Diazotrophic Plant Growth-Promoting Microorganisms and Green Energy – Volume I

Rajesh Kumar Singh · Pratiksha Singh
Editors

Diazotrophic Plant Growth-Promoting Microorganisms and Green Energy – Volume I

Nitrogen Fixation and Crop Productivity

Springer

Editors
Rajesh Kumar Singh
Agromaxx Industry
Bhopal, Madhya Pradesh, India

Pratiksha Singh
School of Agriculture Science
LNCT University
Bhopal, Madhya Pradesh, India

ISBN 978-981-95-9681-2 ISBN 978-981-95-9682-9 (eBook)
https://doi.org/10.1007/978-981-95-9682-9

This Springer imprint is published by the registered company Springer Nature Singapore Pte Ltd.
The registered company address is: 152 Beach Road, #21-01/04 Gateway East, Singapore 189721, Singapore

Preface

Innovative, sustainable, and biologically based solutions are required to address the growing problems of global food insecurity, climate change, and the depletion of fossil fuel supplies. Diazotrophic plant growth-promoting microorganisms (PGPMs) have become essential players in the nexus of renewable energy systems, agriculture, and environmental sustainability. They are positioned as essential elements of future green technologies due to their special capacity to fix atmospheric nitrogen, increase plant production, and support environmentally favorable bioenergy pathways.

Diazotrophic microorganisms including free-living, associative, and symbiotic bacteria and archaea play a fundamental role in natural and managed ecosystems by converting inert atmospheric nitrogen into bioavailable forms. These microbes have a variety of characteristics that promote plant growth in addition to biological nitrogen fixation, including the synthesis of phytohormones, the solubilization of nutrients, the induction of stress tolerance, and the inhibition of phytopathogens. Utilizing these multipurpose qualities presents a viable substitute for energy-intensive chemical fertilizers, lowering greenhouse gas emissions and enhancing soil health. The importance of diazotrophic PGPMs in the production of green energy has received more attention in recent years. Their potential to connect sustainable agriculture with renewable energy projects is shown by their involvement in waste-to-energy systems, microbial biohydrogen generation, biogas production, biofuel crop sustainability, and biomass augmentation. Incorporating diazotrophic microorganisms into circular bio-economies and bioenergy cropping systems is consistent with international efforts to attain carbon neutrality and sustainable development objectives.

The goal of this book is to present a thorough and multidisciplinary overview of diazotrophic plant growth-promoting microorganisms and their developing relevance in green energy applications. This volume's chapters address basic features of microbial nitrogen fixation, rhizosphere molecular and ecological interactions, developments in microbial inoculant technology, and creative uses in the generation of bioenergy. Recent advancements in synthetic microbial consortia, omics-based methods, and climate-resilient agricultural systems are also highlighted.

This book is a useful resource for scientists, students, policymakers, and practitioners working in the fields of microbiology, agronomy, biotechnology, and renewable energy because it compiles contributions from top academics, academicians, and industry experts. We hope that this collection will accelerate the conversion of microbial breakthroughs into sustainable agricultural and energy solutions for a greener future, encourage interdisciplinary collaboration, and drive additional research.

We would like to sincerely thank each and every contributing author for their insightful chapters, dedication to scholarship, and prompt collaboration during the creation of this book. This volume has been greatly influenced by their knowledge, commitment, and zeal. We really appreciate the reviewers' critical assessments and helpful recommendations, which greatly improved the chapters' scientific caliber and coherence. We also thank the publishing team for their help and advice, whose expertise and support made this work possible.

We express our sincere gratitude to our respective institutions, research partners, and funding organizations for their help in providing the academic setting and resources required to successfully complete this work. Lastly, we would like to express our sincere gratitude to our families for their tolerance, inspiration, and steadfast support throughout this project. We sincerely thank **Ms. Swati Sharma** and **Ms. Thesneem Sulthana** of Springer, India, for their kind assistance, direction, and cooperation in completing this book.

Bhopal, Madhya Pradesh, India — Rajesh Kumar Singh
Bhopal, Madhya Pradesh, India — Pratiksha Singh

Contents

1 Nitrogen Fixation in Agriculture, Reducing the Dependency on Chemical Fertilizers

Krishan K. Verma, Xiu-Peng Song, Qiang Liang, Lin Xu, Hai-Rong Huang, Changning Li and Yang-Rui Li

Abstract

Nitrogen fixation represents a cornerstone of sustainable agriculture, offering an eco-friendly alternative to synthetic fertilizers. While symbiotic fixation in legumes is well-established, growing evidence reveals that interconnective nitrogen fixation in cereal crops can contribute 30–50 kg N/ha/year, complementing legume inputs. This chapter synthesizes recent advances in this field, examining the diversity and mechanisms of nitrogen-fixing bacteria, the regulatory challenges of oxygen sensitivity, nitrogen feedback inhibition, and energy supply. We further evaluate innovative strategies from microbial and plant genetic engineering to optimized agronomic practices designed to enhance fixation efficiency. In the context of climate change, regional fertilizer restrictions, and the urgent need for sustainable intensification, we highlight how emerging technologies in microbiome engineering and synthetic biology are poised to unlock nitrogen fixation for a new era of climate-resilient cropping systems in years to come.

Keywords

Nitrogen fixation · Crop plants · Microbial responses and mechanisms · Sustainable agriculture ·

K. K. Verma · X. Song equally contributed of this work.

K. K. Verma · X. Song · Q. Liang · L. Xu · H. Huang · C. Li · Y. Li (✉)
Sugarcane Research Institute, Guangxi Academy of Agricultural Sciences/Key Laboratory of Sugarcane Biotechnology and Genetic Improvement (Guangxi), Ministry of Agriculture and Rural Affairs/Guangxi Key Laboratory of Sugarcane Genetic Improvement, Nanning, Guangxi, China
e-mail: liyr@gxaas.net

R. K. Singh and P. Singh (eds.), *Diazotrophic Plant Growth-Promoting Microorganisms and Green Energy – Volume I*, https://doi.org/10.1007/978-981-95-9682-9_1

1.1 Introduction

Nitrogen (N) is indispensable for plant growth, development, and overall productivity, serving as a key limiting factor in agricultural and natural ecosystems (Santi et al., 2013). It is integral to fundamental physiological functions, such as the synthesis of proteins, nucleic acids, and chlorophyll, as well as the production of plant hormones and the regulation of energy metabolism. Despite molecular nitrogen (N_2) constituting 79% of the atmosphere, the majority of plants are unable to assimilate it directly and must rely on combined nitrogen sources from the soil (Priyadarshini et al., 2021). Biological nitrogen fixation (NF), a microbial-driven process in which specialized microorganisms enzymatically reduce atmospheric N_2 to ammonia thus plays a pivotal role in the global nitrogen cycle. Notably, the symbiotic relationship between legumes and rhizobia stands out as the most efficient nitrogen-fixing system in nature, contributing approximately 55 million tons of fixed nitrogen each year. This represents about 55% of terrestrial biological NF and is essential for sustaining nitrogen balance across ecosystems (Kistner and Parniske, 2002; Magori and Kawaguchi, 2009).

Nitrogen fixation is generally grouped into three main types, i.e., free-living (FNF), symbiotic (SNF), and associative (ANF) (Xu and Wang, 2023). Free-living nitrogen fixation is performed independently by specific prokaryotes, such as cyanobacteria and *Nostoc* that possess both photosynthetic and nitrogen-fixing abilities. By generating energy via photosynthesis, these microorganisms fuel the nitrogen fixation process, supporting their own growth without plant partner (Pankievicz et al., 2019; Soumare et al., 2020). Symbiotic NF involves a highly specialized mutualism between rhizobia and leguminous plants. Rhizobia recognize specific legume hosts and trigger the formation of root nodules, inside which they differentiate into bacteroids capable of fixing nitrogen. In exchange for fixed nitrogen, the host plant supplies photosynthesis-derived carbon to sustain the bacteroids (Hsieh et al., 2009; Zhong et al., 2023).

In contrast, associative nitrogen fixation is mediated by diazotrophic bacteria that colonize the rhizosphere of a wide range of cereal crops, such as *Oryza sativa*, *Zea mays*, and *Triticum aestivum*. These bacteria utilize root exudates for nutrition and fix nitrogen without forming specialized symbiotic structures (Pankievicz et al., 2019; Soumare et al., 2020). Unlike the tightly regulated nodule-based symbiosis in legumes, NF represents a more flexible and context-dependent interaction. Its efficiency is strongly shaped by host genotype, soil conditions, and microbial community composition. This complexity makes NF more challenging to harness, yet it holds considerable promise for enhancing nitrogen sustainability in agricultural cropping systems. The world's population, currently around 8.2 billion, is projected to peak at approximately 10.3 billion by the mid-2080s. This growth intensifies the global demand for food and water supply. Given that the availability of arable land cannot be substantially increased, boosting crop productivity has become essential to meet future needs. While inorganic fertilizers can enhance yields, their overuse leads to soil degradation and environmental pollution (Ahmed et al., 2017). In this context, biological nitrogen fixation represents a promising and sustainable alternative for nitrogen supplementation.

Although cereal crops fix nitrogen less efficiently than legumes, their vast cultivated area implies that their overall NF contribution may be greater than currently recognized (Luo et al., 2023; Yan et al., 2022; Zhang et al., 2022; Aasfar et al., 2024). Advances in omics technologies, including metagenomics, transcriptomics, and metabolomics, have uncovered a hidden diversity of diazotrophs associated with cereals and elucidated novel pathways that regulate nitrogenase activity and plant–microbe interactions (Solanki et al., 2020; Malviya et al., 2021; Jain et al., 2024; Xiong et al., 2024). These insights highlight the importance of integrative approaches to study nitrogen fixation across multiple biological levels. Among these approaches, synthetic microbial communities (SynComs) have emerged as powerful tools for dissecting complex plant–microbe relationships and designing microbial inoculants with more predictable effects. Unlike single-strain inoculants, SynComs can reconstruct core diazotrophic assemblages under controlled conditions, providing reproducible insights into colonization dynamics, nitrogenase activity, and cross-kingdom signaling. Shayanthan et al. (2022). emphasize the potential of SynComs to support crop production by stabilizing microbiome function via synergistic interactions among members.

Parallel advances in synthetic biology have achieved notable proof-of-concept milestones, such as the functional expression of nitrogenase components in transgenic rice (Baysal et al., 2022; He et al., 2022). As highlighted in a recent review, these converging innovations underscore a growing trend toward developing "N-self-fertilizing" cereals capable of autonomous nitrogen fixation (Guo et al., 2023a). Major international initiatives, including the C_4 Rice Project and the Bill & Melinda Gates Foundation, are making substantial research investments to advance nitrogen fixation in staple crops (https://c4rice.com/). These initiatives target a reduction in synthetic fertilizer reliance, increased yield resilience under low-nitrogen conditions, and support for climate-smart agricultural systems. Elucidating the mechanisms and regulation of NF in cereals is essential to counteracting fertilizer-driven environmental degradation and promoting sustainable ecosystem management.

This chapter assesses the associative nitrogen-fixing bacteria in key cereal crops. It explores bacterial diversity, functional niches, regulatory processes, and available microbial resources for agricultural applications. Ultimately, the article aims to outline pathways for enhancing biological NF in cereals, while also discussing present constraints, ongoing challenges, and promising future research priorities.

1.1.1 Nitrogen-Fixing Bacteria groups

The discovery of biological nitrogen fixation in cereal crops dates to 1961, when Dobereiner first isolated the diazotroph *Beijerinckia* from sugarcane. This finding provided seminal evidence for associative nitrogen fixation in a major cereal system (Dobereiner, 1961). The symbiotic relationship between cereals and diazotrophic bacteria delivers nitrogen directly to the plant and, through multiple metabolic pathways, regulates the conversion and metabolism of applied nitrogen fertilizers. This dual action enhances overall nitrogen uptake and improves nitrogen use efficiency (NUE) (Yan et al., 2022; Liu et al., 2024). Based on their physiological and ecological niches, cereal-associated diazotrophs are broadly categorized as rhizospheric or endophytic. Notably, the composition of these diazotrophic communities appears to be influenced more by environmental factors than by host plant selection, often dominated by keystone genera, such as *Azospirillum* and *Methylobacterium* (Xiong et al., 2024).

Metagenomic studies have further uncovered novel metagenome-assembled genomes (MAGs) harboring diverse nitrogen metabolism genes, including complete denitrification pathways, thereby expanding our knowledge of nitrogen cycling in the phyllosphere. Rhizosphere diazotrophs promote plant growth not only by fixing nitrogen but also by producing phytohormones that modify root architecture. Key early examples include *Azotobacter paspali* and *Beijerinckia* spp. Subsequent research has identified other prominent associative nitrogen-fixing genera in cereals, such as *Pseudomonas*, *Klebsiella*, and *Enterobacter* (Tran et al., 2024). Recent advances demonstrate the potent application of Synthetic Microbial Communities (SynComs) as biofertilizers. For instance, a designed SynCom containing *Azospirillum* sp. NL1, *Bradyrhizobium* sp. NL2, and *Azotobacter* sp. NL3 increased the proportion of N derived from the atmosphere (%Ndfa) in rice by about 52%. This inoculant also enhanced root colonization and improved plant architecture, including shoot ramification and tillering. This underscores the dual role of SynComs as both tools for ecological research and effective functional biofertilizers in agriculture. In recent years, the diversity and functional mechanisms of nitrogen-fixing bacteria in cereals have been increasingly clarified. The role and functions of NF bacterium are discussed in Table 1.1.

1.1.2 Regulation of Nitrogen Fixing Bacteria

The symbiotic relationship between cereal crops and NF bacteria is intricately regulated, with nitrogenase serving as the catalytic core that converts atmospheric N_2 into plant-usable ammonia (Rutledge et al., 2022). This enzyme requires substantial energy inputs, specifically 16 ATP molecules and eight electrons to reduce one molecule of N_2 into two NH_3 molecules, while releasing H_2 as a byproduct. Consequently, modulating nitrogenase activity represents a critical strategy for enhancing NF efficiency (Kuypers et al., 2018). However, this process faces significant constraints, including oxygen sensitivity, nitrogen repression, and high metabolic demand. To mitigate oxygen inhibition, certain diazotrophs employ specialized adaptive mechanisms. For instance, in sugarcane, *Gluconacetobacter diazotrophicus* reduces local O_2 levels through elevated respiration and forms protective complexes with FeSII proteins to shield nitrogenase from oxidative damage (Flores-Encarnación et al., 1999; Ureta and Nordlund, 2002). This bacterium also produces levansucrase, an enzyme that hydrolyzes sucrose into monosaccharides while synthesizing levan, an extracellular polysaccharide that acts as a diffusion barrier against oxygen. Molybdenum (Mo) availability further influences nitrogenase functionality, as it is an essential cofactor; increased Mo supply has been shown to enhance enzymatic activity in various systems. Host plants actively participate in regulating this symbiosis by secreting rhizochemicals, such as flavonoids, coumarins, and organic acids, which attract and activate diazotrophic partners. In maize, specific flavonoid exudates not only recruit NF bacteria but also upregulate

Table 1.1 Summary of associated N-fixing bacteria and beneficial impact

Bacterium/group	Functional mechanism	Source
Pseudomonas stutzeri	NF and engineered ammonium excretion, IAA production, PGP traits	Jiang et al. (2022)
Mucilage-associated microbiota (*Sierra Mixe*)	Carbon-rich mucilage hosts diazotrophs, anaerobic/low-O_2 micro-niche supports optimum nitrogenase activity	Van Deynze et al. (2018)
Maize landraces with aerial roots and mucilage enabling NF	Natural aerial root-mediated NF via mucilage-interactive diazotrophs	Connolly et al. (2025)
Alfalfa-maize intercropping effects on diazotrophic community	Intercropping-mediated structural shifts in diazotroph diversity	Zhang et al. (2024a)
Azotobacter	Regulated N assimilation	Nida et al. (2024)
Engineered/remodeled soil diazotroph community	Upregulated biological N_2 fixation	Martinez-Feria et al. (2024)
Diazotrophic inoculants	Stimulates root growth and development; enhanced N acquisition	Carvalho et al. (2023)
Azospirillum brasilense	Hormone induction, modulated N metabolism, increased root structure	Diaz-Zorita & Fernandez-Canigia (2009); Zaheer et al. (2022); Galindo et al. (2022)
Paenibacillus beijingensis (BJ-18)	N fixation + phosphate solubilization, rhizosphere + endophytic colonization	Li et al. (2019); Li et al. (2020)
Combination (B1 + B2) *Brucella oryzae* (B1); *B. ciceri* (B2);	Interactive N fixation, optimum IAA, improved soil C and N, increased enzymatic activities	Saniya et al. (2024)
Azospirillum brasilense, Bacillus subtilis, Pseudomonas fluorescens	Diazotrophic N_2 fixation, phytohormone production (IAA, cytokinin), Zn solubilization, root colonization	Jalal et al. (2022)
Spirillum spp. (historical / root-interaction group; related to *Azospirillum*)	Root restructuring and modulation of nitrate transporters (NAXT/NRT family gene responses reported in colonized roots)	Camilios-Neto et al. (2014); Sigalas et al. (2023)
Azospirillum baldaniorum/A. brasilense	Ammonia scavenging to improve N budget	Dias et al. (2024)
Free-living N-fixers + biostimulant	Biological N fixation; synergy with commercial biostimulant	Buisset et al. (2025)
Salt-tolerant diazotrophs (PGP)	N fixation plus PGP traits under saline/alkaline stress	Kumar et al. (2024)
BNF consortium (varied regimes)	Biological N_2 fixation under contrasting N & P	Aasfar et al. (2024)
Endophytic *Rahnella* spp.	Endophytic N fixation and PGP	Rana et al. (2024)
Azospirillum brasilense (PGPR)	Boosts AE, AFR, N uptake, biomass, and grain N	Galindo et al. (2022)

Bacterium/group	Functional mechanism	Source
Rhizobia strains (*Ensifer meliloti*, *E. mexicanus*, *Rhizobium tropici*) used as biofertilizer	BNF and growth promotion under lab conditions	Serazetdinova et al. (2025)
Stenotrophomonas sp. + NH_4^+ -excreting mutant	BNF with extracellular ammonium release	Zhou et al. (2025)
Herbaspirillum seropedicae	Diazotrophic N_2 fixation, repression of plant defense genes, activation of iron uptake via phytosiderophore synthesis, modulation of root transcriptional responses	Brusamarello-Santos et al. (2019)
Methylobacterium spp. (phyllosphere)	Phyllosphere colonization, plant hormone modulation, and possible enhancement of photosynthesis/growth	Sanjenbam et al. (2022)
Azospirillum spp.	Diazotrophic N_2 fixation, rhizosphere colonization, phytohormone (IAA, cytokinin) production, stimulation of tillering, and nutrient uptake	Sarkodee-Addo et al. (2021)
Clostridium spp. (anaerobic rhizosphere)	Anaerobic heterotrophic BNF in flooded soils; Clostridia enriched under anoxic paddy conditions	Xiao et al. (2022)
Herbaspirillum R3 strain (N-fixing bacterium)	Increased nitrate NH_4^+ and P in the rhizosphere; altered N-fixer community; upregulated N-uptake genes	Li et al. (2024)
Rhodobacter sphaeroides	Phototrophic N fixation in extreme saline acid sulfate soil	Anh et al. (2025)
Engineered *Rhodopseudomonas palustris* (Fe-only nitrogenase, *NifA** regulated)	Constitutive expression of Fe-only nitrogenase even with soil NH_4^+	Zeng et al. (2024)
Terrestrial N-fixing cyanobacteria	Cyanobacterial N_2 fixation supporting rice	Wu et al. (2024)
Klebsiella variicola (DX120E)	Endophytic colonization, influences polyamine metabolism and PGP hormone pathways	Qin et al. (2024)
Enterobacter roggenkampii (ED5)	GS/GDH and other ammonium-assimilation pathway effects; N-metabolism enzyme modulation	Guo et al. (2023b)
Acidithiobacillus (rock solubilizer) + *Beijerinckia indica* (diazotroph) +	Rock phosphate solubilization, N fixation, organic matter decomposition, nutrient release (C, N, P, K, Ca, Mg)	Nunes Oliveira et al. (2017)
Endophytic diazotroph *Klebsiella variicola* DX120E	N_2-fixation, enhanced C & N metabolism, IAA, phosphate solubilization, siderophore production	Qin et al. (2022)
Azospirillum brasilense inoculation	Biological N_2-fixation, phytohormone production, phosphate solubilization	de Souza et al. (2025)
Bacillus megaterium (CY5) and *Bacillus mycoides* (CA1)	*nifH* expression, biological N fixation	Singh et al. (2020)

bacterial *nifH* and nodulation genes, stimulating nitrogenase production (Dong et al., 2002; Velázquez-Hernández et al., 2011).

Similarly, in wheat, naringenin and soyasaponins promote colonization by the endophyte *Azorhizobium caulinodans* ORS571, thereby alleviating ammonia-mediated feedback inhibition. Recent findings also indicate that poplar-derived flavonoids can reshape root-associated *Pseudomonas* communities, improving NUE in these crop plants (Yu et al., 2021a). Intercropping systems reveal additional cross-kingdom signaling: cereal root exudates can induce nodulation-related gene expression in legume-associated bacteria, accelerating nodule formation and nitrogen fixation (Liu et al., 2022). For example, under nitrogen-limited conditions, maize root secretions in a faba bean rhizosphere activate early nodulation genes, initiating symbiosis more rapidly. Once nodules mature, key symbiotic genes remain highly expressed, sustaining elevated NF capacity (Gough et al., 1997; Wu et al., 2025). Root exudates of cereal crops indirectly stimulate nitrogen-fixing bacterial activity through reshaping the rhizosphere microbial community (Liu et al., 2024; Yang et al., 2015; Jauregui et al., 2023). These exudates can suppress non-N-fixing bacteria while secreting phytohormones, such as indole-3-acetic acid (IAA), which promotes root growth, enhances nutrient uptake, and upregulates key genes in N-fixing bacteria. This dual action thereby boosts biological NF through direct and indirect pathways (Ali et al., 2021).

Notably, these interactions can generate a positive feedback cycle, like ammonia derived from bacterial fixation is absorbed and assimilated by the host plants, which in turn further modifies the rhizosphere microbiome and amplifies N-fixation activity. The genotype of the host plant plays a critical role in shaping the assembly of NF bacterial communities. For instance, comparative studies reveal that the rhizosphere of wild rice supports a significantly greater abundance of *Methylobacterium* than cultivated varieties, indicating that specific plant genotypes can actively recruit beneficial diazotrophic bacteria via genetic signaling pathways (Yin et al., 2023). Under conditions of nutrient limitation, host plants further modulate these recruitment processes by integrating environmental stress signals to enrich for microorganisms that enhance nitrogen acquisition (Yu et al., 2021b; Meier et al., 2022; Zhang et al., 2024b). While the molecular dialog governing legume–rhizobia symbiosis is well characterized (Meier et al., 2022), the signaling and regulatory networks underlying cereal diazotroph interactions remain largely unresolved. Advancing this knowledge requires an integrated approach focused on the dynamic interplay between plant genotype, phenotypic expression, and the composition of associative nitrogen-fixing communities. Such research will provide essential theoretical insights and practical strategies for optimizing nitrogen fixation in cereal crop systems.

1.1.3 Synergistic Regulation Between Microorganisms and Abiotic Factors

Interactions among microorganisms are crucial in shaping the functional capacity of diazotrophs (nitrogen-fixing microbes), creating intricate networks with other bacteria, fungi, and viruses. For instance, Zhang et al. (2022). isolated two diazotrophic bacteria and 12 non-diazotrophic bacteria from maize stem xylem sap, finding that the latter significantly enhanced nitrogenase activity in the diazotrophs. Genomic analysis suggested that these non-fixing bacteria likely support NF by regulating local nitrogen and oxygen levels within the xylem microenvironment. Further research has revealed synergistic relationships between diazotrophs and fungi in specialized niches, such as aerial root mucilage. In Marcgravia species, the mucilage of aerial roots selectively recruits abundant diazotrophic bacteria alongside specific fungal partners. These fungi not only display broad-spectrum antimicrobial properties inhibiting numerous plant pathogens but also selectively stimulate the growth of diazotrophs, thereby sustaining efficient nitrogen fixation (Yu et al., 2021c; Xiao et al., 2023).

Research leveraging microbial network analysis has established that synergistic interactions between arbuscular mycorrhizal fungi (AMF) and N-fixing bacteria substantially enhance NUE. In legumes, AMF secrete signaling molecules, including lipo-chitooligosaccharides (LCOs) and chitin oligosaccharides (COs), that initiate host interactions and facilitate N-fixing bacteria recruitment for nodule formation (Luo et al., 2023; Sun et al., 2015). Notably, recent evidence indicates that cereals like rice and barley can also perceive LCO/CO signals under low-nutrient conditions, suggesting this interaction mechanism may extend to non-leguminous crops (Li et al., 2022). The precise regulatory pathways governing this synergy in cereals, however, remain to be fully elucidated. A promising approach to dissecting these complex interactions is the use of SynComs. By providing a controlled, reductionist framework, SynComs act as experimental "testbeds" to decode the molecular dialog underlying plant-microbe and microbe-microbe communication. Fonseca-García et al. (2024) demonstrate how defined SynComs can unravel cooperative behaviors like nutrient exchange and signaling, which are pivotal for regulating N-fixation synergies.

Viral communities are key drivers of microbial composition and function, critically influencing ecosystem processes through mechanisms, such as horizontal gene transfer, including that of auxiliary metabolic genes, and thereby directly participating in carbon and nitrogen cycling (Lopez-Simon et al., 2023; Tong et al., 2023). For example, under nitrogen-limited conditions, bacteriophage-resistant cyanobacterial strains show marked growth and metabolic deficiencies compared to wild-type counterparts, implying that viral resistance may come at a cost to NF capacity (Kolan et al., 2024). Altogether, while the interactions between viruses and NF bacteria and their subsequent effects on nitrogen availability in cereal cropping systems remain an emerging field, existing evidence underscores their potential importance in agricultural nutrient dynamics.

Oxygen and nitrogen are key regulators of biological nitrogen fixation. Nitrogenase activity is directly suppressed by both elements, and the fixation process itself demands substantial amounts of energy. Nitrogenase is particularly oxygen-sensitive, functioning optimally only under microaerobic conditions (Kuypers et al., 2018). For instance, ^{15}N labeling studies have shown that N fixation rates in rice paddies of eastern China are markedly higher than in dryland soil. This difference is likely attributed to oxygen fluctuations resulting from intermittent irrigation and prolonged flooding (Li et al., 2023). Additionally, NF is highly energy-intensive, consuming eight high-energy electrons and 16 ATP molecules to fix each N_2 molecule. In cereal crops, the symbiotic association with diazotrophic bacteria is less developed compared to the specialized root nodules of legumes, which continuously supply carbon and energy to their microbial partners. Consequently, N-fixing bacteria in cereal systems often rely more on soil-derived N, thereby reducing their fixation activity and contributing to lower NF efficiency.

Agricultural reliance on inorganic N fertilizers has long-term suppressive effects on biological NF. This practice impairs the function of NF bacteria, leading to a marked decline in the abundance and genetic diversity of nitrogen fixation (*nif*) genes in the soil microbiome (Norman and Friesen, 2017; Fan et al., 2019; Chen et al., 2021). Empirical evidence from a 35-year wheat-soybean rotation trial demonstrates a reduction in soil NF efficiency of up to 50% (Fan et al., 2019). Similarly, in maize monocultures, the abundance of the *nifH* gene decreased by approximately 54% to 80% following only 5 years of consistent fertilizer application (Chen et al., 2021). In contrast, the application of organic fertilizers, including manure, biochar, and straw, generally promotes biological NF efficiency. This enhancement is attributed to positive changes in soil organic carbon levels, iron mineral composition, pH balance, and microbial community structure (Yu et al., 2021c; Shi et al., 2021). A global meta-analysis supports this, indicating that elevated soil organic carbon can alleviate the suppressive effect of N on NF (Zheng et al., 2023). For example, Yu et al. (2021c) demonstrated that long-term organic fertilizer use substantially increased short-range ordered iron minerals in soil, raising total molybdenum content by 30%. These shifts corresponded with 14% rise in *nifH* gene abundance and 60% increase in nitrogenase activity. Additionally, organic amendments have been shown to counteract the inhibitory impact of low pH on nitrogenase function (Yu et al., 2021b; Shi et al., 2021). Recent research further reveals that diurnal temperature variation plays a significant role in modulating N fixation. Tang et al. (2023) engineered nitrogen-fixing bacterial strains, such as *Escherichia coli* and *Klebsiella oxytoca,* capable of releasing plant-available fixed nitrogen during cooler nighttime periods (23°C), offering novel strategies for temporal regulation of NF (Tang et al., 2023). In summary, oxygen availability, N fertilizer input, and temperature interact within a complex regulatory network that directly governs nitrogenase activity and overall NF efficiency.

1.1.4 Improve the Efficiency of Nitrogen Fixation

1.1.4.1 Engineering Modification of Nitrogen-Fixing Microorganisms

Because directly engineering plants for NF is so complex, a viable alternative is to engineer their associated bacteria. This strategy enhances fixation in key crops or grants the ability to new bacterial hosts, compensating for inadequate natural populations in soil. Initial successes in expressing nitrogenase in *E. coli* proved the principle is genetically portable (Wang et al., 2013). However, to move beyond proof-of-concept, synthetic biology is needed to optimize these gene systems for higher activity, as already demonstrated in bacteria like *Klebsiella oxytoca*. This work lays the essential foundation for the practical microbial engineering (Li et al., 2016; Smanski et al., 2014).

Moreover, NF bacteria are highly responsive to environmental nitrogen levels, which directly regulate their fixation activity. When external nitrogen is plentiful, these bacteria often halt fixation to optimize energy use. Thus, to maintain a consistent and effective supply of plant-accessible nitrogen, it is beneficial to engineer NF strains that remain active irrespective of surrounding nitrogen concentrations. The *nif* gene cluster is mainly controlled by two regulatory proteins, such as *NifA*, an activator that stimulates transcription, and NifL, a repressor that curtails *NifA* function under high nitrogen or aerobic conditions (Ambrosio et al., 2017). By modifying these genes, engineered bacteria can continue fixing nitrogen even in N-rich settings. Furthermore, ammonium uptake in bacteria is mediated by the transporter AmtB. Disrupting *amtB* expression can inhibit bacterial ammonium assimilation, thereby increasing the release of fixed ammonium into the rhizosphere for plant uptake (Zhang et al., 2012).

Similarly, engineering approaches that increase ammonium efflux can further boost NF efficiency (Pankievicz et al., 2019). Nitrogenase converts atmospheric N_2 into bioavailable ammonia (NH_3), yet the enzyme is highly oxygen-sensitive and easily deactivated upon exposure. To address this constraint, diazotrophs have evolved diverse protective strategies, most notably the small iron-sulfur cluster protein *Shethna protein II (FeSII)*, which safeguards nitrogenase via conformational shifts (Narehood et al., 2025). The expression of oxygen-protective proteins, such as Shethna protein in NF bacteria, could therefore alleviate oxygen inhibition of nitrogenase and improve overall fixation activity.

The ability of NF bacteria to colonize host plant roots is critical for effective symbiosis. Central to this process is bacterial chemotaxis, a key function governed by the two-component signaling system. Here, the histidine kinase *CheA* undergoes autophosphorylation and transfers a phosphoryl group to the response regulator *CheY*, which in turn modulates flagellar rotation and directs bacterial movement. In maize, for instance, root exudates strongly upregulate expression of the *cheA* gene in the degradative bacterium *Hansschlegelia zhihuaiae* S113, stimulating its chemotactic accumulation in the rhizosphere and improving root colonization (Zhang et al., 2021). Additionally, extracellular polysaccharides (EPS) synthesized by NF bacteria help shield cells from oxidative stress, both in vitro and during colonization in rice (Meneses et al., 2017). Targeted engineering of key genes, such as those involved in chemotaxis (*cheA*) or biofilm formation (*eps*), offers a promising strategy to enhance bacterial adhesion and root colonization efficiency.

1.1.4.2 Engineering Nitrogen Fixation in Cereal Crops

Introducing NF into cereal crops represents a promising avenue for improving agricultural N efficiency. However, engineering autonomous biological NF in non-legumes faces significant biological hurdles, primarily due to the challenges of heterologously expressing the nitrogenase enzyme system. As an oxygen-sensitive metalloprotein complex, functional nitrogenase expression requires overcoming several interconnected barriers: maintaining the structural stability of the enzyme complex, balancing its oxygen sensitivity with host energy metabolism, and precisely coordinating the expression of numerous essential genes. Current research strategies to address these challenges focus on three principal approaches, i.e., optimizing subcellular localization, refining genetic engineering techniques, and modifying plant signaling pathways to support symbiotic or endophytic associations. Given nitrogenase's extreme sensitivity to oxygen, organelles with reduced oxygen environments, such as mitochondria and chloroplasts, have been identified as preferred sites for its potential localization (Soto et al., 2013; Bennett et al., 2020).

Mitochondrial genome editing, though theoretically possible, faces significant practical constraints due to its low transformation efficiency. As a result, the prevailing method involves integrating target genes into the nuclear genome, followed by peptide-mediated transport into mitochondria. Burén et al. (2017) successfully targeted nine NF (*nif*) genes to yeast mitochondria using the SU9 mitochondrial transit peptide. Similarly, Allen et al. (2017) employed the pFAγ leader sequence to direct 16 *Nif* proteins into yeast mitochondria, demonstrating the feasibility of cross-organellar protein targeting. A persistent challenge in these systems is the instability of the key component *NifD*. Its degradation is associated with specific residues (R98 and Y100) and the activity of mitochondrial processing enzymes. Studies have shown that introducing R98P and Y100Q mutations significantly improves *NifD* stability (Xiang et al., 2020). Another major limitation in plant mitochondria is their restricted capacity for [Fe-S] cluster biosynthesis, which impairs the functionality of *NifH*. This bottleneck can be addressed by using thermostable *NifH* variants from thermophilic bacteria, supplemented with the scaffold protein *NifU* to enhance [Fe-S] cluster assembly (Jiang et al., 2021).

The potential of the plastid system for expressing polycistronic *nif* gene clusters has driven significant research. Initial efforts yielded low nitrogenase activity, but breakthroughs using synthetic biology to assemble and target bacterial *nif* operons to chloroplasts have demonstrated markedly higher function (Ivleva et al., 2016; Eseverri et al., 2020). To overcome inherent challenges, such as the oxygen sensitivity of nitrogenase, strategies like spatiotemporally regulated (e.g., hypoxia-inducible) promoters have been employed (Bothe et al., 2010). Compatibility has been further improved by optimizing the source of *nif* genes and simplifying clusters to their minimal components (Burén et al., 2017). A key insight has been that achieving the correct balance of *nif* protein components is essential for full nitrogenase functionality (Yao et al., 2021; Temme et al., 2012).

Wang et al. (2013) and Yang et al. (2014, 2018) simplified regulation by using gene fusion and polyprotein designs, thereby circumventing the need for Mo cofactors. In transgenic *Arabidopsis*, an engineered *nif* gene cluster demonstrated both high nitrogenase activity and enhanced biomass production under N-limited conditions (Li et al., 2016; Yao et al., 2021), supporting the viability of this strategy. Nevertheless, epigenetic silencing triggered by multigene insertion continues to pose a major obstacle, highlighting the necessity for innovative expression platforms resistant to silencing or for the incorporation of inducible control systems. Rhizosphere metabolite engineering offers a promising approach to enhance NF. For example, disrupting *CYP75B3/4* in rice elevated apigenin secretion, which robustly stimulated colonization by diazotrophic bacteria (Yan et al., 2022). Likewise, engineering barley to produce rhizobactin directly activated bacterial nitrogenase expression (Haskett et al., 2022). Future efforts should systematically map how host-derived metabolites optimize bacterial NF, with particular attention to specific, selective interactions between metabolites and microbial partners.

1.1.4.3 Plant-Mediated Approaches for Microbiome Engineering

The strategic use of plant genetics to position beneficial microbiomes, particularly diazotrophic bacteria in cereal cropping systems, represents a rapidly advancing field. Driven by convergence in genomics, molecular biology, and breeding, this approach allows for the targeted enrichment and manipulation of microbial consortia to improve NF, nutrient acquisition, and resilience in major crops. These plant-mediated strategies provide a sustainable solution to the environmental contamination and resource damage caused by excessive synthetic fertilizer use. Methodologically, the field integrates genome-wide association studies (GWAS), RNAi, CRISPR/Cas genome editing, and cytological visualization to identify and optimize host genetic factors controlling microbiome assembly. Notably, GWAS in maize has elucidated how allelic variation in root exudation and immune pathways directly influences the recruitment and persistence of key diazotrophs like *Herbaspirillum* and *Azospirillum* in the root environment (Meier et al., 2022).

GWAS is proving instrumental in decoding the genetic dialogue between plants and diazotrophs. In rice, interactions between root/metabolite QTLs and the enrichment of *Burkholderia* and *Gluconacetobacter* demonstrate how host genetic variation in pathways like flavonoid biosynthesis directs microbial colonization. Meanwhile, maize GWAS reveals that selection for high NUE has co-selected for distinct diazotrophic communities (Pang et al., 2023). This implies that plant breeding has unknowingly shaped microbiome assembly, an effect that can now be deliberately harnessed. Thus, GWAS provides a direct pathway to marker-assisted selection for genotypes genetically predisposed to recruit efficient NF microbiomes, bridging host genetics and targeted microbiome engineering (Massahiro Yassue et al., 2023).

RNA interference (RNAi) serves as a powerful reverse-genetics tool for analyzing and engineering plant pathways that control diazotroph colonization and NF. For instance, silencing host defense genes in rice via RNAi revealed that suppressing specific receptor-like kinases promotes *Herbaspirillum seropedicae* colonization, demonstrating how innate immunity components regulate symbiotic associations (Rosman et al., 2024). Similarly, in maize, RNAi knockdown of flavonoid biosynthesis alters root exudates and impairs recruitment of *Azospirillum brasilense*, underscoring the role of targeted gene suppression in rhizosphere signaling (Yan et al., 2022). In sugarcane, RNAi-mediated reduction of lignin biosynthesis facilitates deeper endophytic penetration by diazotrophs, boosting nitrogenase activity and ammonia assimilation in nitrogen-limited conditions. Collectively, these studies highlight RNAi's utility for functionally dissecting and modulating plant traits that shape microbiome assembly, especially in cereal-diazotroph interactions (James and Olivares, 1998). Meanwhile, the emergence of CRISPR/Cas genome editing has transformed plant-mediated microbiome engineering by allowing precise genetic modifications in host and microbial genomes.

1.1.4.4 Selective Breeding for Microbiome Compatibility

Selective breeding continues to be a fundamental strategy for enhancing crops, providing a sustainable pathway to increase biological NF in crops. Long-term field studies in maize, for instance, indicate that specific traditional varieties naturally enrich their root environments with beneficial diazotrophs, such as *Azospirillum* and *Herbaspirillum*, which substantially support the plant's nitrogen needs (Wallace et al., 2018). Similarly, in rice, native landraces adapted to nitrogen-limited regions in Asia consistently host elevated populations of NF bacteria like *Burkholderia* and *Gluconacetobacter*. Genome-wide association studies have further linked these microbial associations to specific genetic variants involved in flavonoid biosynthesis. In wheat, breeding efforts have uncovered notable genetic diversity in the recruitment of NF endophytes, with wild relatives supporting more diverse and abundant diazotrophic communities compared to modern high-yielding varieties (Huang et al., 2014; Mitter et al., 2017). Advances in genomic selection are now accelerating this progress, where microbial traits, such as the abundance of the *nifH* gene in maize, are treated as heritable components. This approach enables the simultaneous selection of beneficial plant–microbe partnerships along with agronomic traits like yield and stress resilience capacity (Wallace et al., 2018).

1.1.5 Future Research Directions and Challenges

The synergistic integration of GWAS, RNAi, CRISPR, cytological visualization, selective breeding, and microbiome engineering provides a powerful, multi-pronged strategy to boost associative NF in cereal crops. For instance, coupling multiomics with GWAS in maize can predict functional outcomes like nitrogenase activity. Concurrently, microbiome engineering is evolving beyond single strains toward sophisticated SynComs, often incorporating "helper" strains to improve diazotroph colonization and persistence. Reducing reliance on synthetic NFs through enhanced biological NF offers significant environmental benefits, including the mitigation of soil acidification, groundwater pollution, and greenhouse gas emissions, a core goal of initiatives like the C_4 Rice Project. The ecological stability of introduced SynComs and their long-term safety are pressing concerns, as field applications are frequently challenged by environmental variability, displacement by native microbiota, and regulatory barriers. Recent research underscores the value of using SynComs as tractable models to decipher microbial interactions, a critical step toward engineering resilient field communities.

A promising path forward is the integration of host genotype selection with multiomics-guided SynCom design, aiming to create tailored, robust plant–microbe partnerships that deliver consistent N fixation. Realizing this potential, however, requires navigating complex plant–microbe–environment interactions, potential yield trade-offs, and evolving regulatory frameworks, demanding sustained multidisciplinary collaboration and rigorous field validation. Current research faces several fundamental gaps, such as the catalog of associative diazotrophs and their functions is incomplete;

the genetic basis of compatibility is poorly understood; and interactions between diazotrophs and other microbiome members (bacteria, fungi, viruses) are largely uncharted. Furthermore, the colonization dynamics of diazotrophs in different plant compartments and the mechanisms governing these interactions under varying host genotypes and abiotic stresses remain unclear.

Future efforts should prioritize: (1) elucidating host-microbe genetic networks to enable molecular breeding for improved symbiosis; (2) verifying the field stability and ecological safety of engineered microbial consortia; and (3) developing cost-effective agronomic practices for large-scale implementation. Ultimately, plant-mediated microbiome manipulation represents a holistic approach to enhance diazotroph-driven NF in crop plants. By leveraging model and target crops, such as rice, wheat, maize, and sugarcane, we can design less fertilizer-dependent, more productive, and resilient cropping systems. Advancing this vision will depend on multiomics-guided precision design, novel technologies for in situ monitoring of N fixation, and the development of optimized, region-specific management strategies to translate theoretical promise into field-level impact.

Acknowledgments The authors would like to thank the Guangxi Academy of Agricultural Sciences, Nanning, Guangxi, China, for providing the necessary facilities for this study.

Funding This research was financially supported by the Guangxi Natural Science Foundation (2021GXNSFAA220022; 2023GXNSFAA026459; 2023GXNSFAA026460), Guangxi Innovation Teams of Modern Agriculture Technology (nycytxgxcxtd-2021-03), Guangxi Characteristic Crop Experimental Station (GTS2022022), National Key Research and Development Project (2022YFD2301102-07), The National Natural Science Foundation of China (31760415; 32060468), Fund of Guangxi Academy of Agricultural Sciences (2021YT011), Science and Technology Major Project of Guangxi (Guike AA22117002-1), Fundamental Research Fund of Guangxi Academy of Agriculture Sciences (2023YM55), Key Research and Development Program of Nanning (20232060), and Guangxi Key Laboratory of Sugarcane Genetic Improvement Project (21-238-16-K-03-09).

References

Aasfar, A., Meftah Kadmiri, I., Azaroual, S. E., Lemriss, S., Mernissi, N. E., Bargaz, A., Zeroual, Y., & Hilali, A. (2024). Agronomic advantage of bacterial biological nitrogen fixation on wheat plant growth under contrasting nitrogen and phosphorus regimes. *Frontiers in Plant Science*, *15*, 1388775. https://doi.org/10.3389/fpls.2024.1388775.

Ahmed, M., Rauf, M., Mukhtar, Z., & Saeed, N. A. (2017). Excessive use of nitrogenous fertilizers: An unawareness causing serious threats to environment and human health. *Environmental Science and Pollution Research*, *24*(35), 26983–26987. https://doi.org/10.1007/s11356-017-0589-7.

Ali, M., Ali, Q., Sohail, M. A., Ashraf, M. F., Saleem, M. H., Hussain, S., & Zhou, L. (2021). Diversity and taxonomic distribution of endophytic bacterial community in the rice plant and its prospective. *International Journal of Molecular Sciences*, *22*(18), 10165. https://doi.org/10.3390/ijms221810165.

Allen, R. S., Tilbrook, K., Warden, A. C., Campbell, P. C., Rolland, V., Singh, S. P., & Wood, C. C. (2017). Expression of 16 nitrogenase proteins within the plant mitochondrial matrix. *Frontiers in Plant Science*, *8*, 287. https://doi.org/10.3389/fpls.2017.00287.

Ambrosio, R., Ortiz-Marquez, J. C. F., & Curatti, L. (2017). Metabolic engineering of a diazotrophic bacterium improves ammonium release and biofertilization of plants and microalgae. *Metabolic Engineering*, *40*, 59–68. https://doi.org/10.1016/j.ymben.2017.01.002.

Anh, N. H., Phat, C. T., Nhut, L. M., Thu, L. T. M., Trong, N. D., Quang, L. T., Xuan, L. N. T., Nhan, T. C., Phong, N. T., & Khuong, N. Q. (2025). Effectiveness of nitrogen-fixing bacteria *Rhodobacter sphaeroides* in soil-plant nitrogen and rice performance in extremely saline acid sulfate soil over two consecutive seasons. *Sustainability*, *17*(5), 2228. https://doi.org/10.3390/su17052228.

Baysal, C., Burén, S., He, W., Jiang, X., Capell, T., Rubio, L. M., & Christou, P. (2022). Functional expression of the nitrogenase Fe protein in transgenic rice. *Communications Biology*, *5*(1), 1006. https://doi.org/10.1038/s42003-022-03921-9.

Bennett, A. B., Pankievicz, V. C., & Ané, J.-M. (2020). A model for nitrogen fixation in cereal crops. *Trends in Plant Science*, *25*(3), 226–235. https://doi.org/10.1016/j.tplants.2019.12.004.

Bothe, H., Schmitz, O., Yates, M. G., & Newton, W. E. (2010). Nitrogen fixation and hydrogen metabolism in cyanobacteria. *Microbiology and Molecular Biology Reviews*, *74*(4), 529–551. https://doi.org/10.1128/mmbr.00033-10.

Brusamarello-Santos, L. C. C., Alberton, D., Valdameri, G., Camilios-Neto, D., Covre, R., Lopes, K. D. P., Zibetti Tadra-Sfeir, M., Faoro, H., Adele Monteiro, R., Barbosa-Silva, A., John Broughton, W., Oliveira Pedrosa, F., Wassem, R., & Souza, E. M. (2019). Modulation of defence and iron homeostasis genes in rice roots by the diazotrophic endophyte *Herbaspirillum seropedicae*. *Scientific Reports*, *9*(1), 10573. https://doi.org/10.1038/s41598-019-45866-w.

Buisset, E., Soust, M., & Scott, P. T. (2025). The isolation of free-living nitrogen-fixing bacteria and the assessment of their potential to enhance plant growth in combination with a commercial biostimulant. *Microbiological Research*, *16*(3), 69. https://doi.org/10.3390/microbiolres16030069.

Burén, S., Jiang, X., & López-Torrejón, G. (2017). Purification and *in vitro* activity of mitochondria targeted nitrogenase cofactor maturase nifB. *Frontiers in Plant Science*, *8*, 1567. https://doi.org/10.3389/fpls.2017.01567.

Camilios-Neto, D., Bonato, P., Wassem, R., Tadra-Sfeir, M. Z., Brusamarello-Santos, L. C., Valdameri, G., Donatti, L., Faoro, H., Weiss, V. A., Chubatsu, L. S., Pedrosa, F. O., & Souza, E. M. (2014). Dual RNA-seq transcriptional analysis of wheat roots colonized by *Azospirillum brasilense* reveals up-regulation of nutrient acquisition and cell cycle genes. *BMC Genomics*, *15*(1), 378. https://doi.org/10.1186/1471-2164-15-378.

Carvalho, A. D., Alves, G. C., Alves, B. J. R., Santos, L. A., & Reis, V. M. (2023). Root development and nitrogen acquisition of maize inoculated with two species of diazotrophs. *Bragantia*, *82*, e20230060. https://doi.org/10.1590/1678-4499.20230060.

Chen, L., Li, K. K., Shi, W. J., Wang, X. L., Wang, E. T., Liu, J. F., Sui, X. H., Mi, G. H., Tian, C. F., & Chen, W. X. (2021). Negative impacts of excessive nitrogen fertilization on the abundance and diversity of diazotrophs in black soil under maize monocropping. *Geoderma*, *393*, 114999. https://doi.org/10.1016/j.geoderma.2021.114999.

Connolly, L. N., Lorenz, N., Maleki, K., Kayafas, N., Dick, R. P., & Mercer, K. L. (2025). Nitrogen fixation rates and aerial root production among maize landraces. *Frontiers in Plant Science*, *16*, 1502884. https://doi.org/10.3389/fpls.2025.1502884.

de Souza, J. A. L., Teixeira, L.dS., da Silva Freitas, G., da Silva Alves, L., da Silva, M. B. P., da Silva, J. F., Galindo, F. S., Bonini, C.dS. B., de Oliveira, C. L. B., & Heinrichs, R. (2025). *Azospirillum brasilense* in the planting furrow of sugarcane to minimize the use of N fertilizer. *Plants*, *14*(11), 1599. https://doi.org/10.3390/plants14111599.

Dias, T., Azmaliyev, K., Melo, J., Santos, A. M., Correia, P., & Cruz, C. (2024). The diazotrophic bacteria *Azospirillum baldaniorum* and *A. brasilense* improve wheat seedlings' nitrogen budget through ammonia scavenging. *Applied Soil Ecology*, *204*, 105737. https://doi.org/10.1016/j.apsoil.2024.105737.

Díaz-Zorita, M., & Fernández-Canigia, M. V. (2009). Field performance of a liquid formulation of *Azospirillum brasilense* on dryland wheat productivity. *European Journal of Soil Biology, 45*(1), 3–11. https://doi.org/10.1016/j.ejsobi.2008.07.001.

Dobereiner, J. (1961). Nitrogen-fixing bacteria of the genus *Beijerinckia* Derx in the rhizosphere of sugarcane. *Plant and Soil, 15*(3), 211–216. https://doi.org/10.1007/BF01400455.

Dong, Z., Zelmer, C., Canny, M., McCully, M., Luit, B., Pan, B., Faustino, R., Pierce, G., & Vessey, J. (2002). Evidence for protection of nitrogenase from O_2 by colony structure in the aerobic diazotroph *Gluconacetobacter diazotrophicus*. *Microbiology, 148*(8), 2293–2298. https://doi.org/10.1099/00221287-148-8-2293.

Eseverri, Á, López-Torrejón, G., Jiang, X., Burén, S., Rubio, L. M., & Caro, E. (2020). Use of synthetic biology tools to optimize the production of active nitrogenase Fe protein in chloroplasts of tobacco leaf cells. *Plant Biotechnology Journal, 18*(9), 1882–1896. https://doi.org/10.1111/pbi.13347.

Fan, K., Delgado-Baquerizo, M., Guo, X., Wang, D., Wu, Y., Zhu, M., Yu, W., Yao, H., Zhu, Y. G., & Chu, H. (2019). Suppressed N fixation and diazotrophs after four decades of fertilization. *Microbiome, 7*(1), 143. https://doi.org/10.1186/s40168-019-0757-8.

Flores-Encarnación, M., Contreras-Zentella, M., Soto-Urzua, L., Aguilar, G., Baca, B., & Escamilla, J. (1999). The respiratory system and diazotrophic activity of *Acetobacter diazotrophicus* PAL5. *Journal of Bacteriology, 181*(22), 6987–6995. https://doi.org/10.1128/JB.181.22.6987-6995.

Fonseca-García, C., Pettinga, D., Wilson, A., Elmore, J. R., McClure, R., Atim, J., Pedraza, J., Hutmacher, R., Turumtay, H., Tian, Y., Eudes, A., Scheller, H. V., Egbert, R. G., & Coleman-Derr, D. (2024). Defined synthetic microbial communities colonize and benefit field-grown sorghum. *The ISME Journal, 18*(1), wrae126. https://doi.org/10.1093/ismejo/wrae126.

Galindo, F. S., Pagliari, P. H., Fernandes, G. C., Rodrigues, W. L., Boleta, E. H. M., Jalal, A., Céu, E. G. O., Lima, B. H. D., Lavres, J., & Teixeira Filho, M. C. M. (2022). Improving sustainable field-grown wheat production with *Azospirillum brasilense* under tropical conditions: a potential tool for improving nitrogen management. *Frontiers in Environmental Science, 10*, 821628. https://doi.org/10.3389/fenvs.2022.821628.

Gough, C., Galera, C., Vasse, J., Webster, G., Cocking, E. C., & Dénarié, J. (1997). Specific flavonoids promote intercellular root colonization of *Arabidopsis thaliana* by *Azorhizobium caulinodans* ORS571. *Molecular Plant-Microbe Interactions, 10*(5), 560–570. https://doi.org/10.1094/MPMI.1997.10.5.560.

Guo, D. J., Li, D. P., Yang, B., Verma, K. K., Singh, R. K., Singh, P., Khan, Q., Sharma, A., Qin, Y., Zhang, B. Q., Song, X. P., & Li, Y. R. (2023a). Effect of endophytic diazotroph *Enterobacter roggenkampii* ED5 on nitrogen-metabolism-related microecology in the sugarcane rhizosphere at different nitrogen levels. *Frontiers in Microbiology, 14*, 1132016. https://doi.org/10.3389/fmicb.2023.1132016.

Guo, D. J., Singh, P., Yang, B., Singh, R. K., Verma, K. K., Sharma, A., Khan, Q., Qin, Y., Chen, T. S., Song, X. P., Zhang, B. Q., Li, D. P., & Li, Y. R. (2023b). Complete genome analysis of sugarcane root associated endophytic diazotroph *Pseudomonas aeruginosa* DJ06 revealing versatile molecular mechanism involved in sugarcane development. *Frontiers in Microbiology, 14*, 1096754. https://doi.org/10.3389/fmicb.2023.1096754.

Guo, K., Yang, J., Yu, N., Luo, L., Wang, E. (2023a). Biological nitrogen fixation in cereal crops: Progress, strategies, and perspectives. *Plant Communications, 4* (2). https://doi.org/10.1016/j.xplc.2022.100499

Haskett, T. L., Paramasivan, P., Mendes, M. D., Green, P., Geddes, B. A., Knights, H. E., Jorrin, B., Ryu, M.-H., Brett, P., Voigt, C. A., Oldroyd, G. E. D., & Poole, P. S. (2022). Engineered plant control of associative nitrogen fixation. *Proceedings of the National Academy of Sciences of the United States of America, 119*(16), e2117465119. https://doi.org/10.1073/pnas.2117465119.

He, W., Burén, S., Baysal, C., Jiang, X., Capell, T., Christou, P., & Rubio, L. M. (2022). Nitrogenase cofactor maturase NifB isolated from transgenic rice is active in FeMo-co synthesis. *ACS Synthetic Biology, 11*(9), 3028–3036. https://doi.org/10.1021/acssynbio.2c00194.

Hsieh, L. C., Lin, S. I., Shih, A. C. C., Chen, J. W., Lin, W. Y., Tseng, C. Y., Li, W. H., & Chiou, T. J. (2009). Uncovering small RNA-mediated responses to phosphate deficiency in *Arabidopsis* by deep sequencing. *Plant Physiology*, *151*(4), 2120–2132. https://doi.org/10.1104/pp.109.147280.

Huang, X. F., Chaparro, J. M., Reardon, K. F., Zhang, R., Shen, Q., & Vivanco, J. M. (2014). Rhizosphere interactions: Root exudates, microbes, and microbial communities. *Botany*, *92*(4), 267–275. https://doi.org/10.1139/cjb-2013-0225.

Ivleva, N. B., Groat, J., Staub, J. M., & Stephens, M. (2016). Expression of active subunit of nitrogenase via integration into plant organelle genome. *PLoS ONE*, *11*(8), e0160951. https://doi.org/10.1371/journal.pone.0160951.

Jain, A., Sarsaiya, S., Singh, R., Gong, Q., Wu, Q., & Shi, J. (2024). Omics approaches in understanding the benefits of plant-microbe interactions. *Frontiers in Microbiology*, *15*, 1391059. https://doi.org/10.3389/fmicb.2024.1391059.

Jalal, A., da Silva Oliveira, C. E., Freitas, L. A., Galindo, F. S., Lima, B. H., Boleta, E. H. M., Da Silva, E. C., do Nascimento, V., Nogueira, T. A. R., Buzetti, S., & Teixeira Filho, M. C. M. (2022). Agronomic biofortification and productivity of wheat with soil zinc and diazotrophic bacteria in tropical savannah. *Crop and Pasture Science*, *73*, 817–830. https://doi.org/10.1071/CP21457.

James, E. K., & Olivares, F. L. (1998). Infection and colonization of sugarcane and other graminaceous plants by endophytic diazotrophs. *Critical Reviews in Plant Sciences*, *17*(1), 77–119. https://doi.org/10.1080/07352689891304195.

Jauregui, I., Vega-Mas, I., Delaplace, P., Vanderschuren, H., & Thonar, C. (2023). An optimized hydroponic pipeline for large-scale identification of wheat genotypes with resilient biological nitrification inhibition activity. *The New Phytologist*, *238*(4), 1711–1721. https://doi.org/10.1111/nph.18807.

Jiang, S., Li, J., Wang, Q., Yin, C., Zhan, Y., Yan, Y., Lin, M., & Ke, X. (2022). Maize growth promotion by inoculation with an engineered ammonium-excreting strain of nitrogen-fixing *Pseudomonas stutzeri*. *Microorganisms*, *10*(10), 1986. https://doi.org/10.3390/microorganisms10101986.

Jiang, X., Payá-Tormo, L., Coroian, D., García-Rubio, I., Castellanos-Rueda, R., Eseverri, Á, López-Torrejón, G., Burén, S., & Rubio, L. M. (2021). Exploiting genetic diversity and gene synthesis to identify superior nitrogenase *NifH* protein variants to engineer N_2-fixation in plants. *Communications Biology*, *4*(1), 4. https://doi.org/10.1038/s42003-020-01536-6.

Kistner, C., & Parniske, M. (2002). Evolution of signal transduction in intracellular symbiosis. *Trends in Plant Science*, *7*(11), 511–518. https://doi.org/10.1016/s1360-1385(02)02356-7.

Kolan, D., Cattan-Tsaushu, E., Enav, H., Freiman, Z., Malinsky-Rushansky, N., Ninio, S., & Avrani, S. (2024). Tradeoffs between phage resistance and nitrogen fixation drive the evolution of genes essential for cyanobacterial heterocyst functionality. *The ISME Journal*, *18*(1), wrad008. https://doi.org/10.1093/ismejo/wrad008.

Kumar, S. B., Kalwasińska, A., Świątczak, J., Brzezinska, M. S., & Kesy, J. (2024). Characterization of salt-tolerant diazotrophs with plant growth-promoting potential isolated from soda industry-affected technosoils. *Plant and Soil*, *503*(1), 645–658. https://doi.org/10.1007/s11104-024-06612-y.

Kuypers, M. M., Marchant, H. K., & Kartal, B. (2018). The microbial nitrogen-cycling network. *Nature Reviews Microbiology*, *16*(5), 263–276. https://doi.org/10.1038/nrmicro.2018.9.

Li, P., Tian, Y., Yang, K., Tian, M., Zhu, Y., Chen, X., Hu, R., Qin, T., Liu, Y., Peng, S., Yi, Z., Liu, Z., Ao, H., & Li, J. (2024). Mechanism of microbial action of the inoculated nitrogen-fixing bacterium for growth promotion and yield enhancement in rice (*Oryza sativa* L.). *Advances in Biotechnology*, *2*(4), 32. https://doi.org/10.1007/s44307-024-00038-4.

Li, X., Luo, X., Wang, A., Chen, W., & Huang, Q. (2023). Control factors of soil diazotrophic community assembly and nitrogen fixation rate across eastern China. *Geoderma*, *432*, 116410. https://doi.org/10.1016/j.geoderma.2023.116410.

Li, X. R., Sun, J., Albinsky, D., Zarrabian, D., Hull, R., Lee, T., Jarratt-Barnham, E., Chiu, C. H., Jacobsen, A., Soumpourou, E., Albanese, A., Kohlen, W., Luginbuehl, L. H., Guillotin,

B., Lawrensen, T., Lin, H., Murray, J., Wallington, E., Harwood, W., Choi, J., Paszkowski, U., & Oldroyd, G. E. D. (2022). Nutrient regulation of lipochitooligosaccharide recognition in plants via *NSP1* and *NSP2*. *Nature Communications*, *13*(1), 6421. https://doi.org/10.1038/s41467-022-33908-3.

Li, X. X., Liu, Q., Liu, X. M., Shi, H. W., & Chen, S. F. (2016). Using synthetic biology to increase nitrogenase activity. *Microbial Cell Factories*, *15*(1), 43. https://doi.org/10.1186/s12934-016-0442-6.

Li, Y., Li, Q., Guan, G., & Chen, S. (2020). Phosphate solubilizing bacteria stimulate wheat rhizosphere and endosphere biological nitrogen fixation by improving phosphorus content. *Peer J*, *8*, e9062. https://doi.org/10.7717/peerj.9062.

Li, Y., Li, Y., Zhang, H., Wang, M., & Chen, S. (2019). Diazotrophic *Paenibacillus beijingensis* BJ-18 provides nitrogen for plant and promotes plant growth, nitrogen uptake and metabolism. *Frontiers in Microbiology*, *10*, 1119. https://doi.org/10.3389/fmicb.2019.01119.

Liu, W., Xu, H., Zhou, J., He, Y., Zou, D., Bai, Y., & Zhang, J. (2024). The plant microbiota: From theoretical advances to applications. *Fundamental Research*, https://doi.org/10.1016/j.fmre.2024.04.016.

Liu, Z., Li, S., Liu, N., Huang, G., & Zhou, Q. (2022). Soil microbial community driven by soil moisture and nitrogen in milk vetch (*Astragalus sinicus* L.)-rapeseed (*Brassica napus* L.) intercropping. *Agriculture*, *12*(10), 1538. https://doi.org/10.3390/agriculture12101538.

Lopez-Simon, J., Vila-Nistal, M., Rosenova, A., De Corte, D., Baltar, F., & Martinez-Garcia, M. (2023). Viruses under the Antarctic ice shelf are active and potentially involved in global nutrient cycles. *Nature Communications*, *14*(1), 8295. https://doi.org/10.1038/s41467-023-44028-x.

Luo, T., Li, C. N., Yan, R., Huang, K., Li, Y. R., Liu, X. Y., & Lakshmanan, P. (2023). Physiological and molecular insights into the resilience of biological nitrogen fixation to applied nitrogen in *Saccharum spontaneum*, wild progenitor of sugarcane. *Frontiers in Plant Science*, *13*, 1099701. https://doi.org/10.3389/fpls.2022.1099701.

Magori, S., & Kawaguchi, M. (2009). Long-distance control of nodulation: Molecules and models. *Molecules and Cells*, *27*(2), 129–134. https://doi.org/10.1007/s10059-009-0016-0.

Malviya, M. K., Solanki, M. K., Li, C.-N., Htun, R., Singh, R. K., Singh, P., & Li, Y.-R. (2021). Sugarcane microbiome: Role in sustainable production. In *Microbiomes and Plant Health* . (pp. 225–242). Elsevier. https://doi.org/10.1016/B978-0-12-819715-8.00007-0.

Martinez-Feria, R., Simmonds, M. B., Ozaydin, B., Lewis, S., Schwartz, A., Pluchino, A., McKellar, M., Gottlieb, S. S., Kayatsky, T., Vital, R., Mehlman, S. E., Caron, Z., Colaianni, N. R., Ané, J. M., Maeda, J., Infante, V., Karlsson, B. H., McLimans, C., … Temme, K. (2024). Genetic remodeling of soil diazotrophs enables partial replacement of synthetic nitrogen fertilizer with biological nitrogen fixation in maize. *Scientific Reports*, *14*(1), 27754. https://doi.org/10.1038/s41598-024-78243-3.

Massahiro Yassue, R., Galli, G., James Chen, C. P., Fritsche-Neto, R., & Morota, G. (2023). Genome-wide association analysis of hyperspectral reflectance data to dissect the genetic architecture of growth-related traits in maize under plant growth-promoting bacteria inoculation. *Plant Direct*, *7*(4), e492. https://doi.org/10.1002/pld3.492.

Meier, M. A., Xu, G., Lopez-Guerrero, M. G., Li, G., Smith, C., Sigmon, B., Herr, J. R., Alfano, J. R., Ge, Y., Schnable, J. C., & Yang, J. (2022). Association analyses of host genetics, root-colonizing microbes, and plant phenotypes under different nitrogen conditions in maize. *eLife*, *11*, e75790. https://doi.org/10.7554/eLife.75790.

Meneses, C., Gonçalves, T., Alquéres, S., Rouws, L., Serrato, R., Vidal, M., & Baldani, J. (2017). *Gluconacetobacter diazotrophicus* exopolysaccharide protects bacterial cells against oxidative stress in vitro and during rice plant colonization. *Plant and Soil*, *416*(1), 133–147. https://doi.org/10.1007/s11104-017-3201-5.

Mitter, B., Pfaffenbichler, N., Flavell, R., Compant, S., Antonielli, L., Petric, A., Berninger, T., Naveed, M., Sheibani-Tezerji, R., Von Maltzahn, G., & Sessitsch, A. (2017). A new approach to modify plant microbiomes and traits by introducing beneficial bacteria at flowering into progeny seeds. *Frontiers in Microbiology*, *8*, 11. https://doi.org/10.3389/fmicb.2017.00011.

Narehood, S. M., Cook, B. D., Srisantitham, S., Eng, V. H., Shiau, A. A., McGuire, K. L., Britt, R. D., Herzik, Jr, M. A., & Tezcan, F. A. (2025). Structural basis for the conformational protection of nitrogenase from O_2. *Nature*, *637*(8047), 991–997. https://doi.org/10.1038/s41586-024-08311-1.

Nida, K., Siddiqui, Z. S., Siddiqui, M. H., Salman, Z. A., & Umar, M. (2024). *Azotobacter* modulate nitrogen assimilation, sustain light harvesting efficiency and photosynthetic performance of maize cultivar in a saline soil. *Journal of Plant Nutrition and Soil Science*, *24*(3), 4624–4640. https://doi.org/10.1007/s42729-024-01859-x.

Norman, J. S., & Friesen, M. L. (2017). Complex N acquisition by soil diazotrophs: How the ability to release exoenzymes affects N fixation by terrestrial free-living diazotrophs. *The ISME Journal*, *11*(2), 315–326. https://doi.org/10.1038/ismej.2016.127.

Nunes Oliveira, F. L., Silva Oliveira, W., Pereira Stamford, N., Nova Silva, E. V., & Santiago Freitas, A. D. (2017). Effectiveness of biofertilizer enriched in N by *Beijerinckia indica* on sugarcane grown on an Ultisol and the interactive effects between biofertilizer and sugarcane filter cake. *Journal of Soil Science and Plant Nutrition*, *17*(4), 1040–1057. https://doi.org/10.4067/S0718-95162017000400015.

Pang, Z., Mao, X., Zhou, S., Yu, S., Liu, G., Lu, C., Wan, J., Hu, L., & Xu, P. (2023). Microbiota-mediated nitrogen fixation and microhabitat homeostasis in aerial root-mucilage. *Microbiome*, *11*(1), 85. https://doi.org/10.1186/s40168-023-01525-x.

Pankievicz, V. C., Irving, T. B., Maia, L. G., & Ané, J. M. (2019). Are we there yet? The long walk towards the development of efficient symbiotic associations between nitrogen-fixing bacteria and non-leguminous crops. *BMC Biology*, *17*(1), 99. https://doi.org/10.1186/s12915-019-0710-0.

Priyadarshini, P., Choudhury, S., Tilgam, J., Bharati, A., & Sreeshma, N. (2021). Nitrogen fixing cereal: A rising hero towards meeting food security. *Plant Physiology and Biochemistry*, *167*, 912–920. https://doi.org/10.1016/j.plaphy.2021.09.012.

Qin, Y., Khan, Q., Yan, J. W., Wang, Y. Y., Pan, Y. F., Huang, Y., Wei, J. L., Guo, D. J., Li, Y. R., Dong, D. F., & Xing, Y. X. (2024). Molecular mechanism of endophytic bacteria DX120E regulating polyamine metabolism and promoting plant growth in sugarcane. *Frontiers in Plant Science*, *15*, 1334907. https://doi.org/10.3389/fpls.2024.1334907.

Qin, Y., Xie, X. Q., Khan, Q., Wei, J. L., Sun, A. N., Su, Y. M., Guo, D. J., Li, Y. R., & Xing, Y. X. (2022). Endophytic nitrogen-fixing bacteria DX120E inoculation altered the carbon and nitrogen metabolism in sugarcane. *Frontiers in Microbiology*, *13*, 1000033. https://doi.org/10.3389/fmicb.2022.1000033.

Rana, K. L., Negi, R., Sharma, B., Yadav, A., Devi, R., Kaur, T., Shreaz, S., Rustagi, S., Rai, A. K., Singh, S., Kour, D., & Yadav, A. N. (2024). Potential effect of novel endophytic nitrogen fixing diverse species of *Rahnella* on growth promotion of wheat (*Triticum aestivum* L.). *Journal of Crop Science and Biotechnology*, *27*(5), 605–615. https://doi.org/10.1007/s12892-024-00254-3.

Rosman, A. C., Urquiaga, M. C.dO., Thiebaut, F., Ballesteros, H. G. F., de Oliveira, E. A. G., & Hemerly, A. S. (2024). Dual RNA-seq of maize and *H. seropedicae* ZAE94 association, in different doses of nitrate, reveals novel insights into plant-PGPB-environment relationship. *Frontiers in Plant Science*, *15*, 1346523. https://doi.org/10.3389/fpls.2024.1346523.

Rutledge, H. L., Cook, B. D., Nguyen, H. P., Herzik, Jr, M. A., & Tezcan, F. A. (2022). Structures of the nitrogenase complex prepared under catalytic turnover conditions. *Science*, *377*(6608), 865–869. https://doi.org/10.1126/science.abq7641.

Saniya, T. K., Priya, H., Kokila, V., Nivedha, R. M., Singh, R., Reddy, K. S., Kumar, A., Shivay, Y. S., & Prasanna, R. (2024). Evaluating the promise of diazotrophs from acid soils for improving early vigour and soil nutrient availability in wheat crop grown in contrasting pH environments. *Discover Soil*, *1*(1), 24. https://doi.org/10.1007/s44378-024-00024-1.

Sanjenbam, P., Shivaprasad, P., Agashe, D. (2022). Impact of phyllosphere *Methylobacterium* on host rice landraces. *Applied and Industrial Microbiology*, *10*(4), e0081022. https://doi.org/10.1128/spectrum.00810-22.

Santi, C., Bogusz, D., & Franche, C. (2013). Biological nitrogen fixation in non-legume plants. *Annals of Botany, 111*(5), 743–767. https://doi.org/10.1093/aob/mct048.

Sarkodee-Addo, E., Tokiwa, C., Bonney, P., Aboagye, D. A., Yeboah, A., Abebrese, S. O., Bam, R., Nartey, E. K., Okazaki, S., & Yasuda, M. (2021). Biofertilizer activity of *Azospirillum* sp. B510 on the rice productivity in Ghana. *Microorganisms, 9*(9), 2000. https://doi.org/10.3390/microorganisms9092000.

Serazetdinova, Y. R., Borodina, E. E., Fotina, N. V., Naik, A., Mudgal, G., & Asyakina, L. K. (2025). Rhizobia as complex biofertilizers for wheat: Biological nitrogen fixation and plant growth promotion. *Foods and Raw Materials, 14*(1), 214–227. https://doi.org/10.21603/2308-4057-2026-1-669.

Shayanthan, A., Ordoñez, P. A. C., & Oresnik, I. J. (2022). The role of synthetic microbial communities (SynCom) in sustainable agriculture. *Frontiers in Agronomy, 4*, 896307. https://doi.org/10.3389/fagro.2022.896307.

Shi, W., Zhao, H. Y., Chen, Y., Wang, J. S., Han, B., Li, C. P., Lu, J. Y., & Zhang, L. M. (2021). Organic manure rather than phosphorus fertilization primarily determined asymbiotic nitrogen fixation rate and the stability of diazotrophic community in an upland red soil. *Agriculture, Ecosystems & Environment, 319*, 107535. https://doi.org/10.1016/j.agee.2021.107535.

Sigalas, P. P., Buchner, P., Kröper, A., & Hawkesford, M. J. (2023). The functional diversity of the high-affinity nitrate transporter gene family in hexaploid wheat: Insights from distinct expression profiles. *International Journal of Molecular Sciences, 25*(1), 509. https://doi.org/10.3390/ijms25010509.

Singh, R. K., Singh, P., Li, H. B., Song, Q. Q., Guo, D. J., Solanki, M. K., Verma, K. K., Malviya, M. K., Song, X. P., Lakshmanan, P., Yang, L. T., & Li, Y. R. (2020). Diversity of nitrogen-fixing rhizobacteria associated with sugarcane: A comprehensive study of plant-microbe interactions for growth enhancement in *Saccharum* spp. *BMC Plant Biology, 20*(1), 220. https://doi.org/10.1186/s12870-020-02400-9.

Smanski, M. J., Bhatia, S., Zhao, D., Park, Y., BA Woodruff, L., Giannoukos, G., Ciulla, D., Busby, M., Calderon, J., Nicol, R., Gordon, D. B., Densmore, D., & Voigt, C. A. (2014). Functional optimization of gene clusters by combinatorial design and assembly. *Nature Biotechnology, 32*(12), 1241–1249. https://doi.org/10.1038/nbt.3063.

Solanki, M. K., Wang, Z., Wang, F. Y., Li, C. N., Gupta, C. L., Singh, R. K., Malviya, M. K., Singh, P., Yang, L. T., & Li, Y. R. (2020). Assessment of diazotrophic *proteobacteria* in sugarcane rhizosphere when intercropped with legumes (peanut and soybean) in the field. *Frontiers in Microbiology, 11*, 1814. https://doi.org/10.3389/fmicb.2020.01814.

Soto, G., Fox, A. R., & Ayub, N. D. (2013). Exploring the intrinsic limits of nitrogenase transfer from bacteria to eukaryotes. *Journal of Molecular Evolution, 77*(1), 3–7. https://doi.org/10.1007/s00239-013-9578-8.

Soumare, A., Diedhiou, A. G., Thuita, M., Hafidi, M., Ouhdouch, Y., Gopalakrishnan, S., & Kouisni, L. (2020). Exploiting biological nitrogen fixation: A route towards a sustainable agriculture. *Plants (Basel), 9*(8), 1011. https://doi.org/10.3390/plants9081011.

Sun, J., Miller, J. B., Granqvist, E., Wiley-Kalil, A., Gobbato, E., Maillet, F., Cottaz, S., Samain, E., Venkateshwaran, M., Fort, S., Morris, R. J., Ané, J. M., Dénarié, J., & Oldroyd, G. E. (2015). Activation of symbiosis signaling by arbuscular mycorrhizal fungi in legumes and rice. *The Plant Cell, 27*(3), 823–838. https://doi.org/10.1105/tpc.114.131326.

Tang, Y., Qin, D., Tian, Z., Chen, W., Ma, Y., Wang, J., Yang, J., Yan, D., Dixon, R., & Wang, Y. P. (2023). Diurnal switches in diazotrophic lifestyle increase nitrogen contribution to cereals. *Nature Communications, 14*(1), 7516. https://doi.org/10.1038/s41467-023-43370-4.

Temme, K., Zhao, D., & Voigt, C. A. (2012). Refactoring the nitrogen fixation gene cluster from *Klebsiella oxytoca. Proceedings of the National Academy of Sciences of the United States of America, 109*(18), 7085–7090. https://doi.org/10.1073/pnas.1120788109.

Tong, D., Wang, Y., Yu, H., Shen, H., Dahlgren, R. A., & Xu, J. (2023). Viral lysing can alleviate microbial nutrient limitations and accumulate recalcitrant dissolved organic matter components in soil. *The ISME Journal, 17*(8), 1247–1256. https://doi.org/10.1038/s41396-023-01438-5.

Tran, V. C., Nguyen, T. H., Hoang, T. L. A., Pham, T. T. M., Nguyen, T. M., Tran, T. D., Nguyen, T. H. G., Han, V. C., Tran, A. T., & La, V. H. (2024). Harnessing synthetic microbial communities with nitrogen-fixing activity to promote rice growth. *Open Agriculture*, *9*(1), 20220392. https://doi.org/10.1515/opag-2022-0392.

Ureta, A., & Nordlund, S. (2002). Evidence for conformational protection of nitrogenase against oxygen in *Gluconacetobacter diazotrophicus* by a putative FeSII protein. *Journal of Bacteriology*, *184*(20), 5805–5809. https://doi.org/10.1128/JB.184.20.5805-5809.2002.

Van Deynze, A., Zamora, P., Delaux, P. M., Heitmann, C., Jayaraman, D., Rajasekar, S., Graham, D., Maeda, J., Gibson, D., Schwartz, K. D., Berry, A. M., Bhatnagar, S., Jospin, G., Darling, A., Jeannotte, R., Lopez, J., Weimer, B. C., Eisen, J. A., Shapiro, H. Y., Ané, J. M., & Bennett, A. B. (2018). Nitrogen fixation in a landrace of maize is supported by a mucilage-associated diazotrophic microbiota. *PLoS Biology*, *16*(8), e2006352. https://doi.org/10.1371/journal.pbio.2006352.

Velázquez-Hernández, M. L., Baizabal-Aguirre, V. M., Cruz-Vázquez, F., Trejo-Contreras, M. J., Fuentes-Ramírez, L. E., Bravo-Patiño, A., Cajero-Juárez, M., Chávez-Moctezuma, M. P., & Valdez-Alarcón, J. J. (2011). *Gluconacetobacter diazotrophicus* levansucrase is involved in tolerance to NaCl, sucrose and desiccation, and in biofilm formation. *Archives of Microbiology*, *193*(2), 137–149. https://doi.org/10.1007/s00203-010-0651-z.

Wallace, J. G., Kremling, K. A., Kovar, L. L., & Buckler, E. S. (2018). Quantitative genetics of the maize leaf microbiome. *Phytobiomes Journal*, *2*(4), 208–224. https://doi.org/10.1094/PBIOMES-02-18-0008-R.

Wang, L., Zhang, L., Liu, Z., Zhao, D., Liu, X., Zhang, B., Xie, J., Hong, Y., Li, P., & Chen, S. (2013). A minimal nitrogen fixation gene cluster from *Paenibacillus* sp. WLY78 enables expression of active nitrogenase in *Escherichia coli*. *PLoS Genetics*, *9*(10), e1003865. https://doi.org/10.1371/journal.pgen.1003865.

Wu, J., Liu, S., Zhang, H., Chen, S., Si, J., Liu, L., Wang, Y., Tan, S., Du, Y., Jin, Z., Xie, J., & Zhang, D. (2025). Flavones enrich rhizosphere *Pseudomonas* to enhance nitrogen utilization and secondary root growth in *Populus*. *Nature Communications*, *16*(1), 1461. https://doi.org/10.1038/s41467-025-56226-w.

Wu, L., Dong, J., Song, J., Zhu, Y., Che, S., Qin, X., Xu, Y., Tian, S., Wang, D., Tian, P., Wei, X., Yang, X., Yang, M., & Wu, Z. (2024). The nitrogen fixation characteristics of terrestrial nitrogen-fixing cyanobacteria and their role in promoting rice growth. *Agronomy*, *15*(1), 62. https://doi.org/10.3390/agronomy15010062.

Xiang, N., Guo, C., Liu, J., Xu, H., Dixon, R., Yang, J., & Wang, Y. P. (2020). Using synthetic biology to overcome barriers to stable expression of nitrogenase in eukaryotic organelles. *Proceedings of the National Academy of Sciences of the United States of America*, *117*(28), 16537–16545. https://doi.org/10.1073/pnas.2002307117.

Xiao, B., Huang, J., Guo, J., Lu, X., Zhu, L., Wang, J., & Zhou, C. (2022). Flooding-induced rhizosphere *Clostridium* assemblage prevents root-to-shoot cadmium translocation in rice by promoting the formation of root apoplastic barriers. *Journal of Hazardous Materials*, *439*, 129619. https://doi.org/10.1016/j.jhazmat.2022.129619.

Xiao, D., He, X., Xu, Z., Bai, S. H., Zhang, W., Hu, P., Chen, M., & Wang, K. (2023). Strong cooperations among diazotroph and arbuscular mycorrhizal fungi taxa promote free-living nitrogen fixation at soil-rock mixing layer. *Geoderma*, *437*, 116600. https://doi.org/10.1016/j.geoderma.2023.116600.

Xiong, C., Singh, B. K., Zhu, Y. G., Hu, H. W., Li, P. P., Han, Y. L., Han, L. L., Zhang, Q. B., Wang, J. T., Liu, S. Y., Wu, C. F., Ge, A. H., Zhang, L. M., & He, J. Z. (2024). Microbial species pool-mediated diazotrophic community assembly in crop microbiomes during plant development. mSystems, 9(4), e0105523. https://doi.org/10.1128/msystems.01055-23.

Xu, P., & Wang, E. (2023). Diversity and regulation of symbiotic nitrogen fixation in plants. *Current Biology*, *33*(11), R543–R559. https://doi.org/10.1016/j.cub.2023.04.053.

Yan, D., Tajima, H., Cline, L. C., Fong, R. Y., Ottaviani, J. I., Shapiro, H. Y., & Blumwald, E. (2022). Genetic modification of flavone biosynthesis in rice enhances biofilm formation of soil diazotrophic bacteria and biological nitrogen fixation. *Plant Biotechnology Journal, 20*(11), 2135–2148. https://doi.org/10.1111/pbi.13894.

Yang, B., Wang, X. M., Ma, H. Y., Yang, T., Jia, Y., Zhou, J., & Dai, C. C. (2015). Fungal endophyte *Phomopsis liquidambari* affects nitrogen transformation processes and related microorganisms in the rice rhizosphere. *Frontiers in Microbiology, 6*, 982. https://doi.org/10.3389/fmicb.2015.00982.

Yang, J., Xie, X., Wang, X., Dixon, R., & Wang, Y. P. (2014). Reconstruction and minimal gene requirements for the alternative iron-only nitrogenase in *Escherichia coli. Proceedings of the National Academy of Sciences of the United States of America, 111*(35), E3718–E3725. https://doi.org/10.1073/pnas.1411185111.

Yang, J., Xie, X., Xiang, N., Tian, Z. X., Dixon, R., & Wang, Y. P. (2018). Polyprotein strategy for stoichiometric assembly of nitrogen fixation components for synthetic biology. *Proceedings of the National Academy of Sciences of the United States of America, 115*(36), E8509–E8517. https://doi.org/10.1073/pnas.1804992115.

Yao, Q. H., Peng, R. H., Wang, B., Tian, Y. S., Zhu, Y. M., Gao, J. J., Li, Z. J., Fu, X. Y., Xu, J., Han, H. J., Wang, L. J., Wang, Y., Zhang, F. J., & Deng, Y. D. (2021). Endowing plants with the capacity for autogenic nitrogen fixation. *Research Square*, https://doi.org/10.21203/rs.3.rs-436726/v1. [Preprint].

Yin, Y., Wang, Y. F., Cui, H. L., Zhou, R., Li, L., Duan, G. L., & Zhu, Y. G. (2023). Distinctive structure and assembly of phyllosphere microbial communities between wild and cultivated rice. *Microbiology Spectrum, 11*(1), e0437122. https://doi.org/10.1128/spectrum.04371-22.

Yu, G.-H., Kuzyakov, Y., Luo, Y., Goodman, B. A., Kappler, A., Liu, F. F., & Sun, F. S. (2021a). Molybdenum bioavailability and asymbiotic nitrogen fixation in soils are raised by iron (oxyhydr) oxide-mediated free radical production. *Environmental Science & Technology, 55*(21), 14979–14989. https://doi.org/10.1021/acs.est.1c04240.

Yu, H., Liu, X., Yang, C., Peng, Y., Yu, X., Gu, H., Zheng, X., Wang, C., Xiao, F., Shu, L., He, Z., Wu, B., & Yan, Q. (2021c). Co-symbiosis of arbuscular mycorrhizal fungi (AMF) and diazotrophs promote biological nitrogen fixation in mangrove ecosystems. *Soil Biology & Biochemistry, 161*, 108382. https://doi.org/10.1016/j.soilbio.2021.108382.

Yu, P., He, X., Baer, M., Beirinckx, S., Tian, T., Moya, Y. A., Zhang, X., Deichmann, M., Frey, F. P., Bresgen, V., Li, C., Razavi, B. S., Schaaf, G., von Wirén, N., Su, Z., Bucher, M., Tsuda, K., Goormachtig, S., Chen, X., & Hochholdinger, F. (2021b). Plant flavones enrich rhizosphere Oxalobacteraceae to improve maize performance under nitrogen deprivation. *Nature Plants, 7*(4), 481–499. https://doi.org/10.1038/s41477-021-00897-y.

Zaheer, M. S., Ali, H. H., Iqbal, M. A., Erinle, K. O., Javed, T., Iqbal, J., Hashmi, M. I. U., Mumtaz, M. Z., Salama, E. A., Kalaji, H. M., Wróbel, J., & Dessoky, E. S. (2022). Cytokinin production by *Azospirillum brasilense* contributes to increase in growth, yield, antioxidant, and physiological systems of wheat (*Triticum aestivum* L.). *Frontiers in Microbiology, 13*, 886041. https://doi.org/10.3389/fmicb.2022.886041.

Zeng, Y., Wang, M., Yu, Y., Wang, L., Cui, L., Li, C., Liu, Y., & Zheng, Y. (2024). Rice N-biofertilization by inoculation with an engineered photosynthetic diazotroph. *World Journal of Microbiology and Biotechnology, 40*(5), 136. https://doi.org/10.1007/s11274-024-03956-6.

Zhang, H., Chen, Q. f, Shang, N., Li, N., Niu, Q. h, Hong, Q., & Huang, X. (2021). The enhanced mechanisms of *Hansschlegelia zhihuaiae* S113 degrading bensulfuron-methyl in maize rhizosphere by three organic acids in root exudates. *Ecotoxicology and Environmental Safety, 223*, 112622. https://doi.org/10.1016/j.ecoenv.2021.112622.

Zhang, J., Wu, B., Wang, G., Zhang, J., & Jia, C. (2024a). Responses of diazotrophic network structure and community diversity to alfalfa-maize intercropping are soil property-dependent. *Frontiers in Microbiology, 15*, 1425898. https://doi.org/10.3389/fmicb.2024.1425898.

Zhang, L., Yuan, L., Wen, Y., Zhang, M., Huang, S., Wang, S., Zhao, Y., Hao, X., Li, L., Gao, Q., Wang, Y., Zhang, S., Huang, S., Liu, K., Yu, X., Li, D., Xu, J., Zhao, B., Zhang, L., Zhang, H., Zhou, W., & Ai, C. (2024b). Maize functional requirements drive the selection of rhizobacteria under long-term fertilization practices. *The New Phytologist*, *242*(3), 1275–1288. https://doi.org/10.1111/nph.19653.

Zhang, L., Zhang, M., Huang, S., Li, L., Gao, Q., Wang, Y., Zhang, S., Huang, S., Yuan, L., Wen, Y., Liu, K., Yu, X., Li, D., Zhang, L., Xu, X., Wei, H., He, P., Zhou, W., Philippot, L., & Ai, C. (2022). A highly conserved core bacterial microbiota with nitrogen-fixation capacity inhabits the xylem sap in maize plants. *Nature Communications*, *13*(1), 3361. https://doi.org/10.1038/s41467-022-31113-w.

Zheng, M., Xu, M., Li, D., Deng, Q., & Mo, J. (2023). Negative responses of terrestrial nitrogen fixation to nitrogen addition weaken across increased soil organic carbon levels. *The Science of the Total Environment*, *877*, 162965. https://doi.org/10.1016/j.scitotenv.2023.162965.

Zhang, T., Yan, Y., He, S., Ping, S., Alam, K. M., Han, Y., Liu, X., Lu, W., Zhang, W., Chen, M., Xiang, W., Wang, X., & Lin, M. (2012). Involvement of the ammonium transporter AmtB in nitrogenase regulation and ammonium excretion in *Pseudomonas stutzeri* A1501. *Research in Microbiology*, *163*(5), 332–339. https://doi.org/10.1016/j.resmic.2012.05.002.

Zhong, Y., Tian, J., Li, X., & Liao, H. (2023). Cooperative interactions between nitrogen fixation and phosphorus nutrition in legumes. *The New Phytologist*, *237*(3), 734–745. https://doi.org/10.1111/nph.18593.

Zhou, M., Wang, J., Yang, R., Xu, X., Lian, D., Xu, Y., Shen, H., Zhang, H., Xu, J., & Liang, M. (2025). *Stenotrophomonas sp.* SI-NJAU-1 and its mutant strain with excretion-ammonium capability promote plant growth through biological nitrogen fixation. *Journal of Agricultural and Food Chemistry*, *73*(7), 3874–3886. https://doi.org/10.1021/acs.jafc.4c08697.

Biological Nitrogen Fixation: From Rhizobia to Synthetic Approaches

2

Jitendra Kumar Sharma, Anjali Pathak,
Meenakshi Shrivastav and Mahendra Kumar Gupta

Abstract

Nitrogen has been known to be an essential macronutrient which significantly participates in the metabolic processes of plants, but it often represents the limiting factor responsible for arresting agricultural productivity, mainly due to the poor bioavailability of the nutrient within most soil textures. Biological nitrogen fixation (BNF), represents the process catalyzed by the widespread and heterogeneous class of prokaryotic bacterial microbes, which represents an environment friendly alternative to the energy-intensive processes of chemical synthesis presently utilized for the purpose of nitrogen fixation. Through the action of the enzyme complex known as nitrogenase, these microbes have the ability to convert the gaseous atmospheric dinitrogen (N2) to the ammonia, (NH3), which represents the assimilable and plant-usable form of the nutrient required for plant growth and productive development. In this writing, the evolution, the molecular mechanisms underlying the resultant of biological nitrogen fixation, and the ecological significance thereof, including the codominant relationships between the rhizobia and the legumes and the associative processes of the non-legume plants, have been examined, with special focus placed on the associative processes of the latter plants. Furthermore, this study focused on the new synthetic biology technologies which have come to be developed, all towards the purpose of the insertion of the property of the nitrogen-fixing ability into the most farmed food

J. K. Sharma · A. Pathak · M. K. Gupta (✉)
Microbiology Research Laboratory, School of Studies in Botany, Jiwaji University, Gwalior, Madhya Pradesh, India
e-mail: mkgsac@yahoo.com

M. Shrivastav
School of Studies in Microbiology, Jiwaji University, Gwalior, Madhya Pradesh, India

R. K. Singh and P. Singh (eds.), *Diazotrophic Plant Growth-Promoting Microorganisms and Green Energy – Volume I*, https://doi.org/10.1007/978-981-95-9682-9_2

grain crops. Through the integration of the latest developments of genomics, microbiological engineering, and microbial ecologies, these new technologies have the ability of reducing the worldwide use of chemical fertilizers, improving the soil quality, and ensuring the promotion of the sustainable agricultural production, particularly where the effects of the presently experienced problems of the loss of the greenhouse have been sensed.

Keywords

Biological nitrogen fixation · Nitrogenase · Biotechnology · Agriculture · Nitrogen · Rhizobia ·

2.1 Introduction

Nitrogen is a key factor in the stability and productivity of aquatic and terrestrial ecosystems (Fig. 2.1), but the gaseous atmospheric form, N_2, is inaccessible to most organisms because of its strong triple bond. Therefore, the presence of fixed nitrogen compounds, including ammonium and nitrate, tends to limit the growth of plants and agricultural productivity (Ladha et al., 2022). The industrialized societies are increasingly depending on gaseous nitrogen that is industrially fixed through the Haber-Bosch process, which has led to an unprecedented increase in food production. This convenience is, however, coupled with astronomical environmental and economic costs. Presently, the Haber-Bosch process of producing ammonia, which is energy-intensive, relies on fossil hydrogen, producing large volumes of greenhouse gases. In addition, fertilizer wastes are a cause of eutrophication, nitrate increase in drinking water aquifers, and nitrous oxide emission in soils (Ghavam et al., 2021; Lawrence et al., 2021). Environmental externalities are economically costly in terms of loss of ecosystem services, cost of water treatment, and losses related to climate effects. Moreover, the market price of manufactured nitrogen, although volatile, is prohibitively high and is the largest input for farmers all over the world (Palmero et al., 2022). This paradox, which is well exemplified, emphasizes the deep contradiction of the presence of a lot of atmospheric gaseous N and the ubiquitous lack of nitrogen; to gain access to atmospheric N_2, it is necessary to either use specialized microbial biochemical reactions or to undergo energetically costly industrial changes (Ladha et al., 2022).

The history of the development of biological nitrogen fixation serves as a conceptual transition between traditional natural history and modern molecular biology. The first observations made about root nodules and legume productivity eventually showed that symbiotic bacteria provide lower nitrogen to plants. The following decades of

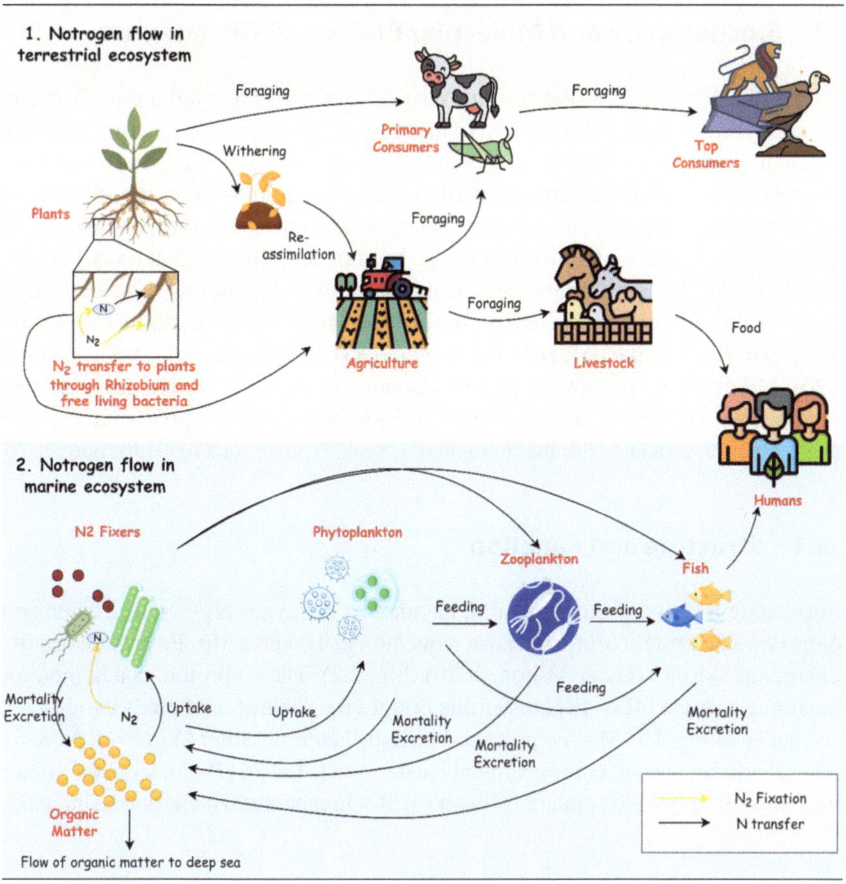

Fig. 2.1 Nitrogen fixation and flow in terrestrial and marine ecosystems

genetics, biochemistry, and imaging advances have further elucidated the nodulation, nitrogenase activity, and nutrient exchange mechanisms (Roy et al., 2019). New methodological approaches such as isotopic ^{15}N tracing, high-throughput omics, and ecosystem-scale modelling have improved quantitative measurements of biological nitrogen fixation and shed light on its spatial distribution in a wide range of crops, natural vegetation, and aquatic ecosystems (Palmero et al., 2022; Wirth et al., 2024). Recent syntheses suggest that biological processes, including symbiotic and free-living diazotrophs, fix a large proportion of fixed nitrogen in the biosphere and are an important source of nitrogen in a vast array of agroecosystems. This observation supports the urgent need to focus on the biological fixation of nitrogen to achieve sustainable intensification, soil carbon sequestration, and the creation of climate-smart food systems (Palmero et al., 2022; Guo et al., 2022; Wirth et al., 2024).

2.2 Biochemistry and Molecular Biology of Nitrogenase

The enzyme nitrogenase is the key enzyme of biological nitrogen fixation (BNF) and processes the atmospheric reduction of elemental dinitrogen (N_2) to ammonia (NH_3) to yield life's major metabolites, e.g., nucleotides and amino acids. Three isoforms of nitrogenase are found in nature that are distinguished based on the composition of the heterometal in their active site cofactor. The molybdenum [(Mo) nitrogenase (encoded by *nifHDK*) is found in all diazotrophs, i.e., bacteria and Archaea that fix N_2] is found irrespective of the environment they inhabit. In contrast to this, only a small group of diazotrophic microorganisms harbor, besides the Mo, also the vanadium (V; encoded by *vnfHDGK*) or iron [(Fe); encoded by *anfHDGK*] nitrogenases (Jasniewski et al., 2020). Mo nitrogenases have a higher efficiency in reducing N_2 and are selectively expressed with respect to the alternate V or Fe nitrogenases that are hypothesized to have a function as a fail-safe mechanism during starvation for Mo (Harwood, 2020).

2.2.1 Structure and Function

Nitrogenase is a two-protein metalloenzyme that catalyzes $N_2 \rightarrow NH_3$. The enzyme comprises a Fe protein (dinitrogenase reductase, *nifH*) and a Mo–Fe protein (dinitrogenase, *nifD*+*nifK*) (Gu & Milton, 2020) (Fig. 2.2). The Fe protein is a homodimer containing a single [4Fe–4S] cluster that bridges the subunits and binds two Mg ATP (one per subunit). The Mo–Fe protein is an $\alpha_2\beta_2$ heterotetramer (*NifD* = α, *NifK* = β) that each $\alpha\beta$ carries two complex metal clusters: an [8Fe–7S] P-cluster at the subunit interface and the Fe–Mo cofactor [(7Fe–9S–1Mo–homocitrate) including an interstitial

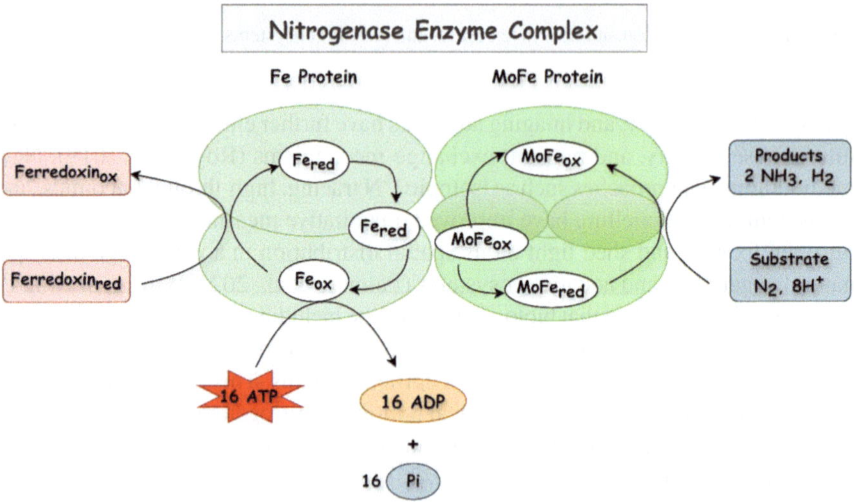

Fig. 2.2 Schematic representation of the electron flow in nitrogen fixation

carbon] in the α subunit (Gu & Milton, 2020). Electron transfer proceeds from the Fe protein's [4Fe–4S] cluster to the P-cluster and then to the Fe–Mo cofactor, with Mg ATP hydrolysis on the Fe protein driving each two-electron transfer (Gu & Milton, 2020; Burén et al., 2020). Overall, eight electrons (supplied by the Fe protein) and 16 ATP are consumed per N_2 reduced, yielding two NH_3 and one H_2 ($N_2 + 8H^+ + 8e^- + 16ATP \rightarrow 2NH_3 + H_2 + 16ADP + 16Pi$). Thus, N_2 fixation by nitrogenase is extremely energy-intensive (Gu & Milton, 2020; Bothe et al., 2010).

2.2.2 Alternative Nitrogenases

Many diazotrophs also harbor alternative nitrogenase systems. In addition to the canonical Mo-nitrogenase, some microbes encode a V-dependent enzyme (*vnfHDGK*) and/or an Fe-only enzyme (*anfHDGK*) (Gu & Milton, 2020; Chanderban et al., 2023). These alternative systems are similarly $\alpha_2\beta_2$ (with an extra δ subunit, making $\alpha_2\beta_2\delta_2$) and share the P-cluster motif, but incorporate a V or all Fe cofactor instead of Mo (Gu & Milton, 2020; Bothe et al., 2010). Generally, Mo-nitrogenase is the preferred enzyme, under Mo-replete conditions microbes express *nif* genes and grow fastest, whereas expression of *vnf* and *anf* genes (and thus V-/Fe-nitrogenase activity) occurs only when Mo (and combined N) are scarce (Chanderban et al., 2023). Growth and transcriptional studies in a methanogen show the highest N_2-fixing growth rate with Mo available and much slower growth with only V or Fe; correspondingly, *vnf/anf* transcription is strongly upregulated when Mo and fixed N are removed (Chanderban et al., 2023). The V- and Fe-nitrogenases are less efficient at N_2 reduction (they produce more H_2 per NH_3) than the Mo enzyme, consistent with these slower growth phenotypes (Chanderban et al., 2023).

2.2.3 Oxygen Sensitivity and Protection Mechanisms

A major challenge for all nitrogenases is O_2, because the [Fe–S] clusters are irreversibly oxidized by O_2 (Bothe et al., 2010). Diazotrophs, therefore, use spatial or temporal segregation to protect nitrogenase. Many filamentous cyanobacteria differentiate O_2-evolving vegetative cells from N_2-fixing heterocysts, degrade Photosystem II, build thick glycolipid envelopes to limit gas diffusion, and maintain very high respiration rates so that residual O_2 is consumed (Bothe et al., 2010). Unicellular cyanobacteria typically fix N_2 at night, using a circadian clock to separate oxygenic photosynthesis (day) from nitrogen fixation (night) (Kumar et al., 2010). In plant symbioses, legume nodules produce leghemoglobin, a high-affinity O_2 carrier that buffers free O_2: it delivers enough O_2 for bacteroid respiration while keeping the free O_2 concentration low enough for nitrogenase to function.

2.2.4 Genetic Organization of Nitrogen Fixation

The *nif* (and *vnf/anf*) genes are typically organized in clusters. For example, the model *Aeromonas/Azotobacter* major nif operon contains *nifH, nifD, nifK* plus ~30 accessory genes (*nifE, nifN, nifB, nifV, nifS, nifU*, etc.) needed for cofactor assembly and regulation. A minimal essential set of genes for Mo-nitrogenase activity is considered *nifH, nifD, nifK, nifE,* and *nifN* (with *nifB* on a separate cassette) (Burén et al., 2020). These clusters often transfer horizontally between bacteria; genes for biosynthesis of the P– and Fe–Mo cofactor (e.g., *nifB, nifV, nifE, nifN, nifS, nifU*) and regulators (*nifA, nifL*) are co-localized with the structural genes. Similarly, *vnf* and *anf* genes form their own operons when present. In all cases, the linked gene organization and transcriptional regulation ensure coordinated expression of nitrogenase structural subunits and the many auxiliary proteins required for cofactor construction and enzyme assembly (Burén et al., 2020).

2.3 Rhizobial Diversity and Classification

Legume-nodulating rhizobia span many genera across two major proteobacterial classes. Recent analyses report over 18 rhizobial genera in the α- and β -Proteobacteria (e.g., *Rhizobium, Sinorhizobium, Bradyrhizobium, Mesorhizobium, Azorhizobium in Alphaproteobacteria; Cupriavidus, Paraburkholderia, Trinickia in Betaproteobacteria*, etc.) (Masson-Boivin et al., 2009; Tang & Capela, 2020). Ongoing genomics has revealed cryptic clades and prompted reclassification: for instance, the historically defined *Mesorhizobium* was shown to be polyphyletic, leading to a recent proposal splitting it into 15 genera (including novel genera, such as *Allomeso rhizobium, Neomeso rhizobium*, etc.) (Li et al., 2024). In practice, surveys of different legumes find that distinct rhizobial assemblages associate with different hosts and soils. For example, invasive *Mimosa* species in China are nodulated by multiple genera: *Cupriavidus* dominates in neutral–alkaline, low-nutrient soils, while *Paraburkholderia* prevails in more acidic, fertile soils (Liu et al., 2020). Such host–environment selection means that even the same legume can form nodules with very different rhizobia in different regions, reflecting local soil pH, nutrient, and climate conditions (Liu et al., 2020). In sum, rhizobial taxonomy is rapidly expanding—hundreds of new *Bradyrhizobium, Rhizobium,* and other species have been described recently—and their geographic distribution is structured by both host plant and environmental factors (pH, salinity, etc.) (Liu et al., 2020; Li et al., 2024).

Host specificity and ecology intertwine in shaping rhizobial community patterns. At the molecular level, legume roots exude flavonoids that induce strain-specific nod genes, which synthesize lipo-chitooligosaccharide Nod factors. These Nod factors, whose structures vary by strain, must precisely match plant receptors to trigger nodule formation (Agudelo et al., 2023). Consequently, the complement of *nod* and *nif* genes largely determines a strain's host range: closely related legumes often share compatible *rhizobia*, whereas transfer of symbiosis genes (via plasmids or lateral transfer) can expand or shift compatibility (Aguilar et al., 2022). Classic "cross-inoculation groups" (grouping legumes and rhizobia by compatibility) have proven limited: many rhizobia nodulate beyond their traditional hosts and vice versa (Mendoza-Suárez et al. 2020, Pankievicz et al. 2019). Current surveys distinguish narrow host range from promiscuous strains based on their ability to nodulate numerous legumes. Free-living rhizobia in soil experience competition and stressing environs. They endure through the creation of biofilms, synthesizing osmoprotectants and competing for niches, frequently having interactions with other microbiota in the rhizosphere (Agudelo et al., 2023). Physical stresses (drought, salinity, alkaline or acidic environments, etc.) considerably impact strain permanence. Acid-tolerant *Burkholderia/Cupriavidus*, for instance, dominate stressed, poor, infertile soils while *Ensifer/Rhizobium* prefer neutral–alkaline soils (Liu et al., 2020). Even larger-scale environmental surveys validate climate and soil pH as principal determinants of worldwide nitrogen-fixer diversity, implying that the biogeography of rhizobia results from adaptation to localized legume symbiont ecosystems (ecological niches) (Liu et al., 2020). Finally, general trends from new research emphasize the fact that classification of rhizobia and their diversity go intimately with host specificity and legume microbiology, so that across the globe, there results an intricate symbiont–legume association mosaic (Li et al., 2024; Agudelo et al., 2023).

2.4 Legume-Rhizobia Symbiosis: Molecular Dialogue

Legumes and their Rhizobia have a highly refined molecular conversation. Flavonoids are released by legume roots and stimulate the rhizobial transcription factor NodD to promote expression of nodule-forming genes and lipo-chito-oligosaccharide Nod factor synthesis (Haskett et al., 2025). These Nod factors carry specific acyl or carbamate substitutions that define host range, and are the main determinants of partner specificity (rather than generic surface polysaccharides) (Tsyganova et al., 2021). Nod factors are perceived by plant LysM-domain receptor kinases (e.g., NFR1/NFR5) on root hairs, triggering intracellular Ca^{2+} oscillations and symbiotic signaling (Haskett et al., 2025; Tsyganova et al., 2021). Compatible rhizobia attach to root hairs and produce Nod factors locally, causing root hair deformation and intense curling that "traps" the bacteria. The resulting microcolony generates a high local Nod factor concentration, which acts as a morphogenetic cue: targeted secretion of cell-wall materials and reorientation of the actin/microtubule cytoskeleton drive the formation of an inward-growing infection thread (Tsyganova et al., 2021). Cytoskeletal dynamics are essential, for example, mutants disrupting fine F-actin bundles or ROP GTPases in root hairs fail to sustain infection-thread growth. Meanwhile, rhizobia modulate host immunity to promote compatibility: Nod-factor signaling triggers an early reactive-oxygen burst that aids infection-thread development, but suppresses a later "defense" ROS wave, and elicits nitric oxide scavenging by non-symbiotic hemoglobins (Tsyganova et al., 2021).

Simultaneously, cortical and pericycle cells in the root begin dividing to form a nodule primordium, synchronized with the epidermal infection. In outer cortex cells, the nucleus migrates toward the presumptive infection site, forming a cytoplasmic "pre-infection thread" that guides the infection thread inward (Oldroyd & Downie, 2008; Tsyganova et al., 2021). The nascent nodule may be indeterminate (with a persistent apical meristem, as in Medicago) or determinate (a transient meristem, as in soybean), but in either case, vascular tissues rapidly differentiate to link the nodule to the root's vasculature (ensuring nutrient exchange) (Madsen et al., 2010; Tsyganova et al., 2021). Infection threads enter the primordium and release rhizobia by endocytosis into host cells, each bacterium enveloped by a plant-derived symbiosome membrane. Inside these symbiosomes, bacteria differentiate into nitrogen-fixing bacteroids. In legumes of the inverted-repeat-lacking clade (e.g., pea, Medicago, etc.), host nodule-specific cysteine-rich (NCR) peptides drive terminal bacteroid differentiation, inducing genome endoreduplication, cell enlargement, and high nitrogenase activity. In tandem, plant metabolism shifts to support nitrogen fixation (for example, via leghemoglobin-mediated oxygen buffering), completing the transition from infection to a fully functional nitrogen-fixing nodule (Zorin et al., 2022).

2.5 Regulation of Nitrogen Fixation

Regulation of biological nitrogen fixation is controlled by multilevel microbial and plant controls that combine environmental clues, host status, and cellular energy to adjust *nif* expression and nodule function with fine precision.

2.5.1 Transcriptional Control of *nif* Genes

Nif gene expression in bacteria is coordinated by the *NifA–NifL* system and its σ^{54} (RpoN) cofactor. Nitrogenase assembly is metabolically expensive (~20% of cell mass; ~40 ATP per NH_4^+ fixed), so *nif* operons are repressed when oxygen or fixed nitrogen is abundant (Dixon & Kahn, 2004; Ryu et al., 2019). *NifL* (often with PII proteins) senses O_2 and NH_3 and binds *NifA* to block its activity under non-permissive conditions. Conversely, under microaerobic, nitrogen-starved conditions, *NifA* is released and, together with σ^{54}/RpoN, activates high-level transcription of *nifHDK* and other structural genes (Ryu et al., 2019). Alternative σ-factors (e.g., heat-shock σ^{32}/RpoH or stationary-phase σ^{S}/RpoS) coordinate broader stress responses and can indirectly influence *nif* expression under adverse conditions. Recent studies also suggest post-transcriptional layers of control, for example, small regulatory RNAs have been found in some diazotrophs that stabilize or repress *nif* mRNAs, fine-tuning nitrogenase production to the cell's energy status.

2.5.2 Plant-Mediated Regulation

Legumes impose additional control on N_2 fixation via systemic signaling. A key mechanism is Autoregulation of Nodulation (AON), upon rhizobial infection or high soil nitrate, roots induce CLE-family peptides (e.g., CLE-RS1/2/3 in Lotus, CLE12/13 in Medicago, RIC1/2 in soybean) under the control of the NODULE INCEPTION (NIN) transcription factor (Okuma & Kawaguchi, 2021). These CLE peptides (often post-translationally arabinosylated) travel to the shoot and bind LRR-receptor kinases (HAR1/SUNN/NARK), which generate a shoot-derived inhibitor that raises root expression of the TML F-box protein, thereby suppressing further nodule initiation. High nitrate triggers an NLP (NIN-like) transcription factor (e.g., LjNRSYM1) that directly upregulates CLE-RS2, linking external N sensing to AON and inhibiting nodulation under ample N (Nishida et al., 2018; Okuma & Kawaguchi, 2021). Plant hormones intersect these pathways: rhizobial infection induces cytokinin biosynthesis (via IPT3) in roots, yet experimental cytokinin feeding reduces nodule number in a dose-dependent manner (Okuma & Kawaguchi, 2021). Auxin is the notable exception among hormones; it accumulates at infection sites and promotes infection-thread formation and cortical cell division, whereas gibberellins generally inhibit early infection stages (but can enhance later cortical organogenesis) (Velandia et al., 2022). Thus, cytokinins, auxins, and gibberellins each modulate nodulation: for example, elevated cytokinin in shoots or roots contributes to systemic feedback, while auxin and GA balance infection versus organogenesis (Velandia et al., 2022).

2.5.3 Carbon–Nitrogen Balance

Carbon supply to nodules is critical for supporting nitrogenase. Legumes export photosynthate (sucrose) to nodules, where nodule sucrose synthase (SuSy) cleaves it into hexoses; disruption of this pathway halts nodule growth and N-fixation capacity (Gordon et al., 1999; Wang et al., 2025). The hexoses are further metabolized into organic acids (chiefly malate) that bacteroids oxidize as their principal carbon source (Udvardi & Poole, 2013; Wang et al., 2025). Because each $N_2 \rightarrow 2NH_4^+$ conversion costs on the order of ~40 ATP, plants must tightly coordinate carbon input with nitrogen demand (Dixon & Kahn, 2004; Ryu et al., 2019). For instance, nitrate application reduces nodule sucrose and malate content, whereas exogenous malate can partially rescue nitrogenase activity under nitrate stress (Heckmann et al., 1989; Wang et al., 2025). In practice, legumes adjust source-sink relations to maintain an optimal C/N ratio: abundant sugar supply and active organic-acid metabolism drive robust fixation, while carbon limitation (or ample soil nitrogen) triggers feedback that down-regulates nodular activity. This dynamic balance ensures that nodules are supplied with sufficient energy and reduce power when fixation is needed, but do not drain carbon when nitrogen is plentiful. The contents of different nutrients and C:N ratios found in the above-ground portions of various important legume green manure crops are shown in Table 2.1.

Table 2.1 The contents of nutrient and C:N ratios of above-ground portions of some important green manure crops

Green manure crops	Total nutrient concentration (% dry weight)						
	N	P	K	S	Ca	Mg	C:N ratio
Sesbaria (*Sesbania aculeata*)	2.62	0.32	1.48	0.19	1.40	1.62	16.4
Sonnhemp (*Crotalariajuncea*)	2.86	0.34	–	–	–	–	16.1
Cowpea (*Vigna unguiculata*)	2.69	058	2.26	028	1.50	1.73	17.1
Cluster bean (*C. tetragonoloba*)	2.80	–	–	–	–	–	17.3
Mung bean (*Vigna radiata*)	2.21	–	–	–	–	–	16.1
Subabul (*L. leucocephala*)	3.15	0.20	1.73	–	1.88	0.41	12.7
Glliricidia (*Gliricidia maculata*)	3.49	0.22	2.44	–	1.89	0.43	10.42

(Source: Ghosh et al., 2007).

2.5.4 Environmental Control Mechanisms

N_2 fixation is also gated by environmental factors. Phosphorus deficiency strongly suppresses nodulation via the plant's phosphate-starvation response, low Pi activates AON through PHR-family transcription factors, reducing NIN and related symbiotic genes and thereby limiting nodule development (Singh et al., 2025; Gao & Chiu, 2025). Micronutrients, such as iron (Fe) and molybdenum (Mo), are indispensable cofactors in nitrogenase and leghemoglobin. Fe deficiency (or excess) in soybean severely impairs infection-thread development and NIN expression via an Fe-sensing BTS–NSP1 regulatory pathway; this reflects Fe's essential role in nitrogenase and leghemoglobin assembly (Singh et al., 2025; Gao & Chiu, 2025). Mo scarcity likewise blocks Fe–Mo cofactor biosynthesis. Soil pH extremes can denature the oxygen-sensitive nitrogenase and limit nutrient uptake, reducing fixation efficiency (Singh et al., 2025; Gao & Chiu, 2025). Water stress (drought, salinity) also curtails nitrogenase: stress-induced declines in nodule SuSy activity and malate availability strongly restrict carbon flow to bacteroids, and drought-driven hormonal changes (e.g., ABA) further inhibit nodule function. Collectively, these nutrient and abiotic signals ensure that legumes activate symbiotic nitrogen fixation only under favorable conditions that balance carbon supply, energy cost, and plant health (Wang et al., 2025).

2.6 Non-Legume Nitrogen-Fixing Associations

The non-legume nitrogen-fixing associations include a wide range of interactions that enhance the advantages associated with biological nitrogen fixation beyond the more traditional legume–rhizobia interactions.

2.6.1 Actinorhizal Symbiosis

In actinorhizal symbioses, filamentous actinobacteria of the genus *Frankia* associate with woody plants [e.g., alder (*Alnus*), casuarina (*Casuarina*), sea-buckthorn (*Hippophae*)] to form nitrogen-fixing root nodules. These nodules are typically coralloid, multilobed structures derived from modified lateral roots, containing internal vesicles that protect the *Frankia* nitrogenase from oxygen. Because *Frankia* fixes N_2 within these structures, actinorhizal plants can thrive on N poor soils and often serve as pioneer species, supplying bioavailable nitrogen that enriches forest soils and aids ecological succession (Astorga-Eló et al., 2021). Unlike legume–*rhizobia* systems, actinorhizal nodulation involves different molecular signals, *Frankia* genomes lack the canonical nod genes found in *rhizobia*, indicating a distinct infection mechanism (though some plant kinases are conserved) (Kucho et al., 2010). Thus, actinorhizal and rhizobial symbioses both fix nitrogen but differ markedly in host range, nodule anatomy, and genetic regulation.

2.6.2 Associative Nitrogen Fixation

Associative diazotrophic bacteria colonize roots or internal tissues without forming true nodules. For example, *Azospirillum brasilense* inhabits cereal rhizospheres (maize, wheat, rice) and fixes N_2 while producing auxins (IAA), gibberellins, and cytokinins that drive lateral root proliferation and nutrient uptake (Giri et al., 2025). It also generates siderophores and organic acids that solubilize phosphorus and micronutrients for plant use (Giri et al., 2025; Renganathan et al., 2025). In sugarcane, nitrogen-fixing endophytes like *Gluconacetobacter diazotrophicus* and *Herbaspirillum seropedicae* colonize internal tissues and can significantly boost plant N content under low-fertilizer conditions (Renganathan et al., 2025). In addition to direct N_2 fixation, these bacteria promote growth via hormone secretion, stress alleviation, and nutrient mobilization (e.g., ACC deaminase, phosphate solubilization) (Renganathan et al., 2025). Together, these traits often translate into improved crop performance and nitrogen use efficiency. In fact, *Azospirillum* inoculants are widely applied to cereals (Giri et al., 2025). However, quantifying the exact fraction of plant N derived from associative fixation remains challenging in field settings (Khan et al., 2021).

2.6.3 Free-Living Diazotrophs

Free-living diazotrophs fix N_2 independently of plants. Soil bacteria, such as *Azotobacter* spp. (strict aerobes) and *Clostridium* spp. (obligate anaerobes), can fix substantial N_2 on their own, though high energy costs and O_2 sensitivity limit their activity (Reed et al., 2010; Khan et al., 2021). In marine environments, filamentous cyanobacteria like *Trichodesmium* form massive blooms that supply a large fraction of oceanic nitrogen fixation. Overall, free-living fixation provides modest but meaningful N inputs: for example, some tropical forest studies estimate ~6 kg N·ha^{-1}·yr^{-1} from free-living soil fixers (litter, rhizosphere) versus ~4.5 kg from symbiotic nodules. Environmental factors, especially O_2 concentration, organic carbon availability, moisture, pH, and trace-metal supply (Mo, Fe), strongly regulate free-living nitrogenase activity. Recent studies emphasize that, despite measurement challenges, free-living diazotrophs (soil heterotrophs and aquatic cyanobacteria) can contribute a significant component of ecosystem nitrogen budgets (Barron et al., 2010; Khan et al., 2021).

2.7 Synthetic Biology Approaches

Developing synthetic biology methods for improving nitrogen fixation is one of the most challenging goals in today's agricultural science. This area has been identified as an important way to solve the global nitrogen problem and minimize the use of the Haber-Bosch process, which is highly demanding in terms of energy. Synthesis methods involve various strategies, such as the use of nitrogenase parts in heterologous organisms and the entire engineering of nitrogen-fixing organism genomes (Fig. 2.3). They all try to deal with the challenges that relate to biological nitrogen fixation using novel design theories and guided evolvability methods. Furthermore, advanced genetic tools are used for designing novel and improved nitrogen-fixing systems.

2.7.1 Engineering Nitrogenase in Heterologous Hosts

Synthetic-biology efforts to reconstitute nitrogenase outside its native diazotrophs have made tangible progress: robust expression of individual *nif* components and even fully active nitrogenase subunits in *Escherichia coli* has been demonstrated, proving that heterologous bacterial chassis can support cofactor assembly and electron transfer when accessory proteins are provided (Solomon et al., 2023; Yang et al., 2014). These bacterial successes, however, contrast with the steep hurdles encountered in eukaryotic systems, where oxygen sensitivity, complex cofactor biosynthesis, and subcellular trafficking conspire to limit activity. Targeting nitrogenase components to low-oxygen organelles, most prominently the mitochondrion in *Saccharomyces cerevisiae*, has emerged as a pragmatic strategy because mitochondria supply ATP and reducing power while providing a more reducing microenvironment; several labs have shown partial cofactor assembly and functional Fe-protein accumulation in yeast mitochondria using organelle-directed polyprotein designs and targeted expression of Nif accessory factors (Dobrzyń ska et al., 2024; Yang et al., 2023). Parallel work defining minimal *nif* gene sets (e.g., core *nifHDK* plus a defined suite of assembly genes, such as *nifB, nifE, nifN*) has clarified the genetic payload needed for a portable nitrogenase and guided modular cloning strategies for both prokaryotic and eukaryotic hosts (Bennett et al., 2023; Lucía Payá-Tormo et al., 2024).

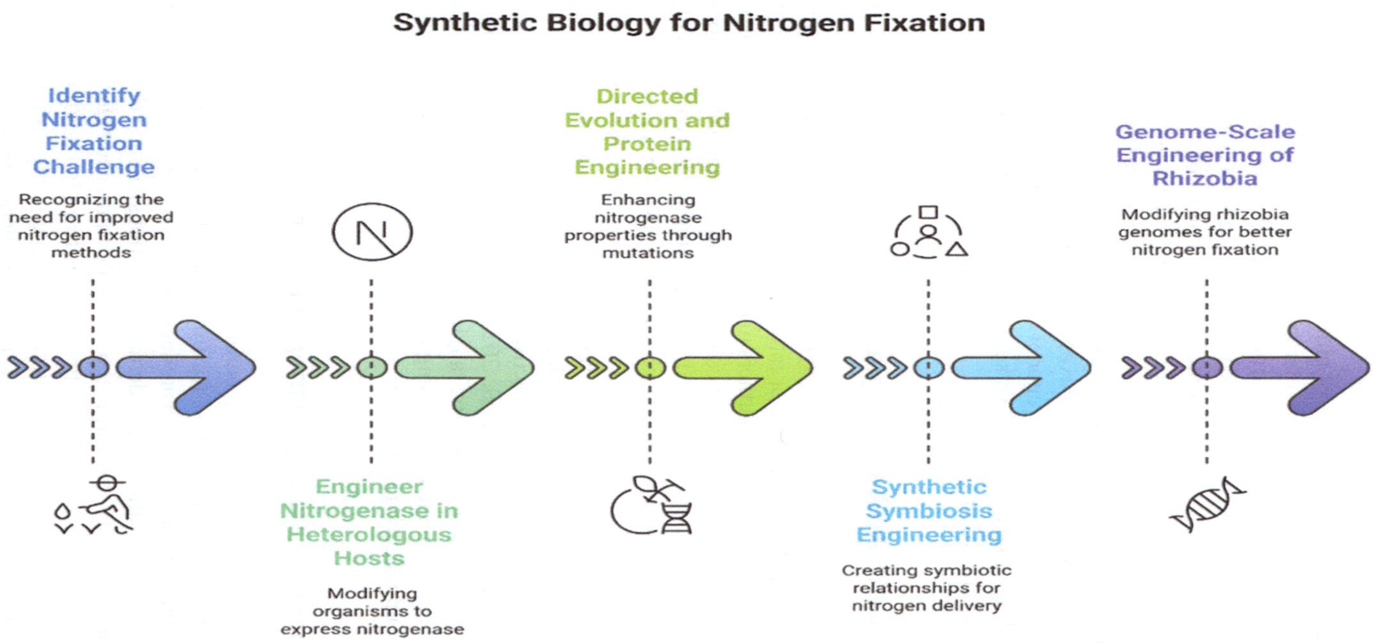

Fig. 2.3 Major synthetic biology approaches to biological nitrogen fixation

2.7.2 Directed Evolution and Protein Engineering

To expand the operational envelope of nitrogenases, directed evolution and protein engineering aim to improve oxygen tolerance, catalytic turnover, and cofactor flexibility. Mutagenesis campaigns and structure-guided substitutions combined with high-throughput screening in *E. coli* or engineered microbial consortia have generated variants with altered kinetics and substrate scope; complementary cofactor engineering studies (iron-only and vanadium cofactor work) highlight routes to alternative-metal active sites that may offer distinct tolerance or catalytic profiles (Bardi Benediktsson & Ragnar Bjornsson, 2020; Quechol et al., 2023; Schmidt et al., 2023). High-throughput selection platforms and screening toolboxes developed for nitrogen-fixing soil microbes accelerate these efforts by coupling reporter outputs or growth readouts to nitrogenase function, enabling rapid cycles of diversification and selection (Venkataraman et al., 2023; Bennett et al., 2023).

2.7.3 Synthetic Symbiosis Engineering

Moving from enzymes to ecosystems, synthetic symbiosis engineering pursues two related goals: (1) enable associative or intracellular nitrogen fixation in non-legume crops and (2) program microbial partners to deliver N on demand. Proof-of-concept studies have shown that plant genetic control over microbial partners (for example, through engineered root exudate signaling or receptor tuning) can bias colonization and restrict benefits to target crops, reducing ecological spillover (Haskett et al., 2022). CRISPR-based editing (including precise base editors) now allows targeted modulation of plant loci that influence root microbiome composition and nodulation signals, while parallel CRISPR and synthetic-circuit strategies in microbes enable inducible *nif* expression or biosensor-mediated activation so that N release can be gated to crop needs or environmental cues (Liu et al., 2024; Wang et al., 2021; Venkataraman et al., 2023). These programmable circuits, ranging from ligand-responsive promoters to kill-switches and metabolic load-sensing modules, are designed to improve efficacy while limiting unintended persistence.

2.7.4 Genome-Scale Engineering of Rhizobia

Complementary genome-scale engineering in rhizobia uses CRISPR–Cas systems to enhance competitiveness, stress resilience, and fixation phenotypes: targeted deletions or promoter rewiring have produced strains with improved colonization or partially deregulated fixation under defined conditions, though deregulation raises ecological and biosafety questions that must be managed (Oussama Bellahsen et al., 2025; Ito et al., 2024). As the field pushes from lab demonstration to field deployment, rigorous containment strategies (genetic biocontainment, ecological dependence designs), multi-tiered risk assessment, and long-term ecological

Table 2.2 Summarized molecular modifications of rhizobial strains and explored improvements in survival and nitrogen fixation characteristics.

Molecular modification	Strain	How implemented	Reported/explored improvements	References
acdS (ACC-deaminase) overexpression	*Bradyrhizobium* sp. (peanut isolate, "Strain 9")	Genomic promoter/RBS engineering and targeted genome editing (*acdS* OE vs KO)	↑ ACC-deaminase activity (≈5×), improved plant growth and stress tolerance in inoculated peanut without loss of nodulation or N-fixation capacity.	(Wang et al., 2025)
otsA (trehalose biosynthesis) overexpression	*Rhizobium* CCBAU25338 (peanut symbiont)	Heterologous/ overexpression of trehalose pathway gene(s) (otsA ± otsB)	Increased cellular trehalose, enhanced salt/ desiccation tolerance, higher nodule numbers, and improved N-fixation and host performance under salinity.	(Liu et al., 2025)
Introduction of *Hup* (uptake hydrogenase) genes	*Rhizobium leguminosarum*, *Mesorhizobium loti*	Mobilization of hup cluster (minitransposon TnHB100 or plasmid) into *Hup*− strains	Hydrogen recycling in nodules → reduced H_2 evolution; in controlled tests increased N incorporation and sometimes higher plant N, effect dependent on host and strain background.	(Sotelo et al., 2021)
parA overexpression (cell-cycle regulator)	*Azorhizobium caulinodans* ORS571	Chromosomal overexpression of *parA*	Alters bacteroid morphotype (swollen/ polyploid) and symbiosome organization; suggests cell-cycle manipulation can change bacteroid differentiation and potentially fixation traits.	(Chien et al., 2019)
NifA/regulatory modulation	Various rhizobia (reviewed examples)	Mutations or expression changes in *nifA/nifL* or regulatory circuits	Alters timing/level of *nif* expression; can decouple nitrogenase expression from strict host/environmental cues—useful but requires careful control because of energy costs and oxygen sensitivity.	(Wang et al., 2025)
Antioxidant/ ROS-scavenging enzyme modulation (e.g., Gpx, catalase, GR)	*Rhizobium leguminosarum* and others	Overexpression or altered regulation of peroxidases/ glutathione systems	Improved oxidative-stress tolerance in vitro and evidence for enhanced survival during infection; may protect nitrogenase and infection thread development under stress.	(Hu et al., 2021) (De Carvalho-Niebel et al., 2024)

Molecular modification	Strain	How implemented	Reported/explored improvements	References
Osmoprotectant biosynthesis/ uptake (e.g., glycine-betaine pathways, choline uptake)	*Sinorhizobium/ Ensifer*, other rhizobia	Introduce or upregulate betaine synthesis/uptake genes or choline dehydrogenases	Improved survival under saline/osmotic stress and better nodulation in marginal soils in greenhouse/field proxies.	(Goszcz et al., 2025)
Symbiosis island/*nod*+*nif* horizontal transfer (synthetic transfer or natural mobilization)	Various recipients (β- and α-rhizobia)	Conjugation or natural transfer of symbiotic genomic islands/plasmids	Enables gain of nodulation and N-fixation traits in new backgrounds; used to expand host range or create novel symbionts, though ecological and regulatory issues remain.	(Xiaohan et al., 2023)

monitoring are essential to balance innovation with stewardship. Taken together, the synthetic-biology toolkit modular *nif* assemblies, directed evolution, programmable symbioses, and genome-scale editing of microbial partners offer a credible pathway to crop-relevant nitrogen inputs, but realization will depend on integrating molecular advances with ecological design and transparent biosafety governance. Table 2.2 summarizes the extensive, long-term efforts that have driven advances in molecular modification of rhizobial strains and documents their demonstrated impacts on host–microbe symbiotic processes.

2.8 Biotechnological Applications and Crop Improvement

Biotechnological strategies to harness biological nitrogen fixation (BNF) for crop improvement bridge tried-and-true microbiological practice and frontier synthetic biology. Agronomic integration and consistent inoculants are practical benefits to farmer livelihoods, while longer-term transformation of cropping systems extending fixation to cereals or producing ammonia in bioreactors will require coordinated progress on strain design, delivery technologies, host genetics, and techno-economic viability. The sections below summarize recent advances in inoculant science, plant- and microbe-directed improvement of fixation efficiency, efforts to extend BNF to non-legumes, and nascent industrial uses of nitrogenase-derived ammonia, highlighting promising results and persistent bottlenecks.

2.8.1 Inoculant Development and Optimization

Effective inoculant programs start with careful selection and testing of rhizobial strains to find those that work well and are competitive in local soils. Now, using molecular and phenotypic screening methods has become common for choosing the best strains that fit the region (Irisarri et al., 2019; Kebede et al., 2021). New formulation technologies, such as polymer and composite carriers, liquid forms, and mixed biopolymer blends, help keep cells alive longer and make them easier to handle. These technologies also allow for quality checks (like shelf-life, CFU counts, and stability) that can predict how well they will perform in the field (Khan et al., 2023; Saberi-Riseh et al., 2021; De Almeida et al., 2025). Strong quality-control systems, along with tests done in different locations and it make sure the strains last under local conditions and support the rules and methods for successful use.

2.8.2 Enhancing Nitrogen Fixation Efficiency

Complementing enhancements come from the plant end breeding for better nodulation traits, and utilising genetic markers for symbiotic efficiency can increase the response to inoculation by the host and stabilize performance across environments (Weith et al., 2022; Oladzad et al., 2020). Multistrain inoculants and co-culture approaches are increasingly utilised to expand functional niches (stress tolerance, P-solubilising, hormonal modulation) and buffer variability from fluctuant soils, whereas ratio-optimisation and co-fermentation technologies increase scalability and consistencies during manufacture (Farda et al., 2024). Integration with precision-agriculture technologies (soil-plant sensors, spatial diagnostics, and DSSs) allows targeted application (time, location, monitoring) that optimises N supply biologically while constraining risk from poor establishment or incongruity with locale agro-conditions (Soussi et al., 2024). Together, these approaches offer a systems-level means to regularly increase fixation efficiency among commercial crop-production packages.

2.8.3 Extending Nitrogen Fixation to Cereals and Industrial Applications of Nitrogenase

Extending biological N benefits beyond legumes via engineered nodulation traits, strengthened cereal–diazotroph associations, or modulation of root exudates to recruit beneficial diazotrophs represents one of the field's most ambitious aims. Review and experimental work document measurable progress (e.g., natural cereal–diazotroph assemblies, such as sorghum mucilage that supply substantial plant N) and outline routes for metabolic and microbiome engineering to increase recruitment and retention of nitrogen-fixing microbes (Guo et al., 2022). At the industrial scale, renewed interest in nitrogenase-based or bio-electrochemical reactors explores ambient-condition ammonia synthesis as a lower-carbon alternative to Haber–Bosch; recent enzyme engineering, heterologous expression, and bio-electrocatalysis studies point to technical feasibility, although rates, oxygen sensitivity, ATP demand, and cofactor management remain key scale-up barriers (Solomon et al., 2024; Rapson & Wood., 2022; Watanabe et al., 2023; Ghavam et al., 2021). Economic comparisons suggest that biological or hybrid pathways could compete with conventional ammonia production when coupled with renewable energy and process intensification, but practical commercial deployment will depend on overcoming productivity and stability challenges.

2.9 Environmental and Ecological Impacts

Biological nitrogen fixation (BNF) delivers a suite of ecosystem services that extend well beyond crop yield, strengthening soil function, biodiversity, and watershed health (Hou, 2023). By converting atmospheric N_2 into plant-available forms, legume–microbe partnerships and free-living diazotrophs enhance soil fertility and contribute to organic matter buildup. That supports soil structure, microbial diversity, and longer-term nutrient retention. Soil health and carbon sequestration studies show that systems that incorporate BNF. Especially when it is coupled with agroforestry or cover-cropping can increase soil organic carbon stocks and help stabilize ecosystem functions (Nazir et al., 2023). Regional restoration projects and agroecological interventions that reintroduce N-fixing plants or inoculants have sustainably enhanced habitat quality and promoted plant and microbial diversity (Nsikani et al., 2018). Targeted reductions in synthetic N inputs reduce downstream nitrate loading and associated eutrophication risks in receiving waters (Liu et al., 2021).

2.9.1 Ecosystem Services from Biological Nitrogen Fixation

Biological nitrogen fixation (BNF) underpins a suite of ecosystem services that extend well beyond simple N inputs to crops: by converting atmospheric N_2 into plant-available forms, symbiotic and free-living diazotrophs improve soil fertility, accelerate organic matter accumulation, and catalyze positive feedbacks in soil microbial communities that sustain nutrient cycling and aggregate stability (Ladha et al., 2022; Soumare et al., 2020). In restoration and agroecological contexts, the deliberate reintroduction or encouragement of N-fixing species (legumes, actinorhizal plants, and N-fixing trees/shrubs) has been shown to increase plant community richness, facilitate colonization by later-successional species, and accelerate recovery of degraded soils through increases in soil organic carbon (SOC) and mineral- associated organic matter pools. These processes carry co-benefits for carbon sequestration because greater N availability usually increases plant productivity and litter inputs, and shifts soil microbial processing toward forms of organic matter that persist longer in soil matrices (Xu et al., 2024). At the watershed scale, legume cover, riparian restoration with N-fixing shrubs, and wetland conservation can reduce the need for synthetic fertilizer and buffer nutrient export, thereby lowering downstream eutrophication risk and protecting water quality, especially when BNF is integrated into landscape designs that emphasize nutrient retention and denitrifying wetlands (Golden et al., 2019; Yan et al., 2025).

2.9.2 Climate-Change Interactions

BNF is very sensitive to climate factors. Both the direct effects of temperature on nitrogenase activity and the indirect effects through plant health and soil moisture greatly influence fixation rates and their effects on ecosystems (Deutsch et al., 2024). Studies show that many nitrogen-fixing bacteria have a temperature range where they work best. Warming can increase fixation up to a certain limit, but it may quickly lower rates beyond specific limits for each species, with important differences in how regions respond (Deutsch et al., 2024; Wirth et al., 2024). On a cellular level, how well nodules handle heat and other stress is connected to the production of heat shock proteins (HSPs) and other helpers that support the bacteria's metabolism and maintain nitrogenase function. Recent research in soybean and pea nodules shows that small HSPs (like GmHSP17.9) are vital for nodule growth and steady N_2 fixation during short periods of stress (Yang et al., 2021; Domingo-Serrano et al., 2025). Drought makes these effects worse by changing how plants manage their resources, cutting carbon supply to nodules, and slowing down nitrogen transfer from the symbiosis. These responses can happen quickly and take time to reverse when conditions improve (Tang et al., 2024). Because of these sensitivities, adaptation strategies that include breeding heat- and drought-tolerant legume varieties, choosing or engineering heat-tolerant rhizobia, and using landscape practices to keep soil moisture and temperature stable (like mulches, agroforestry, and mixed species) provide practical ways to sustain BNF services as climates change (Zhang et al., 2024).

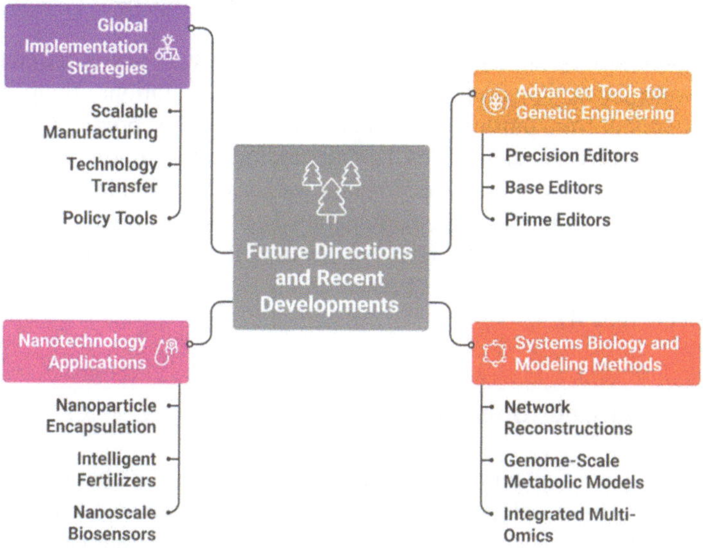

Fig. 2.4 Overview of future research directions and latest technological advances in biological nitrogen fixing research, focusing on genetic engineering innovations, nanotechnology applications, systems biology integration, and global implementation strategies

2.9.3 Biosafety and Risk Assessment

As interest grows in applying engineered diazotrophs and genetically modified microbial inoculants to raise BNF rates in agroecosystems, robust biosafety assessment becomes imperative. Environmental release of engineered organisms raises concerns about persistence, unintended ecological effects on native microbial assemblages, and the potential for horizontal gene transfer (HGT) of engineered traits to non-target taxa; recent reviews emphasize that these risks are context-dependent and require case-by-case empirical evaluation (Lensch et al., 2023; Rafeeq et al., 2023). HGT pathways in soil and rhizosphere environments are complex but documented; risk assessments therefore focus on the mobility of genetic elements, selective advantages conferred by engineered traits, and the ecological consequences if transfer occurs (Azad et al., 2025). Ecosystem disruption potential, such as altered N cycling, shifts in plant–microbe interactions, or downstream nutrient pulses, must be assessed alongside human and animal health endpoints. Consequently, multidisciplinary environmental risk assessments (ERA), layered biocontainment strategies (genetic safeguards, kill switches, ecological containment), post-release monitoring, and adaptive regulatory frameworks that balance innovation with precaution are recommended by recent policy and technical analyses (Turnbull et al., 2021). Strengthening transparency, harmonizing oversight, and investing in long-term field trials with rigorous ecological endpoints will be essential to realize BNF-based technologies safely and equitably.

2.10 Future Directions and Recent Developments

New technologies, computer simulations, and materials chemistry are converging on the next generation of strong, efficient, and scalable solutions. Temporary outcomes will result from putting together accurate gene edits, predictive biological systems, and delivery technologies that take ecological complexity into account. Longer-term outcomes will require economic avenues that provide opportunities for technology diffusion and farmer adoption globally (Pfau et al., 2018; Venkataraman et al., 2023; Maaz et al., 2025). Here we discuss the most encouraging tools and methods, and identify the areas of research that will most likely yield the next big impact (Fig. 2.4).

2.10.1 Advanced Tools for Genetic Engineering

The advent of precision editors, base editors, and prime editors has changed gene-level work away from making gross excisions to making programmable, subtle base alterations and small sequence insertions. This allows for fine-tuning of nodulation receptors, transporters, and regulatory regions with less collateral damage (Tian et al., 2025). Methods and tools that work at the level of chromatin also join the repertoire by changing gene expression without changing the DNA sequence (Ye et al., 2025). This is a hopeful way to transiently increase nodulation or stress responses. Multigene engineering approaches, like receptor, metabolic, and regulatory component stacking, become manageable as modular cloning and synthetic biology tools gain maturity for the rhizobia and their related microorganisms (Venkataraman et al., 2023). Importantly, machine learning and computer-assisted intelligence are also becoming ways to assess genomes, predict desirable edits, and prioritize targets to edit, along with strain designs that consider function and environmental safety (Zhao et al., 2023). Overall, these tools make the construction of programmed increments in symbiotic effectiveness and compatibility with partners easier.

2.10.2 Systems Biology and Modeling Methods

Systems-level approaches, including network reconstructions, genome-scale metabolic models, and integrated multi-omics, hold the key to unravelling the flow and trade-offs that limit nodulated nitrogenase activity. They can predict on the computer which pairs of plant and microbe will have the optimum performance under a specified environment. Genome-scale models and studies on metabolic coupling have clarified how oxygen levels, availability of carbon substrates, and amino acid exchanges control rates of fixation (Pfau et al., 2018; Schulte et al., 2021). By combining such models with transcriptomics, proteomics, and metabolomics, researchers can test their hypotheses on computers before conducting field experiments (Ye et al., 2025). Prediction pipelines that integrate mechanistic models with machine learning can point to the optimal plant–microbe pairs under specified soils and climate regimes (Inomura et al., 2020). This can accelerate the process of taking laboratory findings to farmable levels.

2.10.3 Nanotechnology Applications

Nanoscience offers efficient methods for delivery and monitoring. Encapsulation with nanoparticles and targeted carriers can preserve rhizobia during storage and with seeds, releasing them at controlled rates to assist with colonization (Karnwal et al., 2023). In the same manner, intelligent fertilizers and responsive coatings can release fertilizer elements at times when plants and microbes require them (De Souza-Torres et al., 2020). Nanoscale biosensors that are either electrochemical or optical provide in-time information on root exudates, ammonia, and nitrate in the soil. This enables efficient management whereby inoculants, irrigation, and fertilizer are optimized side by side (Zaman et al., 2025). Recent studies on developing root-microbe interfaces at the nano-scale, e.g., shape of surface, presentation of the ligands, and micro-encapsulation, hold promise for enhancing the prevalence of beneficial diazotrophs. But we must verify environmental safety and cumulative effects on life cycles (Rosa et al., 2025).

2.10.4 Global Implementation Strategies

A technological advantage is not much good without an adoption pathway. We require scalable methods for manufacturing inoculants. Applications must be cheap, simple, and convenient to deploy at a local level, and we require reasonable incentives for farmers to reduce the use of synthetic nitrogen fertilizers (Pudasaini & Raizada, 2024). Technology transfer must be complemented with capacity building quality control facilities, extension services and participatory demonstrations so that small-scale and commercial farmers can acquire, assess and adopt BNF technologies (Zhao et al., 2023). Harmonized standards and regulatory clarity for biofertilizers, nanomaterials, and engineered microorganisms will speed up private-sector investment while safeguarding ecosystems (Asad & Ahmed, 2024). Last, policy tools (subsidies, payments for ecosystem services, procurement specifications) that value the climate and water-quality contributions of biological N solutions can put farmer interests into alignment with wider sustainability objectives (Malusá & Vassilev, 2014)

2.11 Conclusion

Biological fixation of nitrogen provides a viable and climate-friendly solution to the increasing food needs and decreasing environmental degradation. BNF can be used to integrate both time-tested methods like legume rotations and conservation agriculture with newer technologies like high-quality local inoculants, seed coatings, predictive matching models, and genomic or microbial engineering when carefully incorporated into farming systems. Its practical uses are extensive: better inoculants and delivery systems can boost yields in both smallholder and commercial sectors; engineered or selected symbioses can extend the benefits of fixation to nonlegumes; widespread implementation of BNF-based rotations and integrated nutrient management can reduce dependence on fertilizers, enhance soil carbon, and rehabilitate degraded landscapes. To achieve these advantages, it is not only necessary to focus on science and technology but also on supply chains, quality assurance, training farmers, and policy tools that can ensure that adoption is fair and sustainable.

Meanwhile, critical research gaps should be recognized and given priority. We do not yet have a predictive, mechanistic insight into host-microbe signaling in different crop genotypes, soils and climates; the translation of laboratory achievements in multi-gene engineering of fixation into field-stable, competitive strains has not been achieved; and the long-term ecological implications and biosafety of introduced or engineered microbes are not fully understood. There are still delivery and scaling issues, including cost-effective production of inoculants, strong seed or nano-carrier technology, and cold-chain-free storage. Socioeconomic studies are also required to develop incentives and extension models which are locally applicable. To fill these gaps, future research must integrate basic research into the signaling and microbiome ecology with systems-level engineering, rigorous multisite field experiments, transparent safety and regulatory science, and participatory research with farmers. It is only with interdisciplinary cooperation that integrates innovation with prudent assessment, governance and capacity building that BNF can be transformed into a promising idea to a common practice that can help to achieve sustainable, equitable food systems.

References

Agudelo, M. G., Ruiz, B., Capela, D., & Remigi, P. (2023). The role of microbial interactions on rhizobial fitness. *Frontiers in Plant Science*, 14, 1277262. https://doi.org/10.3389/fpls.2023.1277262.

Aguilar, O. M., Collavino, M. M., & Mancini, U. (2022). Nodulation competitiveness and diversification of symbiosis genes in common beans from the American centers of domestication. *Scientific Reports*, *12*, 4591. https://doi.org/10.1038/s41598-022-08720-0.

Asad, M., & Ahmed, S. Y. (2024). Utilizing biofertilizer for achieving sustainable agriculture and rural development strategy towards vision 2040, Oman. *Sustainability*, *16*(10), 4015. https://doi.org/10.3390/su16104015.

Astorga-Eló, M., Gonzalez, S., Acuña, J. J., Sadowsky, M. J., & Jorquera, M. A. (2021). Rhizobacteria from 'flowering desert' events contribute to the mitigation of water scarcity stress during tomato seedling germination and growth. *Scientific Reports*, 11(1), 13745. https://doi.org/10.1038/s41598-021-93303-8.

Azad, R. B., Kasfy, S. H., Molla, K., & Islam, T. (2025). Horizontal gene transfer in plants and implications for biotechnology. *Plant-Environment Interactions*, *6*(5), e70087.

Barron, A. R., Purves, D. W., & Hedin, L. O. (2010). Facultative nitrogen fixation by canopy legumes in a lowland tropical forest. *Oecologia*, 165(2), 511–520. https://doi.org/10.1007/s00442-010-1838-3.

Bellahsen, O., Díaz-Méndez, R., & Romero, D. (2025). Genomic engineering in Rhizobium etli: implementation and evaluation of systems based on dCas9. *Frontiers in Microbiology*, 16, 1604430.

Benediktsson, B., & Bjornsson., R. (2020). Quantum mechanics/molecular mechanics study of resting-state vanadium nitrogenase: Molecular and electronic structure of the iron–vanadium cofactor. *Inorganic Chemistry*, *59*(16), 11514–11527. https://doi.org/10.1021/acs.inorgchem.0c01320.

Bennett, E. M., Murray, J. W., & Isalan, M. (2023). Engineering nitrogenases for synthetic nitrogen fixation: From pathway engineering to directed evolution. *BioDesign Research*, 5, 0005.

Bothe, H., Schmitz, O., Yates, M. G., & Newton, W. E. (2010). Nitrogen fixation and hydrogen metabolism in cyanobacteria. *Microbiology and Molecular Biology Reviews*, *74*(4), 529–551. https://doi.org/10.1128/mmbr.00033-10.

Burén, S., Jiménez-Vicente, E., Echavarri-Erasun, C., & Rubio, L. M. (2020). Biosynthesis of nitrogenase cofactors. *Chemical Reviews*, *120*(12), 4921–4968. https://doi.org/10.1021/acs.chemrev.9b00489.

De Carvalho-Niebel, F., Fournier, J., Becker, A., & Arancibia, M. M. (2024). Cellular insights into legume root infection by rhizobia. *Current Opinion in Plant Biology*, *81*, 102597. https://doi.org/10.1016/j.pbi.2024.102597.

Chanderban, M., Hill, C. A., Dhamad, A. E., & Lessner, D. J. (2023). Expression of V-nitrogenase and Fenitrogenase in Methanosarcina acetivorans is controlled by molybdenum, fixed nitrogen, and the expression of Mo-nitrogenase. *Applied and environmental microbiology*, 89(9), e01033-23.

Chien, H. L., Huang, W. Z., Tsai, M. Y., Cheng, C. H., & Liu, C. T. (2019). Overexpression of the chromosome partitioning gene parA in Azorhizobium caulinodans ORS571 alters the bacteroid morphotype in Sesbania rostrata stem nodules. *Frontiers in Microbiology*, 10, 2422.

Deutsch, C., Inomura, K., Luo, Y.-W., & Wang, Y.-P. (2024). Projecting global biological N_2 fixation under climate warming across land and ocean. *Trends in Microbiology*, *32*(6), 546–553. https://doi.org/10.1016/j.tim.2023.12.007.

De Almeida, J. C., Favero, V. O., Rouws, J. R. C., De Sant'Anna Freitas, C., De Sousa, ÉB., Da Rocha, J. F., Chantre, N. C. S., Xavier, G. R., De Oliveira, P. J., & Rumjanek, N. G. (2025). Inoculant formulation for *Bradyrhizobium* spp.: Optimizing CMC/starch blends for improved performance. *Agriculture*, *15*(9), 1010. https://doi.org/10.3390/agriculture15091010.

Dixon, R., & Kahn, D. (2004). Genetic regulation of biological nitrogen fixation. *Nature Reviews Microbiology*, *2*(8), 621–631. https://doi.org/10.1038/nrmicro954.

Dobrzyńska, K., Pérez-González, A., Echavarri-Erasun, C., Coroian, D., Salinero-Lanzarote, A., & Veldhuizen, M., et al. (2024). Nitrogenase cofactor biosynthesis using proteins produced in mitochondria of Saccharomyces cerevisiae. *MBio*, 15(2), e03088-23.

Domingo-Serrano, L., Sanchis-López, C., Alejandre, C., Soldek, J., Palacios, J. M., & Albareda, M. (2025). A micro-aerobically induced small heat shock protein contributes to *Rhizobium leguminosarum/Pisum sativum* symbiosis and interacts with a wide range of bacteroid proteins. *Applied and Environmental Microbiology*, *91*(1), e01385-24.

Farda, B., Pasquarelli, F., Djebaili, R., Spera, D. M., Del Gallo, M., & Pellegrini, M. (2024). Co-culturing a multistrain Gram-negative inoculant useful in sustainable agriculture. *Frontiers in Industrial Microbiology*, 2, 1380037.

Gao, J., & Chiu, C. H. (2025). Micronutrients: Minor yet crucial for symbiotic nitrogen fixation. *Plant Communications*, 6(5), 101345. https://doi.org/10.1016/j.xplc.2025.101345.

Ghavam, S., Vahdati, M., Wilson, I. G., & Styring, P. (2021). Sustainable ammonia production processes. *Frontiers in energy research*, 9, 580808.

Ghosh, P. K., Bandyopadhyay, K. K., Wanjari, R. H., Manna, M. C., Misra, A. K., Mohanty, M., et al. (2007). Legume effect for enhancing productivity and nutrient use-efficiency in major cropping systems–an Indian perspective: A review. *Journal of Sustainable Agriculture*, *30*, 59–86. https://doi.org/10.1300/J064v30n01_07.

Giri, A., Verma, R., & Mehta, S. (2025). Biofertilizers: A Sustainable Solution for Enhanced Crop Yield and Soil Health in Modern Agriculture. *Int. J. Plant Soil Sci*, 37, 562–576.

Golden, H. E., Rajib, A., Lane, C. R., Christensen, J. R., Wu, Q., & Mengistu, S. (2019). Non-floodplain wetlands affect watershed nutrient dynamics: A critical review. *Environmental Science & Technology*, *53*(13), 7203–7214. https://doi.org/10.1021/acs.est.8b07270.

Gordon, A. C., Minchin, F. R., James, C. L., & Komina, O. (1999). Sucrose synthase in legume nodules is essential for nitrogen fixation1. *Plant Physiology*, *120*(3), 867–878. https://doi.org/10.1104/pp.120.3.867.

Goszcz, A., Furtak, K., Stasiuk, R., Wójtowicz, J., Musiałowski, M., Schiavon, M., & Dębiec-Andrzejewska, K. (2025). Bacterial osmoprotectants—a way to survive in saline conditions and potential crop allies. *FEMS Microbiology Reviews*, 49, fuaf020.

Gu, W., & Milton, R. D. (2020). Natural and engineered electron transfer of nitrogenase. *Chemistry*, *2*(2), 322–346.

Guo, K., Yang, J., Yu, N., Luo, L., & Wang, E. (2022). Biological nitrogen fixation in cereal crops: Progress, strategies, and perspectives. *Plant Communications*, *4*(2), 100499. https://doi.org/10.1016/j.xplc.2022.100499.

Harwood, C. S. (2020). Iron-only and vanadium nitrogenases: Fail-safe enzymes or something more. *Annual Review of Microbiology*, *74*(1), 247–266. https://doi.org/10.1146/annurev-micro-022620-014338.

Haskett, T. L., Cooke, L., Green, P., & Poole, P. S. (2025). Regulation of rhizobial nodulation genes by flavonoidindependent NodD supports nitrogen-fixing symbioses with legumes. *Environmental Microbiology*, 27(1), e70014.

Haskett, T. L., Karunakaran, R., Bueno Batista, M., Dixon, R., & Poole, P. S. (2022). Control of nitrogen fixation and ammonia excretion in *Azorhizobium caulinodans*. PLoS genetics, 18(6), e1010276. https://doi.org/10.1371/journal.pgen.1010276.

Haskett, T. L., Paramasivan, P., Mendes, M. D., Green, P., Geddes, B. A., & Knights, H. E., et al. (2022). Engineered plant control of associative nitrogen fixation. *Proceedings of the National Academy of Sciences*, 119(16), e2117465119.

Heckmann, M., Drevon, J., Saglio, P., & Salsac, L. (1989). Effect of oxygen and malate on NO3– inhibition of nitrogenase in soybean nodules. *PLANT PHYSIOLOGY*, 90(1), 224–229. https://doi.org/10.1104/pp.90.1.224.

Hou, D. (2023). Soil health and ecosystem services. *Soil Use and Management*, *39*(4), 1259–1266. https://doi.org/10.1111/sum.12945.

Hu, A., Chen, X., Luo, S., Zou, Q., Xie, J., & He, D., et al. (2021). Rhizobium leguminosarum glutathione peroxidase is essential for oxidative stress resistance and efficient nodulation. *Frontiers in Microbiology*, 12, 627562.

Inomura, K., Deutsch, C., Masuda, T., Prášil, O., & Follows, M. J. (2020). Quantitative models of nitrogen-fixing organisms. *Computational and Structural Biotechnology Journal*, *18*, 3905–3924. https://doi.org/10.1016/j.csbj.2020.11.022.

Irisarri, P., Cardozo, G., Tartaglia, C., Reyno, R., Gutiérrez, P., & Lattanzi, F. A., et al. (2019). Selection of competitive and efficient rhizobia strains for white clover. *Frontiers in microbiology*, 10, 768.

Ito, Y., Yoshidome, D. aisuke, Hidaka, M., Araki, Y., Ito, K., Kosono, S. aori, & Nishiyama, M. (2024). Improvement of the nitrogenase activity in *Escherichia coli* that expresses the nitrogen fixation-related genes from *Azotobacter vinelandii*. *Biochemical and Biophysical Research Communications*, *728*, 150345–150345. https://doi.org/10.1016/j.bbrc.2024.150345.

Jasniewski, A. J., Lee, C. C., Ribbe, M. W., & Hu, Y. (2020). Reactivity, mechanism, and assembly of the alternative nitrogenases. *Chemical Reviews*, *120*(12), 5107–5157. https://doi.org/10.1021/acs.chemrev.9b00704.

Karnwal, A., Dohroo, A., & Malik, T. (2023). Unveiling the potential of bioinoculants and nanoparticles in sustainable agriculture for enhanced plant growth and food security. *BioMed research international*, 2023(1), 6911851.

Kebede, E. (2021). Competency of rhizobial inoculation in sustainable agricultural production and biocontrol of plant diseases. *Frontiers in Sustainable Food Systems*, 5, 728014.

Khan, A., Singh, A. V., Gautam, S. S., Agarwal, A., Punetha, A., & Upadhayay, V. K., et al. (2023). Microbial bioformulation: a microbial assisted biostimulating fertilization technique for sustainable agriculture. *Frontiers in Plant Science*, 14, 1270039.

Khan, S., Nadir, S., Iqbal, S., Xu, J., Gui, H., Khan, A., & Ye, L. (2021). Towards a comprehensive understanding of free-living nitrogen fixation. *Circular Agricultural Systems*, 1(1), 1–11. https://doi.org/10.48130/CAS-2021-0013.

Kucho, K., Hay, A., & Normand, P. (2010). The determinants of the actinorhizal symbiosis. *Microbes and Environments*, 25(4), 241–252. https://doi.org/10.1264/jsme2.me10143.

Kumar, K., Mella-Herrera, R. A., & Golden, J. W. (2010). Cyanobacterial heterocysts. *Cold Spring Harbor Perspectives in Biology*, *2*(4), a000315. https://doi.org/10.1101/cshperspect.a000315.

Ladha, J. K., Peoples, M. B., Reddy, P. M., Biswas, J. C., Bennett, A., Jat, M. L., & Krupnik, T. J. (2022). Biological nitrogen fixation and prospects for ecological intensification in cereal-based cropping systems. *Field Crops Research*, *283*, 108541. https://doi.org/10.1016/j.fcr.2022.108541.

Lawrence, N. C., Tenesaca, C. G., VanLoocke, A., & Hall, S. J. (2021). Nitrous oxide emissions from agricultural soils challenge climate sustainability in the US Corn Belt. *Proceedings of the National Academy of Sciences*, *118*(46), e2112108118.

Lensch, A., Lindfors, H. A., Duwenig, E., Fleischmann, T., Hjort, C., Kärenlampi, S. O., McMurtry, L., Melton, E., Andersen, M. R., Skinner, R., Wyss, M., & Van Kranenburg, R. (2023). Safety aspects of microorganisms deliberately released into the environment. *EFB Bioeconomy Journal*, 4, 100061. https://doi.org/10.1016/j.bioeco.2023.100061.

Li, Y., Guo, T., Sun, L., Wang, E. T., Young, J. P. W., & Tian, C. F. (2024). Phylogenomic analyses and reclassification of the Mesorhizobium complex: proposal for 9 novel genera and reclassification of 15 species. *BMC genomics*, 25(1), 419.

Liu, J., Yu, X., Qin, Q., Dinkins, R. D., & Zhu, H. (2020a). The impacts of domestication and breeding on nitrogen fixation symbiosis in legumes. *Frontiers in genetics*, 11, 00973.
Liu, J., Wang, D., Tong, R., Ye, S., Zhao, Y., Wu, J., & Gan, Y. (2025). Engineered rhizobia with trehalose-producing genes enhance peanut growth under salinity stress. *Agronomy*, *15*(4), 974. https://doi.org/10.3390/agronomy15040974.
Liu, W., Yuan, Y., & Koropeckyj-Cox, L. (2021). Effectiveness of nutrient management on water quality improvement: A synthesis on nitrate-nitrogen loss from subsurface drainage. *Transactions of the ASABE*, *64*(2), 675–689. https://doi.org/10.13031/trans.14078.
Liu, X., You, S., Liu, H., Yuan, B., Wang, H., & James, E. K., et al. (2020b). Diversity and geographic distribution of microsymbionts associated with invasive Mimosa species in Southern China. *Frontiers in microbiology*, 11, 563389.
Liu, X., Li, P., Wang, H., Han, L.L., Yang, K., Wang, Y., Jiang, Z., Cui, L., & Kao, S.J. (2023). Nitrogen fixation and diazotroph diversity in groundwater systems. *The ISME Journal*, 17(11), 2023–2034. https://doi.org/10.1038/s41396-023-01513-x.
Liu, X., Dong, H., Wang, H., Ren, X., Yang, X., Li, T., Fu, G., Xia, M., Fang, H., Du, G., Jin, Z., & Zhang, D. (2024). Recent advances in genetic engineering strategies of *Sinorhizobium meliloti*. *ACS Synthetic Biology*, *13*(11), 3497–3506. https://doi.org/10.1021/acssynbio.4c00348.
Maaz, T. M., Dobermann, A., Lyons, S. E., & Thomson, A. M. (2025). Review of research and innovation on novel fertilizers for crop nutrition. *npj Sustainable Agriculture*, 3(1), 25.
Madsen, L. H., Tirichine, L., Jurkiewicz, A., Sullivan, J. T., Heckmann, A. B., & Bek, A. S., et al. (2010). The molecular network governing nodule organogenesis and infection in the model legume Lotus japonicus. *Nature communications*, 1(1), 10.
Malusá, E., & Vassilev, N. (2014). A contribution to set a legal framework for biofertilisers. *Applied Microbiology and Biotechnology*, *98*(15), 6599–6607. https://doi.org/10.1007/s00253-014-5828-y.
Masson-Boivin, C., Giraud, E., Perret, X., & Batut, J. (2009). Establishing nitrogen-fixing symbiosis with legumes: How many rhizobium recipes?. *Trends in Microbiology*, *17*(10), 458–466.
Mendoza-Suárez, M. A., Geddes, B. A., Sánchez-Cañizares, C., Ramírez-González, R. H., Kirchhelle, C., Jorrin, B., & Poole, P. S. (2020). Optimizing Rhizobium-legume symbioses by simultaneous measurement of rhizobial competitiveness and N_2 fixation in nodules. *Proceedings of the National Academy of Sciences*, *117*, 9822–9831. https://doi.org/10.1073/pnas.1921225117.
Nazir, M. J., Li, G., Nazir, M. M., Zulfiqar, F., Siddique, K. H., Iqbal, B., & Du, D. (2023). Harnessing soil carbon sequestration to address climate change challenges in agriculture. *Soil and Tillage Research*, *237*, 105959. https://doi.org/10.1016/j.still.2023.105959.
Nishida, H., Tanaka, S., Handa, Y., Ito, M., Sakamoto, Y., & Matsunaga, S., et al. (2018). A NIN-LIKE PROTEIN mediates nitrate-induced control of root nodule symbiosis in Lotus japonicus. *Nature communications*, 9(1), 499.
Nsikani, M. M., van Wilgen, B. W., & Gaertner, M. (2018). Barriers to ecosystem restoration presented by soil legacy effects of invasive alien N_2-fixing woody species: Implications for ecological restoration. *Restoration Ecology*, *26*(2), 235–244.
Okuma, N., & Kawaguchi, M. (2021). Systemic optimization of legume nodulation: A shoot-derived regulator, miR2111. *Frontiers in Plant Science*, *12*. https://doi.org/10.3389/fpls.2021.682486.
Oladzad, A., González, A., Macchiavelli, R., De Jensen, C. E., Beaver, J., Porch, T., & McClean, P. (2020). Genetic factors associated with nodulation and nitrogen derived from atmosphere in a Middle American common bean panel. *Frontiers in Plant Science*, *11*. https://doi.org/10.3389/fpls.2020.576078.
Oldroyd, G. E., & Downie, J. A. (2008). Coordinating nodule morphogenesis with rhizobial infection in legumes. *Annual Review of Plant Biology*, *59*(1), 519–546. https://doi.org/10.1146/annurev.arplant.59.032607.092839.
Palmero, F., Fernandez, J. A., Garcia, F. O., Haro, R. J., Prasad, P. V., Salvagiotti, F., & Ciampitti, I. A. (2022). A quantitative review into the contributions of biological nitrogen fixation to

agricultural systems by grain legumes. *European Journal of Agronomy*, *136*, 126514. https://doi.org/10.1016/j.eja.2022.126514.

Pankievicz, V. C. S., Irving, T. B., Maia, L. G. S., & Ané, J.-M. (2019). Are we there yet? The long walk towards the development of efficient symbiotic associations between nitrogen-fixing bacteria and non-leguminous crops. *BMC Biology*, *17*, 99. https://doi.org/10.1186/s12915-019-0710-0.

Payá-Tormo, L., Echavarri-Erasun, C., Makarovsky-Saavedra, N., Pérez-González, A., Yang, Z. Y., & Guo, Y., et al. (2024). Iron-molybdenum cofactor synthesis by a thermophilic nitrogenase devoid of the scaffold NifEN. *Proceedings of the National Academy of Sciences*, 121(46), e2406198121.

Pfau, T., Christian, N., Masakapalli, S. K., Sweetlove, L. J., Poolman, M. G., & Ebenhöh, O. (2018). The intertwined metabolism during symbiotic nitrogen fixation elucidated by metabolic modelling. *Scientific Reports*, 8(1), 12504.

Pudasaini, R., & Raizada, M. N. (2024). Nodule crushing: a novel technique to decentralize rhizobia inoculant technology and empower small-scale farmers to enhance legume production and income. *Frontiers in Sustainable Food Systems*, 8, 1423997.

Quechol, R., Solomon, J. B., Liu, Y. A., Lee, C. C., Jasniewski, A. J., & Górecki, K., et al. (2023). Heterologous synthesis of the complex homometallic cores of nitrogenase P-and M-clusters in Escherichia coli. *Proceedings of the National Academy of Sciences*, 120(44), e2314788120.

Rafeeq, H., Afsheen, N., Rafique, S., Arshad, A., Intisar, M., Hussain, A., Bilal, M., & Iqbal, H. M. N. (2023). Genetically engineered microorganisms for environmental remediation. *Chemosphere*, *310*, 136751. https://doi.org/10.1016/j.chemosphere.2022.136751.

Rapson, T. D., & Wood, C. C. (2022). Analysis of the ammonia production rates by nitrogenase. *Catalysts*, *12*(8), 844. https://doi.org/10.3390/catal12080844.

Reed, S. C., Townsend, A. R., Cleveland, C. C., & Nemergut, D. R. (2010). Microbial community shifts influence patterns in tropical forest nitrogen fixation. *Oecologia*, 164(2), 521–531. https://doi.org/10.1007/s00442-010-1649-6.

Renganathan, P., Astorga-Eló, M., Gaysina, L. A., Puente, E. O. R., & Sainz-Hernández, J. C. (2025). Nitrogen fixation by diazotrophs: a sustainable alternative to synthetic fertilizers in hydroponic cultivation. *Sustainability*, 17(13), 5922. https://doi.org/10.3390/su17135922.

Rosa, A. R. T., do Valle Jr, R. F., da Silva, M. V., Perini, H. F., Oliveira, C. J. F., & Rosa, R. C., et al. (2025). Nano-and microstructured systems for controlled release of agricultural inputs: innovations for efficiency and sustainability. *Journal of Agricultural and Food Chemistry*, 73(18), 10681–10691.

Roy, S., Liu, W., Nandety, R. S., Crook, A., Mysore, K. S., Pislariu, C. I., Frugoli, J., Dickstein, R., & Udvardi, M. K. (2019). Celebrating 20 years of genetic discoveries in legume nodulation and symbiotic nitrogen fixation. *The Plant Cell*, 32(1), 15–41. https://doi.org/10.1105/tpc.19.00279.

Ryu, M.-H., Zhang, J., Toth, T., Khokhani, D., Geddes, B. A., Mus, F., Garcia-Costas, A., Peters, J. W., Poole, P. S., Ané, J.-M., & Voigt, C. A. (2019). Control of nitrogen fixation in bacteria that associate with cereals. *Nature Microbiology*, *5*(2), 314–330. https://doi.org/10.1038/s41564-019-0631-2.

Saberi-Riseh, R., Moradi-Pour, M., Mohammadinejad, R., & Thakur, V. K. (2021). Biopolymers for biological control of plant pathogens: Advances in microencapsulation of beneficial microorganisms. *Polymers*, *13*(12), 1938. https://doi.org/10.3390/polym13121938.

Schmidt, F. V., Schulz, L., Zarzycki, J., Prinz, S., Oehlmann, N. N., Erb, T. J., & Rebelein, J. G. (2023). Structural insights into the iron nitrogenase complex. *Nature Structural & Molecular Biology*, *31*(1), 150–158. https://doi.org/10.1038/s41594-023-01124-2.

Schulte, C. C., Borah, K., Wheatley, R. M., Terpolilli, J. J., Saalbach, G., & Crang, N., et al. (2021). Metabolic control of nitrogen fixation in rhizobium-legume symbioses. *Science Advances*, 7(31), eabh2433.

Singh, M., Bhutani, S., & Dinkar, N. (2025). Rhizospheric microbiome: A key player in non-leguminous plant dynamics. *Environment Conservation Journal*, 26(3), 1090–1099. https://doi.org/10.36953/ecj.31113004.

Solomon, J. B., Liu, Y. A., Górecki, K., Quechol, R., Lee, C. C., & Jasniewski, A. J., et al. (2023). Heterologous expression of a fully active Azotobacter vinelandii nitrogenase Fe protein in Escherichia coli. *MBio*, 14(6), e02572-23.

Solomon, J. B., Lee, C. C., Liu, Y. A., Duffin, C., Ribbe, M. W., & Hu, Y. (2024). Ammonia synthesis via an engineered nitrogenase assembly pathway in *Escherichia coli. Nature Catalysis*, *7*(10), 1130–1141. https://doi.org/10.1038/s41929-024-01229-x.

Sotelo, M., Ureta, A. C., Muñoz, S., Sanjuán, J., Monza, J., & Palacios, J. (2021). Introduction of H_2-uptake hydrogenase genes into rhizobial strains improves symbiotic nitrogen fixation in *Vicia sativa* and *Lotus corniculatus* forage legumes. *Frontiers in Agronomy*, *3*, 661534.

Soumare, A., Diedhiou, A. G., Thuita, M., Hafidi, M., Ouhdouch, Y., Gopalakrishnan, S., & Kouisni, L. (2020). Exploiting biological nitrogen fixation: A route towards a sustainable agriculture. *Plants*, *9*(8), 1011–1011. https://doi.org/10.3390/plants9081011.

Soussi, A., Zero, E., Sacile, R., Trinchero, D., & Fossa, M. (2024). Smart sensors and smart data for precision agriculture: A review. *Sensors (Basel, Switzerland)*, *24*(8), 2647. https://doi.org/10.3390/s24082647.

De Souza-Torres, A., Govea-Alcaide, E., Gómez-Padilla, E., Masunaga, S. H., Effenberger, F. B., Rossi, L. M., López-Sánchez, R., & Jardim, R. F. (2020). Fe_3O_4 nanoparticles and Rhizobium inoculation enhance nodulation, nitrogen fixation and growth of common bean plants grown in soil. *Rhizosphere*, *17*, 100275. https://doi.org/10.1016/j.rhisph.2020.100275.

Tang, M., & Capela, D. (2020). Rhizobium diversity in the light of evolution. In Frendo, P., Frugier, F., Masson-Boivin, C. (Eds.)Eds. Advances in Botanical Research. (pp. 251–288). Academic Press. 10.1016/bs.abr.2019.09.006Eds..

Tang, Y., Wang, L., Zhou, Y., Geng, X., Deng, X., & Zhang, Z. (2024). Drought decreases symbiotic nitrogen fixation and nitrogen transfer in a *Populus tomentosa-Hippophae rhamnoides* mixed plantation on the semiarid Loess Plateau, China. *Forest Ecology and Management*, *558*, 121815. https://doi.org/10.1016/j.foreco.2024.121815.

Tian, S., Yao, L., Zhang, Y., Rao, X., & Zhu, H. (2025). Prime editing for crop improvement: A systematic review of optimization strategies and advanced applications. *Genes*, *16*(8), 965. https://doi.org/10.3390/genes16080965.

Tsyganova, A. V., Brewin, N. J., & Tsyganov, V. E. (2021). Structure and development of the legume-rhizobial symbiotic interface in infection threads. *Cells*, *10*(5), 1050. https://doi.org/10.3390/cells10051050.

Turnbull, C., Lillemo, M., & Hvoslef-Eide, T. A. (2021). Global regulation of genetically modified crops amid the gene edited crop boom–a review. *Frontiers in plant science*, 12, 630396.

Udvardi, M., & Poole, P. S. (2013). Transport and metabolism in legume-rhizobia symbioses. *Annual Review of Plant Biology*, *64*(1), 781–805. https://doi.org/10.1146/annurev-arplant-050312-120235.

Velandia, K., Reid, J. B., & Foo, E. (2022). Right time, right place: The dynamic role of hormones in rhizobial infection and nodulation of legumes. Plant Communications, 3(5).

Venkataraman, M., Yñigez-Gutierrez, A., Infante, V., MacIntyre, A., Fernandes-Júnior, P. I., Ané, J., & Pfleger, B. (2023). Synthetic biology toolbox for nitrogen-fixing soil microbes. *ACS Synthetic Biology*, *12*(12), 3623–3634. https://doi.org/10.1021/acssynbio.3c00414.

Wang, L., Xiao, Y., Wei, X., Pan, J., & Duanmu, D. (2021). Highly efficient CRISPR-mediated base editing in Sinorhizobium meliloti. *Frontiers in Microbiology*, 12, 686008.

Wang, S., Wang, X., Adeniji, O. D., Batchelor, W. D., Wang, Y., Blersch, D., Higgins, B. T., Liles, M. R., Luo, W., Chen, C. Y., Feng, Y., & Wang, Y. (2025). Targeted genome editing of the ACC deaminase gene in *Bradyrhizobium*: Toward enhanced plant growth and stress tolerance. *Biotechnology and Bioengineering*, https://doi.org/10.1002/bit.70064.

Watanabe, Y., Aoki, W., & Ueda, M. (2023). Ammonia production using bacteria and yeast toward a sustainable society. *Bioengineering*, *10*(1), 82. https://doi.org/10.3390/bioengineering10010082.

Weith, S. K., Jahufer, M. Z., Hofmann, R. W., Anderson, C. B., Luo, D., & Ehoche, O. G., et al. (2022). Quantitative genetic analysis reveals potential to breed for improved white clover growth in symbiosis with nitrogen-fixing Rhizobium bacteria. *Frontiers in Plant Science*, 13, 953400.

Wirth, S. B., Braun, J., Heinke, J., Ostberg, S., Rolinski, S., Schaphoff, S., Stenzel, F., Von Bloh, W., Taube, F., & Müller, C. (2024). Biological nitrogen fixation of natural and agricultural vegetation simulated with LPJmL 5.7.9. *Geoscientific Model Development*, *17*(21), 7889–7914. https://doi.org/10.5194/gmd-17-7889-2024.

Xu, M., Fan, L., Li, A., Liu, Q., Yu, G., & Wang, S., et al. (2024). Plant and microbial carbon are important drivers of free-living nitrogen fixation in tropical forest soils: A new discovery of carbon-driven nitrogen input. *Geophysical Research Letters*, 51(20), e2024GL111238.

Xu, Z. X., Zeng, B., Chen, S., Xiao, S., Jiang, L. G., & Li, X., et al. (2024). Soil microbial community composition and nitrogen enrichment responses to the operation of electric power substation. *Frontiers in Microbiology*, 15, 1453162.

Yan, X., Xia, Y., Zhao, X., Ti, C., Xia, L., Chang, S. X., & Yan, X. (2025). Coupling nitrogen removal and watershed management to improve global lake water quality. *Nature Communications*, 16(1), 2182.

Yang, J., Xie, X., Wang, X., Dixon, R., & Wang, Y. P. (2014). Reconstruction and minimal gene requirements for the alternative iron-only nitrogenase in Escherichia coli. *Proceedings of the National Academy of Sciences*, 111(35), E3718–E3725.

Yang, J., Xiang, N., Liu, Y., Guo, C., Li, C., & Li, H., et al. (2023). Organelle-dependent polyprotein designs enable stoichiometric expression of nitrogen fixation components targeted to mitochondria. *Proceedings of the National Academy of Sciences*, 120(34), e2305142120.

Yang, Z., Du, H., Xing, X., Li, W., Kong, Y., Li, X., & Zhang, C. (2021). A small heat shock protein, GmHSP17.9, from nodule confers symbiotic nitrogen fixation and seed yield in soybean. *Plant Biotechnology Journal*, *20*(1), 103–115. https://doi.org/10.1111/pbi.13698.

Ye, K., Zheng, J., Dong, Z., Wang, S., & Huang, S. (2025). Harnessing omics to decode the mechanisms of symbiotic nitrogen fixation. *Abiotech*, 6(4), 602–617.

Zaman, W., Khalil, A. A. K., Amin, A., & Ali, S. (2025). Nanofabrication techniques for enhancing plant–microbe interactions in sustainable agriculture. *Nanomaterials*, *15*(14), 1086. https://doi.org/10.3390/nano15141086.

Zhang, Y., Ku, Y.-S., Cheung, T.-Y., Cheng, S.-S., Xin, D., Gombeau, K., Cai, Y., Lam, H.-M., & Chan, T.-F. (2024). Challenges to rhizobial adaptability in a changing climate: Genetic engineering solutions for stress tolerance. *Microbiological Research*, *288*, 127886. https://doi.org/10.1016/j.micres.2024.127886.

Zhao, G., Zhu, X., Zheng, G., Meng, G., Dong, Z., Baek, J. H., Jeon, C. O., Yao, Y., Xuan, Y. H., Zhang, J., & Jia, B. (2023). Development of biofertilizers for sustainable agriculture over four decades (1980–2022). *Geography and Sustainability*, *5*(1), 19–28. https://doi.org/10.1016/j.geosus.2023.09.006.

Zhao, L., Walkowiak, S., & Fernando, W. G. D. (2023). Artificial intelligence: A promising tool in exploring the phytomicrobiome in managing disease and promoting plant health. *Plants*, *12*(9), 1852. https://doi.org/10.3390/plants12091852.

Zorin, E. A., Kliukova, M. S., Afonin, A. M., Gribchenko, E. S., Gordon, M. L., & Sulima, A. S., et al. (2022). A variable gene family encoding nodule-specific cysteine-rich peptides in pea (Pisum sativum L.). *Frontiers in plant science*, 13, 884726.

Microbial Consortia of Diazotrophs in Agriculture: Synergistic Approaches for Soil and Plant Health

3

Kusum Sharma and Manoher Saryam

Abstract

The excessive application of chemical pesticides and fertilizers spurred on by agricultural intensification has a detrimental impact on crop quality, nutritional value, soil health, and eventually human health because of the buildup of harmful residues. A dynamic microhabitat full of advantageous microbial communities, the plant rhizosphere aids in nutrient uptake, disease prevention, stress reduction, and plant growth promotion. Among these, diazotrophic bacteria—microbes that can fix nitrogen from the atmosphere—are essential to plant nutrition and the natural cycle of nitrogen. When incorporated into microbial consortia, diazotrophs function in collaboration with mycorrhizal fungi and other plant growth-promoting rhizobacteria to improve soil health and structure and promote hormone production. These partnerships serve as potent bioformulations that can be applied to non-legume and cereal crops, providing a sustainable substitute for chemical inputs. The ecological roles and mechanisms of diazotrophic microbial consortia in the rhizosphere are examined in this chapter, with a focus on how they enhance soil fertility, support long-term agricultural sustainability, and foster plant health. This chapter attempts to shed light on the creation of efficient microbial formulations for robust and fruitful agroecosystems by unraveling the interacting networks found in the plant microbiome, particularly in the rhizosphere.

K. Sharma (✉)
Department of Life Science, Faculty of Science, Rabindranath Tagore University, Raisen, Bhopal, Madhya Pradesh, India
e-mail: kusum.sharma15@gmail.com

M. Saryam
Faculty of Agriculture, Rabindranath Tagore University, Raisen, Bhopal, Madhya Pradesh, India

R. K. Singh and P. Singh (eds.), *Diazotrophic Plant Growth-Promoting Microorganisms and Green Energy – Volume I*, https://doi.org/10.1007/978-981-95-9682-9_3

Keywords
Diazotroph · PGPR · modern agriculture · soil fertility and plant health

3.1 Introduction

Modern agriculture has achieved significant improvements in output, but in order to feed the expanding population, it has become more and more reliant on artificial fertilizers and pesticides. However, there are significant issues with crop quality, soil fertility, human health, and environmental sustainability with this chemical-heavy approach. Among the hazardous materials that accumulate as a result of extended use of synthetic inputs are heavy metals and chemical residues. These contaminants introduce hazardous compounds into the food chain, harm soil structure, and reduce microbial diversity. These detrimental effects not only raise production but also emphasize the urgent need for sustainable, nature-based solutions that support better soil and plant health (Jaishankar et al., 2014).

The incorporation of microbial consortia, composed of diazotrophs that can transform atmospheric nitrogen into forms that are bioavailable is one of the most promising approaches in this regard (Llamas et al., 2023). Beneficial microorganisms, such as diazotrophs, interact with plants and with one another in the rhizosphere, the thin layer of soil that surrounds plant roots. Important functions such as stress tolerance, disease suppression, phytohormone synthesis, and nutrient solubilization are made possible by these kinds of interactions. Together with other plant growth-promoting rhizobacteria (PGPR) and arbuscular mycorrhizal fungi (AMF), diazotrophs can form synergistic consortia that can greatly enhance soil health, plant growth, and nutrient-use efficiency (Varinderpal-Singh et al., 2020).

Recent research indicates that microbial consortia outperform single-strain inoculants due to their collaborative nature and functional diversity. Most commercially available products are based on single species and have proven success in attaining specified targets following validation, as evidenced by numerous recent meta-analyses (Thilakarathna & Raizada, 2017; Schütz et al., 2018; Tufail et al., 2021). However, each species has constraints in terms of its functions and niches. The idea of blending various species with complementary qualities, known as a consortium, was advanced in the recent decade (Santos et al., 2019).

Bacterial consortia have been demonstrated to be significantly more advantageous than single bacterial isolates in agricultural applications. In compensating for the shortcomings of single bacterial isolates, a consortium of several bacterial strains can increase total efficacy. Broader spectrums of characteristics that promote plant growth are made possible by the synergistic interaction between many strains, which enhances soil fertility and plant production. Consortia of isolates that can produce indole-3-acetic acid (IAA) and solubilize inorganic phosphate, for example, have been found to be more successful at promoting plant development than single strains used alone. By increasing nutrient availability, promoting root growth, and fortifying plant resistance to environmental stress, bacterial consortia represent a viable and sustainable approach to raising agricultural output (Vishwakarma et al., 2020; Anandakumar et al., 2024). Second, members within a consortium can facilitate the establishment and functioning of target strains through synergistic cooperation (Deter & Lu, 2022). This interaction of consortium was also observed between PGPB and AMF, where plants could achieve a higher salt-stress resistance (Sabah et al., 2025) and better organic phosphorus mineralization (Liu et al., 2024) compared to inoculated with either microorganism alone. Interactions between inoculants and indigenous species are more likely mediated through quorum sensing and antibiotic release within bacterial consortia (Mukherjee et al., 2021). These aspects support using microbial consortia to achieve more stable and effective outcomes (Cao et al., 2022; Jiang et al., 2023; Wu et al., 2023). Despite the advantages mentioned, the application of microbial consortia still needs to be explored, with similar disadvantages as single species approaches (Lahiri et al., 2022). One concern is that microbial inoculation represents a deliberate biotic disturbance, also known as microbial invasion, for the native soil microbiome. This can trigger complex interactions between inoculated and native species, leading to unpredictable and unreliable outcomes (Liu et al., 2022). Additionally, the consortium's characteristics and application strategies such as soil inoculation or seed and plant-based inoculation play a crucial role in determining the inoculation effect. Therefore, more effort should be put into experiments that explore these effects and suitable application approaches when developing commercial consortia. However, the majority of microbial inoculants are developed by laboratory screening and combination. Since soils contain living resident microorganisms, it remains difficult to determine if and how to make microbial inoculants survive and work steadily in the natural environment (Pareek & Soni, 2025; Cornell et al., 2021). In order to evaluate the effectiveness of microbial consortia in living soil and investigate the variables that can affect the inoculation impact, systematic and quantitative research is therefore desperately needed. This information would direct the subsequent creation and use of microbial inoculants. These collaborations offer a sustainable and ecologically friendly means of reducing dependency on chemical fertilizers, particularly for non-legume cereal crops that are generally high consumers of nitrogen, phosphorus, and potassium. In addition to promoting plant nutrition, diazotrophic consortia-based bioformulations enhance soil biological activity, organic matter dynamics, and overall agroecosystem resilience.

A case study conducted by Singh et al. (2022a) demonstrated the effectiveness of microbial consortia containing diazotrophs such as *Azospirillum brasilense*, *Azotobacter chroococcum*, and *Rhizobium leguminosarum* in enhancing soil fertility and crop productivity in maize-based systems. The study revealed that the consortium not only improved nitrogen fixation efficiency but also promoted root growth, chlorophyll content, and yield by 18%–25% compared to chemical nitrogen fertilization alone. Moreover, the application of this microbial mix enhanced soil enzymatic activities and microbial biomass, indicating long-term improvement in soil health. The findings emphasized that integrating such diazotrophic consortia into existing nutrient management practices could significantly reduce chemical fertilizer use while maintaining crop productivity and ecological balance. This case highlights the potential of microbial synergy as a sustainable strategy for improving agricultural productivity and soil resilience.

The aim of this chapter is to provide a thorough review of the functions and workings of diazotrophic microbial consortia in agriculture. It emphasizes the use of bioinoculants to support sustainable crop production, the importance of microbial diversity in the rhizosphere, and the synergistic interactions that propel plant-microbe partnerships. Through an examination of the aboveground and belowground aspects of the plant microbiome, this chapter highlights the revolutionary potential of microbial consortia in forming ecologically conscious, future-ready farming methods.

3.2 Diazotrophic Microorganisms: An Overview

Diazotrophic bacteria are a broad category of prokaryotes that use nitrogenase's enzymatic activity to fix atmospheric nitrogen (N_2) into accessible ammonia. In particular, as an alternative to manufactured nitrogen fertilizers, the biological nitrogen fixation (BNF) process is essential to sustainable agriculture. Numerous ecological niches, including bulk soil, rhizosphere, plant roots, and even inside plant tissues, are inhabited by diazotrophs, which comprise both bacteria and archaea. In general, these microorganisms can be divided into four groups according to how they interact with plants: free-living, associative, endophytic, and symbiotic.

Free-living diazotrophs that may fix nitrogen without direct plant connection include organisms from the genera *Azotobacter, Clostridium, Klebsiella, Beijerinckia*, and *cyanobacteria* like *Anabaena* and *Nostoc* (Sun et al., 2021; Pahari et al., 2021). As per the study, it has been described that *Azotobacter* is an obligatory aerobe that fixes nitrogen and releases compounds that promote plant growth, such as siderophores and indole acetic acid (IAA), in addition to creating protective cysts and exopolysaccharides. Cyanobacteria play a significant role in the availability of nitrogen during floods and are especially significant in rice fields. More recently, metagenomic research has shown that uncultured taxa like *Geobacter* and *Anaeromyxobacter* have the ability to fix nitrogen, particularly in anaerobic or moist conditions (Masuda et al., 2024).

Associative and endophytic diazotrophs more effectively increase nitrogen availability by developing closer bonds with internal plant tissues or roots. Cereal crops like maize, rice, and wheat are frequently colonized by genera like *Azospirillum, Herbaspirillum, Gluconacetobacter, Enterobacter, Burkholderia*, and *Pseudomonas* (Sood & Kaushal, 2021). Initially identified from sugarcane, *Gluconacetobacter* diazotrophicus is well-known for its potent capacity to fix nitrogen in plant tissues and its resistance to acid, which frequently lessens the requirement for nitrogen fertilizers. Similarly, strains of *Herbaspirillum* and *Burkholderia* have demonstrated exceptional promise in raising the nitrogen content and plant biomass of graminaceous crops (Liu-Xu et al., 2024; Imran et al., 2021). In legumes and many woody non-leguminous plants, symbiotic diazotrophs, most often *Rhizobium, Bradyrhizobium*, and *Frankia,* form root nodules (Patra & Mandal, 2023). These nodules offer a low-oxygen environment that is necessary for nitrogenase activity. In legume-based cropping systems, symbiotic diazotrophs are essential nitrogen fixers that frequently supply more than

50% of the nitrogen needed by the plant. These microorganisms assist the host plant as well as the next crop in the rotation by enriching the soil with leftover nitrogen.

Functionally, diazotrophs offer many characteristics that promote plant growth in addition to fixing nitrogen. They generate phytohormones that strengthen root architecture and increase stress tolerance, including auxins, cytokinins, gibberellins, and ethylene modulators such ACC deaminase (Imran et al., 2021). They also release siderophores to chelate iron, solubilize other vital elements like potassium and phosphorus, and produce antimicrobial chemicals to control soil-borne diseases. This diverse function is particularly noticeable when diazotrophs are paired with other advantageous microorganisms in consortia, which enhances their capacity for colonization, their involvement in nutrient cycling, and their overall advantages for plant growth. Organic matter, crop genotype, soil pH, moisture content, and management techniques are some of the biotic and abiotic variables that affect diazotroph function (Sneha et al., 2021). It is well recognized that the structure and activity of diazotrophic communities are impacted by the long-term use of chemical fertilizers, tillage, crop rotation, and biochar amendments. To get the most out of integrating diazotrophs into integrated nutrition management techniques, it is necessary to fully comprehend these interactions.

Diazotrophs are being utilized more and more in the creation of microbial consortia and biofertilizers for a range of crops due to their agronomic and ecological adaptability (Xavier et al., 2023). Their use in traditional farming systems improves soil health, promotes sustainable agriculture, and lessens the need for chemical inputs. The identification, characterization, and enhancement of diazotrophic strains are being made possible by developments in genomes, metagenomics, and synthetic biology. This opens the door for the development of next-generation bioinoculants that are suited to certain agro-ecological zones and crop kinds (Table 3.1).

3.3 Microbial Consortia: Synergistic Interactions

Microbial consortia are a harmonious grouping of many microbial species that work together to support soil health and plant growth, particularly in stressful environmental settings. Diazotrophic bacteria are essential to the plant microbiome because they fix atmospheric nitrogen and provide it in a form that plants can use (Imran et al., 2021). However, when they work as a part of a well-organized microbial consortium, their effectiveness and persistence in the rhizosphere are greatly increased. The consortium members' dynamic relationships, which range from commensalism to mutualism and protocooperation, are essential to maintaining their combined advantages (Sarsan et al., 2021). In addition to improving nitrogen fixation, positive interactions between diazotrophs and other plant growth-promoting bacteria (PGPB) encourage cross-feeding, co-metabolism, and the sharing of growth-stimulating molecules. These interactions also improve nutrient cycling, disease suppression, and the development of root systems.

In these beneficial assemblages, mutualistic connections are particularly significant since species depend on each other for metabolic support and survival.

Table 3.1 Key characteristics of different diazotrophic microbial consortia

Microbial cnsortium	Major components	Key characteristics / functional traits	Target crop(s)	Reference
Consortium I	*Azotobacter chroococcum* + *Azospirillum brasilense*	Enhances nitrogen fixation, promotes root elongation, increases chlorophyll content and yield by 20%–25%	Maize (*Zea mays*)	Singh et al. (2022b)
Consortium II	*Rhizobium leguminosarum* + *Bradyrhizobium japonicum* + *Bacillus subtilis*	Improves nodule formation, nitrogen fixation efficiency, and stress tolerance in legumes	Soybean (*Glycine max*)	Kumar et al. (2021)
Consortium III	*Azospirillum lipoferum* + *Pseudomonas fluorescens* + *Trichoderma harzianum*	Improves plant growth, disease resistance, and nutrient uptake through synergistic interactions	Rice (*Oryza sativa*)	Meena and Singh (2020)
Consortium IV	*Azotobacter vinelandii* + *Phosphate-solubilizing bacteria (PSB)* + *Mycorrhiza*	Enhances nitrogen and phosphorus availability, root colonization, and soil aggregation	Wheat (*Triticum aestivum*)	Pandey and Thakur (2023)
Consortium V	*Beijerinckia indica* + *Azospirillum spp.* + *Bacillus megaterium*	Improves nitrogen fixation and nutrient solubilization, enhances drought tolerance	Sorghum (*Sorghum bicolor*)	Gupta and Singh (2019)

Protocooperation enriches all parties without requiring mandatory dependency, whereas commensalism allows one to profit without affecting the other. These cooperative mechanisms are especially helpful in agricultural consortia because of their ability to maintain stability over an extended period of time and their consistent performance in a range of contexts. On the other hand, harmful interactions such as competition, parasitism, and amensalism can reduce the effectiveness of microbial consortia (Sarsan et al., 2021). These happen when members battle for limited resources or release compounds that are antagonistic to one another, highlighting how important compatibility is when creating microbial consortia.

An important molecular communication mechanism that controls microbial compatibility and functional synchronization within consortia is quorum sensing (Zeng et al., 2023). Auto-inducers, such as acyl-homoserine lactones (AHLs), are released by bacteria and, when they reach a threshold concentration, alter gene expression across the population (Laj et al., 2022). These changes make it possible for consortia to function by moving from individual activity at low cell densities to coordinated activities at larger densities.

Quorum sensing is used by diazotrophs such as *Serratia liquefaciens, Sinorhizobium fredii*, and *Pantoea ananatis* to create biofilms, control root growth, and improve colonization in rice and bean crops (Ghosh & Mandal, 2022). Furthermore, another family of bacterial signaling molecules called volatile organic compounds (VOCs) is involved in both plant-microbe interactions and communication between and across species (Sharifi et al., 2022). These volatile organic compounds (VOCs) stimulate plant defense systems and hormone signaling pathways while also influencing microbial competitiveness, stress resistance, virulence, and biofilm formation (Montejano-Ramírez et al., 2024).

Identifying constituent strains accurately, documenting consortium assembly clearly, comprehending soil physicochemical interactions, and maintaining viable microbial populations at application are all critical to the formation and effectiveness of microbial consortia. Microbial consortia, particularly those designed to integrate with diazotrophs to decrease the use of chemical fertilizers and increase sustainable crop output, can only function reliably in field conditions when these factors are fully taken into consideration. Since inter-microbial harmony is in line with plant health, productivity, and ecological resilience, synergistic interactions among microbial consortia serve as the foundation for next-generation biofertilizer solutions (Table 3.2).

Table 3.2 Complex interactions of microorganisms in the rhizosphere

Type of interaction	Microbial partners	Synergistic effect	References
Nitrogen Fixation + Phosphate Solubilization	*Azospirillum* + *Pseudomonas*, *Bacillus* + *Rhizobium*	Enhanced nitrogen fixation and increased phosphorus availability	Janati et al. (2021), Timofeeva et al. (2023)
IAA Production + ACC Deaminase Activity	*Gluconacetobacter* + *Enterobacter*	Promotes root elongation and reduces ethylene stress	Ahmed et al. (2021), Kour et al. (2024), Singh et al. (2024)
Siderophore Production + HCN Production	*Pseudomonas fluorescens* + *Bacillus subtilis*	Suppresses phytopathogens through iron competition and antibiosis	Khaing et al. (2021), Gade and Koche (2022).
Biofilm Formation + Motility	*Azotobacter* + *Bacillus megaterium*	Improved root colonization and microbial persistence	Roy et al. (2024), Sowmiya et al., (2024
Halotolerance + PGPR Traits	*Halomonas* + *Rhizobium*	Enhanced growth under saline conditions	Oliva et al. (2023), Shahid et al. (2024).
Organic Acid Production + Zn Solubilization	*Pseudomonas putida* + *Bacillus cereus*	Increased micronutrient solubility and uptake	Masood et al. (2022), Rawat et al. (2021)

3.4 Mechanisms of Plant Growth Promotion by Diazotrophs

Diazotrophs use a combination of direct and indirect methods to greatly increase plant growth and output. Their primary function is biological nitrogen fixation (BNF), in which nitrogenase enzymatically transforms atmospheric nitrogen (N_2) into ammonia (NH_3), so providing plants with usable nitrogen (Mukherjee & Sen, 2021). This procedure lessens the need for artificial nitrogen fertilizers, which is particularly important in soils that lack nitrogen. It is commonly known that diazotrophic bacteria, whether free-living or associative, have a high capacity for fixing nitrogen such as in *Azospirillum, Azotobacter, Herbaspirillum*, and *Gluconacetobacter* (Jehani et al., 2023). In order to create a low-oxygen environment that is conducive to nitrogenase activity and to provide the host with fixed nitrogen, symbiotic diazotrophs such as *Rhizobium* and *Bradyrhizobium* form nodular associations with legumes (Goyal et al., 2021).

In addition to fixing nitrogen, diazotrophs also stimulate plant development by producing and controlling phytohormones. They are known to generate auxins, like indole-3-acetic acid (IAA), which promote the production of lateral roots and root elongation, improving the uptake of nutrients and water (Roychoudhry & Kepinski, 2022). Gibberellins and cytokinins, which encourage cell division and shoot growth, are also produced by certain strains (Prasad, 2022). Furthermore, 1-aminocyclopropane-1-carboxylate (ACC) deaminase (Figs. 3.1a and 3.1b), an enzyme found in many diazotrophs, lowers plant ethylene levels in response to stress (Gamalero et al., 2023). ACC deaminase activity helps counteract this inhibitory impact and promotes plant vigor since ethylene, although necessary for plant development, can limit root growth when produced in excess during stress (Gamalero et al., 2023).

The solubilization of nutrients that are not readily available, including potassium and phosphorus (Fig. 3.1b), is another significant mechanism (Rawat et al., 2021). Mineral-bound phosphates are chelated and released into the soil solution by the organic acids secreted by diazotrophic bacteria, such as gluconic, citric, or oxalic acid.

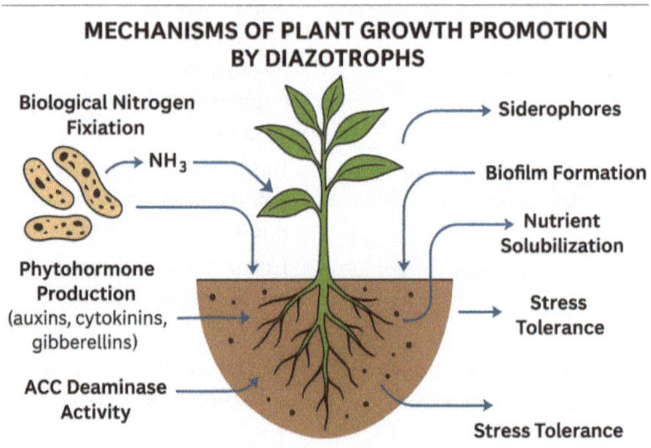

Fig. 3.1a Multiple mechanisms of promote plant growth by diazotrophs

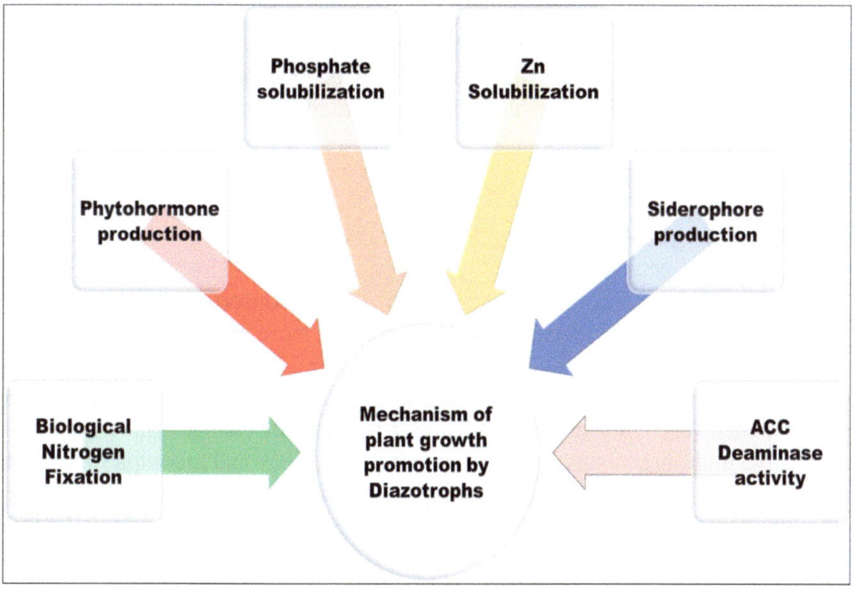

Fig. 3.1b Multiple mechanisms of plant growth promotion by diazotrophic consortia

Azospirillum brasilense and *Burkholderia* spp., for instance, have been demonstrated to enhance the amount of phosphorus in soil, which supports plant root development and energy metabolism (Sun et al., 2022; Ribeiro et al., 2022). Likewise, potassium-solubilizing diazotrophs increase the efficiency of nutrient intake by mobilizing K from insoluble sources.

Additionally, diazotrophs are essential for improving plant immunity and stress tolerance. They operate as biocontrol agents by producing siderophores (Fig. 3.1a), which are iron-chelating chemicals that restrict harmful microorganisms' access to iron (Swarnalatha et al., 2022). In order to reduce soil-borne infections and enhance plant health, several diazotrophs also release hydrogen cyanide (HCN), lytic enzymes (such as chitinase and glucanase), and antibiotics (Abdelaziz et al., 2023). Additionally, diazotrophs engage the plant's defense mechanisms by causing systemic resistance (ISR), readying it for biotic or abiotic stressors in the future (Verma et al., 2024).

Another crucial characteristic of many diazotrophs is the production of biofilms, which promotes root surface colonization and rhizosphere persistence. In order to maintain a long-term relationship with the plant host, biofilms shield bacterial cells against desiccation, predators, and harmful substances. Plant biomass and health are favorably correlated with the biofilm-mediated colonization of plant roots by strains of *Pseudomonas*, *Sinorhizobium*, and *Gluconacetobacter* (Khan et al., 2024).

These impacts on plant growth are further enhanced by the synergistic interactions that occur when diazotrophic bacteria are included into microbial consortia. Together with other helpful microorganisms, diazotrophs provide rhizospheric stability, nutrient cycling, and increased crop yields. Diazotrophs are positioned as important players in sustainable agriculture by these complex processes, which are bolstered by developments in microbial ecology, omics technologies, and bioformulation techniques (Pallucchini, 2023). They provide an affordable, environmentally responsible substitute for chemical fertilizers while maintaining food security in the face of shifting environmental conditions.

3.5 Challenges and Future Perspectives

A multitude of hurdles hinder diazotrophic microbial consortia from being widely used in agriculture, even though they are becoming more widely acknowledged as sustainable substitutes for synthetic nitrogen fertilizers. One major issue is the inconsistent field performance of inoculated diazotrophs, which is frequently brought on by environmental variability such as pH, salinity, temperature changes, and competition with native microorganisms. Poor colonization efficiency and incompatibility with the current soil microbiota frequently impair the effectiveness of diazotroph-based inoculants (Berde et al., 2022). Furthermore, antagonism between introduced microorganisms might interfere with nitrogen fixation and plant growth promotion capabilities; the synergistic interaction within multi-strain consortia is largely strain-specific (Benmrid et al., 2023). The scalability of bioformulation technologies that guarantee the delivery, survival, and shelf-life of viable microbial consortia in diverse agro-climatic zones is another significant hurdle.'

The intricacy of diazotrophic communities has been deciphered by molecular and metagenomic techniques, however there is still a gap in applying this understanding in the field. The creation of artificial microbial consortia containing carefully chosen diazotrophs that have been designed or screened for host-specific root colonization, abiotic stress tolerance, and mutual compatibility will determine future prospects. The preparation of functional consortia that replicate natural plant-microbe interactions in the field can be accelerated by combining bioinformatics with omics-based technologies (genomics, proteomics, and metabolomics) (Jain et al., 2024). The survivability and usefulness of these consortia can also be improved by the creation of smart carriers (such as nanomaterials, charcoal, or alginate beads) for efficient microbial delivery Najim et al. (2024).

Additionally, many nations still lack adequate regulatory and quality control systems for commercial diazotrophic formulations, which results in inconsistent product quality and farmer acceptance. In order to address this, a multidisciplinary strategy comprising soil microbiologists, agronomists, and policymakers is needed to establish crop-oriented, region-specific consortia backed by strong guidelines. In order to measure the environmental impact of microbial consortia on soil health and nutrient cycling, future research must also concentrate on ecological studies and long-term field trials. In the end, including diazotroph microbial consortia into comprehensive agroecological techniques can greatly support low-input and climate-resilient farming systems, which is consistent with the objectives of sustainable agriculture and global food security.

3.6 Conclusion

Diazotroph microbial consortia represent a viable, environmentally responsible solution to improve soil fertility and plant health. These synergistic combinations are more successful than single-strain inoculants at improving nitrogen fixation, nutritional solubilization, and stress tolerance. According to research, co-inoculation of suitable diazotrophs improves soil structure, increases crop yields, and lessens reliance on chemical fertilizers. Assuring strain compatibility, formulation stability, and reliable field performance are still difficult tasks, nonetheless. Creating crop- and region-specific consortia and incorporating them into sustainable farming methods should be the main goal of future initiatives. All things considered, the development of robust and fruitful agricultural systems depends on diazotrophic consortia. Future research on microbial consortia of diazotrophs should focus on exploring their molecular diversity, synergistic interactions, and field-level performance to enhance soil and plant health sustainably. Advanced omics tools can help identify efficient nitrogen-fixing strains and their functional roles, while crop- and soil-specific consortia combining diazotrophs with other beneficial microbes can improve nutrient uptake, stress tolerance, and productivity. Long-term field trials are needed to validate their effectiveness under diverse agro-climatic conditions. Practical applications should emphasize integrating these bioinoculants into sustainable farming systems, such as organic and integrated nutrient management, supported by improved formulation technologies for better microbial viability. Moreover, policy support, quality assurance, and farmer training are essential to promote large-scale adoption and commercialization of diazotrophic consortia, ultimately reducing chemical fertilizer dependency and ensuring long-term soil fertility and ecosystem resilience.

References

Abdelaziz, A. M., Hashem, A. H., El-Sayyad, G. S., El-Wakil, D. A., Selim, S., Alkhalifah, D. H., & Attia, M. S. (2023). Biocontrol of soil borne diseases by plant growth promoting rhizobacteria. *Tropical Plant Pathology*, *48*(2), 105–127.

Ahmed, B., Shahid, M., Syed, A., Rajput, V. D., Elgorban, A. M., Minkina, T., & Lee, J. (2021). Drought tolerant *Enterobacter* sp./ *Leclercia adecarboxylata* secretes indole-3-acetic acid and other biomolecules and enhances the biological attributes of Vigna radiata (L.) R. Wilczek in water deficit conditions. *Biology*, *10*(11), 1149.

Anandakumar, S., Senthamilselvi, D., & Kalaiselvi, T. (2024). Microbial consortium with multifunctional plant growth-promoting traits and its significant contribution in sustainable agriculture. In *Progress in soil microbiome research* (pp. 53–75). Springer Nature Switzerland.

Benmrid, B., Ghoulam, C., Zeroual, Y., Kouisni, L., & Bargaz, A. (2023). Bioinoculants as a means of increasing crop tolerance to drought and phosphorus deficiency in legume-cereal intercropping systems. *Communications Biology*, *6*(1), 1016.

Berde, C. V., Bramhachari, P. V., & Berde, V. B. (2022). Plant growth-promoting diazotrophs: Current research and advancements (Chapter 8).

Cao, Z., Yan, W., Ding, M., & Yuan, Y. (2022). Construction of microbial consortia for microbial degradation of complex compounds. *Frontiers in Bioengineering and Biotechnology*, *10*, 1051233.

Cornell, C., Kokkoris, V., Richards, A., Horst, C., Rosa, D., Bennett, J. A., & Hart, M. M. (2021). Do bioinoculants affect resident microbial communities? A meta-analysis. *Frontiers in Agronomy, 3*, 753474.

Deter, H. S., & Lu, T. (2022). Engineering microbial consortia with rationally designed cellular interactions. *Current Opinion in Biotechnology, 76*, 102730.

Gade, R. M., & Koche, M. D. (2022). Enhancing the growth and disease suppression ability of Pseudomonas fluorescens. In *New and future developments in microbial biotechnology and bioengineering* (pp. 351–368). Elsevier.

Gamalero, E., Lingua, G., & Glick, B. R. (2023). Ethylene, ACC, and the plant growth-promoting enzyme ACC deaminase. *Biology, 12*(8), 1043.

Ghosh, R., & Mandal, N. C. (2022). Quorum sensing in rhizosphere engineering. In Rhizosphere engineering. (pp. 355–381). Academic Press.

Goyal, R. K., Mattoo, A. K., & Schmidt, M. A. (2021). Rhizobial–host interactions and symbiotic nitrogen fixation in legume crops toward agriculture sustainability. *Frontiers in Microbiology, 12*, 669404.

Gupta, P., & Singh, K. (2019). Functional diversity of diazotrophic consortia in improving sorghum productivity under stress conditions. *Soil and Plant Health Journal, 7*(2), 99–108. https://doi.org/10.1080/sphj.2019.099.

Imran, A., Hakim, S., Tariq, M., Nawaz, M. S., Laraib, I., Gulzar, U., & Ahmad, M. (2021). Diazotrophs for lowering nitrogen pollution crises: Looking deep into the roots. *Frontiers in Microbiology, 12*, 637815.

Jain, A., Sarsaiya, S., Singh, R., Gong, Q., Wu, Q., & Shi, J. (2024). Omics approaches in understanding the benefits of plant-microbe interactions. *Frontiers in Microbiology, 15*, 1391059.

Jaishankar, M., Tseten, T., Anbalagan, N., Mathew, B. B., & Beeregowda, K. N. (2014). Toxicity, mechanism and health effects of some heavy metals. *Interdisciplinary Toxicology, 7*(2), 60–72. https://doi.org/10.2478/intox-2014-0009.

Janati, W., Benmrid, B., Elhaissoufi, W., Zeroual, Y., Nasielski, J., & Bargaz, A. (2021). Will phosphate bio-solubilization stimulate biological nitrogen fixation in grain legumes?. *Frontiers in Agronomy, 3*, 637196.

Jehani, M. D., Singh, S., Kumar, D., & Kumar, G. (2023). *Azospirillum*—a free-living nitrogen-fixing bacterium. In *Rhizobiome.* (pp. 285–308). Academic Press.

Jiang, Y., Wu, R., Zhang, W., Xin, F., & Jiang, M. (2023). Construction of stable microbial consortia for effective biochemical synthesis. *Trends in Biotechnology, 41*(11), 1430–1441.

Khaing, A., Theint, W. T., Thi, O. K., & Fu, P. (2021). Antagonistic activity of indigenous rhizobacteria through biosynthesis of indole-3-acetic acid (IAA), hydrogen cyanide (HCN), and siderophores. *Austin Journal of Biotechnology & Bioengineering, 8*(1), 1110.

Khan, I., Sultan, G., Miskeen, S., Madar, I. H., Najeeb, S., Sivanandan, P. K., & Oh, D. H. (2024). Biobased nanomaterials and their interaction with plant growth-promoting rhizobacteria/blue-green algae/Rhizobium for sustainable plant growth and development. In *Biostimulants in plant protection and performance.* (pp. 33–60). Elsevier.

Kour, D., Khan, S. S., Kour, H., Kaur, T., Devi, R., Rai, A. K., & Yadav, A. N. (2024). ACC deaminase producing phytomicrobiomes for amelioration of abiotic stresses in plants for agricultural sustainability. *Journal of Plant Growth Regulation, 43*(4), 963–985.

Kumar, S., Patel, D., & Sharma, N. (2021). Co-inoculation of diazotrophs and PGPR enhances soybean yield and soil enzyme activity. *Applied Microbial Ecology, 18*(2), 87–95. https://doi.org/10.1080/ame.2021.087.

Lahiri, D., Nag, M., Ghosh, S., Dey, A., & Ray, R. R. (2022). Microbial consortium and crop improvement: Advantages and limitations. In Trends of applied microbiology for sustainable economy. (pp. 109–123). Academic Press.

Laj, N., Elayadeth-Meethal, M., Huxley, V. A. J., Hussain, R. R., Kuruniyan, M. S., & Naseef, P. P. (2022). Quorum-sensing molecules: Sampling, identification and characterization of N-acyl-homoserine lactone in Vibrio sp. *Saudi Journal of Biological Sciences, 29*(4), 2733–2737.

Liu, H., Cui, Y., Zhou, J., Penttinen, P., Liu, J., Zeng, L., Chen, Q., Gu, Y., Zou, L., Zhao, K., Xiang, Q., & Yu, X. (2022). Nickel mine soil is a potential source for soybean plant growth promoting and heavy metal tolerant rhizobia. PeerJ, 10, e13215.

Liu, J., Liu, J., Wang, S., Fu, Q., Baoyin, T., Bao, Z., & Li, F. Y. (2024). Soil biological nitrogen fixation is closely associated with soil ammonium nitrogen content in a mowing semiarid natural grassland. *Applied Soil Ecology*, 203, 105690.

Liu-Xu, L., González-Hernández, A. I., Camañes, G., Vicedo, B., Scalschi, L., & Llorens, E. (2024). Harnessing green helpers: Nitrogen-fixing bacteria and other beneficial microorganisms in plant–microbe interactions for sustainable agriculture. *Horticulturae*, *10*(6), 621.

Llamas, A., Leon-Miranda, E., & Tejada-Jimenez, M. (2023). Microalgal and nitrogen-fixing bacterial consortia: From interaction to biotechnological potential. *Plants*, *12*(13), 2476.

Masood, F., Ahmad, S., & Malik, A. (2022). Role of rhizobacterial bacilli in zinc solubilization. *Microbial biofertilizers and micronutrient availability: The role of zinc in agriculture and human health* (pp. 361–377).

Masuda, Y., Mise, K., Xu, Z., Zhang, Z., Shiratori, Y., Senoo, K., & Itoh, H. (2024). Global soil metagenomics reveals distribution and predominance of deltaproteobacteria in nitrogen-fixing microbiome. *Microbiome*, *12*(1), 95.

Meena, M., & Singh, D. (2020). Integrated use of diazotrophic and biocontrol microbes for sustainable rice production. *Agricultural Microbiology Journal*, *15*(1), 35–42. https://doi.org/10.1016/amj.2020.35.

Montejano-Ramírez, V., Ávila-Oviedo, J. L., Campos-Mendoza, F. J., & Valencia-Cantero, E. (2024). Microbial volatile organic compounds: Insights into plant defense. *Plants*, *13*(15), 2013.

Mukherjee, A., Chouhan, G. K., Gaurav, A. K., Jaiswal, D. K., & Verma, J. P. (2021). Development of indigenous microbial consortium for biocontrol management. In *New and future developments in microbial biotechnology and bioengineering* . (pp. 91–104). Elsevier.

Mukherjee, R., & Sen, S. (2021). Role of biological nitrogen fixation (BNF) in sustainable agriculture: A review. *International Journal of Advancement in Life Sciences Research*, *4*(3), 1–7.

Najim, A. A., Radeef, A. Y., al-Doori, I., & Jabbar, Z. H. (2024). Immobilization: The promising technique to protect and increase the efficiency of microorganisms to remove contaminants. *Journal of Chemical Technology & Biotechnology*, *99*(8), 1707–1733.

Oliva, G., Di Stasio, L., Vigliotta, G., Guarino, F., Cicatelli, A., & Castiglione, S. (2023). Exploring the potential of four novel halotolerant bacterial strains as plant-growth-promoting rhizobacteria (PGPR) under saline conditions. *Applied Sciences*, *13*(7), 4320.

Pahari, A., Nayak, S. K., Banik, A., Lakra, P. B., & Mishra, B. B. (2021). Biological nitrogen fixation mechanism and applications. In *Agriculturally important microorganisms* . (pp. 137–151). CRC Press.

Pallucchini, M. (2023). *An investigation into the mechanisms underlying the plant growth promoting properties of G. diazotrophicus* (Doctoral dissertation, University of Nottingham).

Pandey, A., & Thakur, R. (2023). Impact of multifunctional microbial consortia on nutrient dynamics and wheat yield. *Journal of Sustainable Agriculture*, *11*(4), 210–220. https://doi.org/10.1080/jsa.2023.210.

Pareek, A., & Soni, S. (2025). Soil microbial inoculants for eco-friendly farming: Limitations and opportunities. In *Microbial Inoculants*. (pp. 395–412). Academic Press.

Patra, D., & Mandal, S. (2023). Non-rhizobia are the alternative sustainable solution for growth and development of the nonlegume plants. *Biotechnology and Genetic Engineering Reviews*, *39*(2), 435–464.

Prasad, R. (2022). Cytokinin and its key role to enrich the plant nutrients and growth under adverse conditions-an update. Frontier, *13*, 1–14.

Rawat, P., Das, S., Shankhdhar, D., & Shankhdhar, S. C. (2021). Phosphate-solubilizing microorganisms: Mechanism and their role in phosphate solubilization and uptake. *Journal of Soil Science and Plant Nutrition*, *21*(1), 49–68.

Ribeiro, V. P., Gomes, E. A., de Sousa, S. M., de Paula Lana, U. G., Coelho, A. M., Marriel, I. E., & de Oliveira-Paiva, C. A. (2022). Co-inoculation with tropical strains of Azospirillum and Bacillus is more efficient than single inoculation for improving plant growth and nutrient uptake in maize. *Archives of Microbiology*, *204*(2), 143.

Roy, B., Maitra, D., Biswas, A., Chowdhury, N., Ganguly, S., Bera, M., & Mitra, A. K. (2024). Efficacy of high-altitude biofilm-forming novel Bacillus subtilis species as plant growth-promoting rhizobacteria on Zea mays L. *Applied Biochemistry and Biotechnology*, *196*(2), 643–666.

Roychoudhry, S., & Kepinski, S. (2022). Auxin in root development. *Cold Spring Harbor Perspectives in Biology*, *14*(4), a039933.

Sabah, F. U., Bano, A., Bano, A., & Mufti, R. (2025). Interactive effects of arbuscular mycorrhizal fungi (AMF) and plant growth promoting bacteria (PGPB) on the alleviation of salt stress in wheat. *Pakistan Journal of Botany*, *57*(2), 425–432.

Santos, F., Mucha, A. P., Alexandrino, D. A., Almeida, C. M. R., & Carvalho, M. F. (2019). Biodegradation of enrofloxacin by microbial consortia obtained from rhizosediments of two estuarine plants. *Journal of Environmental Management*, 231, 1145–1153.

Sarsan, S., Pandiyan, A., Rodhe, A. V., & Jagavati, S. (2021). Synergistic interactions among microbial communities. In *Microbes in microbial communities: Ecological and applied perspectives* (pp. 1–37). Springer.

Schütz, L., Gattinger, A., Meier, M., Müller, A., Boller, T., Mäder, P., & Mathimaran, N. (2018). Improving crop yield and nutrient use efficiency via biofertilization—A global meta-analysis. *Frontiers in plant science*, 8, 2204.

Shahid, M., Altaf, M., & Danish, M. (2024). The halotolerant exopolysaccharide-producing Rhizobium azibense increases the salt tolerance mechanism in Phaseolus vulgaris (L.) by improving growth, ion homeostasis, and antioxidant defensive enzymes. *Chemosphere*, *360*, 142431.

Sharifi, R., Jeon, J. S., & Ryu, C. M. (2022). Belowground plant–microbe communications via volatile compounds. *Journal of Experimental Botany*, *73*(2), 463–486.

Singh, R., Verma, P., & Yadav, A. K. (2022b). Synergistic effects of diazotrophic microbial consortia on soil fertility and maize productivity under field conditions. *Journal of Soil Biology and Sustainable Agriculture*, *9*(3), 145–156. https://doi.org/10.1016/jsbsa.2022.145.

Singh, S., Annapurna, K., Shrivastava, N., & Varma, Y.A. (2022a). Symbiotic interplay of Piriformospora indica and Azotobacter chroococcum augments crop productivity and biofortification of Zinc and Iron. *Microbiological research*, 262(), 127075.

Sneha, G. R., Swarnalakshmi, K., Sharma, M., Reddy, K., Bhoumik, A., Suman, A., & Kannepalli, A. (2021). Soil type influence nutrient availability, microbial metabolic diversity, eubacterial and diazotroph abundance in chickpea rhizosphere. *World Journal of Microbiology and Biotechnology*, *37*(10), 167.

Sood, G., & Kaushal, R. (2021). Plant growth promoting rhizobacteria for crop health in wheat-maize cropping systems in northwest himalayas. In *Microbial biotechnology in crop protection* . (pp. 269–287). Springer Singapore.

Sowmiya, K., Aravindharajan, S. T. M., Bhagyashree, K. B., & Manoj, M. (2024). Biofilm-forming capability of Bacillus and its related genera. Applications of Bacillus and Bacillus Derived Genera in Agriculture, *Biotechnology and Beyond*, 71–89.

Sun, N., Huang, L., Zhao, H., Zhang, N., Lin, X., & Sun, C. (2022). Beneficial bacterium Azospirillum brasilense induces morphological, physiological and molecular adaptation to phosphorus deficiency in Arabidopsis. *Plant and Cell Physiology*, *63*(9), 1273–1284.

Sun, W. E. N. L. I., Shahrajabian, M. H., & Cheng, Q. I. (2021). Nitrogen fixation and diazotrophs–a review. *Rom Biotechnol Lett*, *26*(4), 2834–2845.

Swarnalatha, G. V., Goudar, V., Reddy, E. C. R. G. S., Al Tawaha, A. R. M., & Sayyed, R. Z. (2022). Siderophores and their applications in sustainable management of plant diseases. In *Secondary metabolites and volatiles of PGPR in plant-growth promotion* (pp. 289–302).

Thilakarathna, M. S., & Raizada, M. N. (2017). A meta-analysis of the effectiveness of diverse rhizobia inoculants on soybean traits under field conditions. Soil Biology and Biochemistry, 105, 177–196.

Timofeeva, A. M., Galyamova, M. R., & Sedykh, S. E. (2023). Plant growth-promoting soil bacteria: Nitrogen fixation, phosphate solubilization, siderophore production, and other biological activities. *Plants*, *12*(24), 4074.

Tufail, M. A., Bejarano, A., Shakoor, A., Naeem, A., Arif, M. S., Dar, A. A., Farooq, T. H., Pertot, I., & Puopolo, G. (2021). Can bacterial endophytes be used as a promising bio-inoculant for the mitigation of salinity stress in crop plants?—a global meta-analysis of the last decade (2011–2020). Microorganisms, 9(9), 1861.

Varinderpal-Singh, Sharma, S., Kunal, Gosal, S. K., Choudhary, R., Singh, R., & Bijay-Singh (2020). Synergistic use of plant growth-promoting rhizobacteria, arbuscular mycorrhizal fungi, and spectral properties for improving nutrient use efficiencies in wheat (Triticum aestivum L.). *Communications in Soil Science and Plant Analysis*, *51*(1), 14–27.

Verma, P., Bhattacharya, A., Bharti, C., & Arora, N. K. (2024). Diazotrophs for enhancement of productivity of major cereal crops under abiotic stresses. In Microbial biotechnology for sustainable agriculture, volume 2. (pp. 131–158). Springer Nature Singapore..

Vishwakarma, K., Kumar, N., Shandilya, C., Mohapatra, S., Bhayana, S., & Varma, A. (2020). Revisiting plant–microbe interactions and microbial consortia application for enhancing sustainable agriculture: A review. *Frontiers in Microbiology*, *11*, 560406.

Wu, D., Wang, W., Yao, Y., Li, H., Wang, Q., & Niu, B. (2023). Microbial interactions within beneficial consortia promote soil health. *Science of the Total Environment*, *900*, 165801.

Xavier, G. R., Jesus, E. D. C., Dias, A., Coelho, M. R. R., Molina, Y. C., & Rumjanek, N. G. (2023). Contribution of biofertilizers to pulse crops: From single-strain inoculants to new technologies based on microbiomes strategies. *Plants*, *12*(4), 954.

Zeng, X., Zou, Y., Zheng, J., Qiu, S., Liu, L., & Wei, C. (2023). Quorum sensing-mediated microbial interactions: Mechanisms, applications, challenges and perspectives. *Microbiological Research*, *273*, 127414.

Comparative Evaluation of Urea and Integrated PGPR with Nitrogen-Fixing Microbes: and Their Effects on Maize (*Zea Mays*)

4

Ajay Malviya, Pratiksha Singh, Rajesh Kumar Singh and Preetibala Solanki

Abstract

Nitrogen is the only macronutrient that is abundant in the atmosphere; however, plants lack specific mechanisms to utilize it directly. When nitrogen deficiency occurs, plants exhibit several morphological symptoms that negatively affect their growth. To address nitrogen scarcity, the long-term and indiscriminate use of chemical fertilizers has been adopted. Unfortunately, this practice adversely impacts a wide range of environmental factors, including crop health and pollution levels. To minimize the use of chemical fertilizers for nitrogen supplementation, microbiologists have isolated potent nitrogen-fixing bacterial strains from plant rhizospheres and endophytes. These microorganisms not only improve soil structure and biological properties but also exhibit biocontrol potential and plant growth–promoting characteristics. The isolated microorganisms can be used in bio-formulations for applications such as seed treatment and foliar spraying, helping to alleviate nitrogen deficiency and strengthen the plant's immune system against pathogens. This exploratory research concludes that bio-formulation could serve as an excellent alternative to urea and other chemical fertilizers for addressing nitrogen deficiency in maize plants. Furthermore, it contributes to the sustainability of green and healthy agriculture.

A. Malviya (✉) · P. Solanki
Department of Microbiology, Barkatullah University, Bhopal, MP, India
e-mail: ajaymalviya458@gmail.com

P. Singh
School of Agriculture Science, LNCT University, JK Town Sarvadharam C Sector, Bhopal, MP, India

R. K. Singh
Department of Research and Development, Agromaxx Industry, Bhopal, MP, India

R. K. Singh and P. Singh (eds.), *Diazotrophic Plant Growth-Promoting Microorganisms and Green Energy – Volume I*, https://doi.org/10.1007/978-981-95-9682-9_4

Keywords

Nitrogen fixing Microbes · Urea · Rhizosphere · Biocontrol

4.1 Introduction

Maize (*Zea mays*) is a well-known crop recognized for its ability to tolerate both high and low temperatures. It can be cultivated throughout all three planting seasons: kharif, rabi, and spring. Interestingly, maize is grown on every continent except Antarctica. In Asia, it ranks as the second-largest producer of maize; however, in India, maize is the third-largest crop produced, followed by wheat and rice (www.researchandmarkets.com). During the 2023–2024 periods, India exported 1,442,671 metric tons of maize to various other countries (www.Protectourlivehod.com). Due to the ever-increasing population, domestic consumption of maize is projected to rise by 4.6 lakh tons in 2024 compared to the 2022–2023 period (ANGRAU maize outlook report—June to May 2023–2024). Maize is not only a staple food and animal feed, but also a vital source of bioenergy, including bioethanol, biogas, and biodiesel production. This growing demand highlights the need for integrating various technologies into traditional cropping systems. Implementing practices such as using hybrid seeds, seed dressing, and intercropping can enhance both the quantity and quality of maize yield. Additionally, maintaining rich soil quality is essential for optimal maize production, as the maize plant is sensitive to soil conditions, including acidity, waterlogging, and salinity stress. According to the report of food and agriculture organization (FAO) of United Nation–2024, about 6.7 million hectares of crop area in India is suffering from salinity stress, out of which Gujarat is the major state affected by soil salinization with 33.18% of the area, whereas 49 million hectares of agricultural land are affected by acid stress, in which 21 million hectares of agricultural land in the North-Eastern region are highly affected by acidified soil (Kumar & Sharma, 2020). The reason behind the acidic and salt stressed soil is that, indiscriminately apply of pesticides and other chemical fertilizers, this will lead to continually degrading a largest precious farming area and reduce the efficient plant growth with increase valuable money investment of farmer (Machado & Serralheiro, 2017).

Therefore, it is essential for us to explore and identify effective alternative methods in agriculture to safeguard the soil and the environment. Required the process of isolation, screening, and characterizing beneficial microbes that thrive in a variety of environmental conditions is of paramount importance, especially in arid regions where agricultural challenges are prominent (Singh et al., 2017). Microbes, which include bacteria and fungi, possess unique traits that enable them to enhance plant resilience against diseases and pests. Plant growth promoting rhizobacteria (PGPR) is defined as the rhizospheric region, in which a large number of distinct kinds of microorganism founded along with secreting multiple plant growth stimulating substances such as P-solubilizers, N_2-fixers, K-mobilizers, Ammonia, and other phytohormons producers, will play the crucial role in promoting the plant growth. PGPR must be proficient to colonize the root surface they must survive, multiply, and compete with other microbiota, and they must promote plant growth (Singh et al., 2017).

In maize cultivation, nitrogen plays a crucial role as an essential macronutrient that significantly influences the plant growth and its overall productivity. Whereas, Nitrogen deficiency is a common challenge faced by maize plants, which can adversely affect the plant growth and productivity. To combat this deficiency, a diverse range of nonsymbiotic nitrogen-fixing bacterial genera has been identified as effective agents for enhancing nitrogen availability in the soil such as *Azotobacter, Azospirillum,* and pseudomonas, etc., are able to colonize on or around the surface of plant root (Samantaray et al., 2024). In addition there is several endophytic bacteria, such as *Gluconacetobacter diazotrophicus, Herbaspirillum seropedicae,* and *Serratia marcescens*, have been discovered that can reside within the plant tissues and contribute to fix atmospheric nitrogen symbiotically into the plant (Kumar et al., 2024).

Nowadays, the Indian government is giving more attention to the application of fertilizers such as organic fertilizers, vermicompost, and microbiologically developed bioformulations as alternatives to chemically manufactured fertilizers via various governmental schemes, for instance, Paramparagat Krishi Vikash Yojana and Mission Organic Value Chain Development for North Eastern Region (Ministry of Agriculture & Farmers Welfare, 2024).

4.2 The Impact of Nitrogen Deficiency on Maize Plant

In the context of the maize plant growth and their defense or immunity mechanisms against pathogens, both micronutrients and macronutrients play a vital role. Additionally, nitrogen is a precious element for synthesis of protein, amino acid, nucleic acid, enzymes, and enhancing plant growth and helping to prevent susceptibility to diseases. However, the scarcity of Nitrogen in maize plants will showing symptoms to reduce the leaf area, leaf width, beginning of necrosis at the apex of leaf, and Yellowing of leaves starting at the tips and spreading along the edges to the base of the leaf, forming a V-shape (Kumar et al., 2021), which is important for trapping light and absorbing carbon dioxide effectively that will hindering to the rate of photosynthesis and this decline in photosynthetic activity subsequently affects the overall dry matter production (Hu et al., 2020). Moreover, deficiency of major macronutrient nitrogen also decreases stomatal conductance, soluble protein, and chlorophyll content and the activity of key enzymes involved in the Calvin cycle, such as ribulose-1, 5-bisphosphate carboxylase and phosphoenolpyruvate carboxylase, this disparity in metabolic processes during photosynthesis can negatively impact the overall yield and productivity of crops (Wei et al., 2016). Few studies have been reported that several abiotic factors, including temperature, drought, and osmotic stress, are responsible for deficiency of nitrogen in plant. An extreme raise in temperature can disrupt the proteins and enzymes involved in metabolic activity, additionally, enhanced in osmotic stress, can imbalance the ion concentration in soil, whereas drought is another significant abiotic stress that reduces level of water from the soil, which in turn decelerates the accumulation of nutrients including nitrogen and down regulates the activity of gene involved in nitrogen cycle are nitrate reductase and glutamine synthase (Ullah & Ali, 2025). The nutritional requirements of plant depends on the base pillar of the plant known as the root system, and the maize grow from monocotyledons seed belonging to the grass family develops both type of root system including embryonic and post embryonic that will develop adventitious root, which help anchor the plant securely in the soil. However, there are various morphological changes reported in the maize root due to scarcity of nitrogen including an increase in length and diameter of primary, seminal and crown root, to seek nitrogen from the deeper soil layer consequentially the dry mass of root is also getting raise (Lynch et al., 2024). Followed by lateral root associated with the primary and seminal root tends to become shorter in length and decrease in their number. This nutritional deficiency leads to hindered the growth and development of newly formed lateral roots under the circumstance of nitrogen deficiency (Sen, 2015). In contrast, there was a significant relation occurs in between the root and shoot system, due to low nitrogen acquisition by root. Thus, it has a greatly negative impact on the plant shoot system also, in result of our experiment observed from control treatment as compared to the microbial (*Bacillus megaterium* and *Azospirillum brasilense*) inoculant give betterment in plant height, stem diameter, fresh weight, dry weight, and leaf area and average length of plant leaf (Table 4.1). This decline in growth of the control treatment plant indicates that, lack of availability of nitrogen in sufficient quantity at the right time adversely affects the crop yield of the plant that cannot be corrected by applying nitrogen at later stages (Binder et al., 2000).

4.3 Impact of Inorganic Nitrogen Fertilizer in Different Field

In the soil, silicon, aluminum, and iron are present as major elements, while magnesium, manganese, calcium, nitrogen, phosphorus, and potassium are found in smaller amounts (de Oliveira Morais et al., 2021). Plants require nutrients such as nitrogen, phosphorus, and potassium significantly in larger quantities for healthy growth. Therefore, agricultural lands often need to be supplemented with various micronutrients and macronutrients in the form of fertilizers, which can be organic, inorganic, or biofertilizers. To quickly meet the nutrient needs of plants, many farmers apply inorganic or synthetic chemicals in large quantities. While the initial use of synthetic fertilizers, such as an urea and DAP, can lead to good yields, relying on these fertilizers continuously and indiscriminately can actually harm crop yields over time; currently, there are around twenty urea-making companies present in fourteen states of India (Table 4.4). Research has shown that excessive use of inorganic fertilizers beyond the recommended doses can negatively impact not only crop yields, but also the soil's physical characteristics (pore space, water retention capacity, water precipitation rate, and bulk density) (Haynes & Naidu, 1998), mineral properties (depleting organic minerals, accumulation of toxic ions, altering soil pH, reduce rate of mineral suction by plant) (Chandini, 2019), diversity of agriculturally important living microorganisms (Bacteria, actinomycetes, fungi, and algae), and develop resistance in phytopathogenic strains. Along with this, extert in influence on other enviornmental factors (Fig. 4.1).

Table 4.1 Experiments were carried out to evaluate and compare different physiological parameters of maize inoculated with nitrogen-fixing bacteria

Parameter (s)	Treatments	Replicates (cm)					Average
		R1	R2	R3	R4	R5	
Plant height	Control	93	78	65	71	90	79.40 ± 10.74
	BM	92	98	103	111	106	102.00 ± 6.54
	AB	76	92	94	88	81	86.20 ± 6.76
Stem diameter	Control	0.5	0.6	0.5	0.5	0.49	0.52 ± 0.04
	BM	0.9	0.8	0.9	0.8	0.9	0.86 ± 0.05
	AB	0.6	0.8	0.8	0.8	0.7	0.74 ± 0.08
Average leaf length	Control	22.5	24	18	21.5	21.4	21.48 ± 1.98
	BM	23.5	27	24.6	29.5	26	26.12 ± 2.07
	AB	21	20	23	26	24	22.80 ± 2.14
Leaf area	Control	51.5	69.5	40	45.3	62.5	53.76 ± 10.86
	BM	96.2	89.5	93	85.7	116	96.08 ± 10.56
	AB	60.5	87	72.3	77.8	70.12	73.54 ± 8.75
Dry weight of plant	Control	1.4	2.12	0.82	0.86	0.89	1.22 ± 0.50
	BM	1.84	1.92	2.59	1.3	1.73	1.88 ± 0.42
	AB	1.15	1.7	2.71	1.53	1.31	1.68 ± 0.55
Fresh weight of plant	Control	6.65	12.09	4.64	5.3	6.22	6.98 ± 2.65
	BM	13.76	13.28	15.53	10.73	18.46	14.35 ± 2.56
	AB	8.45	12.51	13.7	12.4	11.8	11.77 ± 1.77

"BM" represents *Bacillus megaterium*, whereas "AB" represents *Azospirillum brasilense*

In the experimental study of 8 years long continues application of inorganic chemical fertilizer and organic manure onto the maize field, used urea as inorganic treatment for nitrogen, on the other hand used farm yard manure (FYM) as an organic treatment. Result was indicated significantly declined the natural chemical properties and fertility of soil in the case of chemical apply. Negatively affected soil parameters including, pH, available phosphorous, alkaline phosphate whereas other parameters such as available potassium, total nitrogen, soil organic carbon, and grain yield are not significantly improved in inorganic chemical compared to control treatment (Table 4.2) (Liu et al., 2010).

In a 36-year experimental study on maize, the use of balanced or recommended doses of inorganic nitrogen fertilizer despite, led to alterations in soil physical properties. The pH ranged from 7.40, soil bulk density was 1.54 Mg $m-^{3}$, cation exchange capacity was 5.2 $cmol^{c}$ $kg-^{1}$ (CEC), and electrical conductivity was 0.20 dS $m-^{1}$ (EC). In comparison, the control treatment had values of 7.90 for pH, 1.59 for soil bulk density, 5.2 for CEC, and 0.20 for EC. The study Conclude that long-term chemical

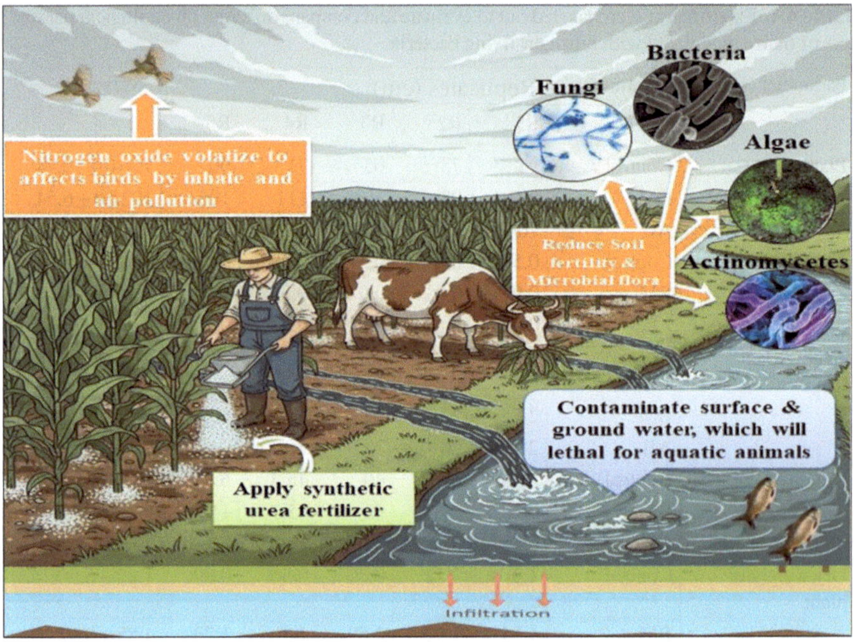

Fig. 4.1 Excess use of synthetic nitrogen fertilizers can severely disrupt environmental conditions by polluting soil, water, and air

fertilization did not result in significant growth or improvement in maize production (Brar et al., 2015).

In a long-term study examining the application of inorganic nitrogen fertilizer (urea) over 13–14 years on a wheat–maize rotation crop, it was found that this practice reduced both exchangeable nutrients (calcium and magnesium) and available form of nutrients (copper and zinc), while increase in exchangeable hydrogen and Aluminum; however, excessive H+ and $Al+^3$ can also detrimental for soil. Additionally, it also contributed to increased soil acidification 0.07–0.12 pH/year (Table 4.3) (Zhang et al., 2008).

Table 4.2 Application of urea affects several soil characteristics, such as acidity, organic matter decomposition, and overall soil fertility

Treatment	Soil pH (H_2O)	AP (mg $kg-^1$)	ALKP (mg PNP $kg-^1$ $h-^1$)	AK (mg kg^{-1})	TN (g kg^{-1})	SOC (g kg^{-1})
Control	8.18	5.76	250	144	0.98	7.52
N_2	7.62	4.98	211	145	1.01	7.67
FYM	8.04	12.66	303	193	1.22	9.98

AP available phosphorous, *ALKP* alkaline phosphate, *AK* available potassium, *TN* total nitrogen, *SOC* soil organic carbon

4.3.1 Effect of Chemical Nitrogen Fertilizer on Soil

1. Long-term application of inorganic nitrogen fertilizer (Urea), affects significant nutrient cycle in the soil such as nitrogen cycle, phosphorous cycle, etc.
2. Minimizing pore size of the soil that will lead to affect the rate of water precipitation and water retention capacity.
3. Deposition of heavy metals for instance copper, lead, and nickel, etc.
4. Reduction in population or diversity of rhizospheric agricultural beneficial microorganism.
5. Decrease in soil organic carbon and soil microbial biomass.
6. Down regulate the soil enzymatic activity of Beta-glucosidase, dehydrogenase, urease, and phosphatase, etc.
7. Disrupt the balance of chief element including nitrogen, phosphorous, and potassium.
8. Imbalance of nutrients content and H+ ions resultant decreasing level of soil pH or increase acidification that may contribute to soil breakdown.
9. Elevate the level of salt concentration in the soil.
10. Stimulate the process of soil compaction through reducing the void space in soil, which will result in reduced oxygen content for plant root.

4.3.2 Effect on Plant Microbes Interaction

1. Prolonged utilization of inorganic nutrients fertilization specially urea for nitrogen that will weakens the interaction among the microorganisms and plants.
2. The diminished interaction also affects the availability of the mineralized nutrients as an insoluble form of potassium is mineralized by microbes for the plants.
3. Plants become more susceptible to phytopathogens due to their weaker interactions with microorganisms.
4. Plants have reduced the ability to withstand in abiotic stress such as drought, salinity, and nutritional starvation, under the influence of low microbial diversity.

Table 4.3 Application of long term urea affects several micronutrients of soil

Treatment	Ca^{2+} (cmol kg $-^{1}$)	Mg^{2+} (cmol kg $^{-1}$)	Cu (mg kg$-^{1}$)	Zn (mg kg$-^{1}$)	H+ (cmol kg^{-1})	Al^{3+} (cmol kg^{-1})
Control	6.48	2.43	2.71	3.6	0.13	0.26
Nitrogen	1.38	0.61	1.83	2.22	0.27	5.66

Ca^{2}+ Calcium, Mg^{2}+Magnesium, *Cu* Cupper, *Zn* Zinc, H^{+}Hydrogen, Al^{3+}Aluminum

5. Long-term practices of excessive inorganic nitrogen fertilizer are the major cause of soil degradation that will also negatively effects on the plant microbes interaction.
6. Excess dose of synthetic nitrogen fertilizer from its recommended dose, which were unassertively influence on the microbial community of plant rhizospheric region.
7. Decreasing microbial diversity from soil resultant greatly impact on the induce systemic resistance in the plant agents phytopathogens.
8. Reduction in the secretion of secondary metabolites called root exudates like flavonoids from plants, which the microorganism utilize as nutrients for their growth and are attracted towards the rhizospheric region of plant.
9. Weakening of plant–microbial interactions significantly impacts overall plant growth and also reduces grain yields.
10. Plant–microbe interactions depend on the soil's fertility status; however, indiscriminate use of chemical nitrogen fertilizers can render agricultural land unfertile, thereby weakening mutualistic relationship between the plants and microorganisms.

4.3.3 Air and Water Pollution: Its Impact on Humans and Animals

1. The environmental impact of synthetic nitrogen manufacturing industries is detrimental.
2. Applying more than the recommended dose of inorganic nitrogen fertilizer in crop, which can be volatilization in the form of nitrous oxide, that may prejudicially affect lungs health upon inhalation.
3. When applied nitrogen is not completely consumed or absorbed by crop consequently volatilized in air into the form nitrogen oxide and nitric oxide after oxidation.
4. Emitted toxic form of nitrogen compound in air that will react with the various atmospheric particles to reducing the quality of air and drop-down in the form of acid rain.
5. Nitrous oxide is considered as the form of greenhouse gases, therefore, its abundant proliferation can deplete the ozone layer, increasing level of UV radiation on earth, which is harmful to both humans and animals.
6. Unmineralized and unutilized synthetic nitrogen fertilizer in soil can leached into deep soil to contaminate groundwater.
7. Unused inorganic nitrogen from soil by plant can runoff into the water streams, polluting them and creating an unstable environment, which may depletes oxygen levels critical for the survival of aquatic life.
8. Gathering of nitrogen fertilizer in soil can lead to evaporation of water indirectly.
9. According to the Indian government, standard rule 45 mg/L nitrate, while the world health organization recommended maximum limit of nitrate contamination in water is 50 mg/L, however, continues practices with synthetically

manufactured nitrogen increases the contamination of nitrate into the groundwater.

10. Drinking nitrate contaminated water will cause serious health issues such as thyroid, high blood pressure, methemoglobinemia, even that there is also chances of cancer

4.4 Role of Entophytes in Plant Health Stimulation

The increasing demand of food and agricultural modern technology has led to a rapid rise in the application of urea fertilizer for nitrogen in maize cultivation. However, this trend is negatively impacting soil fertility. Research in the field of microbiology plays a crucial role to alleviating application of urea from the agriculture, in this way unearthed numbers of nitrogen fixing bacterial genera such as *Rhizobium, Azotobacter, Azospirillum, Pseudomonas, Bradyrhizobium*, and *Bacillus*, (Aquino et al., 2019) etc. these endophytic bacteria isolated from the soil of rhizospheric region, inter part of root, stem and leaf of the plant (Table 4.5).

Table 4.4 List of registered synthetic urea manufacturing industries across different state of India

<table>
<tr><th>S. no.</th><th>Name of the state</th><th>Name of the company</th><th>Name of the urea manufacturing plant</th><th>Brand name</th></tr>
<tr><td>1</td><td rowspan="4">Andhra Pradesh</td><td rowspan="2">Nagarjuna fertilizers and chemicals Ltd.</td><td>Nagarjuna fertilizers and chemicals Ltd. Kakinada-I</td><td rowspan="2">Nagarjuna, Urea</td></tr>
<tr><td>2</td><td>Nagarjuna Fertilizers and Chemicals Ltd. Kakinada-II</td></tr>
<tr><td>3</td><td rowspan="2">Coromandel International Ltd.</td><td>Coromandal International Ltd. Kakinada</td><td rowspan="2">Gromor, Urea</td></tr>
<tr><td>4</td><td>Coromandel International Ltd. Visakhapatnam</td></tr>
<tr><td>5</td><td rowspan="2">Assam</td><td rowspan="2">Brahamaputra Valley Fertilizers Corp. Ltd.</td><td>Brahmaputra valley fertilizers corporation Ltd. Namrup-II</td><td rowspan="2">Urea</td></tr>
<tr><td>6</td><td>Brahmaputra Valley Fertilizers Corporation Ltd. Namrup-III</td></tr>
<tr><td>7</td><td>Goa</td><td>Zuari Agro Chemicals Ltd.</td><td>Zuari Agro Chemicals Ltd. Goa</td><td>Jai Kisaan, Urea</td></tr>
</table>

8	Gujarat	Indian Farmers Fertilizers Co-operative Ltd.	Indian Farmers Fertilizers Co-operative Ltd. Kandla	IFFCO, Urea
9			Indian Farmers Fertilizers Co-operative Ltd. Kalol	
10		Krishak Bharati Cooperative Ltd.	Krishak Bharati Co-operative Ltd. Hazira	KRIBHCO, Urea
11		Gujarat Narmada Valley Fertilizers & Chemicals Ltd.	Gujarat Narmada Valley Fertilizer & Chemicals Ltd. Bharuch	Narmada, Urea
12		Gujarat State Fertilizers & Chemicals Ltd.	Gujarat State Fertilizers & Chemicals Ltd. Vadodara	Neem Urea
13			Gujarat State Fertilizers & Chemicals Ltd. Sikka	
14	Haryana	National Fertilizers Ltd.	National Fertilizers Ltd. Panipat	Kisan, Urea
15	Karnataka	Manglore Chemicals & Fertilizers Ltd.	Manglore Chemicals & Fertilizers Ltd. Manglore	Mangala, Urea
16	Madhya Pradesh	National Fertilizers Ltd	National Fertilizers Ltd. Vijaipur-I	Kisan, Urea
17			National Fertilizers Ltd. Vijaipur-II	
18	Maharashtra	Rashtriya Chemicals & Fertilizers Ltd.	Rashtriya Chemicals & Fertilizers Ltd. Trombay	Ujjwala, Urea
19			Rashtriya Chemicals & Fertilizers Ltd. Thal	
20	Odisha	Paradeep Phosphate Ltd	Paradeep Phosphate Ltd. Paradeep	Jai Kisaan Navratna, Urea
21	Punjab	National Fertilizers Ltd.	National Fertilizers Ltd. Bhatinda	Kisan, Urea
22			National Fertilizers Ltd. Nangal-II	

23	Rajasthan	Chambal Fertilizers and Chemicals Ltd.	Chambal Fertilizers and Chemicals Ltd. Gadepan-I	Uttam Veer, Urea
24			Chambal Fertilizers and Chemicals Ltd. Gadepan-II	
25			Chambal Fertilizers and Chemicals Ltd. Gadepan-III	
26	Tamil Nadu	Madras Fertilizer Ltd.	Madras Fertilizer Ltd. Chennai	Vijay, Urea
27		Southern Petrochemicals Inds. Corpn. Ltd.	Southern Petrochemical Inds.Corpn.Ltd. Tuticorin	SPIC Neem Coated, Urea
28	Uttar Pradesh	Indian Farmers Fertilizers Co-operative Ltd.	Indian Farmers Fertilizer Co-operative Ltd. Phulpur-I	IFFCO, Urea
29			Indian Farmers Fertilizer Co-operative Ltd. Phulpur-II	
30			Indian Farmers Fertilizer Co-operative Ltd. Aonla-I	
31			Indian Farmers Fertilizer Co-operative Ltd. Aonla-II	
32		Yara Fertilizers India Pvt. Ltd.	Yara Fertilizers India Pvt. Ltd./ TCL. Babrala	YaraVera, Urea
33		Kribhco Fertilizers Ltd.	KFL/Kribhco Shyam Fertilizers Ltd. Shahjahanpur	Kribhco, Urea
34		Kanpur Fertilizers & Chemicals Ltd.	Kanpur Fertilizers & Chemicals Ltd. Kanpur	Jaypee Chaand Chhap, Urea
35		Indorama India Private Limited	Indorama India Private Ltd. Jagdishpur	Shaktiman, Urea
36	West Bengal	Matix Fertilizers & Chemicals Ltd.	Matix Fertilizers & Chemicals Ltd. Panagarh	Bharat, Urea

Source: file:///D:/Research%20assistant/N2%20Experiment/urea%20manufacturing%20company.pdf

Table 4.5 Endophytic nitrogen fixing bacterial isolates isolated from different plant parts

S. no.	Plant	Plant organs	Bacterial genera	References
1	Maize	Root	*Herbaspirillum, Sphingomonas, Pantoea, Neorhizobium, Enterobacteriaceae, Stenotrophomonas, Pseudomonas, Peribacillus, Klebsiella, Cellulomonas, Burkholderia, Variovorax, Bradyrhizobium, Phytobacter, Lentzea, Kosakonia, Pseudoxanthomonas, Acromobacter*	Gao et al. (2024), Li et al. (2024), Miranda-Carrazco et al. (2022)
2		Seed	*Stenotrophomonas, Lysinibacillus, Pantoea, Paenibacillus, Pararhizobium, Burkholderia, Pseudomonas, Bacillus, Enterobacteriaceae, Staphylococcus, Lachnospiraceae, Pseudonocardia*	Pal et al. (2022), Gao et al. (2024)
3		Leaf	*Microbispora, Streptomyces, Streptosporangium, Yersinia, Bacillus, Pseudomonas, Enterobacter, Pantoea, Klebsiella, Priestia, Kosakonia, Methylorubrum*	Araújo et al. (2000), Mashiane et al. (2018), Moturu et al. (2023)
4		Stem	*Kitasatospora, Micromonospora, Mycolicibacterium, Burkholderia, Cellulosimicrobium, Pantoea, Streptomyces, Xanthomonas, Streptomyces, Gordonia, Pseudomonas, Stenotrophomonas, Bacillus, Acinetobacter, Serratia, Priestia, Klebsiella*	Miranda-Carrazco et al. (2022), Mashiane et al. (2018), Moturu et al. (2023)

The term nitrogen fixing diazotrophic endophytic bacteria is coined by Johanna dobereiner (Baldani & Baldani, 2005). Endophytic microorganism divides into different category including passenger endophyte, facultative endophyte, and the principal one of this symbiotic interaction within the root and in shoot and leaf, called obligate endophyte (Kandel et al., 2017), depends on the penetration and inhabit mechanism into the root system of plant. This all positive association consequently refers to as Plant–endophyte microbial interaction (Santos et al., 2018), obligate endophyte just like a parasitic pathogen depends on the host, and however, pathogen characteristics to introducing infection within the host system, whereas obligate endophyte does not causing infection to the plant (Azevedo, 1998, Santos et al., 2018). Rather microbial community of obligate endophyte colonized within the host plant to help in fix nitrogen, nutrients sequestration from soil, along with produce various different kinds of secondary metabolites, including antibacterial, antifungal compounds such as flavonoids, chinones, lipopeptides, ecomycin, polymyxins, quinones, fengycin, phenazines, and iturin, etc. and these secondary metabolites collectively develops Induce systemic resistance in plant together with increase the ability of plants agents various phytopathogens, improve the plant's capacity to withstand environmental stress, in this way promotes plant growth to enhance crop yield via two different mechanism directly (promoting plant growth directly by staying in different parts of the plant like in xylem, phloem) and indirectly (evaluate endophytic isolated bacteria). As rhizospheric plant beneficial microorganism, endophytic bacteria is also directly contributes to promoting sustainable plant growth by colonizing within the interior of the plant organs and fixating nitrogen from atmosphere, phosphate solubilization, phytate solubilization (organic source of phosphate), potassium mineralization, zinc solubilization (zinc oxide, zinc sulphate, zinc carbonate, and zinc phosphate), iron chelation, via secreting enzymes phosphatase, phytase, along with making siderophore chelating enzyme for the accession of iron even and producing multiples of different low-molecular-weight organic acids such as citric acid, gluconic acid, propionic acid, keto-D glutrate, mallic acid, succinic acid, and malonic acid that will sequesters both micronutrients and macronutrients including zinc and potassium by lowering the pH of immobilized nutrients existing nearby rhizospheric reign and making it available for the plant, additionally production of crucial phytohormones gibberellin, Auxin, cytokinins. Which can improves the framework of root including the dry and fresh weight, increases proliferation of root surface area (root elongation and branching), consequently, the water and nutrients can be readily acquisition by the root from the soil. The distinct significant diversity of phyllospharic bacterial endophytes inhibit within the tissue of above-ground part of the plant, namely leaf, shoot, fruits, and flowers plays a significant role in plant health. This community of endophytic microorganisms stimulates the plant's ability to endure various biotic and abiotic environmental stressed situations. As a result, crops can achieve improved yields alongside improve better grain nutritive quality. In the study of Pal et al. (2022), isolated 23 endophitic bacteria from the two different varieties of maize seed, which was baby corn and maize K25. Further screening revealed that every isolates are able to fix nitrogen, 70% IAA and 74% isolates solubilizing phosphate alongside among these strains *Burkholderia anthina* (ZM4), *Pseudomonas aeruginosa* (ZM5), and *Bacillus velezensis* (ZMW8), demonstrated the best biocontrol activity upon the phytopathogens *R. solani* and *F. verticillioides*. Thereafter, evaluate for seed priming and pot trials using maize seed ZM1 and ZM2 trial with *Lysinibacillus* sp. and *Paenibacillus dendritiformis* respectively, the result showed significantly improvements in physiological parameters of plant growth including, seed germination, seedling, shoot length, fresh and dry mass of shoot, leaf surface area, chlorophyll content, total count of lateral root, root length, fresh and dry mass of root, compared to the control treatment.

4.5 Path from Nitrogen to Ammonia

Bacterial nitrogen fixation are reductive process in which the transformation of organic nitrogen into the plant-usable form, ammonium. Moreover, those eubacteria and archebacteria capable of fixing nitrogen are collectively called as diazotrophs. Diazotrophs including both free living and symbiotic nitrogen fixers, play the significant role in replenishing the sick of nitrogen for plant via nitrogen cycle (Ladha et al., 2022). Naturally, the process of nitrogen fixation occurs in two ways, biological and nonbiological. Nonbiologically fixation contributes small amount up to 10% during the atmospheric celestial fire and photochemical reactions (https://geo.libretexts.org). Free-living nitrogen fixers can be either aerobic or anaerobic and photosynthetic or nonphotosynthetic. Nitrogenase (encoded by *nif* gene) is the precious intermediate enzyme for nitrogen fixation involved in catalyzes of molecular nitrogen to ammonia. However, nitrogen-fixing activity is diminished in the presence of atmospheric oxygen due to the nitrogenase enzyme's inability to function effectively (York et al., 2016). Therefore, nitrogen-fixing organisms employ various strategies to protect nitrogenase from oxygen exposure, such as perform higher rate of respiration together with carbon consumption and making FSII (small ferredoxin) protein, which will associate or making complex with the nitrogenase to suppress it's activity and further refurnish under the oxygen limited situation (Franke et al., 2025). Enzyme nitrogenase divided into three important metal proteins or metalloproteins namely vanadium type, iron type, and molybdenum type protein (Rucker & Kaçar, 2024). Molebdenium observed commonly in wide range of bacterial species but in some microorganism venedium and iron type found as an alternative, for instance, *Methanosarcina acetivorans*, *Azotobacter vinelandii*, *Rhodopseudomonas palustris* (Addo & Dos Santos, 2020). Further molybdenum nitrogenase divides into two subtype dinitrogenase and dinitrogenase reductase. Dinitogenase is the constituents of molybdenum-iron (MoFe) protein, which is composed by four subunits, $\alpha 2\beta 2$ heterotetrameric structure among this alpha is encoded by *nifD,* whereas beta from *nifK*. Dinitrogenase *nifDK* inclusives two metal clusters "P" and "M." M-cluster found at the structure of alpha subunit and significantly with this considered as the final binding site for molecule of nitrogen occurs here. Iron-protein, dinitrogenase reductase composed of two identical subunits creating homodimer structure of $\gamma 2$ (gama2), encoded by the gene *nifH*. Iron protein contains one F cluster and transfer the required eight electrons which will accept by cluster P to further achieve the nitrogen fixation reaction (Sun et al., 2021).

4.6 Influence of Formulations in Agriculture

4.6.1 Impact of Fertilizer on Soil Microbes

In view of the ever-increasing population and demand for food grains, we often resort to using various types of chemical fertilizers or planting the same crop multiple times to get higher production. However, we tend to overlook the impact this can have on soil health. While Soil is an underlying platform not only for the proper growth of plants, but also for the living biodiversity residing in it and these both factors are depends upon the structure and availability of nutrients within the soil. The existence of giant microbial diversity indicates the bonny soil. Aside from chemical fertilizers, several environmental or abiotic factors may also affect the population of microbial biomass, including temperature (excessive cold and warmth), pH (acidity and alkalinity), and moisture content (https://microbiologyclass.net/factors-that-affect-microbial-population-in-the-soil).

There are number of research based experimental analysis that will determine the quality of soil (Tahat et al., 2020). (1) A technique evolved from RFLP called

Terminal-Restriction fragment length polymorphism used to assess the microbial group in soil (Siqueira Jr et al., 2022). (2) Analysis of soil microbial community via biochemical method called phospholipid fatty acid (Giray et al., 2024 and Li et al., 2020). (3) Ester Linked-Fatty acid Methyl Ester is used to detect the changeable biomass of microbial community from soil (Li et al., 2020). (4) Denaturing gradient gel electrophoresis, based on bands appear on the gel, appraise the availability of soil microbial diversity (Nikolcheva & Bärlocher, 2020). (5) Single-strand-conformation polymorphism, applied SSCP marker to estimate the microbial biomass of soil (Pasookhush et al., 2021). (6) Metagenomics, etc. and culture dependent method like Plate count.

It has been definitely considered that the indiscriminate use of chemical fertilizers and stressed environmental situation on soil consequently reduces microbial diversity in the soil. To restore and maintain a healthy microbial community in agricultural ecosystems, it is essential to incorporate microbial formulations, which is formulate from individual or together combination of two or more living microbes. This formulation can function cooperatively with soil, crop, and environment, to enhancing the properties of soil by solubilizing unmineralized nutrients of soil, as well as increasing microbial biomass to enhance the significant parameters in respect to soil fertility consequently supports the growth of plant (Sing & Pathak, 2024). In case of study, inoculated a mixture of two beneficial microorganisms *Azotobacter chroococcum* and *Trichoderma afroharzianum* in tomato cultivated plot resulted in increase of bacteria by 1.03 log CFU g and fungi by 1.58 log CFU g, which increased the number of microorganisms by 7.73 and 6.40 log CFU/g soil, respectively, compared to control treatment. It was concluded that bio-formulation significantly contributes to increase the living microbial diversity, thereby promoting sustainable living soil ecosystem (Cirillo et al., 2023).

4.6.2 Important of Seed Treatment Biologically

Maize is one of the most important crops grown among the coarse grain varieties, and farmers use various techniques to increase its production. But various types of fungal and bacterial diseases affecting maize affect the crop production a lot. Seed treatment is an effective method which is defined as the coating of seed via insecticide and fungicide (it will be the chemical and biological) to make a shell on seed agent's seed born or soil born pathogen. Because the worrying thing is that due to environmental stress and spread of soil born pathogen, seeds may fail to germinate or, after germination, the plants may not produce well. It was reported that upto 45%–50% annually crop yield loss due to spread of pathogens (Fisher et al., 2018). Hence, it is essential to evaluate treated seeds in agricultural fields to ensure good germination and plant growth while reducing the risk of infection (Simona et al., 2025). Since the sixteenth century, seeds have been treated with chemicals such as brine and copper sulfate. Today, these methods have greatly improved with the use of various fungicides and insecticides for seed coating, including chlorpyrifos, carbendazim, imidacloprid, thiamethoxam, and thiram, among others. In the 2-year experimental study by Ravindra et al., (2022), rice seeds were treated with different doses of the fungicide Carbendazim before sowing. The results of Carbendazim 50% WP 4 g/kg seed showed a 37% improvement in seed germination compared to the untreated control group, which had a germination rate of 51%. Additionally, the treatment mitigated the severity of sheath blight at various growth stages, with reductions of 19.2% during flowering, 35% at the milking stage, and 48.2% at maturity, compared to untreated rates of 49.5%, 57.5%, and 66.2%, respectively. Furthermore, the harvested yield increased by 2.4 tons/hectare, rising from 2.0 tons/hectare in the untreated group. Nevertheless, several outcomes of using chemical fungicides or pesticides in seed treatment can negatively impact human health through direct contact and the inhalation of chemical odors (Lamichhane et al., 2020). Hoesel et al. (2017) reported in wheat that chemically applied treated seed could significantly influence the activity and occurrence of earthworms into the top most layers of soil. Sivparsad et al., (2014) reported that Pesticides including propiconazole, imidacloprid, thiram, triflumizole, and hexaconazole were showing significant incompatibility with the naturally occurring biological control fungus *Trichoderma harzianum*. Pesticides and fungicides help control fungal, bacterial, and pest infections. However, they also negatively impact living microorganisms, animals, and humans, as well as gradually collection of chemicals contributing to the decline of environmental health (Lamichhane et al., 2022).

Therefore, this cannot be ignored anymore, and it is now necessary to change the traditional method of chemical seed treatment and adopt biological methods that are farming-friendly, which can retouch the activity of seed germination to plant growth. In case study of Walid et al. (2022) pot experiment analysis, tomato seeds were treated with the fungal strain *Trichoderma harzianum* and compared with the untreated control plants. The pathogens *Rhizoctonia solani* and *Phytophthora infestans* were artificially introduced on the plants, and with respect to this inoculated pathogen develop disease in tomato called damping off. The results indicated that the biological control efficiency of *T. harzianum* was 49% for *Phytophthora* and 64% for *Rhizoctonia*. The study also stated that *Trichoderma* acts as an effective biocontrol agent against damping-off disease during both pre- and post-seed emergence. Additionally, the treated seeds showed improvements in several important plant physiological parameters, including seed germination, plant height, and number of leaves, enhanced flowering, and both fresh and dry mass of roots. In the study of Wang et al., (2023), watermelon and melon seeds were primed against *Acidovorax citrulli*, which causes a seed-borne disease known as bacterial fruit blotch. The bacterial strains *Bacillus amyloliquefaciens* and *Pseudomonas fluorescence* were both strain mixed separately with a sterilized solid matrix, applied for seed treatment resulting in effective biocontrol, significantly reducing disease incidence. Until over the period of 36 days study, the disease incidence was recorded at 1.8% and 6%–7% for watermelon and melon plant, respectively, treated with *Bacillus amyloliquefaciens*. For both melon and watermelon, the disease incidences were 0.8% with *Pseudomonas fluorescence*. In contrast, the control treatment exhibited much higher disease incidences of 6.5% for watermelon and 16.3% for melon. Additionally, the control treatment showed a reduction in seedling germination energy, while the microbial primed seeds demonstrated improved seed germination energy. There are also several species of plant-parasitic nematodes present in the soil, including *Bursaphelenchus cocophilus, Heterodera glycines, Pratylenchus zeae, Aphelenchoides fragariae*, and the commonly known *Meloidogyne incognita.* These nematodes invade the roots, leaves, and stems of plants, hindering their growth and leading to significant yield losses (https://ohioline.osu.edu/factsheet/plpath-gen-8). In the research of Mohammed & Khan (2021) reveled that Cucumber seeds were coated with microbial strains known for their antagonistic potential and then sown in soil infected with the nematode *Meloidogyne incognita.* The untreated control plants exhibited multiple disease symptoms, including root knot infections, yellowing of foliage, and reductions in chlorophyll and photosynthesis. In contrast, the seeds treated with microbial strains such as *Trichoderma longibrachiatum, Bacillus subtilis, Pseudomonas fluorescence, Trichoderma viride*, and *Trichoderma harzianum* demonstrated a significant increase in nematode mortality percentages of 61, 68, 68, 71, and 73, respectively. Additionally, these treatments stimulates the several physiological traits of the plants, including fresh and dry mass of roots and shoots, root and stem heights, and rates of transpiration, photosynthesis, carotenoid content, and both chlorophyll a and b.

4.7 Conclusions

Maize (*Zea mays* L.) is one of the most important cereal crops worldwide. However, the excessive use of chemical fertilizers can negatively affect maize growth and soil health, underscoring the need for sustainable alternatives such as biofertilizers. The present study demonstrates that the integration of plant growth-promoting rhizobacteria (PGPR) with nitrogen-fixing ability not only enhances nitrogen assimilation, but also promotes overall plant health through additional mechanisms, including phytohormone production, nutrient solubilization, and disease resistance. This approach showed promising results in improving plant biomass, nutrient use efficiency, and sustainable soil fertility, suggesting that it can partially or even completely reduce dependence on synthetic fertilizers.

Hence, combining PGPR and diazotrophs presents a sustainable strategy for maize cultivation, contributing to ecofriendly agriculture while maintaining crop productivity. Future research should focus on optimizing microbial combinations and field-level applications to maximize agronomic benefits, enhancing microbial strains using genomics and synthetic biology, designing effective microbial consortia, and developing optimized bioinoculant formulations for improved agricultural outcomes.

References

Addo, M. A., & Dos Santos, P. C. (2020). Distribution of nitrogen-fixation genes in prokaryotes containing alternative nitrogenases. *European Journal of Chemical Biology, 21*(12), 1749–1759.

Aquino, J. P. A. D., Macedo, F. B. D., Antunes, J. E. L., Figueiredo, M. D. V. B., Alcântara, F. D., & Araujo, A. S. F. D. (2019). Plant growth-promoting endophytic bacteria on maize and sorghum1. *Pesquisa Agropecuária Tropical, 49*, e56241.

Araújo, J. M. D., Silva, A. C. D., & Azevedo, J. L. (2000). Isolation of endophytic actinomycetes from roots and leaves of maize (*Zea mays* L.). *Brazilian Archives of Biology and Technology, 43*, 447–451.

Azevedo, J. L. (1998). Microorganismos endofíticos. In: Ecologia microbiana. Embrapa–CNPMA.

Baldani, J. I., & Baldani, V. L. (2005). History on the biological nitrogen fixation research in graminaceous plants: Special emphasis on the Brazilian experience. *Anais da Academia Brasileira de Ciências, 77*, 549–579.

Binder, D. L., Sander, D. H., & Walters, D. T. (2000). Maize response to time of nitrogen application as affected by level of nitrogen deficiency. *Journal of Agronomy, 92*, 1228–1236.

Brar, B. S., Singh, J., Singh, G., & Kaur, G. (2015). Effects of long term application of inorganic and organic fertilizers on soil organic carbon and physical properties in maize–wheat rotation. *Agronomy, 5*(2), 220–238.

Chandini, K. R., Kumar, R., & Prakash, O. (2019). The impact of chemical fertilizers on our environment and ecosystem. Research trends in environmental Sciences, 2nd ed., pp. 69–86.

Cirillo, V., Romano, I., Woo, S. L., Di Stasio, E., Lombardi, N., Comite, E., & Maggio, A. (2023). Inoculation with a microbial consortium increases soil microbial diversity and improves agronomic traits of tomato under water and nitrogen deficiency. *Frontiers in Plant Science, 14*, 1304627.

de Oliveira, M., Augusto, p, de Souza, D. M., Madari, B. E., & de Oliveira, A. E. (2021). Predicting silicon, aluminum, and iron oxides contents in soil using computer vision and infrared. *Microchemical Journal, 170*, 106669.

Fisher, M. C., Hawkins, N. J., Sanglard, D., & Gurr, S. J. (2018). Worldwide emergence of resistance to antifungal drugs challenges human health and food security. *Science, 2018*(360), 739–742.

Franke, P., Freiberger, S., Zhang, L., & Einsle, O. (2025). Conformational protection of molybdenum nitrogenase by Shethna protein II. *Nature, 637*(8047), 998–1004.

Gao, Y., Zhang, Y., Wang, P., & Zhao, L. (2024). Structure and diversity of endophytic bacteria in maize seeds and germinating roots. *Microorganisms, 12*(7), 1348.

Giray, K., Banfield, C., Piepho, H. P., Joergensen, R. G., Dippold, M., & Wachendorf, C. (2024). Main soil microbial groups assessed by phospholipid fatty acid analysis of temperate alley agroforestry systems on crop-and grassland. *Applied Soil Ecology, 195*, 105277.

Haynes, R., & Naidu, R. (1998). Influence of lime, fertilizer and manure applications on soil organic matter content and soil physical conditions: A review. *Nutrient Cycling in Agroecosystems, 51*, 123–137.

Hu, W., Lu, Z., Meng, F., Li, X., Cong, R., Ren, T. , & Lu, J. (2020). The reduction in leaf area precedes that in photosynthesis under potassium deficiency: The importance of leaf anatomy. *New Phytologist, 227*(6), 1749–1763.

Kandel, S. L., Joubert, P. M., & Doty, S. L. (2017). Bacterial endophyte colonization and distribution within plants. *Microorganisms, 5*(4), 77.

Kumar, A., Jhal, C. K., Singh, S. P., & Kumari, S. (2021). Diagnosing nutrient deficiency symptoms in Maize. Agriculture and Food E-newsletter.

Kumar, A., Meena, S. K., Sinha, S. K., Singh, A. K., & Singh, S. K. (2024). Isolation and biochemical characterization of Endophytic bacterium from native *Gluconacetobacter diazotrophocus* sugarcane cultivar of Middle Gangetic Plains of India. *Indian Journal of Ecology, 51*(1), 104–112.

Kumar, P., & Sharma, P. K. (2020). Soil salinity and food security in India. *Frontiers in Sustainable Food Systems, 4*, 533781.

Ladha, J. K., Peoples, M. B., Reddy, P. M., Biswas, J. C., Bennett, A., Jat, M. L., & Krupnik, T. J. (2022). Biological nitrogen fixation and prospects for ecological intensification in cereal-based cropping systems. *Field Crops Research, 283*, 108541.

Lamichhane, J. R., Corrales, D. C., & Soltani, E. (2022). Biological seed treatments promote crop establishment and yield: a global meta-analysis. *Agronomy for Sustainable Development, 42*(3), 45.

Lamichhane, J. R., You, M. P., Laudinot, V., Barbetti, M. J., & Aubertot, J. N. (2020). Revisiting sustainability of fungicide seed treatments for field crops. *Plant Disease*, *104*(3), 610–623.

Li, C., Cano, A., Acosta-Martinez, V., Veum, K. S., & Moore-Kucera, J. (2020). A comparison between fatty acid methyl ester profiling methods (PLFA and EL-FAME) as soil health indicators. *Soil Science Society of America Journal*, *84*(4), 1153–1169.

Li, G., Shi, M., Wan, W., Wang, Z., Ji, S., Yang, F., & Zhang, J. (2024). Maize endophytic plant growth-promoting bacteria *Peribacillus simplex* can alleviate plant saline and alkaline stress. *International Journal of Molecular Sciences*, *25*(20), 10870.

Liu, E., Yan, C., Mei, X., He, W., Bing, S. H., Ding, L., & Fan, T. (2010). Long-term effect of chemical fertilizer, straw, and manure on soil chemical and biological properties in northwest China. *Geoderma*, *158*(3–4), 173–180.

Lynch, J. P., Galindo-Castañeda, T., Schneider, H. M., Sidhu, J. S., Rangarajan, H., & York, L. M. (2024). Root phenotypes for improved nitrogen capture. *Plant and Soil*, *502*(1), 31–85.

Machado, R. M. A., & Serralheiro, R. P. (2017). Soil salinity: Effect on vegetable crop growth. Management practices to prevent and mitigate soil salinization. *Horticulturae*, *3*(2), 30.

Mashiane, A. R., Adeleke, R. A., Bezuidenhout, C. C., & Chirima, G. J. (2018). Community composition and functions of endophytic bacteria of Bt maize. *South African Journal of Science*, *114*(7–8), 88–97.

Miranda-Carrazco, A., Navarro-Noya, Y. E., Govaerts, B., Verhulst, N., & Dendooven, L. (2022). Nitrogen fertilizer application alters the root endophyte bacterial microbiome in maize plants, but not in the stem or rhizosphere soil. *Microbiology Spectrum*, *10*(6), e01785–22.

Mohammed, R. K. A., & Khan, M. R. (2021). Management of root-knot nematode in cucumber through seed treatment with multifarious beneficial microbes in polyhouse under protected cultivation. *Indian Phytopathology*, *74*(4), 1035–1043.

Nikolcheva, L. G., & Bärlocher, F. (2020). Denaturing gradient gel electrophoresis (DGGE) to estimate fungal diversity. In *Methods to study litter decomposition: A practical guide*. (pp. 319–326). Springer International Publishing.

Pal, G., Kumar, K., Verma, A., & Verma, S. K. (2022). Seed inhabiting bacterial endophytes of maize promote seedling establishment and provide protection against fungal disease. *Microbiological Research*, *255*, 126926.

Pal, G., Kumar, K., Verma, A., & Verma, S. K. (2022). Seed inhabiting bacterial endophytes of maize promote seedling establishment and provide protection against fungal disease. *Microbiological Research*, *255*, 126926.

Pasookhush, P., Usmani, A., Suwannahong, K., Palittapongarnpim, P., Rukseree, K., Ariyachaokun, K., & Ajawatanawong, P. (2021). Single-strand conformation polymorphism fingerprint method for dictyostelids. *Frontiers in Microbiology*, *12*, 708685.

Ravindra, H., HB, N., & Patil, B. J. (2022). Appraise a seed treatment of carbendazim 50% WP against sheath blight disease of rice (*Oryza sativa* L.). *Ecology, Environment and Conservation*, *28*, 497–502.

Rucker, H. R., & Kaçar, B. (2024). Enigmatic evolution of microbial nitrogen fixation: Insights from Earth's past. *Trends in Microbiology*, *32*(6), 554–564.

Samantaray, A., Chattaraj, S., Mitra, D., Ganguly, A., Kumar, R., Gaur, A., & Thatoi, H. (2024). Advances in microbial based bio-inoculum for amelioration of soil health and sustainable crop production. *Current Research in Microbial Sciences*, *7*, 100251.

Santos, M. L. D., Berlitz, D. L., Wiest, S. L. F., Schünemann, R., Knaak, N., & Fiuza, L. M. (2018). Benefits associated with the interaction of endophytic bacteria and plants. *Brazilian Archives of Biology and Technology*, *61*, e18160431.

Sen, S., Smith, M. E., & Setter, T. L. (2015). Effect of low nitrogen stress on various shoot traits of maize (*Zea mays* L.). *International Journal of Modern Botany*, *9*, 18–22.

Simona, P., Domas, B., Kristina, B., & Tomas, Ū (2025). Seed treatment: Importance, application, impact, and opportunities for increasing sustainability. *Agronomy*, *15*(7), 1689.

Sing, P. C., & Pathak, S. G. (2024). Role of microorganisms in soil fertility: Agronomic perspectives. *International Journal of Research in Agronomy*, *7*(10), 409–414.

Singh, R. K., Singh, P., Li, H. B., Yang, L. T., & Li, Y. R. (2017). Soil–plant–microbe interactions: use of nitrogen-fixing bacteria for plant growth and development in sugarcane. In *Plant-microbe interactions in agro-ecological perspectives: Volume 1: Fundamental mechanisms, methods and functions* (pp. 35–59). Springer Singapore.

Siqueira, Jr, J. F., Sakamoto, M., & Rosado, A. S. (2022). Microbial community profiling using terminal restriction fragment length polymorphism (T-RFLP) and denaturing gradient gel electrophoresis (DGGE). In *Oral biology: Molecular techniques and applications* pp. 91–104. Springer US.

Sivparsad, B. J., Chiuraise, N., Laing, M. D., & Morris, M. J. (2014). Negative effect of three commonly used seed treatment chemicals on biocontrol fungus *Trichoderma harzianum*. *African Journal of Agriculture Research*, *9*, 2588–2592.

Sun, W. E. N. L. I., Shahrajabian, M. H., & Cheng, Q. I. (2021). Nitrogen fixation and diazotrophs–a review. *Romanian Biotechnological Letters*, *26*(4), 2834–2845.

Tahat, M. M., Alananbeh, K. A., Othman, Y. A., & Leskovar, D. I. (2020). Soil health and sustainable agriculture. *Sustainability*, *12*(12), 4859.

Ullah, S., & Ali, I. (2025). Understanding the molecular mechanisms of nitrogen assimilation in C3 plants under abiotic stress: A mini review. *Phyton-International Journal of Experimental Botany*, *94*(4), 1029–1045.

Van Hoesel, W., Tiefenbacher, A., König, N., Dorn, V. M., Hagenguth, J. F., Prah, U., Widhalm, T., Wiklicky, V., Koller, R., Bonkowski, M., Lagerlöf, J., Ratzenböck, A., & Zaller, J. G. (2017). Single and combined effects ofpesticide seed dressings and herbicides on earthworms, soil microorganisms, and litter decomposition. *Frontiers in Plant Science*, *8*, 215.

Walid, N., Al-Jaramany, L., Elbenay, A., & Al-Mhethawi, R. (2022). Biological control of tomato damping-off and potato black scurf by seed treatment with *Trichoderma harzianum*. *Jordan Journal of Biological Sciences*, 15(3), 373–480.

Wang, Y., Qiu, Y., Lu, Y., Wang, X., Zhang, H., Wu, P., & Xu, X. (2023). Biocontrol of bacterial fruit blotch by seed priming with *Bacillus amyloliquefaciens* and *Pseudomonas fluorescence*. *Seed Science and Technology*, *51*(2), 191–203.

Wei, S., Wang, X., Shi, D., Li, Y., Zhang, J., Liu, P., & Dong, S. (2016). The mechanisms of low nitrogen induced weakened photosynthesis in summer maize (*Zea mays* L.) under field conditions. *Plant Physiology and Biochemistry*, *105*, 118–128.

York, L. M., Carminati, A., Mooney, S. J., Ritz, K., & Bennett, M. J. (2016). The holistic rhizosphere: Integrating zones, processes, and semantics in the soil influenced by roots. *Journal of Experimental Botany*, *67*, 3629–3643.

Zhang, H., Wang, B., & Xu, M. (2008). Effects of inorganic fertilizer inputs on grain yields and soil properties in a long-term wheat–corn cropping system in South China. *Communications in Soil Science and Plant Analysis*, *39*(11–12), 1583–1599.

Online Source

File:///D:/Research%20assistant/N2%20Experiment/urea%20manufacturing%20company.pdf

https://agritech.tnau.ac.in/seed_certification/seed_treatment_Insecticides%20&%20Fungicides.html#:~:text=T.,or%20polythene%20bags%20are%20used.

https://angrau.ac.in/downloads/AMIC/OutlookReports/2023_24/Maize%20outlook-June-july-2023-24.pdf

https://geo.libretexts.org/Bookshelves/Geography_(Physical)/BioGeoChemistry_(LibreTexts)/12%3A_The_Global_Cycles_of_Nitrogen_Phosphorus_and_Potassium/12.02%3A_The_global_nitrogen_cycle#:~:text=Three%20processes%20are%20responsible%20for,second%20process%20is%20industrial%20fixation

https://www.pib.gov.in/PressReleasePage.aspx?PRID=2037424

https://www.protectourlivelihood.in/crops/maize/

https://www.researchandmarkets.com/report/asia-pacific-corn-seed-market

Nitrogen-Fixing Biofertilizers and Microbial Inoculants: The Future of Sustainable Agriculture

5

Deepti Jaswal and Pooja Nikhanj

Abstract

The overuse of chemical fertilizers in agriculture has resulted in severe environmental consequences, including soil degradation, loss of fertility, and ecological disruption. As a sustainable alternative, nitrogen-fixing biofertilizers and microbial inoculants, particularly plant growth-promoting rhizobacteria (PGPR), offer an eco-friendly approach to improving soil health and crop yields. These beneficial microbes enhance plant growth through biological nitrogen fixation (BNF), solubilization of key nutrients (phosphorus, potassium), phytohormone production, and induced systemic resistance against pathogens. Symbiotic bacteria like *Rhizobium*, *Bradyrhizobium,* and free-living nitrogen fixers, *Azotobacter*, *Azospirillum* convert atmospheric nitrogen into plant-usable forms, reducing reliance on synthetic fertilizers. Beyond nutrient enrichment, microbial inoculants improve plant resilience under abiotic stresses, such as drought, salinity, and heavy metal toxicity, by producing stress-mitigating compounds like ACC deaminase and siderophores. However, challenges, such as variable field performance, limited shelf life, and inconsistent quality standards, restrict their widespread adoption. Future research must focus on optimizing microbial formulations, enhancing strain adaptability, and integrating precision farming techniques for efficient delivery. Advances in genetic engineering and microbiome research hold promise for developing next-generation biofertilizers with improved efficiency. With the global biofertilizer market expanding at 12.1% annually, microbial inoculants represent a transformative solution for sustainable

D. Jaswal · P. Nikhanj (✉)
Department of Microbiology, Punjab Agricultural University, Ludhiana, India
e-mail: poojanikhanj.217@gmail.com

R. K. Singh and P. Singh (eds.), *Diazotrophic Plant Growth-Promoting Microorganisms and Green Energy – Volume I*, https://doi.org/10.1007/978-981-95-9682-9_5

agriculture. By overcoming current limitations, they can significantly reduce agriculture's environmental footprint while ensuring long-term food security and ecosystem balance.

Keywords

Biofertilizers · PGPR · Nitrogen fixation · Sustainable agriculture · Microbial inoculants · Nutrient solubilization · Stress tolerance ·

5.1 Introduction

Biofertilizers represent a sustainable approach to soil fertility management, utilizing beneficial microorganisms, such as nitrogen-fixing bacteria, phosphate solubilizers, and plant growth-promoting rhizobacteria (PGPR) to enhance crop productivity. These microbial inoculants serve as eco-friendly alternatives to chemical fertilizers by improving nutrient availability, stimulating root development, and increasing yields while maintaining long-term soil health (Singh et al., 2024). Unlike synthetic fertilizers, biofertilizers significantly reduce environmental pollution, lower production costs, and support sustainable farming systems (Mahanty et al., 2017). Nitrogen-fixing biofertilizers are particularly valuable in agriculture, converting atmospheric nitrogen into plant-usable forms through natural biological processes. These formulations are broadly classified into solid (carrier-based) and liquid types, each offering distinct advantages. Solid biofertilizers, consisting of microorganisms like *Rhizobium*, *Azotobacter*, and *Azospirillum* immobilized on organic carriers (peat, lignite, or charcoal), are ideal for seed treatment or soil application. *Rhizobium* forms symbiotic relationships with legumes (soybean, chickpea), while *Azospirillum* associates with cereal roots (wheat, maize), providing efficient nitrogen fixation (Firdos et al., 2025). Liquid biofertilizers contain microbial suspensions in nutrient solutions, offering extended shelf life and better bacterial survival rates. These formulations are applied through foliar sprays, soil drenching, or irrigation systems, ensuring uniform distribution and rapid microbial activity. Specialized liquid formulations, such as *Acetobacter* for sugarcane and cyanobacteria (*Anabaena*, *Nostoc*) for rice paddies, enhance nitrogen availability while improving water retention and reducing nutrient leaching (Prasad et al., 2015). The broader context of nitrogen fertilizers includes both chemical and biological types. Chemical fertilizers (urea, ammonium nitrate) provide immediate nitrogen through the energy-intensive Haber-Bosch process but contribute to soil degradation and environmental pollution when overused. In contrast, biological nitrogen fertilizers harness natural processes through microbes, such as *Rhizobium* (legume symbionts), *Azotobacter* (free-living soil bacteria), and cyanobacteria (blue-green algae in rice fields), which convert atmospheric N_2 into plant-available ammonia and nitrates (Koirala and Volker, 2021). The strategic integration of biofertilizers in agriculture delivers multiple benefits: improved soil structure, enhanced nutrient cycling, and reduced environmental impact. By combining appropriate chemical fertilizers with targeted biological nitrogen sources, farmers can optimize productivity while promoting agricultural sustainability. This chapter examines the diverse types of nitrogen fertilizers, their mechanisms of action, and their vital role in developing eco-friendly farming practices for the future.

5.2 Mechanism of Nitrogen Fixation and Plant Growth Promotion

Nitrogen fixation is the process by which atmospheric nitrogen (N_2) is converted into ammonia (NH_3), a form usable by living organisms. This reaction is fundamental to life as nitrogen is a key component of biomolecules, such as proteins and nucleic acids. Biological nitrogen fixation is primarily catalyzed by the oxygen-sensitive enzyme nitrogenase, found in certain bacteria and archaea (Einsle and Rees, 2020). The genetic basis for this system lies in the *nif* gene cluster, whose core catalytic components, the Fe protein (nitrogenase reductase) encoded by *nifH* and the MoFe protein (dinitrogenase) encoded by *nifDK*, are universal (Dobrzyńska et al., 2024). The Fe protein, a homodimer, transfers electrons to the heterotetrameric MoFe protein in an ATP-dependent process. The MoFe protein contains the unique iron-molybdenum cofactor (FeMo-co), the active site for N_2 reduction, whose biosynthesis itself requires auxiliary *nif* genes like *nifB*, *nifE*, and *nifN*. The chemical mechanism involves the sequential reduction of N_2, where the strong N≡N triple bond is weakened. Electrons and protons are transferred, forming intermediates via an alternating pathway (Harris et al., 2018). A critical step involves the Janus state (E_4), where two bridging hydrides facilitate N_2 binding and the obligatory release of H_2, which is essential for activating the cofactor. The entire process consumes at least 16 ATP molecules and eight electrons per molecule of N_2 reduced. Some microbes possess alternative, less efficient nitrogenases encoded by *vnf* (V-Fe) and *anf* (Fe–Fe) genes, which are activated in molybdenum-deficient environments (Narehood et al., 2025). This process is central to the symbiotic relationship between nitrogen-fixing bacteria, such as *Rhizobium*, and plants of the Fabaceae family (legumes), which form root nodules to house the bacteria. To protect the oxygen-sensitive nitrogenase, root nodules contain leghemoglobin, an iron-containing protein that regulates oxygen levels to ensure efficient nitrogen fixation (Rutten & Poole, 2019). The ammonia produced is assimilated by the plant into amino acids and other essential compounds, enhancing plant growth by improving nutrient availability. This natural process reduces the need for chemical fertilizers, making legumes highly valuable in sustainable agriculture. Unlike synthetic fertilizers, which can degrade soil health over time, biological nitrogen fixation maintains soil fertility and minimizes environmental pollution. Furthermore, decaying legume roots release nitrogen into the soil, benefiting subsequent crops. By adopting nitrogen-fixing crops and biofertilizers, farmers can achieve higher yields while preserving soil health for future generations (Fig 5.1).

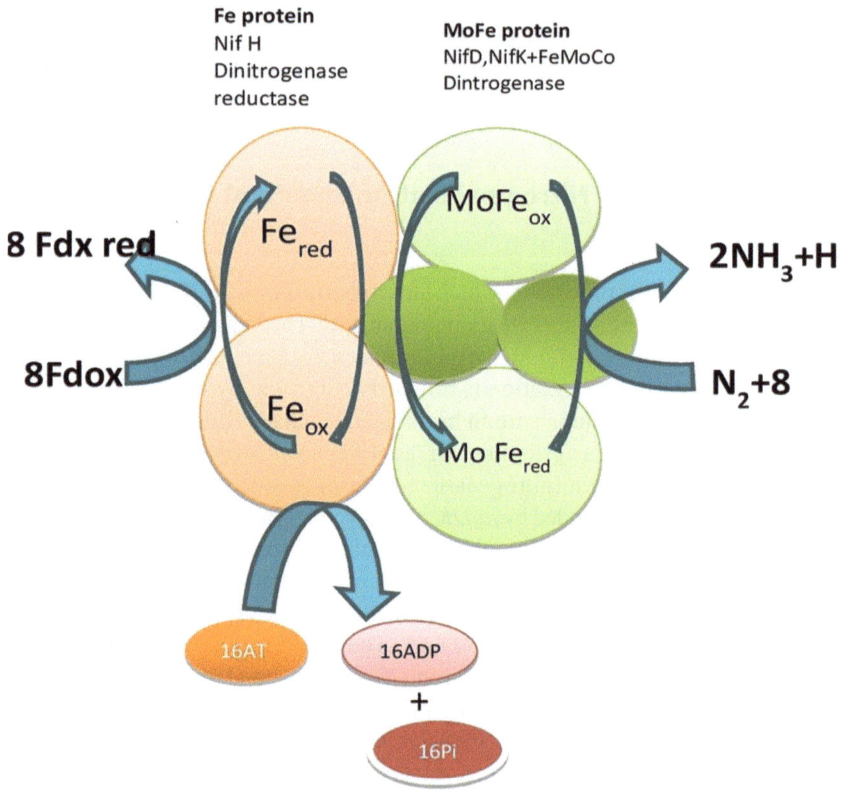

Fig. 5.1 Nitrogen enzyme complex catalyzing nitrogen fixation

5.3 Mechanisms of Microbial Inoculants in Plant Growth and Resilience

Beneficial microbes, known as plant growth-promoting microbes (PGPMs), enhance plant health and crop productivity through a diverse array of direct and indirect mechanisms. These include solubilizing essential nutrients, producing phytohormones, and priming the plant's own defense systems.

5.3.1 Phosphate Solubilization: Unlocking a Vital Resource

Following nitrogen, phosphorus (P) is often the second most limiting macronutrient for global crop production. Despite its presence in soils, up to 95% exists in insoluble, fixed forms that plants cannot absorb. Phosphate-solubilizing microbes (PSMs) serve as a critical biological key to this locked reservoir (Bashir et al., 2024). The primary mechanism involves the secretion of low-molecular-weight organic acids, such as gluconic, citric, and oxalic acids. These acids dissolve mineral phosphates by chelating bound cations (e.g., Ca^{2+}, Al^{3+}, Fe^{3+}) and lowering the pH in the rhizosphere, a process known as acidolysis (Etesami et al., 2021). Concurrently, many PSMs produce extracellular enzymes like acid phosphatases and phytases, which hydrolyze organic phosphorus compounds (e.g., phytate) into plant-available inorganic phosphate, a process called enzymolysis. This dual-action approach makes PSMs invaluable as biofertilizers, with potent strains found in bacterial genera like *Pseudomonas*, *Bacillus,* and *Rhizobium*, and fungal genera like *Aspergillus* and *Penicillium* (Devi et al., 2022). The strategic use of these microbes enhances P-use efficiency and reduces dependency on non-renewable rock phosphate.

5.3.2 Phytohormone Production: Microbial Directors of Plant Growth

PGPMs profoundly influence plant physiology by acting as external hormone regulators. The production of Indole-3-acetic acid (IAA), the most common auxin, is a widespread trait among plant growth-promoting rhizobacteria (PGPR) (Aloo et al., 2023). Bacterial IAA stimulates lateral root formation and root hair elongation, creating a prolific root system that enhances the plant's capacity for water and nutrient uptake, thereby improving vigor and drought tolerance (Sosnowski et al., 2023). PGPMs also synthesize other crucial phytohormones like cytokinins, which promote cell division and delay senescence, and gibberellins, which regulate stem elongation and flowering. (Poveda and Gonzalez- Andres, 2021) Genera, such as *Azospirillum*, *Bacillus*, and *Pseudomonas*, are well-documented producers. By fine-tuning the plant's hormonal balance, these microbes mitigate growth repression caused by environmental stresses. It is a notable paradox that the same hormonal mechanisms can be hijacked by pathogens to cause disease, highlighting the context-dependent nature of these interactions.

5.3.3 Induced Systemic Resistance (ISR): Priming the Plant's Immune System

A key indirect benefit conferred by PGPMs is the activation of Induced Systemic Resistance (ISR), a physiological state of "priming" that enhances the plant's readiness to respond to biotic challenges. When beneficial microbes colonize the rhizosphere, their microbe-associated molecular patterns (MAMPs) and secreted metabolites are recognized by the plant (Van et al., 2008). This triggers a signaling cascade dependent on jasmonic acid (JA) and ethylene (ET), which systemically primes the entire plant without activating defenses constitutively, thus conserving energy (Broekgaarden et al., 2015). Upon subsequent attack, the primed plant mounts a dramatically faster and stronger defensive response, including rapid callose deposition, accumulation of antimicrobial compounds, and activation of pathogenesis-related (PR) genes (Akram et al., 2008). This broad-spectrum resistance makes ISR a highly desirable trait for sustainable crop protection, reducing the need for chemical pesticides (Ayaz et al., 2023).

5.3.4 Volatile Organic Compounds (VOCs): The Invisible Language of Microbial Benefactors

Microbial volatile organic compounds (VOCs) are key mediators of plant-microbe interactions. These small, carbon-based molecules can diffuse through soil pores, acting as long-distance signals. Bacterial VOCs have been demonstrated to directly stimulate plant growth, likely by modulating auxin homeostasis and cell wall expansion (Poveda and Gonzalez-Andres, 2021). Furthermore, they are potent triggers of ISR; VOCs from certain *Bacillus* and *Pseudomonas* strains can prime the plant's immune system without physical contact (Grahovac et al., 2023). VOCs also function in microbial warfare by inhibiting the growth of phytopathogens. The exploration of the "volatilome" opens up exciting possibilities for developing novel, gas-based biocontrol strategies and biofertilizers (Beatty et al., 2021) (Fig. 5.2).

5.4 The Role of Biofertilizers in Sustainable Agriculture

Biofertilizers are gaining global recognition as an eco-friendly alternative to conventional chemical fertilizers, driven by growing environmental concerns and the push for sustainable farming practices. Unlike synthetic fertilizers, which rely on nitrogen, phosphorus, and potassium, biofertilizers consist of biomass-based materials like compost or humic acid, enriched with live or dormant microorganisms (Ijaz et a., 2019). These microbes enhance plant nutrition through natural processes, such as

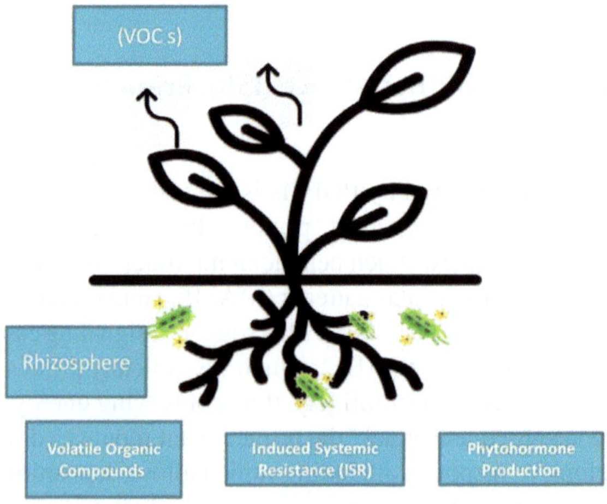

Fig. 5.2 Direct and indirect modes of action of beneficial microbes

nitrogen fixation, phosphate solubilization, cellulolytic degradation, and phytohormone production. Studies have shown that applying biofertilizers to seeds and soil improves nutrient availability, boosts plant growth, and increases crop yields by 10%–20%, all while minimizing environmental harm (Ali et al., 2022). By leveraging natural nutrient cycles, biofertilizers offer a sustainable solution for agriculture without compromising ecological balance. Given their benefits, integrating biofertilizers into modern farming practices can help mitigate the long-term negative impacts of chemical fertilizers. Among the various types available, plant growth-promoting rhizobacteria (PGPR) and vermicompost stand out for their effectiveness in supporting plant health and soil fertility (Al-Obaidi and Abdul, 2021). Their role in sustainable agriculture highlights the importance of shifting toward more natural and environmentally responsible farming methods.

5.4.1 Rhizobium Biofertilizers

Rhizobium-based biofertilizers play a crucial role in sustainable agriculture by enhancing plant growth and crop productivity through symbiotic nitrogen fixation. These bacteria form root nodules in leguminous plants, converting atmospheric nitrogen into ammonia, which is then utilized by the plant for growth. This process reduces the dependency on chemical fertilizers, minimizing environmental pollution and soil degradation (Udvardi and Poole, 2013). Rhizobium inoculants improve soil fertility, promote nutrient uptake, and increase crop yields by 10%–35% in legumes like soybeans, peas, and lentils. The mechanism involves the secretion of nodulation (Nod) factors that trigger root hair curling and infection thread formation, leading to nodule development. Inside the nodules, Rhizobium fixes nitrogen through the nitrogenase enzyme, providing the plant with a sustainable nitrogen source. Additionally, Rhizobium produces phytohormones like auxins and cytokinins, which stimulate root growth and enhance stress tolerance against drought, salinity, and pathogens. Quality control is essential for effective Rhizobium biofertilizers, including viable cell count ($\geq 10^8$ CFU/mL), pH (6.5–7.5), and moisture content (30%–40%). Carrier materials like lignite, peat, or liquid formulations influence shelf life, with liquid biofertilizers offering longer viability (5–6 months) compared to solid carriers (3 months) (Goyal et al., 2021). Government regulations, such as India's Fertilizer Control Order (FCO), mandate strict standards for microbial purity and contamination levels. Application methods include seed coating, root dipping, and soil application, ensuring optimal colonization. Co-inoculation with phosphate-solubilizing bacteria (PSB) further enhances nutrient availability (Table 5.1).

5.4.2 Plant Growth-Promoting Rhizobacteria (PGPR): Multifunctional Allies in Sustainable Agriculture

Plant growth-promoting rhizobacteria (PGPR) are beneficial bacteria that colonize the rhizosphere, the soil zone surrounding plant roots, where they enhance plant growth

Table 5.1 Classification of biofertilizers

Classification group and primary mechanism	Microorganisms	Elaboration on function and recent research insights
1. **Biological Nitrogen Fixers** *Conversion of atmospheric* N_2 *to plant-available ammonia* (NH_3).	**Symbiotic:** • *Rhizobium, Bradyrhizobium, Mesorhizobium* (Legumes) • *Frankia* (Actinorhizal plants)	**Recent Focus:** Engineering host range and enhancing efficiency under stress. Research explores the role of bacterial Nod factors and the plant's innate immune system in nodulation (Roy et al., 2020). Co-inoculation with PGPR like *Pseudomon*as can boost nodulation and N fixations
	Associative/Endophytic: • *Azospirillum* spp. • *Gluconacetobacter diazotrophicus* • *Herbaspirillum* spp.	**Recent Focus:** Their dual role as biofertilizers and phytostimulants. *Azospirillum* is a model for IAA-induced root growth, which significantly improves nutrient and water uptake, synergistically enhancing plant performance beyond N-fixation (Fukami et al., 2018).
	Free-Living and Cyanobacteria: • *Azotobacter* spp. • *Cyanobacteria* (*Anabaena*, *Nostoc*, *Trichodesmium*)	**Recent Focus:** Cyanobacteria are vital for sustainable rice cultivation, contributing N and improving soil structure via exopolysaccharides (Prasanna et al., 2022). Free-living fixers are key players in building the soil N pool in conservation agriculture.
2. **Nutrient solubilizing and mobilizing biofertilizers** *Increasing the bioavailability of bound soil nutrients.*	**Phosphate (P) solubilizers:** • *Pseudomonas* spp. • *Bacillus* spp. • *Penicillium* spp. • *Aspergillus* spp.	**Mechanism:** Production of organic acids (e.g., gluconic, citric) and phosphatases. **Recent Focus:** Omics approaches are identifying the *pqq* and *gcd* genes critical for efficient P solubilization, aiding in the selection of superior strains (Alori et al., 2017).
	Potassium (K) and silicate solubilizers: • *Bacillus mucilaginosus* • *B. edaphicus* • *Arthrobacter* spp.	**Mechanism:** Production of organic acids and chelating substances that degrade K-bearing minerals (e.g., illite). **Recent Focus:** Gaining prominence for reducing dependency on costly potash fertilizers, especially in K-deficient soils (Meena et al., 2022).

Classification group and primary mechanism	Microorganisms	Elaboration on function and recent research insights
	Zinc (Zn) and micronutrient solubilizers: • *Pseudomonas* spp. • *Bacillus* spp. • *Serratia* spp.	**Mechanism:** Acidification and chelation to solubilize Zn oxides and carbonates. **Recent Focus:** Critical for addressing "hidden hunger" (micronutrient deficiencies) in crops, directly impacting human nutrition (Kamran et al., 2021).
3. **Phytostimulants: microbes for direct growth promotion** *Enhancing plant growth via phytohormone modulation and stress mitigation.*	**ACC deaminase producers:** • *Pseudomonas* spp. • *Variovorax paradoxus* • *Bacillus* spp.	**Mechanism:** Lowers plant ethylene levels, especially under stress, promoting root elongation and survival. **Recent Focus:** A key trait sought in microbes for climate-resilient agriculture. The *acdS* gene is a marker for selecting effective stress-mitigating strains (Glick, 2014).
	Phytohormone producers: • *Azospirillum* (IAA) • *Pseudomonas* (IAA, Cytokinins) • *Bacillus* (Gibberellin)	**Mechanism:** Bacterial synthesis of hormones modulates plant development, primarily by stimulating root system architecture, leading to better resource acquisition.
4. **Biocontrol and rhizomediators** Protecting plants and improving health via **indirect mechanisms**.	**Siderophore producers:** • Pseudomonas (Pyoverdine) • Azotobacter • Streptomyces (Desferrioxamine)	**Mechanism:** High-affinity iron chelation sequesters Fe, starving pathogenic fungi and bacteria. **Recent focus:** Understanding the diversity and specificity of siderophores for developing targeted biocontrol strategies (Koskey et al., 2021).

through multiple mechanisms. These microbes improve nutrient availability by fixing atmospheric nitrogen, solubilizing phosphorus and zinc, and facilitating their absorption by plants (Glick, 2014). By increasing nutrient uptake efficiency, PGPR reduce reliance on chemical fertilizers, promoting long-term soil fertility and environmental sustainability. Beyond nutrient enrichment, PGPR also serve as biocontrol agents, suppressing plant diseases by outcompeting pathogens and producing antimicrobial compounds. Their effectiveness has been demonstrated in diverse crops, including maize, wheat, legumes, and vegetables, making them valuable tools in modern agriculture. PGPR interact closely with plant roots, either adhering to root surfaces (rhizoplane) or penetrating root tissues. Successful colonization depends on their ability to compete with other soil microbes for nutrients and root attachment sites. Their growth is often concentrated at root junctions and emerging lateral roots, where they exert maximum influence on plant health. PGPR benefits plants through direct and indirect mechanisms, though these effects often overlap. Direct benefits include phytohormone production like auxins, cytokinins, which stimulate root development and nutrient absorption, while indirect benefits involve disease suppression through antibiotics, siderophores, and induced systemic resistance (ISR) (Table 5.2).

Advantage of PGPR is their ability to synthesize growth-promoting compounds, such as volatile organic compounds (VOCs) and pyrroloquinolinequinone (PQQ), which enhance photosynthesis, stress tolerance, and antioxidant activity. Additionally, PGPR produce siderophoresiron-chelating molecules that improve iron uptake for plants while depriving pathogens of this essential nutrient. They also secrete hydrolytic enzymes (e.g., chitinases, proteases) that degrade fungal cell walls and nematode population, further protecting crops (Kumar et al., 2020). PGPR also mitigates abiotic stresses, such as salinity, drought, and heavy metal toxicity, by modulating ethylene levels through ACC deaminase activity. This enzyme breaks down ethylene precursors, preventing root growth inhibition and enhancing stress resilience. Furthermore, PGPR contribute to rhizoremediation by degrading soil pollutants, including heavy metals and organic toxins, creating a healthier environment for plants and soil organisms.

5.4.3 *Azotobacter* as a Bio-fertilizer

Nitrogen-fixing biofertilizers like *Azotobacter* represent a revolutionary approach to sustainable agriculture, offering multiple benefits that address both crop productivity and environmental concerns (Kizilkaya, 2009). As a free-living, nitrogen-fixing bacterium, *Azotobacter* converts atmospheric nitrogen into plant-available ammonia, contributing 20–40 kg of nitrogen per hectare annually, while its independence from host plants makes it versatile for various crops, including rice, wheat, and cotton. Beyond nitrogen fixation, *Azotobacter* enhances plant growth through production of phytohormones (IAA, gibberellins, cytokinins), vitamins, and siderophores that improve nutrient uptake and provide pathogen protection, while also contributing to organic matter decomposition and soil fertility (Kumar et al., 2020). Different species adapt to varied soil conditions, with *A. chroococcum* thriving in neutral to alkaline soils and *A. beijerinckii* in slightly acidic environments, and though naturally occurring in

Table 5.2 Comparison between Rhizobium biofertilizers and plant growth-promoting rhizobacteria (PGPR)

Feature	Rhizobium biofertilizer	Plant growth-promoting rhizobacteria (PGPR)
Primary function	Nitrogen fixation in legume root nodules (Bahuguna et al., 2025)	Multiple growth-promoting mechanisms (N-fixation, P-solubilization, phytohormones, biocontrol) (Timofeeva et al., 2024)
Host specificity	Mostly symbiotic with legumes (e.g., peas, beans) (Kumar et al.,2021)	Broad host range (legumes and non-legumes) (Gouda et al., 2021)
Mechanism of action	Forms root nodules & fixes atmospheric N_2 (Bahuguna et al., 2025)	Direct & indirect: N-fixation, P-solubilization, siderophores, phytohormones (IAA, cytokinins), pathogen suppression (Vejan et al., 2016)
Nutrient contribution	Primarily nitrogen (50–100 kg N/hectare) (Harman et al.,2021)	Nitrogen, phosphorus, iron, and other micronutrients (Timofeeva et al., 2024)
Common strains	*Rhizobium, Bradyrhizobium, Sinorhizobium*	*Pseudomonas, Bacillus, Azospirillum, Azotobacter*
Application method	Seed coating, root dipping, soil application (Kumar et al.,2021); (Bahuguna et al., 2025)	Seed treatment, soil drenching, foliar spray (Vejan et al., 2016)
Effect on plants	Improves N-uptake, nodulation, and legume yield (Harman et al.,2021), (Harman et al.,2021), (Bahuguna et al., 2025)	Enhances root growth, nutrient uptake, stress tolerance, and disease resistance (Timofeeva et al., 2024)
Environmental role	Reduces synthetic N-fertilizer dependency	Improves soil fertility, reduces chemical inputs, and promotes sustainable farming

low concentrations (10,000–100,000 cells/gram soil), its application as a biofertilizer significantly boosts yields while reducing reliance on chemical fertilizers and associated environmental damage. The bacterium's plant growth promotion occurs through multiple mechanisms, including nitrogen fixation via oxygen-protected nitrogenase enzymes, phytohormone production that stimulates root development and germination, siderophore secretion that enhances iron availability while suppressing pathogens, and phosphate solubilization. (Tejera et al., 2020). Its disease suppression capabilities against pathogens like Fusarium and Rhizoctonia occur through antibiotic production, competitive iron exclusion via siderophores, and induced systemic resistance in plants. When combined with other beneficial microbes like *Rhizobium*, *Azospirillum*, or *arbuscular mycorrhizal* fungi in synergistic consortia, *Azotobacter* effects are amplified, enhancing nodulation in legumes, root growth in cereals, and phosphorus uptake. Modern biotechnological advances are further optimizing *Azotobacter* strains through genetic engineering of *nif* genes for enhanced nitrogen fixation, increased phytohormone production, and development of region-specific biofertilizers.

5.4.4 *Azospirillumas* a Bio-fertilizer

Azospirillum has emerged as a powerful biofertilizer in sustainable agriculture, offering a natural solution to enhance crop productivity while reducing reliance on synthetic fertilizers (Bashan & de-Bashan, 2010). This nitrogen-fixing, plant-growth-promoting bacterium was first identified in the roots of digit grass in Brazil and has since been recognized for its ability to colonize the rhizosphere of important crops like rice, maize, wheat, sorghum, and millets (Fukami et al., 2018). Through its nitrogenase enzyme, Azospirillum fixes atmospheric nitrogen (20–40 kg N/ha) while simultaneously producing growth-promoting phytohormones including auxins (IAA), gibberellins, and cytokinins that stimulate root development, improve nutrient uptake, and boost seed germination—collectively contributing to yield increases of 15%–30%. Beyond nitrogen fixation, *Azospirillum* provides multiple agricultural benefits: it enhances plant stress tolerance against drought, salinity, and heavy metal toxicity; produces siderophores that suppress pathogenic microbes through iron competition; and some strains even solubilize phosphorus to improve plant nutrition (Cassán & Diaz-Zorita, 2016). The bacterium is typically isolated from rhizospheric soil using nitrogen-free semi-solid malate medium under microaerophilic conditions, with *A. lipoferum* and *A. brasilense* being the most widely used species due to their adaptability to various soil conditions (optimal pH 5.6–7.2). Its application is particularly effective in rice cultivation, where it colonizes root surfaces and xylem vessels. The microorganism's plant growth promotion occurs through multiple mechanisms, including nitrogen fixation, hormonal regulation of root architecture, and siderophore-mediated iron uptake. When used in combination with other beneficial microbes like *Azotobacter, Rhizobium,* or phosphate-solubilizing bacteria, its effects are significantly amplified (Santos et al., 2021). As sustainable farming practices gain momentum, *Azospirillum*-based biofertilizers present an environmentally friendly alternative to chemical inputs, with future research directions, including development of high-efficiency strains through genetic engineering, optimization of large-scale production methods, and integration with precision agriculture technologies.

5.4.5 *Bradyrhizobium* as a Bio-fertilizer

The *Bradyrhizobium* isolate Bra6 possesses the ability to fix nitrogen in a free-living state (non-symbiotically) due to the presence and expression of the nif gene cluster, with the *nifH* gene being a central component. As confirmed by the study's PCR amplification and sequencing, the *nifH* gene encodes for the enzyme dinitrogenase reductase, which is a critical subunit of the nitrogenase enzyme complex (Okubo et al., 2016). This complex is responsible for the actual biochemical reduction of inert atmospheric nitrogen gas (N_2) into bioavailable ammonia (NH_3). The process requires a tremendous amount of energy in the form of ATP and a source of electrons. In a free-living state, the bacterium must generate this energy itself through its own metabolism. The nitrogenase complex, comprising dinitrogenase reductase (the "Fe protein" encoded by *nifH*) and dinitrogenase (the "MoFe protein" encoded by *nifD* and *nifK*), is highly sensitive to oxygen. Therefore, Bra6 must maintain a micro-oxic environment, likely through its high respiratory rate, to protect this oxygen-labile enzyme (Einsle and Rees, 2020). The fixed ammonia is then assimilated into amino acids for the bacterium's own growth, and while some may be released into the surrounding environment, the quantities are generally much smaller than those achieved through the highly efficient symbiotic process within root nodules. This free-living nitrogen fixation, as measured by the acetylene reduction assay in the study, represents a baseline capability of the bacterium that is dramatically enhanced when it establishes a symbiotic partnership with the soybean plant, where the plant provides a protected, low-oxygen niche (the nodule) and a direct supply of carbohydrates to fuel the energy-intensive fixation process (Montes-Luz et al., 2023).

5.4.6 *Rhodospirillum* as a Bio-fertilizer

Rhodospirillum rubrum is a prominent example of a free-living, photosynthetic bacterium utilized as a biofertilizer, primarily for its ability to fix atmospheric nitrogen into ammonia without forming root nodules. This process is catalyzed by its molybdenum nitrogenase enzyme complex, encoded by the *nif* gene cluster. The core structural genes include *nifH* (encoding dinitrogenase reductase, the Fe protein), *nifD*, and nifK (encoding the α and β subunits of dinitrogenase, the MoFe protein containing the FeMo-cofactor) (Schleicher et al., 2023). The electron transfer to this nitrogenase is a critical and energy-intensive process, requiring low-potential electrons. In *R. rubrum*, a key electron donor is a specific 2[4Fe-4S] cluster-containing ferredoxin (FdxN), which is phylogenetically related to the essential FdN found in *Rhodobacter capsulatus* (Addison et al., 2024). The reduction of this ferredoxin is powered by the Rhodobacter nitrogen fixation (Rnf) complex, an membrane-bound enzyme that uses the proton motive force to oxidize NADH and reduce ferredoxin. Furthermore, *R. rubrum* possesses a unique regulatory system for its nitrogenase activity, known as "switch-off," where rapid environmental changes like the addition of ammonia or exposure to light/dark cycles trigger the reversible ADP-ribosylation of the *NifH* protein, inactivating the complex to conserve energy. This intricate system, governed by the *draT* (ADP-ribosyltransferase) and *draG* (glycohydrolase) genes, allows the bacterium to fine-tune nitrogen fixation in response to its metabolic status. As a biofertilizer, *Rhodospirillum* not only enriches the soil with fixed nitrogen but also promotes plant growth through the production of phytohormones, siderophores, and vitamins, making it a multifaceted and valuable asset in sustainable agriculture (Aloo et al., 2022).

5.4.7 *Bacillus subtilis*: A Dual-Strategy Biofertilizer for Sustainable Nitrogen Management

Nitrogen, as a fundamental component of chlorophyll, proteins, and nucleic acids, plays a pivotal role in plant growth, yet its atmospheric form (N_2) remains inaccessible to plants. Biological nitrogen fixation (BNF), responsible for approximately 70% of Earth's annually fixed nitrogen (175 million metric tonnes), bridges this gap through diazotrophic bacteria. Among these, *Bacillus subtilis* has emerged as a versatile biofertilizer, demonstrating remarkable potential in sustainable agriculture by addressing two critical challenges: nitrogen fixation and ammonia (NH_3) volatilization mitigation. This Gram-positive, endospore-forming bacterium thrives in diverse environments, colonizing the rhizosphere to fix atmospheric nitrogen via the oxygen-sensitive nitrogenase enzyme complex encoded by *nif* genes (Pankievicz et al., 2021). Unlike symbiotic *Rhizobium*, which partners exclusively with legumes, *B. subtilis* operates as a free-living nitrogen fixer, benefiting non-leguminous crops through associative relationships. Field studies reveal its capacity to reduce chemical nitrogen inputs by 20%–30% while and minimizing greenhouse gas emissions like nitrous oxide (N_2O) (Fan et al., 2022). Beyond nitrogen fixation, *B. subtilis* significantly curtails NH_3 volatilization a major environmental pollutant contributing to haze and soil acidification. In alkaline soils, this biofertilizer reduces NH_3 emissions by 71% compared to chemical fertilizers and 44% compared to organic fertilizers, without compromising crop yields (Xiao et al., 2021). This dual functionality stems from a multi-tiered mechanism: (1) enzymatic regulation, where *B. subtilis* suppresses urease activity, slowing NH_4^+-N production while boosting nitrification via enhanced potential ammonia oxidation (PAO); (2) genetic modulation, through downregulation of the *ureC* gene (associated with NH_4^+-N formation) and upregulation of *amoA* and *comammoxamoA* genes, accelerating NH_4^+-N conversion to plant-available NO_3^--N; and (3) microbial community engineering, enriching ammonia-oxidizing bacteria (AOB) like *Nitrosospira* and *Nitrosomonas* while fostering organic nitrogen mineralization via *Bacillus* and *Clostridium*. By simultaneously diminishing the "source" (NH_4^+-N accumulation) and amplifying the "sink" (nitrification), *B. subtilis* optimizes nitrogen use efficiency, redirecting fertilizer-derived nitrogen toward plant uptake rather than atmospheric loss (Ding et al., 2023). The integration of *B. subtilis* into organic farming systems offers a paradigm shift in sustainable agriculture. Its endospore-forming capability ensures resilience in adverse conditions, while synergies with phosphate-solubilizing bacteria and mycorrhizal fungi enhance soil microbiome robustness (Wan et al., 2023). However, challenges such as competition with native soil microbiota and environmental variability necessitate advanced bioformulations and genomic optimization. Future adoption hinges on field-scale validation, farmer education, and precision agriculture integration.

5.4.8 *Clostridium* as Biofertilizers

Clostridium species are emerging as highly effective biofertilizers due to their remarkable ability to fix atmospheric nitrogen through biological nitrogen fixation (BNF). This diverse genus comprises Gram-positive, mesophilic, and obligately anaerobic bacteria, with recent advancements in taxonomic classification further elucidating their ecological and agricultural significance. These bacteria play crucial roles in various ecosystems, particularly in agroecology, where they enhance plant growth, reduce dependency on synthetic fertilizers, and contribute to sustainable farming practices (Mostafa et al., 2024). Beyond agriculture, *Clostridium* species are industrially valuable, producing biofuels such as biohydrogen, biobutanol, and acetone, organic acids, and enzymes, thereby offering eco-friendly alternatives to petroleum-based chemicals. Their dual role in agriculture and industry highlights their potential in mitigating environmental pollution and promoting circular bioeconomy models (Ponsetto et al., 2024). The genetic machinery enabling *Clostridium* to function as a biofertilizer is complex and involves multiple gene clusters responsible for nitrogen fixation, nutrient solubilization, and stress adaptation. Central to their nitrogen-fixing capability are the *nif* genes, particularly the *nifHDK* operon, which encodes the nitrogenase enzyme complex that catalyzes the conversion of atmospheric nitrogen (N_2) into ammonia (NH_3) (Usmanet et al., 2024). Additional genes, such as *nifENB*, facilitate the assembly and activation of nitrogenase, while regulatory genes like *nifA* and *nifL* modulate expression under varying oxygen and nutrient conditions (Wang et al., 2022). Beyond nitrogen fixation, *Clostridium* species possess genes for phosphate solubilization (*pqq* and *gcd*), siderophore production, and phytohormone synthesis like auxins and cytokinins, all of which enhance soil fertility and plant growth (Charubin et al., 2018). Their metabolic versatility is further supported by genes involved in carbon metabolism (e.g., *hyd* genes for hydrogenase activity) and stress tolerance (e.g., heat shock proteins and oxidative stress regulators), enabling survival in diverse soil conditions. Advancements in genomic and proteomic analyses have deepened our understanding of *Clostridium* plant-growth-promoting mechanisms, paving the way for genetic engineering and microbial consortia development to optimize their efficacy. By leveraging these bacteria as biofertilizers, farmers can improve crop yields, restore degraded soils, and reduce greenhouse gas emissions associated with synthetic fertilizer use. Future research should focus on field applications, strain optimization, and synergistic interactions with other soil microbes to fully unlock *Clostridium*'s potential in sustainable agriculture (Figueiredo et al., 2020) .

5.4.9 Blue-Green Algae as Sustainable Biofertilizers

Blue-green algae, or cyanobacteria, represent one of Earth's oldest photosynthetic prokaryotes, playing a pivotal role in sustainable agriculture, particularly in rice paddies. These microorganisms thrive in warm, moist environments and are uniquely adapted to fix atmospheric nitrogen through specialized cells called heterocysts, converting inert N_2 into plant-available ammonia (NH_3). This biological nitrogen fixation (BNF) contributes 30–40 kg of nitrogen per hectare to rice crops, reducing reliance on synthetic fertilizers. Beyond nitrogen fixation, cyanobacteria enhance soil fertility by solubilizing phosphorus, secreting growth-promoting hormones (auxins, vitamin B12, and ascorbic acid), and improving soil structure through extracellular polysaccharides that increase water retention and organic matter (Prasanna et al., 2022). In rice fields, cyanobacteria form symbiotic relationships with aquatic ferns like *Azolla*, creating a natural biofertilizer system. The *Azolla-Anabaena* symbiosis, for instance, fixes nitrogen efficiently while decomposing to enrich soils with organic nutrients. Field studies demonstrate that cyanobacterial inoculation boosts rice yields by 10%–15% by enhancing nutrient availability and stimulating root development. Their ability to colonize waterlogged soils makes them ideal for paddy ecosystems, where they also suppress weeds and reduce soil salinity through ion exchange (Singh et al., 2021). Cyanobacteria employ two primary nitrogen-fixing mechanisms: heterocystous and non-heterocystous pathways. Heterocystous species (e.g., *Nostoc*, *Anabaena*) isolate nitrogenase enzymes in heterocysts to protect them from oxygen inactivation, while non-heterocystous types *Oscillatoria* fix nitrogen under microaerophilic conditions. Their metabolic versatility allows them to thrive in diverse climates, from tropical rice fields to extreme environments like hot springs. Despite their benefits, large-scale application faces challenges, including competition with native microbiota and sensitivity to agrochemicals. Advances in mass cultivation techniques such as open ponds, polythene-lined pits, and photobioreactors aim to optimize production. Additionally, research focuses on genetic strains with enhanced nitrogenase activity and stress tolerance (Ladha et al., 2022).

5.4.10 *Anabaena* as a Sustainable Biofertilizer

Anabaena is a genus of filamentous, nitrogen-fixing cyanobacteria (blue-green algae) that plays a crucial role in sustainable agriculture, particularly through its symbiotic relationship with the aquatic fern *Azolla* (mosquito fern). This partnership is highly beneficial in rice cultivation, as the *Azolla-Anabaena* system can fix atmospheric nitrogen at an impressive rate of 1.1 kg N ha^{-1} day^{-1}, delivering 20–40 kg N ha^{-1} to rice crops within just 20–25 days. This nitrogen-fixing capability is facilitated by specialized cells called heterocysts, which create an anaerobic environment to protect the oxygen-sensitive nitrogenase enzyme, enabling the conversion of atmospheric nitrogen (N_2) into ammonia (NH_3), a form readily usable by plants (Hakkoum et al., 2025). The biofertilizer potential of *Anabaena* and *Azolla* is significant, offering an eco-friendly alternative to chemical fertilizers by enhancing soil fertility and promoting sustainable farming practices. In tropical paddy fields, these cyanobacteria contribute substantially to carbon and nitrogen cycling, fixing approximately 20–30 kg N/ha/season. While this is lower than the nitrogen fixation achieved by legume-*Rhizobium* symbiosis (150–300 kg N/ha/season), the *Anabaena-Azolla* system remains a valuable resource for rice farming due to its low cost and environmental benefits (Jawad et al., 2023). One notable advancement is the genetic modification of *Anabaena* sp. PCC7120 to overexpress the *hetR* gene, a master regulator of heterocyst formation. This engineering has led to several key improvements: increased heterocyst frequency (12%–18%) compared to wild-type strains (5%–8%), higher nitrogenase activity (1.5–2× increase), and consequently, greater ammonia production. The engineered strain also features stable genomic integration of the *hetR* gene under a light-inducible promoter, ensuring consistent expression without the need for antibiotic resistance markers, thereby reducing the risk of horizontal gene transfer. Hydroponic studies have demonstrated the strain's enhanced compatibility with rice crops, showing improved seedling growth, chlorophyll content, and biomass when treated with the modified *Anabaena* (Hakkoum et al., 2025).

5.4.11 Nostoc as a Sustainable Biofertilizer

Nostoc, a nitrogen-fixing cyanobacterium, has emerged as a promising biofertilizers due to its ability to convert atmospheric nitrogen into plant-available forms while enhancing soil fertility. Thriving in diverse environments from aquatic ecosystems to arid deserts and even Antarctica Nostoc exhibits extreme resilience, particularly through its capacity to survive desiccation and rapidly reactivate upon rehydration. This adaptability makes it well-suited for agricultural use in variable climates. Studies confirm that Nostoc enriches soil by fixing significant amounts of nitrogen (0.029–487.38 kg N/ha/year), potentially replacing synthetic fertilizers in nutrient-deficient soils. Beyond nitrogen, it boosts soil organic carbon by 30%–100% and improves the availability of phosphorus, potassium, and micronutrients, as demonstrated in African semi-arid regions (Naik et al., 2024). In flooded rice paddies, it naturally contributes 20–80 kg N/ha per crop, reducing reliance on chemical fertilizers (Oco et al., 2024). Compared to synthetic alternatives, Nostoc offers ecological advantages: it minimizes nitrogen runoff, acts as a slow-release fertilizer, and improves long-term soil health by regulating pH and organic matter. Despite its potential, challenges remain in scaling production and optimizing field performance. Future research should focus on selecting high-efficiency strains, developing crop-specific formulations, and refining mass cultivation techniques (Shamim et al., 2020).

5.4.12 Azolla as a Sustainable Biofertilizer for Modern Agriculture

Azolla, a small free-floating aquatic fern, has gained recognition as an eco-friendly biofertilizer due to its symbiotic relationship with the nitrogen-fixing cyanobacterium Anabaena azollae. This unique association allows Azolla to fix atmospheric nitrogen at rates comparable to chemical fertilizers, providing 30–60 kg N/ha per rice crop while enhancing soil fertility. Its rapid growth and ability to form dense mats on water surfaces make it an excellent green manure, particularly in flooded rice paddies, where it suppresses weeds by blocking sunlight and reducing germination by up to 70%. Beyond nitrogen fixation, Azolla improves soil structure by increasing organic carbon (30%–100% in some cases) and releasing essential nutrients like phosphorus, potassium, calcium, and micronutrients upon decomposition (Marzouk et al., 2023). One of Azolla's key advantages is its sustainability. Unlike synthetic fertilizers, it prevents nitrogen runoff, reduces ammonia volatilization, and maintains soil pH balance. Studies show that integrating Azolla with rice cultivation can boost yields by 14%–40%, with some cases reporting a 50% reduction in chemical fertilizer use without compromising productivity. Additionally, Azolla enhances microbial activity in soil, promoting nutrient cycling and long-term fertility. Its fast decomposition ensures quick nutrient release, benefiting crops like rice, wheat, and vegetables (Thapa & Poudel, 2021).

5.4.13 Spirulina as a Biofertilizer for Sustainable Agriculture

Spirulina (*Arthrospiraplatensis*) has emerged as a promising biofertilizer and biostimulant, offering a sustainable alternative to chemical inputs in agriculture. As a nitrogen-fixing cyanobacterium, Spirulina enhances soil fertility by converting atmospheric nitrogen into plant-available forms, reducing dependency on synthetic nitrogen fertilizers. Its rich composite, including proteins, amino acids, phytohormones (auxins, cytokinins, gibberellins), vitamins, and micronutrients promotes plant growth, improves nutrient uptake, and boosts stress tolerance. Studies demonstrate that Spirulina-based formulations enhance germination, root development, chlorophyll synthesis, and crop yield in various plants, including rice, wheat, and vegetables. Additionally, Spirulina's polysaccharides and antioxidants elicit natural plant defenses against biotic and abiotic stresses, such as salinity, heavy metals, and pathogens. Unlike macroalgae, Spirulina's controlled cultivation ensures consistent quality, though production costs remain a challenge. Innovations like wastewater-based cultivation and biorefinery approaches (e.g., co-producing phycocyanin and biostimulants) can improve economic viability (Arahou et al., 2023). By integrating Spirulina into agricultural practices, farmers can achieve eco-friendly crop production, reduce chemical use, and enhance soil health, aligning with global sustainability goals. Further research is needed to optimize application methods, dosages, and cost-effective production to fully harness Spirulina's potential as a biofertilizer for sustainable agriculture.

5.4.14 Mycorrhizal Fungi as Biofertilizers: Mode of Action and Nitrogen Fixation

Mycorrhizal fungi enhance soil fertility and plant growth by forming symbiotic associations with plant roots, primarily improving nutrient uptake, especially phosphorus, through an extensive hyphal network. While mycorrhizal fungi themselves do not fix atmospheric nitrogen (N_2), they often facilitate nitrogen acquisition by partnering with nitrogen-fixing bacteria (e.g., *Rhizobia* in legumes or free-living *Azotobacter*). The fungal hyphae act as conduits, transporting nitrogen compounds from the soil or nitrogen-fixing bacteria to the plant in exchange for photosynthetic carbon (Devi et al., 2021). In arbuscular mycorrhizal (AM) fungi, the hyphae extend beyond the root zone, absorbing ammonium (NH_4^+) and nitrate (NO_3^-) from the soil and transferring them to the plant via specialized transporters. This process is crucial in nitrogen-poor soils, where the fungi's enzymatic activity (e.g., proteases) breaks down organic nitrogen into plant-available forms. In ectomycorrhizal (ECM) associations, fungi secrete extracellular enzymes like chitinases and peptidases to mineralize organic nitrogen, enhancing nitrogen cycling in forest ecosystems. Additionally, Mycorrhizae improve soil structure by producing glomalin, a glycoprotein that stabilizes soil aggregates, further promoting nitrogen retention and reducing leaching. The symbiotic relationship also enhances plant resistance to abiotic stresses (drought, salinity) by regulating stress-responsive genes and improving water-use efficiency, indirectly supporting nitrogen metabolism (Karunarathna et al., 2026). Though mycorrhizal fungi do not directly fix nitrogen, their role in nitrogen mobilization and synergy with nitrogen-fixing microbes makes them vital biofertilizers. Their application in agriculture reduces dependency on synthetic fertilizers, promoting sustainable farming. Future research aims to optimize fungal-bacterial consortia for enhanced nitrogen fixation in non-leguminous crops, leveraging genetic and metabolic pathways for eco-friendly soil enrichment.

5.5 Inoculum-Based Biofertilizers

Inoculum-based biofertilizers represent a revolutionary approach to sustainable agriculture, harnessing the power of beneficial microorganisms to enhance soil fertility and plant growth. These living biological products contain carefully selected strains of bacteria, fungi, or algae that form symbiotic relationships with plants or improve soil nutrient availability (Singh et al., 2024). Unlike chemical fertilizers that provide direct nutrient inputs, biofertilizers work through biological processes, making them an eco-friendly alternative that supports long-term soil health and reduces environmental pollution. The science behind these inoculants has evolved significantly, with modern formulations offering targeted solutions for different crop types and soil conditions. Their importance in contemporary agriculture cannot be overstated, particularly in the context of climate change and the global push for sustainable farming practices (Gautam et al., 2021). For legume crops, this typically involves selecting highly efficient rhizobial strains that demonstrate excellent nodulation capacity and nitrogen fixation potential. The isolation process often involves collecting soil samples from healthy, productive fields or from the root nodules of vigorously growing plants. The selected strains are carefully preserved in culture collections to maintain their genetic stability and biological activity (Reddy et al., 2020). Once suitable strains are identified, the production process moves to the culture multiplication phase. This critical step requires sterile laboratory conditions to prevent contamination and ensure pure cultures. Microorganisms are typically grown in specialized liquid or solid media formulations designed to maximize their growth and metabolic activity. For rhizobia, a common medium is yeast-extract mannitol broth, which provides the necessary carbohydrates and nutrients for optimal bacterial growth. The culture conditions, including temperature, pH, aeration, and incubation time, are carefully controlled to promote maximum cell multiplication while maintaining the microorganisms' beneficial properties. Large-scale production may involve fermentation technology, where microorganisms are grown in bioreactors under precisely controlled conditions. The resulting microbial biomass is then harvested at the optimal growth phase, typically during late log phase when cell viability and metabolic activity are highest (Roy et al., 2020).

The formulation of biofertilizers requires careful consideration of carrier materials that will protect and deliver the microorganisms effectively. Sterilized peat has traditionally been the most common carrier due to its excellent water-holding capacity and ability to maintain microbial viability. However, alternative carriers, such as lignite, biochar, clay, vermiculite, or even synthetic polymers, are increasingly being used. These materials must be carefully processed to eliminate competing microorganisms while preserving their physical and chemical properties that support microbial survival. The carrier is typically mixed with the cultured microorganisms at specific ratios to achieve concentrations of 10^8–10^9 viable cells per gram of final product. To enhance adhesion to seeds, formulation scientists may add natural adhesives like gum arabic or synthetic stickers. Some advanced formulations include nutrients that support microbial activity during storage and after application, while others incorporate protective compounds that enhance survival under environmental stress (Mahanty et al., 2017). Quality control represents a critical aspect of biofertilizer production, ensuring that products meet established standards for microbial purity, concentration, and performance. Rigorous testing protocols assess the number of viable cells (typically requiring $>10^8$ CFU/g), the absence of contaminants, and the product's ability to perform its intended function. Accelerated aging tests may be conducted to determine shelf life under various storage conditions. Packaging is carefully designed to protect the microorganisms from heat, moisture, and UV radiation, with some products using vacuum-sealed or oxygen-impermeable materials. Regulatory compliance is essential, with many countries establishing specific standards for biofertilizer registration and quality assurance (Khan et al., 2023). The application of inoculum-based biofertilizers in agricultural systems involves several well-established methods, each with specific advantages depending on the crop and farming system. Seed treatment remains the most common and economical approach, where seeds are coated with a slurry containing the biofertilizer and adhesive compounds. This method ensures that the beneficial microorganisms are positioned exactly where they need to be when the seed germinates (Jacob and Paranthama, 2023). Table 5.3 shows the various types of inoculum-based biofertilizers.

5.6 Enhancing Soil Health and Crop Quality through Microbial Biofertilizers

The application of microbial inoculants, or biofertilizers, represents a transformative approach to sustainable agriculture that moves beyond mere yield enhancement to holistically improve soil ecosystem functioning and the nutritional quality of crops.

These beneficial microbial formulations are pivotal for sustainable intensification, offering eco-friendly alternatives to chemical inputs while addressing both agronomic productivity and human nutritional security (Rehman et al., 2021). A primary mechanism of biofertilizers is biological nitrogen fixation (BNF), which provides a renewable source of plant-available nitrogen. This process is driven by specialized microorganisms categorized into symbiotic systems, such as rhizobia (e.g., *Rhizobium*, *Bradyrhizobium*) that form nodules on legumes, and associative systems, like *Azospirillum* and *Azotobacter*, that associate with the roots of non-leguminous crops (Choudhury and Rajam, 2021) . The nitrogen contribution from these systems is substantial; legume-rhizobia symbioses can fix 50–300 kg N/ha annually, while associative fixation in cereals can contribute 20–50 kg N/ha, potentially replacing 30%–60% of synthetic fertilizer needs without yield penalties (Beleri et al., 2023). This reduction in chemical fertilizer use mitigates environmental issues, such as nitrogen oxide emissions and nitrate leaching, while also avoiding the significant carbon footprint of industrial fertilizer production (Rehmaan et al., 2022). The benefits of biofertilizers extend far beyond nitrogen provision. Their application induces profound, beneficial shifts in soil microbial ecology. Metagenomic analyses reveal that introduced nitrogen-fixing strains stimulate complementary microbial communities, including phosphate-solubilizing bacteria (e.g., *Pseudomonas*, *Bacillus*) and mycorrhizal fungi, creating synergistic networks that enhance overall nutrient cycling and soil organic matter (Otaiku et al., 2022). This creates a healthier, more resilient soil biome, contrasting with chemical fertilizers, which can degrade soil health over time. Concurrently, a promising frontier of biofertilizer research is their ability to enhance crop quality and nutritional value. By improving the uptake and assimilation of essential micronutrients like zinc and iron through mechanisms, such as siderophore production and rhizosphere acidification, microbial inoculants can biofortify staple crops, directly addressing widespread "hidden hunger" and human nutrient deficiencies (Bundela et al., 2025). Furthermore, by modulating the plant's hormonal and defense pathways, PGPMs can influence the synthesis of valuable secondary metabolites. For instance, inoculants that influence ethylene or jasmonic acid pathways can enhance the content of antioxidants (e.g., lycopene, carotenoids, polyphenols), vitamins, and essential oils in fruits and vegetables, which are critical for taste, color, aroma, shelf-life, and nutritional value (Prisa et al., 2023). The future of this technology lies in multi-strain consortia that combine nitrogen-fixers with other plant growth-promoting microbes, creating comprehensive packages that address multiple plant needs simultaneously. Despite challenges related to environmental variability and microbial competition, ongoing research in strain selection, nano-formulations, and precision application is optimizing their efficacy (Saini et al., 2021). As global agriculture strives to balance productivity with environmental stewardship, biofertilizers stand as a cornerstone technology for enhancing both the health of our soils and the nutritional quality of our food.

Table 5.3 Types of inoculum-based biofertilizers

Type of biofertilizer	Microorganisms	PGP genes and traits	Mechanism of action	Target crops and application notes
Rhizobial inoculants (Symbiotic N_2 -fixers)	*Bradyrhizobium japonicum*, *Rhizobium leguminosarum*, *Sinorhizobium meliloti*, *Mesorhizobium ciceri*	*nif* gene cluster (N_2 fixation), *nod* genes (nodulation), ACC deaminase (*acdS*)	Form root nodules, convert atmospheric N_2 to ammonia for the plant. Produce phytohormones (IAA) and volatile organic compounds (VOCs) that stimulate root growth.	**Soybean, Pea, Lentil, Chickpea, Alfalfa**. • **Application:** Seed coating is most effective. Strain specificity is critical; must match the crop species. (Compant et al., 2019)
***Azospirillum* inoculants (Associative N_2 -fixers)**	*Azospirillum brasilense*, *A. lipoferum*	*nif* genes, ACC deaminase (*acdS*), IAA production (IPDC pathway)	Associative nitrogen fixation in the rhizosphere. Major growth promotion is via production of auxins (IAA), which profoundly alters root architecture (increased hair density and length), enhancing water and nutrient uptake.	**Maize, Wheat, Rice, Sugarcane**. • **Application:** Seed treatment or soil application. Highly effective in co-inoculation with rhizobia or PSB. (Fukami et al., 2018)
Azotobacter inoculants (Free-living N_2-fixers)	*Azotobacter chroococcum*, *A. vinelandii*	*nif* genes, P-solubilization (gcd), siderophore production	Non-symbiotic N_2 fixation. Solubilizes phosphates and produces siderophores, antibiotics, and phytohormones. Improves soil health by enriching organic matter.	**Cotton, Sugarcane, Vegetables (Onion, Tomato), Rice**. • **Application:** Suitable for a wide range of crops. Used as seed treatment, soil application, or for compost enrichment. (Mahanty et al., 2017)
Phosphate-solubilizing inoculants (PSM)	**Bacteria:** *Pseudomonas putida*, *Bacillus megaterium* • **Fungi:** *Penicillium* spp., *Aspergillus* spp.	P-solubilization (*pqq*, *gcd* genes for gluconic acid production), Phytase (*phy*)	Secretion of organic acids (e.g., gluconic, citric) and phosphatases that dissolve insoluble mineral phosphates (e.g., tricalcium phosphate) and mineralize organic phosphorus, converting it to plant-available orthophosphate.	Cereals, Legumes, Oilseeds, Forest seedlings. • **Application:** Crucial in P-fixing acidic and alkaline soils. Granular formulations are common for soil applications (Alori et al., 2017); (Singh et al., 2024).

Type of biofertilizer	Microorganisms	PGP genes and traits	Mechanism of action	Target crops and application notes
Pseudomonas-based PGPR inoculants (Multifunctional)	*Pseudomonas fluorescens*, *P. putida*, *P. aeruginosa* (non-pathogenic strains)	**Nutrient Solubilization:***pqq*, *gcd* • **Biocontrol:***phl* (2,4-DAPG), *prn* (pyrrolnitrin) • **Stress Mitigation:***acdS* (ACC deaminase) • **Siderophores:***pvd* (pyoverdine)	1. **Direct Growth Promotion:** P & K solubilization; IAA production. 2. **Biocontrol:** Antibiosis (DAPG, phenazines); Siderophore-mediated iron competition; Lytic enzymes (chitinases); Induced Systemic Resistance (ISR). 3. **Abiotic Stress Mitigation:** ACC deaminase reduces stress ethylene; exopolysaccharide production aids in drought tolerance.	**Tomato, Wheat, Chickpea, Cucumber, Potato**. • **Application:** Highly versatile. Used for disease suppression (e.g., wilt, damping-off), stress alleviation (drought, salinity), and general growth promotion. Formulations include wettable powders and liquids. (Singh et al., 2024)
Potassium and silicate solubilizers	*Bacillus mucilaginosus*, *Frateuria aurantia*, *Pseudomonas* spp.	Organic acid production genes	Weather silicate minerals (e.g., orthoclase, biotite) by excreting organic acids and polysaccharides, releasing soluble potassium and silicon. Silicon deposition enhances plant resistance to pests and diseases.	**Sugarcane, Rice, Banana, Cotton**. • **Application:** Particularly beneficial in K-deficient soils. Often applied as a co-inoculant with other biofertilizers. (Meena et al., 2022)
Biocontrol and phytostimulation fungi	*Trichoderma harzianum*, *T. viride*	Cell wall-degrading enzymes (*ech42* - chitinase), ISR elicitors	1. **Mycoparasitism:** Coils around and lyses pathogenic fungi. 2. **Antibiosis:** Produces antifungal metabolites. 3. **Competition:** For space and nutrients. 4. **Phytostimulation:** Enhances root growth and nutrient uptake efficiency.	**Vegetables, Pulses, Cereals**. • **Application:** Used for soil-borne disease management (e.g., *Fusarium* wilt, *Rhizoctonia* root rot). Applied as seed treatment, soil drench, or in nursery mixes. (Singh et al., 2017)
Arbuscular Mycorrhizal Fungi (AMF) inoculants	*Rhizophagus irregularis*, *Funneliformis mosseae*	Symbiotic pathway genes (e.g., *RAM2*)	Form extensive hyphal networks (mycelium) that act as an extension of the root system, dramatically increasing the surface area for absorption of immobile nutrients like Phosphorus, Zinc, and Copper. Also, improve water uptake and soil structure.	**Most agricultural crops (except Brassicaceae)**. • **Application:** Inoculation is most critical in soils low in native AMF. Applied as spores/infected root fragments in a carrier near the seed during planting. (Mahmud et al., 2021)

5.7 The Future of Biofertilizers: Integrating Microbiome Research, Precision Engineering, and Market Growth

The next frontier for biofertilizers lies in the synergistic integration of advanced microbiome science, precision genetic engineering, and robust market development to create a new generation of highly effective and reliable "smart" biofertilizers. This multifaceted approach is essential to meet the growing global demand for sustainable agricultural solutions that enhance food security while reducing the environmental footprint of farming (Fatimah et al., 2021). The foundation of this evolution is a deep, systems-level understanding of the plant microbiome. Advanced metagenomics and metabolomics are moving beyond simply cataloging microbial diversity to revealing the functional networks and communication within these complex communities (Diwan et al., 2022). This knowledge provides the blueprint for designing superior microbial consortia. CRISPR–Cas technology emerges as a quintessential tool for translating this blueprint into reality. Its programmability via a simple guide RNA allows for rapid, multiplexed, and precise genome editing, surpassing the limitations of older tools, such as ZFNs and TALENs. This enables not just gene knock-outs but also the fine-tuning of gene expression, the insertion of entire biosynthetic pathways, and the installation of synthetic genetic circuits. For example, bacteria can be engineered to express nitrogen-fixing genes only upon detecting root exudates, conserving energy, or to produce antifungal compounds specifically in response to pathogen signatures (Dudeja et al., 2025). This shift from simply applying microbes to *programming* them for reliable, context-aware functions represents a paradigm shift, paving the way for bespoke microbial solutions that deliver targeted benefits in nutrient solubilization, stress tolerance, and biocontrol (Amen et al., 2025).

Concurrent with these biological advances, technological innovations are revolutionizing product formulation and reliability. Modern encapsulation techniques protect sensitive nitrogen-fixing bacteria, significantly extending shelf life and field efficacy. Advanced fermentation processes yield higher concentrations of viable microorganisms, while novel carrier materials enhance microbial survival in diverse soil conditions, directly addressing historical concerns about inconsistent performance (Thakur and Kaur, 2024). Furthermore, the future market will be characterized by rigorous quality assurance. Progressive manufacturers are implementing comprehensive quality control systems that monitor microbial viability, nutrient content, and purity throughout the production chain, with third-party certification programs emerging to verify product claims and build user trust (Yadav and Yadav, 2024). By combining ecological insights from microbiome studies with the surgical precision of genetic engineering and the reliability of advanced formulations, the biofertilizer industry is poised for transformative growth. This integrated approach will create powerful biological tools that are highly effective, resilient, and capable of supporting a new paradigm of sustainable agriculture rooted in biological precision, ultimately improving soil health, maintaining productivity, and mitigating environmental impact.

5.8 Conclusion

Nitrogen-fixing biofertilizers and microbial inoculants are pivotal in transitioning toward sustainable agriculture. By harnessing the power of beneficial microorganisms, such as *Rhizobium*, *Azospirillum*, and *Azotobacter*, these biofertilizers enhance soil fertility, reduce chemical fertilizer dependency, and mitigate environmental pollution. Their ability to fix atmospheric nitrogen, solubilize phosphorus, and promote plant growth through phytohormones makes them indispensable for eco-friendly farming. However, challenges, such as strain competition, variable field efficacy, and storage stability, must be addressed through advanced biotechnological interventions, including microbial consortia, nano-encapsulation, and genetic engineering. Precision agriculture and farmer education will further optimize biofertilizer application, ensuring consistent results. As global demand for sustainable food production rises, nitrogen-fixing biofertilizers offer a scalable solution to enhance crop yields while maintaining soil health. With continued innovation and policy support, biofertilizers can revolutionize agriculture, aligning productivity with environmental conservation for future generations (Hwang et al., 2020; AlAli et al., 2021; Burgess and Lowe, 1996; Chaudhary et al., 2022; Dobrzyńska et al., 2024; Dudeja et al., 2025; Fincheira and Quiroz, 2018; Gauri et al., 2012; Hoffman et al., 2014; Jama et al., 2023; Karunarathna, 2024; Khan et al., 2022; Kholssi et al., 2022; Mazhar et al., 2013; Pal et al., 2023; Patel et al., 2021; Rehman et al., 2022; Roy, 2020; Samantaray et al., 2024; Seefeldt et al., 2009; Vasseur Coronado, 2021; Zhang and Yang, 2023; Zia et al., 2021).

References

Addison, H., Glatter, T., Hochberg, G. K. A., & Rebelein, J. G. (2024). Two distinct ferredoxins are essential for nitrogen fixation by the iron nitrogenase in *Rhodobacter capsulatus*. *MBio*, *15*(3), 14–23. https://doi.org/10.1128/mbio.03314-23.

Akram, A., Ongena, M., Duby, F., Dommes, J., & Thonart, P. (2008). Systemic resistance and lipoxygenase-related defence response induced in tomato by *Pseudomonas putida* strain BTP1. *BMC Plant Biology*, *8*, 110–113. https://doi.org/10.1186/1471-2229-8-113.

AlAli, H. A., Khalifa, A., & Almalki, M. (2021). Plant growth-promoting rhizobacteria from *Ocimum basilicum* improve growth of *Phaseolus vulgaris* and *Abelmoschus esculentus*. *South African Journal of Botany*, *139*, 200–209. https://doi.org/10.1016/j.sajb.2021.02.019.

Ali, A., Iftikhar, Y., Mubeen, M., Haider, A., Zeshan, M., Asad, Z., Zafar-ul-Hye, M., Hussain, S., Sohail, M. A., & Rizwan, M. (2022). Antagonistic potential of bacterial species against fungal plant pathogens (FPP) and their role in plant growth promotion (PGP): A review. *Phyton*, *91*(9), 1859–1877.

Al-Obaidi, S. M. J., & Abdul-Ratha, H. A. (2021). Evaluation of the combination of bacterial biofertilizer and vermicompost in the availability of N, P, K and some of plant parameters of beans (*Phaseolus vulgaris* L.). *Iraqi Journal of Agricultural Sciences*, *52*(4), 1024–1033.

Aloo, B. N., Tripathi, V., Makumba, B. A., & Mbega, E. R. (2022). Plant growth-promoting rhizobacterial biofertilizers for crop production: The past, present, and future. *Frontiers in Plant Science*, *13*, 100–248. https://doi.org/10.3389/fpls.2022.1002448.

Aloo, B. N., Dessureault-Rompré, J., Tripathi, V., Nyongesa, B. O., & Were, B. A. (2023). Signaling and crosstalk of rhizobacterial and plant hormones that mediate abiotic stress tolerance in plants. *Frontiers in Microbiology*, *14*, 117–104. https://doi.org/10.3389/fmicb.2023.1171104.

Alori, E. T., Glick, B. R., & Babalola, O. O. (2017). Microbial phosphorus solubilization and its potential for use in sustainable agriculture. *Frontiers in Microbiology*, *8*, 35–71. https://doi.org/10.3389/fmicb.2017.00971.

Amen, R. A., Hassan, Y. M., Essmat, R. A., Ahmed, R. H., Azab, M. M., Shehata, N. R., Elgazzar, M. M., & El-Sayed, W. M. (2025). Harnessing the microbiome: CRISPR-based gene editing and antimicrobial peptides in combating antibiotic resistance and cancer. *Probiotics and Antimicrobial Proteins*, *17*(4), 1938–1968. https://doi.org/10.1007/s12602-025-10573-8.

Arahou, F., Lijassi, I., Wahby, A., Rhazi, L., Arahou, M., & Wahby, I. (2023). Spirulina-based biostimulants for sustainable agriculture: Yield improvement and market trends. *BioEnergy Research*, *16*(3), 1401–1416. https://doi.org/10.1007/s12155-023-10614-6.

Ayaz, M., Li, C. H., Ali, Q., Zhao, W., Chi, Y. K., Shafiq, M., Ali, F., Yu, X. Y., Yu, Q., Zhao, J. T., Yu, J. W., Qi, R. D., & Huang, W. K. (2023). Bacterial and fungal biocontrol agents for plant disease protection: Journey from lab to field, current status, challenges, and global perspectives. *Molecules*, *28*(18), 63–75. https://doi.org/10.3390/molecules28186735.

Bahuguna, V., Matura, R., Farswan, A. S., Naqvi, S. S., Sharma, N., & Chaudhary, M. (2025). Rhizobium as a potential biofertilizer and its quality control analysis for sustainable agriculture. *Journal of Applied Biology & Biotechnology*, *13*(3), 97–105.

Bashan, Y., & de-Bashan, L. E. (2010). How the plant growth-promoting bacterium Azospirillum promotes plant growth—A critical assessment. Advances in Agronomy, 108, 77–136. https://doi.org/10.1016/S0065-2113(10)08002-8.

Bashir, Z., Hamid, B., Yatoo, A. M., Nisa, M., Sultan, Z., & Popescu, S. M. (2024). Phosphorus solubilizing microorganisms: An eco-friendly approach for sustainable plant health and bioremediation. Journal of Soil Science and Plant Nutrition, 24(4), 6838–6854. https://doi.org/10.1007/s42729-024-02007-1.

Beatty, D. S., Aoki, L. R., Graham, O. J., & Yang, B. (2021). The future is big-and small: Remote sensing enables cross-scale comparisons of microbiome dynamics and ecological consequences. *MSystems*, *6*(6), 06–21. https://doi.org/10.1128/mSystems.01106-21.

Beleri, P. S. (2023). Microbial solutions to soil health: The role of biofertilizers in sustainable agriculture. *Environmental Reports*, *5*(1), 45–60.

Broekgaarden, C., Caarls, L., Vos, I. A., Pieterse, C. M. J., & Van Wees, S. C. M. (2015). Ethylene: Traffic controller on hormonal crossroads to defense. *Plant Physiology*, *169*(4), 2371–2379. https://doi.org/10.1104/pp.15.01020.

Bundela, V., Saini, N., Kumar, A., Singh, A., & Ranjan, P. (2025). Microbial synergy for zinc and iron biofortification: Enhancing crop nutrition through plant-microbe interactions. *Journal of Advances in Biology & Biotechnology*, *28*, 898–918. https://doi.org/10.9734/jabb/2025/v28i72607.

Burgess, B. K., & Lowe, D. J. (1996). Mechanism of molybdenum nitrogenase. *Chem Rev*, *96*(7), 2983–3012. https://doi.org/10.1021/cr950055x.

Cassán, F., & Diaz-Zorita, M. (2016). *Azospirillum* sp. in current agriculture: From the laboratory to the field. *Soil Biology and Biochemistry*, *103*, 117–130. https://doi.org/10.1016/j.soilbio.2016.08.020.

Charubin, K., Bennett, R. K., Fast, A. G., & Papoutsakis, E. T. (2018). Engineering *Clostridium* organisms as microbial cell-factories: Challenges & opportunities. *Metabolic Engineering*, *50*, 173–191. https://doi.org/10.1016/j.ymben.2018.07.011.

Chaudhary, P., et al. (2022). Overview of biofertilizers in crop production and stress management for sustainable agriculture. *Frontiers in Plant Science*, *13*, 230–340. https://doi.org/10.3389/fpls.2022.930340.

Choudhury, A., & Rajam, M. V. (2021). Genetic transformation of legumes: An update. Plant Cell Reports, 40(10), 1813–1830. https://doi.org/10.1007/s00299-021-02749-7.

Compant, S., Samad, A., Faist, H., & Sessitsch, A. (2019). A review on the plant microbiome: Ecology, functions, and emerging trends in microbial application. Journal of Advanced Research, 19, 29–37. https://doi.org/10.1016/j.jare.2019.03.004.

Devi, R., Kaur, T., Kour, D., Yadav, A., Yadav, A. N., & Suman, A., et al. (2022). Minerals solubilizing and mobilizing microbiomes: A sustainable approach for managing minerals' deficiency in agricultural soil. Journal of Applied Microbiology, 133(3), 1245–1272. https://doi.org/10.1111/jam.15627.

Devi, S. H., Bhupenchandra, I., Sinyorita, S., Chongtham, S. K., & Devi, E. L. (2021). Mycorrhizal fungi and sustainable agriculture. Nitrogen in Agriculture– Physiological, Agricultural and Ecological Aspects,225-228. 1–19. https://doi.org/10.5772/intechopen.99262

Ding, Y., Wang, J., Liu, Y., & Chen, S. (2023). Biofertilizers alter nitrogen use efficiency and gaseous losses in cereal-legume intercropping systems. *Agriculture Ecosystems & Environment*, *342*, 108–216. https://doi.org/10.1016/j.agee.2022.108216.

Diwan, D., Rashid, M. M., & Vaishnav, A. (2022). Current understanding of plant-microbe interaction through the lenses of multi-omics approaches and their benefits in sustainable agriculture. *Microbiological Research*, *265*, 280. https://doi.org/10.1016/j.micres.2022.127180.

Dobrzyńska, K., Pérez-González, A., Echavarri-Erasun, C., Coroian, D., Salinero-Lanzarote, A., & Veldhuizen, M., et al. (2024). Nitrogenase cofactor biosynthesis using proteins produced in mitochondria of Saccharomyces cerevisiae. mBio, 15(2), e03088–23. https://doi.org/10.1128/mbio.03088-23.

Dudeja, C., Mishra, A., Ali, A., Singh, P. P., & Jaiswal, A. K. (2025). Microbial genome editing with CRISPR–Cas9: Recent advances and emerging applications across sectors. *Fermentation*, *11*(7), 410. https://doi.org/10.3390/fermentation11070410.

Einsle, O., & Rees, D. C. (2020). Structural enzymology of nitrogenase enzymes. *Frontiers in Microbiology*, *11*, 603529. https://doi.org/10.3389/fmicb.2020.603529.

Etesami, H., Jeong, B. R., & Glick, B. R. (2021). Contribution of arbuscular mycorrhizal fungi, phosphate–solubilizing bacteria, and silicon to P uptake by plant. *Frontiers in Plant Science*, *12*, 699618. https://doi.org/10.3389/fpls.2021.699618.

Fan, D., Schwinghamer, T., Smith, D. L., & Xu, J. (2022). *Bacillus subtilis* strain L1 promotes nitrate assimilation by *Arabidopsis* via stimulating nitrate uptake and inhibiting nitrogen volatilization. *Science of The Total Environment*, *806*, 150–771. https://doi.org/10.1016/j.scitotenv.2021.150771.

Fatimah, N., Dar, S. A., Ashraf, S., Rashid, S., Mukhtar, Y., & Mir, S. H. (2021). Biofertilizers for sustainable agriculture: An overview. *Journal of Sustainable Agriculture*, *10*(6), 45–58.

Figueiredo, G. G. O., Lopes, V. R., Romano, T., & Camara, M. C. (2020). Clostridium. In Beneficial Microbes in Agro-Ecology Academic Press 477–491. https://doi.org/10.1016/B978-0-12-823414-3.00022-8

Fincheira, P., & Quiroz, A. (2018). Microbial volatiles as plant growth inducers. *Microbiological Research*, *208*, 63–75. https://doi.org/10.1016/j.micres.2018.01.002.

Firdos, S., Niranjan Raj, S. N., & Baker, S. (2025). *Nitrogen fixing microbes Azospirillum as bio fertilizer* (pp 1–44). Department of Studies in Microbiology, Karnataka State Open University.

Fukami, J., Cerezini, P., & Hungria, M. (2018). Azospirillum: Benefits that go far beyond biological nitrogen fixation. *AMB Express*, *8*(1), 73. https://doi.org/10.1186/s13568-018-0608-1.

Gauri, S. S., Mandal, S. M., & Pati, B. R. (2012). Impact of *Azotobacter* exopolysaccharides on sustainable agriculture. *Applied Microbiology and Biotechnology*, *95*(2), 331–338. https://doi.org/10.1007/s00253-012-4159-0.

Gautam, K., Sirohi, C., Singh, N. R., Thakur, Y., Jatav, S. S., Rana, K., & Parihar, M. (2021). Microbial biofertilizer: Types, applications, and current challenges for sustainable agricultural production. In, *Biofertilizers* (pp. 3–19). Woodhead Publishing.

Glick, B. R. (2014). Bacteria with ACC deaminase can promote plant growth and help to feed the world. *Microbiological Research*, *169*(1), 30–39. https://doi.org/10.1016/j.micres.2013.09.009.

Gouda, Y., El-Meniawy, S., Abd Elhady, S. A. E., & Ragab, M. E. (2021). Performance of grafted and non-grafted cantaloupe plants under different fertilization rates of nitrogen, phosphorus, and potassium. Egyptian Journal of Horticulture, 48(2), 277–291. https://doi.org/10.21608/ejoh.2021.78036.1175.

Goyal, R. K., Schmidt, M. A., & Hynes, M. F. (2021). Legume-*Rhizobium* symbiosis: Advances and perspectives. *Frontiers in Microbiology, 12*, 634–687. https://doi.org/10.3389/fmicb.2021.634687.

Grahovac, J., Pajčin, I., & Vlajkov, V. (2023). *Bacillus* VOCs in the context of biological control. *Antibiotics, 12*(3), 50–81. https://doi.org/10.3390/antibiotics12030581.

Hakkoum, Z., Minaoui, F., Chabili, A., Douma, M., Mouhri, K., & Loudiki, M. (2025). Biofertilizing effect of soil cyanobacterium *Anabaena cylindrica*–based formulations on wheat growth, physiology, and soil fertility. *Agriculture, 15*(2), 189. https://doi.org/10.3390/agriculture15020189.

Harman, G., Khadka, R., Doni, F., & Uphoff, N. (2021). Benefits to plant health and productivity from enhancing plant microbial symbionts. *Frontiers in Plant Science, 11*, 61–65. https://doi.org/10.3389/fpls.2020.610065.

Harris, D. F., Lukoyanov, D. A., Shaw, S., Compton, P., Tokmina-Lukaszewska, M., Bothner, B., Kelleher, N., Dean, D. R., Hoffman, B. M., & Seefeldt, L. C. (2018). Mechanism of N_2 reduction catalyzed by Fe-nitrogenase involves reductive elimination of H_2. *Biochemistry, 57*(5), 701–710. https://doi.org/10.1021/acs.biochem.7b01142.

Hoffman, B. M., Lukoyanov, D., Yang, Z. Y., Dean, D. R., & Seefeldt, L. C. (2014). Mechanism of nitrogen fixation by nitrogenase: The next stage. *Chemical Reviews, 114*(8), 4041–4062. https://doi.org/10.1021/cr400641x.

Hwang, H. -H, Wang, M. -H, Lee, Y. -L., Tsai, Y. -L., Li, Y. -H., Yang, F. -J., Liao, Y. -C., Lin, S. -K., & Lai, E. -M. (2010). *Agrobacterium*-produced and exogenous cytokinin-modulated *Agrobacterium*-mediated plant transformation. *Molecular Plant Pathology11*(5), 677–690. https://doi.org/10.1111/j.1364-3703.2010.00637.x.

Ijaz, M., Shahzadi, R., Rahman, M. U., & Iqbal, M. (2019). Molecular approaches to study plant growth-promoting rhizobacteria (PGPRs). In, *Plant growth promoting Rhizobacteria for agricultural sustainability: From theory to practices* (pp. 215–232). Springer.

Jacob, S. M., & Paranthaman, S. (2023). Biofertilizers: An advent for eco-friendly and sustainable agriculture development. *Vegetos, 36*(4), 1141–1153. https://doi.org/10.1007/s42535-023-00609-1.

Jama, A., Widiastuti, D. P., Gafur, S., & Davis, J. G. (2023). *Azolla* biofertilizer is an effective replacement for urea fertilizer in vegetable crops. *Sustainability, 15*(7), 11–45. https://doi.org/10.3390/su15076045.

Jawad, M., Jabeen, A., Gurmani, A. R., & Rafique, M. (2023). Phosphate-mobilizing mycorrhizal-based biofertilizers. In Biofertilizers for Sustainable Soil Management . CRC Press 89–106. https://doi.org/10.1201/9781003286233

Kamran, M., Huang, L., Nie, J., Geng, M., Lu, Y., & Liao, Y., et al. (2021). Effect of reduced mineral fertilization (NPK) combined with green manure on aggregate stability and soil organic carbon fractions in a fluvo-aquic paddy soil. Soil and Tillage Research, 211, 105005. https://doi.org/10.1016/j.still.2021.105005.

Karunarathna, S. C. (2024). Mycorrhizal fungi in sustainable agriculture and land restoration. *Agronomy, 14*(4), 32–43. https://doi.org/10.3390/agronomy14040743.

Karunarathna, S. C., Tibpromma, S., Karunarathna, B. S., Dai, D.-Q., Kumla, J., Lu, W., Perera, R. H., et al. (2026). Mushrooms in climate change mitigation: A comprehensive review 20–46. https://doi.org/10.3389/fmicb.2025.1727022

Khan, H., Akbar, W. A., Shah, Z., Rahim, H. U., Taj, A., & Alatalo, J. M. (2022). Coupling phosphate-solubilizing bacteria (PSB) with inorganic phosphorus fertilizer improves mungbean (*Vigna radiata*) phosphorus acquisition, nitrogen fixation, and yield in alkaline-calcareous soil. *Heliyon, 8*(3), 61–81. https://doi.org/10.1016/j.heliyon.2022.e09081.

Khan, A., Panthari, D., Sharma, R. S., Punetha, A., Singh, A. V., & Upadhayay, V. K. (2023). Biofertilizers: A microbial-assisted strategy to improve plant growth and soil health. In, *Advanced microbial techniques in agriculture, environment, and health management* (pp. 97–118). Academic Press.

Kholssi, R., Marks, E. A. N., Miñón, J., Montero, O., Lorentz, J. F., Debdoubi, A., & Rad, C. (2022). Biofertilizing effects of *Anabaena cylindrica* biomass on the growth and nitrogen uptake of wheat. *Communications in Soil Science and Plant Analysis*, *53*(9), 1216–1225. https://doi.org/10.1080/00103624.2022.2052361.

Kizilkaya, R. (2009). Nitrogen fixation capacity of *Azotobacter* spp. strains isolated from soils in different ecosystems and relationship between them and the microbiological properties of soils. *Journal of Environmental Biology*, *30*(1), 73–82.

Koirala, A., & Brözel, V. S. (2021). Phylogeny of nitrogenase structural and assembly proteins: Implications for the evolution of biological nitrogen fixation. *mBio*, *12*(4), 20–21. https://doi.org/10.1128/mBio.00254-21.

Koskey, G., Mburu, S. W., Awino, R., Njeru, E. M., & Maingi, J. M. (2021). Potential use of beneficial microorganisms for soil amelioration, phytopathogen biocontrol, and sustainable crop production in smallholder agroecosystems. *Frontiers in Sustainable Food Systems*, *5*, 303–308. https://doi.org/10.3389/fsufs.2021.606308.

Kumar, A., Patel, J. S., & Meena, V. S. (2020). *Azotobacter chroococcum*: A potential biofertilizer for sustainable agriculture. *Sustainability*, *12*(15), 12–20. https://doi.org/10.3390/su12156020.

Kumar, S., Diksha, Sindhu, S. S., & Kumar, R. (2021). Biofertilizers: An eco-friendly technology for nutrient recycling and environmental sustainability. *Current Research in Microbial Sciences*, *3*, 71–94. https://doi.org/10.1016/j.crmicr.2021.100094.

Ladha, J. K., Peoples, M. B., Reddy, P. M., Biswas, J. C., Bennett, A., Jat, M. L., & Krupnik, T. J. (2022). Biological nitrogen fixation and prospects for ecological intensification in cereal-based cropping systems. *Field Crops Res*, *35*, 403–415. https://doi.org/10.1016/j.fcr.2022.108541.

Mahanty, T., Bhattacharjee, S., Goswami, M., Bhattacharyya, P., Das, B., Ghosh, A., & Tribedi, P. (2017). Biofertilizers: A potential approach for sustainable agriculture development. *Environmental Science and Pollution Research*, *24*(4), 3315–3335. https://doi.org/10.1007/s11356-016-8104-0.

Mahmud, A. A., Upadhyay, S. K., Srivastava, A. K., & Bhojiya, A. A. (2021). Biofertilizers: A Nexus between soil fertility and crop productivity under abiotic stress. *Current Research in Environmental Sustainability*, *3*, 55–63. https://doi.org/10.1016/j.crsust.2021.100063.

Mahmud, K., Missaoui, A. M., Lee, K., Ghimire, B., Presley, H. W., & Makaju, S. (2021). Rhizosphere microbiome manipulation for sustainable crop production. *Current Plant Biology*, *27*, 200–210. https://doi.org/10.1016/j.cpb.2021.100210.

Marzouk, S. H., Tindwa, H. J., Amuri, N. A., & Semoka, J. M. (2023). An overview of underutilized benefits derived from *Azolla* as a promising biofertilizer in lowland rice production. *Heliyon*, *9*(1), 128–145. https://doi.org/10.1016/j.heliyon.2023.e12845.

Mazhar, S., Cohen, J. D., & Hasnain, S. (2013). Auxin producing non-heterocystous cyanobacteria and their impact on the growth and endogenous auxin homeostasis of wheat. *Journal of Basic Microbiology*, *53*(12), 996–1003. https://doi.org/10.1002/jobm.201200723.

Meena, V. S., Bahadur, I., Maurya, B. R., Kumar, A., Meena, R. K., Meena, S. K., & Verma, J. P. (2022). Potassium-solubilizing microorganisms for sustainable agriculture. In, *Potassium solubilizing microorganisms for sustainable agriculture* (pp. 1–25). Springer. https://doi.org/10.1007/978-81-322-2776-2_1

Montes-Luz, B., Conrado, A. C., Ellingsen, J., Monteiro, R. A., Souza, E. M., & Stacey, G. (2023). Acetylene reduction assay: A measure of nitrogenase activity in plants and bacteria. *Current Protocols*, *3*(1), 66–76. https://doi.org/10.1002/cpz1.766.

Mostafa, S. S., Fares, C. N., Bishara, M. M., Azzam, C. R., Awad, A. A., Elgaml, N. M., & Mostafa, M. S. (2024). Microbiomes–plant interactions and K-humate application for salinity stress mitigation and yield enhancement in wheat and faba bean in Egypt's Northeastern Delta. International Journal of Plant Biology, 15(4), 1077–1107. https://doi.org/10.3390/ijpb15040076.

Naik, S. S. (2024). Promotive effect of different carrier based *Nostoc commune* biofertilizers (pp. 1–50). Dissertation, Goa University.

Narehood, S. M., Cook, B. D., Srisantitham, S., Eng, V. H., Shiau, A. A., McGuire, K. L., Britt, R. D., Herzik, Jr, M. A., & Tezcan, F. A. (2025). Structural basis for the conformational protection of nitrogenase from O_2. *Nature*, *637*, 991–997. https://doi.org/10.1038/s41586-024-08311-1.

Oco, R. G., Devanadera, M. K., & De Grano, R. V. R. (2024). Utilization of *Nostocpiscinale* as potential biofertilizer to the growth and development of *Oryzasativa* L. *Caraka Tani: Journal of Sustainable Agriculture, 39*(1), 22–37. https://doi.org/10.20961/carakatani.v39i1.76543.

Okubo, T., Piromyou, P., Tittabutr, P., Teaumroong, N., & Minamisawa, K. (2016). Origin and evolution of nitrogen fixation genes on symbiosis islands and plasmid in Bradyrhizobium. *Microbes and Environments, 31*(3), 260–267. https://doi.org/10.1264/jsme2.ME15159.

Otaiku, A. A., Soretire, A. A., & Mmom, P. C. (2022). Biofertilizer impacts on soybean (*Glycine max L.*) cultivation, humid tropics: Biological nitrogen fixation, yield, soil health and smart agriculture framework. *International Journal of Agricultural Extesion and Rural Development Studies, 9*(1), 138–139.

Pal, P., Ansari, S., Jalil, S. U., & Ansari, M. I. (2023). Regulatory role of phytohormones in plant growth and development. In, Awasthi, L. P. (Ed.), *Molecular aspects of plant beneficial microbes in agriculture* (pp. 197–208). *Academic Press*. https://doi.org/10.1016/B978-0-323-91886-2.00016-1

Pankievicz, V. C. S., do Amaral, F. P., Ane, J. M., & Stacey, G. (2021). Diazotrophic bacteria and their mechanisms to interact and benefit cereals. *Molecular Plant–Microbe Interactions, 34*(5), 491–498. https://doi.org/10.1094/MPMI-11-20-0316-FI.

Patel, S. K., Gupta, R. K., & Das, D. (2021). *Clostridium*-based biofertilizers: A sustainable approach to enhance soil fertility and crop productivity. *Journal of Cleaner Production, 320*, 12–57. https://doi.org/10.1016/j.jclepro.2021.128857.

Ponsetto, P., Sasal, E. M., Mazzoli, R., Valetti, F., & Gilardi, G. (2024). The potential of native and engineered Clostridia for biomass biorefining. *Frontiers in Bioengineering and Biotechnology, 12*, 14–35. https://doi.org/10.3389/fbioe.2024.1423935.

Poveda, J., & González-Andrés, F. (2021). Bacillus as a source of phytohormones for use in agriculture. Applied Microbiology and Biotechnology, 105(23), 8629–8645. https://doi.org/10.1007/s00253-021-11492-8.

Prasad, R., Kumar, M., & Varma, A. (2015). Role of PGPR in soil fertility and plant health. In, *Plant-growth-promoting rhizobacteria (PGPR) and medicinal plants* (pp. 247–260). Springer. https://doi.org/10.1007/978-3-319-13401-7_12

Prasanna, R., Kanchan, A., Ramakrishnan, B., Ranjan, K., & Venkatachalam, S. (2022). Cyanobacteria-based biofertilizers: Innovations and future perspectives. *Microbiological Research, 254*, 26–91. https://doi.org/10.1016/j.micres.2021.126891.

Prasanna, R., Babu, S., Bidyarani, N., Kumar, A., Triveni, S., & Monga, D. (2022). Prospecting cyanobacteria-fortified composts as plant growth promoting and biocontrol agents in cotton. *Indian Journal of Microbiology, 62*(1), 135–144. https://doi.org/10.1007/s12088-021-00985-9.

Prisa, D., Fresco, R., & Spagnuolo, D. (2023). Microbial biofertilisers in plant production and resistance: A review. *Agriculture, 13*(9), 61–66. https://doi.org/10.3390/agriculture13091666.

Reddy, G. C., Goyal, R. K., Puranik, S., Waghmar, V., Vikram, K. V., & Sruthy, K. S. (2020). Biofertilizers toward sustainable agricultural development. In, *Plant microbe symbiosis* (pp. 115–128). Springer. https://doi.org/10.1007/978-3-030-36248-57.

Rehmaan, I. U., Jan, B., Khan, N. F., Islam, T., Rehman, S., Sheergojri, I. A., & Wani, A. H. (2022). Nitrogen biofertilizers: Role in sustainable agriculture. In, *Advances in plant nitrogen metabolism* (pp. 170–184). CRC Press.

Rehman, I. U., Islam, T., Wani, A. H., Rashid, I., Sheergojri, I. A., Bandh, M. M., & Rehman, S. (2021). Biofertilizers: The role in sustainable agriculture. In Sustainable Agriculture: Technical Progressions and Transitions (pp. 25–38). https://doi.org/10.1007/978-3-030-83066-3_2

Rehman, I. U., Islam, T., Wani, A. H., Rashid, I., Sheergojri, I. A., Bandh, M. M., & Rehman, S. (2022). Biofertilizers: The role in sustainable agriculture. In, Sustainable agriculture: Technical progressions and transitions (pp. 25–38). Springer.

Roy, S., Liu, W., Nandety, R. S., Crook, A., Mysore, K. S., Pislariu, C. I., Frugoli, J., Dickstein, R., & Udvardi, M. K. (2020). Celebrating 20 years of genetic discoveries in legume nodulation and symbiosis. *The Plant Cell, 32*(1), 15–41. https://doi.org/10.1105/tpc.19.00279.

Roy, A. (2020). Biofertilizers for agricultural sustainability: Current status and future challenges. In, Current trends in microbial biotechnology for sustainable agriculture (pp. 525–553). Springer. https://doi.org/10.1007/978-981-15-6949-421

Rutten, P. J., & Poole, P. S. (2019). Oxygen regulatory mechanisms of nitrogen fixation in rhizobia. *Advances in Microbial Physiology*, *75*, 325–389. https://doi.org/10.1016/bs.ampbs.2019.08.001.

Saini, P. K., Prasad, S., Bhati, J., & Ram, D. (2021). Use of biofertilizer for increasing soil health and sustainable agriculture production. *Biotic Research Today*, *3*(12), 1119–1122.

Samantaray, A., Chattaraj, S., Mitra, D., Ganguly, A., Kumar, R., Gaur, A., Das Mohapatra, P. K., de los Santos-Villalobos, S., Rani, A., & Thatoi, H. (2024). Advances in microbial based bio-inoculum for amelioration of soil health and sustainable crop production. *Current Research in Microbial Sciences*, *7*, 200–251. https://doi.org/10.1016/j.crmicr.2024.100251.

Santos, M. S., Nogueira, M. A., & Hungria, M. (2021). Microbial inoculants: Reviewing the past, discussing the present and previewing an outstanding future for the use of beneficial bacteria in agriculture. *AMB Express*, *11*(1), 205–207. https://doi.org/10.1186/s13568-021-01320-7.

Schleicher, E., Heidinger, L., Perez, K., Spatzal, T., Einsle, O., Weber, S., & Rees, D. (2023). Analysis of early intermediate states of the nitrogenase reaction by Se incorporation and regularization of EPR spectra 11–45. https://doi.org/10.21203/rs.3.rs-3120611/v1

Seefeldt, L. C., Hoffman, B. M., & Dean, D. R. (2009). Mechanism of Mo-dependent nitrogenase. *Annual Review of Biochemistry*, *78*, 701–722. https://doi.org/10.1146/annurev.biochem.78.070907.103812.

Shamim, A., Mahfooz, S., Hussain, A., & Farooqui, A. (2020). Ability of acclimatized immobilized *Nostoc muscorum* to combat abiotic stress and its potential as a biofertilizer. *Journal of Pure and Applied Microbiolody*, *14*(2), 1377–1386. https://doi.org/10.22207/JPAM.14.2.45.

Singh, J. S., Kumar, A., Rai, A. N., & Singh, D. P. (2021). Cyanobacteria: A precious bioresource for sustainable agriculture. *Applied Soil Ecology*, *158*, 103–31. https://doi.org/10.1016/j.apsoil.2020.103791.

Singh, S., Devi, N. B., Divya, D., Kumar, A., Singh, S., Tyagi, G., & Kumar, A. (2024). Soil fertility management: Role of organic amendments and bio-fertilizers: A review. International Journal of Research in Agronomy, 7(12), 766–772. https://doi.org/10.33545/2618060X.2024.v7.i12j.2253.

Singh, R. K., Singh, P., Li, H. B., Yang, L. T. & Li, Y. R. (2017). Soil–plant–microbe interactions: Use of nitrogen-fixing bacteria for plant growth and development in sugarcane. In, Plant-microbe interactions in agro-ecological perspectives (pp. 35–39). Springer. https://doi.org/10.1007/978-981-10-5813-43

Sosnowski, J., Truba, M., & Vasileva, V. (2023). The impact of auxin and cytokinin on the growth and development of selected crops. Agriculture, 13(3), 724. https://doi.org/10.3390/agriculture13030724.

Tejera, N., Lluch, C., Martínez-Toledo, M. V., & González-López, J. (2020). Isolation and characterization of *Azotobacter* and *Azospirillum* strains from the sugarcane rhizosphere. *Plant and Soil*, *267*(1–2), 223–232. https://doi.org/10.1007/s11104-005-5349-9.

Thakur, B., & Kaur, S. (2024). The role of artificial intelligence in biofertilizer development. In, Metabolomics, proteomics and gene editing approaches in biofertilizer industry (vol, II, pp. 157–176). Springer Nature.

Thapa, P., & Poudel, K. (2021). *Azolla*: Potential biofertilizer for increasing rice productivity, and government policy for implementation. *Journal of Wastes and Biomass Management*, *3*(2), 62–68.

Timofeeva, A. M., Galyamova, M. R., & Sedykh, S. E. (2024). Plant growth-promoting soil bacteria: Nitrogen fixation, phosphate solubilization, siderophore production, and other biological activities. *Plant Soil and Environment*, *70*(1), 1–15.

Udvardi, M., & Poole, P. S. (2013). Transport and metabolism in legume-rhizobia symbioses. *Annual Review of Plant Biology*, *64*, 781–805. https://doi.org/10.1146/annurev-arplant-050312-120235.

Usman, N., & Wali, M. (2024). Nitrogen fixation by rhizobacterial Nif mechanism: An advanced genetic perspective. In, *Nitrogen Fixation* (vol. 54, pp. 75–87). IntechOpen. 10.5772/intechopen.1004087.

Vasseur Coronado, M. F. (2021). Strategies to design a new generation of biofertilisers for a more sustainable agriculture. *Plant and Soil, 18*, 09–22.

Vejan, P., Abdullah, R., Khadiran, T., Ismail, S., & Boyce, A. N. (2016). Role of plant growth promoting *Rhizobacteria* in agricultural sustainability: A review. *Molecules, 21*(5), 573–85. https://doi.org/10.3390/molecules21050573.

Wan, Q., Li, L., Liu, B., Zhang, Z., Liu, Y., & Xie, M. (2023). Different and unified responses of soil bacterial and fungal community composition and predicted functional potential to 3 years' drought stress in a semiarid alpine grassland. Frontiers in Microbiology, 14, 1104944. https://doi.org/10.3389/fmicb.2023.1104944.

Wang, Q., Zhang, W., & Yang, H. (2022). *Clostridium* as a versatile nitrogen-fixing bacterium: Genetic insights and agricultural applications. *Applied and Environmental Microbiology, 88*(15), 47–22. https://doi.org/10.1128/aem.00847-22.

Van Wees, S. C. M., Van der Ent, S., & Pieterse, C. M. J. (2008). Plant immune responses triggered by beneficial microbes. *Current Opinion in Plant Biology, 11*(4), 443–448. https://doi.org/10.1016/j.pbi.2008.05.005.

Xiao, Y., Wang, X., Chen, W., & Huang, Q. (2021). Isolation and identification of three kinds of heterotrophic nitrifying aerobic bacteria and their nitrogen removal characteristics. *Environmental Science and Pollution Reearchs, 28*, 18637–18649. https://doi.org/10.1007/s11356-021-13034-1.

Yadav, A., & Yadav, K. (2024). Challenges and opportunities in biofertilizer commercialization. *SVOA Microbiology, 5*(1), 1–14.

Zhang, L., & Yang, X. (2023). Genetic engineering of *Clostridium* for improved nitrogen fixation: Challenges and opportunities. *Biotechnology Advances, 64*, 108–121. https://doi.org/10.1016/j.synbio.2023.12.001.

Zia, R., Nawaz, M. S., Siddique, M. J., Hakim, S., & Imran, A. (2021). Plant survival under drought stress: Implications, adaptive responses, and integrated rhizosphere management strategy for stress mitigation. *Microbiological Research, 242*, 20–26. https://doi.org/10.1016/j.micres.2020.126626.

6 Biological Nitrogen Fixation is a Revolution of Unlocking Sustainable Agriculture

Priyanka Jindal, Pooja Nikhanj, Mohini Prabha Singh and Pratiksha Singh

Abstract

There is broad scientific consensus that nitrogen-based fertilizers are nearly essential for modern agriculture. Yet, their escalating use and the resulting environmental challenges have intensified the search for more sustainable alternatives. Nitrogen fixation offers a practical and ecoconscious method for tackling a major agricultural challenge: maintaining high crop productivity while protecting the environment. By employing nitrogen-fixing organisms and embracing modern farming techniques, farmers can lessen their dependence on chemical fertilizers, boost soil fertility, and reduce environmental harm. As the global demand for food continues to grow, incorporating nitrogen fixation into conventional farming practices is essential for achieving long-term sustainability and resilience in food production. Among these, biological nitrogen fixation (BNF) stands out as one of the most ecofriendly methods for meeting crop nitrogen needs. However, like synthetic fertilizers, BNF is not without limitations. A key issue is the highly variable and often limited transfer of fixed nitrogen to nearby non-nitrogen-fixing plants ranging from 0% to 70% depending on various conditions. This chapter examines current developments in BNF across selected legume crops, its global economic impact, and recent findings on nitrogen transfer in agricultural systems. It also explores the key drivers of nitrogen transfer and outlines management approaches to enhance both fixation and interplant nitrogen sharing.

P. Jindal · P. Nikhanj (✉) · M. P. Singh
Department of Microbiology, Punjab Agricultural University, Ludhiana, India
email: poojanikhanj.217@gmail.com

P. Singh
School of Agriculture Science, LNCT University, JK Town Sarvadharam C Sector, Bhopal, MP, India

R. K. Singh and P. Singh (eds.), *Diazotrophic Plant Growth-Promoting Microorganisms and Green Energy – Volume I*, https://doi.org/10.1007/978-981-95-9682-9_6

Keywords

Nitrogen fixation · Nitrogen fixing organisms · BNF · Agriculture · Biofertilizers

6.1 Introduction

Nitrogen (N) is an essential macronutrient that plants require in considerable quantities for proper growth and reproduction. It is identified as the third most influential factor in crop development (Thilakarathna et al., 2016), emphasizing its crucial role in sustaining agricultural output. The importance of nitrogen in global food production is clearly demonstrated by the ongoing rise in nitrogen fertilizer usage. Reducing the adverse impacts of agrochemicals while meeting the increasing food demands of a growing population requires research into sustainable bioemerging strategies (Singh et al., 2019). This increase is primarily driven by the growing need to ensure sufficient food production for a global population now exceeding 7 billion. While nitrogen fertilizers are broadly acknowledged by researchers as critical to contemporary farming systems, their use raises significant environmental issues. These include the pollution of water resources, the release of greenhouse gases like nitrous oxide (N_2O), and low nitrogen use efficiency (NUE).

Consequently, managing nitrogen inputs effectively requires an integrated and strategic approach. Although synthetic fertilizers have enhanced crop productivity, there is an increasing emphasis on more environmentally sound practices such as climate-smart agriculture and sustainable intensification to fulfill crop nitrogen demands responsibly (Vorholt et al., 2017). Through a systematic process of literature review, data collection, and processing (Fig. 6.1), a dataset containing 32 variables is assembled. These variables encompass site location, International Geosphere-Biosphere Program (IGBP) land cover classification, N-fixing niche (including root-nodulating N-fixing trees, shrubs, and herbs; free-living BNF in soil, leaf litter, woody litter, dead wood; and BNF linked to mosses, lichens, and biocrusts), sampling year, quantification method, BNF rate (kg N ha^{-1} $year^{-1}$), proportion of nitrogen derived from the atmosphere (%Ndfa), abundance of N fixers or N-fixing substrates, BNF rate per unit of N fixer abundance, and species identity. Altogether, the dataset compiles information from 376 studies on biological nitrogen fixation (BNF), totaling 1359 measurements—1207 BNF rates and 152 %Ndfa values (Carl et al., 2025). The process of nitrogen fixation in legumes begins with the development of nodules on their roots. Rhizobia bacteria present in the soil infect the root hairs and reproduce within the root's cortical cells. The plant provides the bacteria with the nutrients and energy that they require to thrive. About a week after infection, small nodules can be seen without magnification. In field settings, these nodules usually become visible 2–3 weeks after planting, depending on the legume species and environmental factors influencing germination. Young nodules that have not yet started nitrogen fixation are typically white or gray inside. As they mature, they take on a pink or reddish hue, signaling active nitrogen fixation. This color results from leghemoglobin—a compound similar to hemoglobin in blood—that

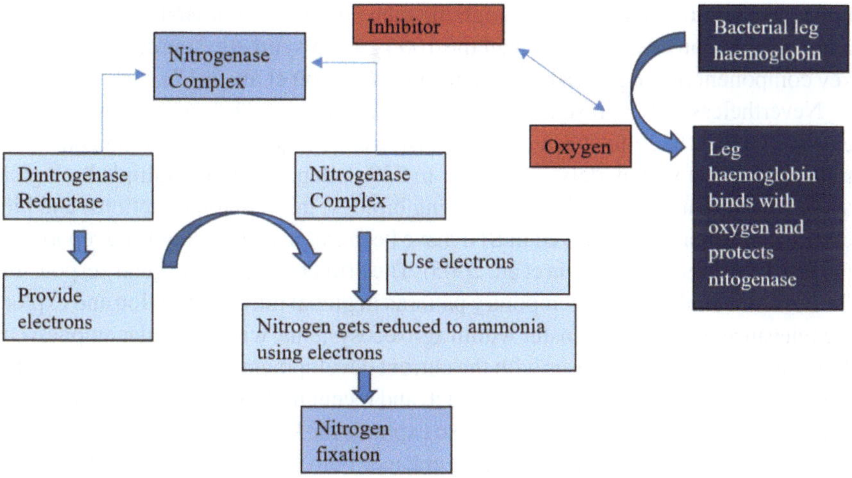

Fig. 6.1 Molecular mechanism of nitrogen fixation

manages the oxygen supply available to the bacteria. Perennial legumes such as alfalfa and clover commonly form finger-like nodules. When fully developed, these nodules may resemble a small hand, with a central portion (the palm) and several narrow extensions (the fingers), although they rarely exceed ½ inch in diameter. These nodules are long-lived and remain active in nitrogen fixation throughout the growing season as long as conditions are suitable. On mature alfalfa plants, roughly 10–50 nodules are typically found clustered near the taproot. Annual legumes—including beans, peanuts, and soybeans—produce rounded nodules that may grow to the size of a large pea. Unlike those on perennials, these nodules are short-lived and are continuously replaced during the growing season. As the plant enters the pod-filling stage, nitrogen fixation decreases because the plant redirects its energy toward seed development. The number of nodules varies among species: bean plants usually have fewer than 100, soybeans can develop several hundred, and well-established peanut plants may produce more than 1000 nodules (Amack et al., 2020).

Prior to the Industrial Revolution, biological nitrogen fixation (BNF) served as the dominant natural process through which plants obtained nitrogen. This process is carried out by certain microorganisms that, via the nitrogenase enzyme, convert inert atmospheric nitrogen (N_2) into forms that plants can absorb and use. Today, BNF is considered one of the most sustainable methods for supplying nitrogen to crops. Research shows that nitrogen use efficiency tends to increase significantly with higher levels of biologically fixed nitrogen, while it declines with excessive application of synthetic fertilizers. Despite its sustainability, accurately assessing the contribution of BNF remains a complex task. According to conservative estimates based on 2005 FAOSTAT data, pulses and oilseed crops fixed approximately 2.95 Tg and 18.5 Tg of nitrogen per year, respectively (Herridge et al., 2008). Of this, soybean *Glycine max* (L.) alone accounted for 16.4 Tg, representing 77% of the total nitrogen fixed by legumes during that year. Although the total contribution of BNF—around 25 Tg annually—is considerably less than the 100 Tg from synthetic sources, it remains a key component of the global nitrogen balance (Vorholt et al., 2017).

Nevertheless, BNF presents its own set of challenges. One major concern is the extent to which fixed nitrogen can be transferred to adjacent nonfixing plants. This transfer rate can vary widely—from 0% to 73%—depending on multiple biological and environmental factors. The underlying biology, microbial interactions, and biochemical mechanisms involved in BNF have been extensively studied and reported in the scientific literature (Sylvia et al., 2005). The primary microbial agents responsible for BNF and analyzing contemporary patterns in global nitrogen fixation and explore the relevance of nitrogen transfer within agro ecosystems, with particular emphasis on *Rhizobium-legume* symbioses with the current developments in BNF across selected legume crops, its global economic impact, and recent findings on nitrogen transfer in agricultural systems are discussed. It also explores the key drivers of nitrogen transfer and outlines management approaches to enhance both fixation and interplant nitrogen sharing. An interesting strategy that offers an alternative to traditional fertilizers is the use of microorganisms, particularly those that have the ability to fix nitrogen biologically, or BNF. Along with helping agricultural plants increase their capacity for N uptake and assimilation, these bacteria may also create phytohormones, enhance stress tolerance, and promote general soil health (Cassim et al., 2024).

6.2 Nitrogen Fixation

Although nitrogen gas (N_2) accounts for nearly 80% of the Earth's atmosphere, it frequently limits primary productivity in many ecosystems. This is because atmospheric nitrogen is not directly usable by plants and animals. For nitrogen to be incorporated into vital biomolecules such as proteins and DNA, it must first be converted into a bioavailable form through nitrogen fixation. The N_2 molecule is extremely stable due to a strong triple bond between its atoms, making it energetically difficult to break. This conversion process requires the input of eight electrons and a minimum of 16 ATP molecules. Consequently, only specific prokaryotic organisms are capable of performing this demanding biochemical reaction. While most nitrogen fixation occurs biologically, a smaller portion is fixed abiotically through natural events like lightning or through human activities such as fossil fuel combustion (Varley et al., 2015). Certain nitrogen-fixing organisms operate independently, while others rely on a symbiotic relationship with host plants to facilitate nitrogen fixation. These partnerships are often highly specific and involve intricate biological processes to maintain their function. For example, legume species such as soybeans, clover, and peas release root exudates that serve as chemical cues, attracting *Rhizobium* bacteria. This interaction sets off a complex series of events that allow the bacteria to penetrate the roots and form nodules—specialized structures where nitrogen fixation occurs (Betancourt et al., 2024).

Nitrogen-fixing bacteria display a wide spectrum of physiological characteristics. Some are aerobic and thrive in oxygen-rich environments, while others are anaerobic and function in the absence of oxygen. They also differ in their energy sources—some are phototrophic, relying on light, while others are chemotrophic, using chemical compounds for energy (Table 6.1). Despite their diversity, all nitrogen-fixing organisms utilize a common enzyme complex called nitrogenase, which facilitates the transformation of atmospheric nitrogen (N_2) into ammonia (NH_3), a usable form of nitrogen for living systems. However, nitrogenase is highly sensitive to oxygen and becomes inactive when exposed to it. This presents a challenge for aerobic and photosynthetic nitrogen-fixers that either depend on or produce oxygen. To overcome this, different organisms have evolved protective strategies. For instance, certain cyanobacteria develop heterocysts specialized cells that provide a low-oxygen environment suitable for nitrogen fixation. Others restrict nitrogen fixation to nighttime hours when photosynthetic oxygen production is not occurring.

Nitrogen compounds have long been essential in agriculture as fertilizers. During the nineteenth century, scientists increasingly recognized the importance of fixed nitrogen for healthy plant growth. This awareness led to the recovery and use of ammonia produced during coal coking as a fertilizer, along with natural sodium nitrate (saltpetre) from Chile. In areas where farming was intensive, the soil's nitrogen content was

Table 6.1 Representative prokaryotes to carry out nitrogen fixation

Genus	Phylogenetic affiliation	Lifestyle	References
Pseudomonas, Azotobacter, Methylomonas	Bacteria	Free-living, aerobic, chemoorganotrophic	Galloway et al. (2020)
Thiobacillus, Alcaligenes	Bacteria	Free-living, aerobic, chemolithotrophic	Howarth (2008)
Chromatium, Chlorobium	Bacteria	Free-living, anaerobic, phototrophic	Johnson et al. (2010)
Rhizobium, Frankia	Bacteria	Symbiotic, aerobic, chemoorganotrophic	Könneke et al. (2005)
Desulfovibrio, Clostridium	Bacteria	Free-living, anaerobic, chemoorganotrophic	Könneke et al. (2005)
Methanosarcina, Methanococcus	Archaea	Free-living, anaerobic, phototrophic	Johnson et al. (2010)
Nostoc, Anabaena	Cyanobacteria	Free-living, aerobic, phototrophic	Johnson et al. (2010)
Psuedomonas	Bacteria	Aerobic	Taule et al. (2012)
Aulosira	Blue green algae	Free-living, aerobic	Johnson et al. (2010)
Plectridium	Bacteria	Anaerobic	Johnson et al. (2010)
Chlorobacterium	Bacteria	Anaerobic	–
Cyclindrospermum	Blue green algae	Free-living, aerobic	–

often insufficient, necessitating additional nitrogen inputs. Meanwhile, the growing demand for Chilean saltpetre in the production of gunpowder spurred a global search for alternative natural nitrogen sources. By the end of the 1800s, it became clear that neither ammonia from coal nor imported Chilean nitrates could meet the expanding agricultural and industrial requirements. There were also fears that if a nation were cut off from Chilean nitrate during wartime, it would face serious difficulties in producing munitions. To solve this problem, early 20th-century scientists turned their focus to industrial nitrogen fixation (Varley et al., 2015). Three major processes were developed: combining nitrogen with oxygen, reacting nitrogen with calcium carbide, and bonding nitrogen with hydrogen. The first method involved heating air or another nitrogen-oxygen mixture to extremely high temperatures, which produced small amounts of nitric oxide gas that could then be processed into nitrates for fertilizer use. Although this approach was tested in 1902 using electric generators at Niagara Falls, it was not commercially viable. However, in 1904, Norwegian researchers Christian Birkeland and Samuel Eyde successfully used this arc process in a small plant, paving the way for larger, effective facilities in Norway and abroad. The Haber-Bosch process is the most practical and economical method for fixing nitrogen, producing ammonia by combining nitrogen and hydrogen gases. In 1909, German chemist Fritz Haber demonstrated that atmospheric nitrogen could react with hydrogen under high pressure and moderately elevated temperatures in the presence of a catalyst, resulting in a high output of ammonia (Fig. 6.1). This ammonia serves as a crucial raw material for the production of numerous nitrogen-based compounds. Carl Bosch later adapted Haber's method for industrial-scale use, leading to its designation as the Haber-Bosch process, also referred to as the synthetic ammonia process. Germany's effective use of this technique during World War I significantly contributed to its war efforts and spurred rapid industrial expansion. After the war, many other countries adopted the process, building similar facilities. Today, the Haber-Bosch process remains a fundamental part of the global chemical manufacturing industry. Despite its initial promise, the arc method consumed large amounts of energy and was ultimately deemed inefficient. It was soon replaced by more effective alternatives. One such process involved reacting nitrogen with calcium carbide at high temperatures to form calcium cyanamide, which could later be converted into ammonia and urea (Johnson et al., 2022).

Nitrogenase genes have been identified in a wide variety of ecosystems across the globe, ranging from oxygen-rich environments like marine and freshwater systems and soils, to low-oxygen or anaerobic settings such as termite intestines, sediments, hypersaline lakes, microbial mats, and planktonic organisms (Unkovich et al., 2008). This extensive distribution illustrates the ecological flexibility of nitrogen-fixing organisms and underscores the essential role of nitrogen fixation in supporting life on Earth. Nitrogen-fixing microorganisms are vital to the nitrogen cycle, as they help convert atmospheric nitrogen into forms that plants can absorb and use. These microbes fall into two main categories: symbiotic and free-living nitrogen fixers.

- **Symbiotic Nitrogen Fixers:** These organisms form mutually beneficial relationships with specific plants, particularly legumes. A common example is *Rhizobium*, which resides in root nodules. Through this partnership, the bacteria transform nitrogen gas (N_2) into ammonia (NH_3), which the host plant can utilize for growth. In return, the plant provides the bacteria with energy in the form of carbohydrates produced via photosynthesis.

- **Free-Living Nitrogen Fixers:** These bacteria live freely in the soil and do not depend on a host plant. Notable examples include *Azotobacter* and *Clostridium*. Although they do not form direct relationships with plants, their nitrogen-fixing activity enhances soil fertility, thereby supporting plant development indirectly.

Biological nitrogen fixation (BNF) is the primary natural source of new nitrogen in terrestrial ecosystems, crucial for supporting ecosystem productivity, carbon storage, and other key Earth system functions. To improve insights into global BNF patterns, a comprehensive dataset was assembled, gathering field measurements from all major nitrogen-fixing niches across natural terrestrial biomes. Drawing on 376 studies, the dataset includes 32 variables such as site location, biome type, nitrogen-fixing niche, year of sampling, measurement method, BNF rate (kg N ha^{-1} $year^{-1}$), percentage of nitrogen derived from the atmosphere (%Ndfa), abundance of nitrogen fixers or substrates, BNF rate per unit abundance, and species identity. In total, it contains 1207 BNF rate measurements for ecosystem components including trees, shrubs, herbs, soils, leaf litter, woody litter, dead wood, mosses, lichens, and biocrusts, along with 152 herb %Ndfa values, 1005 abundance measurements, and 762 BNF rates normalized by fixer abundance, covering 424 species across 66 countries (Carl et al., 2025). This dataset provides a valuable basis for synthesis, meta-analyses, upscaling studies, and benchmarking models of BNF fluxes at multiple spatial scales.

Furthermore, Biological nitrogen fixation is the natural process by which inert atmospheric nitrogen (N_2) is transformed into biologically usable ammonia. This conversion is exclusively carried out by nitrogen-fixing rhizobial bacteria belonging to the family *Rhizobiaceae* and class *α-Proteobacteria*. Other plants gain indirect advantages from these bacteria either when the microbes decompose or release nitrogen into the soil or through close associations with living bacteria. In legumes and certain other plants, the bacteria reside within specialized root structures known as nodules. Within these nodules, the bacteria perform nitrogen fixation, generating NH_3 that the plant can readily utilize. Consequently, nitrogen fixation in legumes exemplifies a symbiotic partnership between the plant and its bacterial counterpart. Bacteria including *Rhizobium, Bradyrhizobium, Mesorhizobium, Azorhizobium*, and *Sinorhizobium* form symbiotic partnerships with legume plants to convert atmospheric nitrogen into a usable form. This nitrogen fixation process relies on the nitrogenase enzyme, produced by genes in the <id>nif</id> (nitrogen fixation) gene cluster, which catalyzes the conversion of nitrogen gas (N_2) into ammonia (NH_3) inside specialized root nodules.

6.3 Isolation of Nitrogen-Fixing Bacteria and Launch of N-Fixing Biofertilizers

The study of biological nitrogen fixation (BNF) dates back to 1886, when Hellriegel and Wilfarth discovered that certain legumes could absorb atmospheric nitrogen (N_2) through root nodules (Franche et al., 2009). A major breakthrough followed in

1888 when Beijerinck isolated *Rhizobium leguminosarum*, a bacterium capable of fixing nitrogen. Further progress came in 1893 with Winogradsky's identification of *Clostridium pasteurianum* (Winogradsky, 1895), and in 1896, Guthrie introduced the use of *Rhizobium* inoculants to legume farmers in New South Wales, Australia (Gurthie, 1896). In 1901, Beijerinck also identified *Azotobacter*, an aerobic nitrogen-fixing microorganism. Field experiments involving *Rhizobium* and field peas began in 1905 at Hawkesbury Agricultural College in Australia, paving the way for the distribution of rhizobial inoculants to farmers by 1914 (Bullard et al., 2005). In 1953, Johanna Döbereiner expanded the scope of BNF research by discovering new nitrogen-fixing endophytic bacteria, such as *Beijerinckia fluminensis* in sugarcane and *Azotobacter paspali* in Bahia grass (*Paspalum notatum*) (Dobereiner, 1966). A significant methodological advancement occurred in 1975 when Fabio Pedrosa developed a nitrogen-free medium (NFb) designed for isolating *Spirillum* species. This led to the discovery of two nitrogen-fixing *Azospirillum* species *A. lipoferum* and *A. brasilense* which are heterotrophic and nonsymbiotic, typically fixing between 20 and 30 kilograms of nitrogen per hectare per year. In 1979, nitrogen fixation in *cyanobacteria* (formerly called blue–green algae) was confirmed by Steward using the Nitrogen-15 (^15N) isotope tracing method. More recently, nitrogen-fixing abilities were also discovered in archaea, specifically in two methanogenic species: *Methano sarcinabarkeri* and *Methanococcus thermolithotrophicus*. These findings were supported by both ^15N tracing and the acetylene reduction assay (ARA). Table 6.2 outlines these key historical developments in BNF research.

6.4 Biological Nitrogen Fixation Importance and Economics

Biological nitrogen fixation (BNF) plays a crucial role in the global nitrogen cycle, with an estimated annual contribution of 122 million tons of nitrogen. Agricultural crops are responsible for fixing between 55 and 60 million tons of this total (Vitousek et al., 2013). Among these, *Glycine max* (soybean) is the most significant contributor, accounting for approximately 16.4 million tons of nitrogen annually through a symbiotic relationship with *Bradyrhizobium* bacteria.

As sustainable farming practices gain momentum, BNF is increasingly recognized as a viable alternative to synthetic nitrogen fertilizers. This approach is mainly driven by the cultivation of nitrogen-fixing legumes such as soybean, chickpea, cowpea, lupin, and alfalfa, which form mutualistic associations with rhizobia (Aloril et al., 2018). The inoculation of legumes with rhizobia has become one of the most effective applications of biofertilizers. The positive impacts of these microorganisms have led to the expansion of the biofertilizer industry, resulting in a wide array of commercially available nitrogen-fixing inoculants. Many of these products have demonstrated their potential to increase crop productivity and reduce chemical fertilizer dependence. In Brazil, e.g., BNF inoculants helped save over USD 2.5 billion in nitrogen fertilizer costs by 2002 (Bruno et al., 2003), while also enhancing soybean production and elevating the country to the position of the world's second-largest soybean producer, behind only the United States. In the U.S., BNF is estimated to supply between 23%

Table 6.2 Historical developments in BNF research

Date	Events	References
1886	Hellriegel and Wilfarth demonstrated the ability of legumes to convert N_2	Santi et al. (2013)
1888	First *rhizobia* were isolated from nodules	Winogradsky (1895)
1893	Isolation of *Clostridium pasteurianum*	–
1836	Identification of nitrogen as a nutrient for plants	Smil (2001)
1913	Carl Bosch performed Haber's ammonia synthesis on an industrial scale	–
1946	Second commercial inoculant	Timonin (1949)
1972	Isolation of *Spirillum* sp. and demonstration of their nitrogenase activity	Von et al. (1975)
1984	Nitrogen fixation in Methanogens	Belay (1984c)
2011	The European Nitrogen Assessment provides the first integrated and comprehensive look at N use in Europe	Parnell et al. (2016)

and 65% of the nitrogen required by soybean crops. Pastor-Bueis et al. (2019), working in Spain, found that inoculating common beans with *Rhizobium leguminosarum* bv phaseoli LCS0306 using a perlite-biochar carrier produced a yield of 3640 kg/ha—substantially higher than the 3165 kg/ha obtained with synthetic nitrogen fertilizer. In Ghana, Ulzen et al. (2016) compared urea with two commercial inoculants (Biofix and Legumefix) on soybean and cowpea and found the inoculants superior. They doubled nodule dry weight, increased nodule counts by 90%–118%, and boosted grain yields by 12%–19%. Similar outcomes were seen in northern Nigeria, where Ronner et al. (2016) reported a 447 kg/ha increase in soybean yield following rhizobial inoculation. Thuita et al. (2018) also observed that combining Legumefix or Biofix with Sympal, a nutrient formulation compatible with inoculants, could double yields from 2000 to 4000 kg/ha. In low-fertility soils, supplementing with vermicompost alongside Sympal and Legume fix led to even greater improvements.

In Zambia, the economic benefits of rhizobial inoculant use were considerable, amounting to more than USD 23 million over 8 years. In Germany, a refined bacterial inoculant known as "Nitragin" was recently tested on soybeans and other soy products, resulting in three-fold to four-fold yield increases along with elevated protein content in roots and foliage.

Numerous nonleguminous crops, particularly cereals, have evolved mechanisms to alleviate nitrogen deficiency by partnering with nitrogen-fixing (diazotrophic) microorganisms. Some of these microbes have been developed into commercial inoculants for agricultural use. Mexico was among the first to introduce maize seeds treated with *Azospirillum*, with Argentina soon adopting similar practices. Long-term field studies in the Sierra Mixe region of Oaxaca, Mexico, conducted over 5 years using ^15 N natural abundance and enrichment techniques, revealed that biological nitrogen fixation contributed between 29% and 82% of the nitrogen taken up by maize plants.

In Egypt, El-Sayed et al. (2000) reported notable improvements in barley yields following the application of Microbin and Azottein biofertilizers that include nitrogen-fixing and phosphorus-solubilizing bacteria. Yield increases ranged from 24.8% to 27.2% during the first season and from 18.4% to 22.0% in the second, performing similarly to synthetic nitrogen fertilizers. Additional support comes from Rose et al. (2014), who showed that the commercial biofertilizer "BioGro" Table 6.2 could replace 23% to 52% of synthetic nitrogen inputs in rice systems across Southeast Asia without compromising yields. On average, biological nitrogen fixation contributes approximately 30 kg of nitrogen per hectare each year in rice cultivation. In sugarcane farming, Serna-Cock et al. (2011) found that applying a microbial mixture containing *Azospirillum brasilense*, *Azotobacter chroococcum*, and *Trichoderma lignorum*to the CC 934418 variety could satisfy up to 60% of its nitrogen demand. Likewise, Antunes et al. (2019) documented yield increases of between 18% and 57.31% in the RB92579 sugarcane variety after inoculation with *Herbaspirillum seropedicae*, *Pseudomonas* spp., and *Bacillus megaterium*. However, it's worth noting that the nitrogen-fixing performance of these diazotrophs tends to diminish in environments with high levels of synthetic nitrogen fertilizer application.

6.5 Effectiveness of BNF in Different Crops or Regions

Although synthetic fertilizers are often emphasized in discussions about nitrogen inputs, biological nitrogen fixation (BNF) remains the largest natural source of reactive nitrogen globally (Fowler et al., 2015). Broadly, three main ecosystems contribute significantly to BNF: natural (unmanaged) terrestrial systems, aquatic and marine environments, and agricultural lands. Global BNF estimates vary widely across these ecosystems because of major uncertainties and the difficulty of extrapolating limited or inconsistent field data to a global scale. While nitrogen transfer occurs between natural and agricultural ecosystems, this section focuses on BNF inputs in cultivated croplands.

Estimates of BNF from grain legumes have increased notably—from about 10 Tg N in the mid-1990s to 35.5 Tg N by 2018 (Herridge et al., 2022)—a rise largely linked to the expansion of legume cultivation. Soybean accounts for the majority of this contribution, representing approximately 70% of the total nitrogen fixed in 2018. However, a long-term global nitrogen budget assessment covering 1960–2010 by Ladha et al. (2020) estimated that nonsymbiotic BNF in cereal cropping systems contributed about 14.8 Tg N in 2010—distributed as 3.3 Tg for wheat, 5.6 Tg for rice, and 5.9 Tg for maize annually. When Ladha et al.'s (2016) relationships were applied to updated production data, the 2018 global estimate of nonsymbiotic BNF rose slightly to 15.7 Tg N, including 3.1 Tg for wheat, 5.7 Tg for rice, and 6.9 Tg for maize. These adjustments mainly reflect an 18.2% increase in maize harvested area and a 3.7% increase in rice, while wheat experienced a slight 1.3% decline between 2010 and 2018.

6.6 Management Considerations for Improving Biological Nitrogen Fixation

Several management practices play a critical role in regulating biological nitrogen fixation (BNF) within agricultural systems. These include nitrogen fertilizer application, the selection of plant species, genetic factors like cultivar and genotype, and cultivation strategies such as intercropping. Implementing appropriate management practices can significantly boost N_2 fixation efficiency. For example, in mixed swards, competition for soil nitrogen from perennial ryegrass (*Lolium perenne* L.) has been shown to influence nitrogen fixation in legumes such as birdsfoot trefoil (*Lotus corniculatus* L.), alfalfa, and white clover (Rasmussen et al., 2012). Different plant species vary in how much they rely on nitrogen from the atmosphere versus the soil; in a red clover–grass–forb mixture, grasses primarily depended on fixed nitrogen, while forbs sourced more nitrogen from the soil. Choosing well-matched plant species and cultivars can enhance both nitrogen fixation and interplant nitrogen transfer within agricultural systems. For instance, white clover grown alone obtained between 75% and 94% of its nitrogen from BNF, but this figure rose to 85%–97% when cultivated with ryegrass. The increased nitrogen fixation in these mixed systems is likely due to intensified competition for soil nitrogen by nonfixing plants.

Rouquette and Smith (2010) noted that BNF performance in forage legumes can vary widely based on factors such as species and cultivar selection, soil nutrient status, local climate, and environmental conditions. These diverse influences help explain the wide variability in BNF levels reported by different researchers even at the same site. For example, in grazed fields at one location, the proportion of nitrogen from BNF ranged from 12% to 96%. Moreover, the application of nitrogen fertilizers is known to suppress nitrogen fixation in legumes. One study showed that the use of nitrogen fertilizer reduced the proportion of nitrogen fixed from the atmosphere in clover from 77% to 43%.

The effectiveness of nitrogen fixation is strongly determined by the particular *Rhizobia* strain used. These microorganisms are usually highly specific to their host plants, and because of the complex and finely balanced symbiosis between nitrogen-fixing plants and their bacterial partners, any disruption to this relationship can lead to a noticeable decline in nitrogen fixation. Therefore, selecting the correct *Rhizobia* strains for inoculation is vital to achieving optimal nitrogen fixation. In mixed grass-legume pastures, biological nitrogen fixation is commonly limited by three main factors: low total forage production, a small percentage of legumes in the mix, and a reduced reliance of legumes on atmospheric nitrogen sources. Maintaining a robust legume population is essential for effective nitrogen fixation, yet this can be difficult due to challenges such as livestock preferentially grazing on legumes, inadequate soil conditions, and pest or disease pressures. Nonetheless, maintaining appropriate seed mass ratios and following best practices in grazing and hay management can help sustain legume presence in the sward (Adjesiwor et al., 2017).

6.7 Development of New Technologies to Enhance BNF

Recent progress in science and technology is greatly advancing the potential of biological nitrogen fixation (BNF). Biotechnology firms are now engineering microorganisms to improve nitrogen fixation in nonleguminous crops such as wheat, rice, and maize. For example, nitrogen-fixing bacteria are being genetically modified to enhance their efficiency and ability to colonize plant roots. These "engineered microbes" have the potential to significantly decrease dependence on chemical fertilizers, thereby cutting agricultural costs and reducing environmental pollution. Additionally, genetic engineering is being applied to enable nonleguminous crops to form symbiotic relationships with nitrogen-fixing bacteria. By introducing genes from legumes into crops like maize, researchers aim to develop plants capable of directly hosting nitrogen-fixing microorganisms. Advancements in precision agriculture are also transforming nitrogen management by increasing fertilizer efficiency and promoting sustainability. The use of Internet of Things (IoT) technologies and smart soil sensors allows continuous, real-time monitoring of nitrogen levels in the soil. This provides farmers with precise information for targeted fertilizer applications, minimizing waste and preventing excess nutrient runoff into nearby ecosystems. Artificial intelligence (AI) further supports these practices by integrating data from sensors, satellite images, and weather forecasts to create optimized fertilizer application models. These AI-based systems guide farmers on when, where, and how much nitrogen to apply, helping to maximize yields while minimizing environmental harm (Chen et al., 2020). Nanotechnology is also emerging as a key innovation in efficient nitrogen delivery. Nano-fertilizers—composed of nitrogen compounds encapsulated in nanomaterials—release nutrients slowly in synchronization with plant needs, reducing nitrogen losses through leaching and volatilization (Karimi et al., 2015). This slow-release mechanism enhances nitrogen use efficiency, decreases fertilizer requirements, and is particularly beneficial for resource-limited farmers or areas with restricted fertilizer access. Furthermore, nanofertilizers help mitigate environmental issues like eutrophication and reduce the energy consumption associated with fertilizer production and use (Liu et al., 2016). In addition, major breakthroughs in synthetic biology and gene-editing technologies such as CRISPR are reshaping nitrogen fixation research. Scientists are developing synthetic nitrogenase pathways that can be integrated directly into plants, allowing them to fix atmospheric nitrogen autonomously—without the need for bacterial symbionts. CRISPR is also being used to modify genes in crops to enhance nitrogen absorption and utilization. By improving root systems and metabolic processes, these genetically optimized plants can make more efficient use of nitrogen, reducing dependency on synthetic fertilizers. Meanwhile, advancements in renewable energy are offering sustainable alternatives to the traditional, energy-intensive Haber–Bosch process used in fertilizer production (Lassaletta et al., 2014).

6.8 Pathway to Large-Scale Integration of Nitrogen-Fixing Biofertilizers

Modern agriculture continues to rely heavily on synthetic nitrogen fertilizers (Burghardt, 2019). To fully realize the benefits of biological nitrogen fixation (BNF) in improving soil fertility and crop performance, researchers and stakeholders must work together within a coordinated, integrated framework to address key challenges. The following outlines the main research and development priorities needed to promote the large-scale adoption of nitrogen-fixing biofertilizers (Fig. 6.2):

- Enhance our understanding of the biological and molecular mechanisms driving nitrogen fixation, including the selection of microbial partners and the stability of their associations with host plants.
- Broaden the host range of nitrogen-fixing bacteria by reducing symbiotic specificity, informed by deeper knowledge of the genetic and molecular controls governing symbiosis.
- Conduct rigorous field evaluations of inoculant products across various environmental settings and soil types before introducing them commercially.
- Focus on isolating and developing highly effective and competitive strains, particularly mixed-strain consortia, which may outperform single-strain formulations due to complementary traits that improve resilience and function in field conditions.

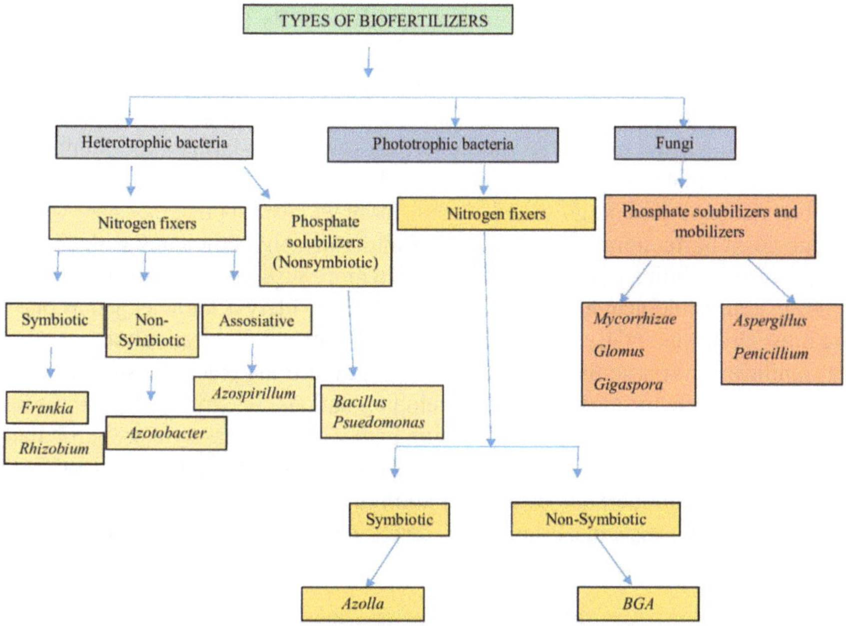

Fig. 6.2 Pathway to large-scale integration of nitrogen-fixing biofertilizers

- Upgrade commercial inoculant standards, emphasizing microbial strain purity, adequate population density, and formulations that protect viability during storage. Strain performance should be a critical measure of product quality.
- Reinforce regulatory oversight, requiring formal product registration and independent third-party assessments to ensure consistent quality and reliability of inoculants.
- Build farmers' knowledge and skills through training initiatives and support networks that disseminate best practices, protocols, and up-to-date information on BNF technologies.
- Ensure timely and widespread availability of quality inoculants to end users, along with the development of innovative and user-friendly inoculation techniques.
- Integrate mycorrhizal fungi with nitrogen-fixing systems, as research has shown that arbuscular mycorrhizae enhance nitrogen transfer across plant species, enabling nonfixing crops to benefit from nearby nitrogen-fixing plants.
- Identify optimal soils for inoculation, prioritizing areas with slightly acidic to neutral pH (5.5–7.2) and low nitrogen content—conditions conducive to nodulation and microbial activity. Mapping nitrogen availability alongside soil pH data have highlighted numerous promising zones, particularly across Africa, for extensive use of nitrogen-fixing biofertilizers.

6.9 Conclusions

The development of advanced nitrogen fixation technologies represents a crucial breakthrough in the evolution of modern agriculture. By lessening dependence on synthetic fertilizers, these innovations help confront major global challenges such as food insecurity, climate change, and ecosystem degradation. Through cutting-edge approaches—including engineered microorganisms, precision farming methods, nanotechnology, and renewable energy–driven solutions—farmers are gaining sustainable tools that boost crop productivity while preserving environmental health. As these technologies progress and become more widely adopted, they are expected to transform agriculture into a system that integrates high efficiency with environmental responsibility, paving the way toward a more sustainable and resilient global food system. Biological nitrogen fixation (BNF) is widely acknowledged as one of the most environmentally sustainable sources of nitrogen for agricultural systems. Its significance is particularly evident in low-input farming practices, such as forage cultivation, and in the agriculture of developing nations. Integrating BNF traits into crop breeding programs offers a promising pathway to increase yields, particularly in regions where access to synthetic nitrogen fertilizers is limited. While challenges remain for the commercial-scale application of nitrogen-fixing biofertilizers, this technology holds substantial promise for supporting more sustainable and ecologically balanced farming systems.

References

Adjesiwor, A. T., Islam, M. A., Zheljazkov, V. D., Ritten, J. P., & Garcia y Garcia, A. (2017). Grass-legume seed mass ratios and nitrogen rates affect forage accumulation, nutritive value, and profitability. *Crop Science*, *57*(1), 1–13. https://doi.org/10.2135/cropsci2016.03.0192.

Alori, E. T., & Babalola, O. O. (2018). Microbial inoculants for improving crop quality and human health in Africa. *Frontiers in Microbiology*, *9*, 2213–2224. https://doi.org/10.3389/fmicb.2018.02213.

Amack, C., & Antunes, S. (2020). CaMV35S promoter–a plant biology and biotechnology workhorse in the era of synthetic biology. *Current Plant Biology*, *24*, 100179.

Antunes, J. E. L., de Freitas, A. D. S., Oliveira, L. M. S., de Lyra, M. D. C. C. P., Fonseca, M. A. C., Santos, C. E. R. S., Oliveira, J. P., de Araújo, A. S. F., & Figueiredo, M. V. B. (2019). Sugarcane inoculated with endophytic diazotrophic bacteria: Effects on yield, biological nitrogen fixation and industrial characteristics. *Anais da Academia Brasileira de Ciências*, *91*(1), e20180990. https://doi.org/10.1590/0001-3765201920180990.

Belay, N., Sparling, R., & Daniels, L. (1984c). Dinitrogen fixation by a thermophilic methanogenic bacterium. Nature, 312(5990), 286–288. https://doi.org/10.1038/312286a0.

Betancourt, M. (2024). Guano and the rise of the American empire. *Socius Sociological Research: A Dynamic World*, *10*, 1–11.

Bruno, J. R. A., Boddey, R. M., & Urquiaga, S. (2003). The success of BNF in soybean in Brazil. In In In plant and soil (vol . 252pp. 1–9). Kluwer Academic Publishers.

Bullard, G. K., Roughley, R. J., & Pulsford, D. J. (2005). The legume inoculant industry and inoculant quality control in Australia: 1953–2003. *Australian Journal of Experimental Agriculture*, *45*(2), 127–134. https://doi.org/10.1071/EA03159.

Burghardt, L. T. (2019). Evolving together, evolving apart: Measuring the fitness of rhizobial bacteria in and out of symbiosis with leguminous plants. *New Phytologist. Advance Online Publication*, https://doi.org/10.1111/nph.16045. , *228*(1), 28–34.

Carl, R., Stevan, S., Cory, C., Menge, S., & Nina, W. (2025). A global dataset of terrestrial biological nitrogen fixation. *Scientific Data*, *12*, 1362.

Cassim, B. M. A. R., Lisboa, I. P., Besen, M. R., Otto, R., Cantarella, H., Inoue, T. T., & Batista, M. A. (2024). Nitrogen: From discovery, plant assimilation, sustainable usage to current enhanced efficiency fertilizers technologies—A review. *Revista Brasileira de Ciência do Solo*, *48*, e0230–037.

Chen, C. C., Dai, L., Ma, L., & Guo, R. T. (2020). Enzymatic degradation of plant biomass and synthetic polymers. *Nature Reviews Chemistry*, *4*(3), 114–126.

Döbereiner, J. (1966). *Azotobacter paspali* sp. nov.: Uma bactériafixadora de nitrogênionarizosfera de Paspalum. *Pesquisa Agropecuária Brasileira*, 1, 357–365.

El-Sayed, A. A., Elenein, R. A., Shalaby, E. E., Shalan, M. A., & Said, M. A. (2000). Response of barley to biofertilizer with N and P application under newly reclaimed areas in Egypt. In Proceedings of the 3rd International Crop Science Congress (ICSC) (pp. 17–22). Hamburg, Germany.

Fowler, N., Cape, J., Dore, M., & Vieno (2015). Effects of global change during the 21st century on the nitrogen cycle. *Atmospheric Chemistry and Physics*, *15*, 13849–13893.

Franche, C., Lindström, K., & Elmerich, C. (2009). Nitrogen-fixing bacteria associated with leguminous and non-leguminous plants. *Plant and Soil*, *321*(1–2), 35–59. https://doi.org/10.1007/s11104-008-9833-8.

Galloway, J. N., Leach, A. M., Bleeker, A., & Erisman, J. W. (2020). A chronology of human understanding of the nitrogen cycle. Philosophical Transactions of the Royal Society B: Biological Sciences, 368(1621), 20130120. https://doi.org/10.1098/rstb.2013.0120.

Gurthie, F. B. (1896). Inoculation of the soil for leguminous crops. *Agricultural Gazette of New South Wales*, *7*, 690–694.

Herridge, D. F., Peoples, M. B., & Boddey, R. M. (2008). Global inputs of biological nitrogen fixation in agricultural systems. *Plant and Soil*, *311*(1–2), 1–18. https://doi.org/10.1007/s11104-008-9668-3.

Herridge, D. F., Giller, K. E., Jensen, E. S., & Peoples, M. B. (2022). Quantifying country-to-global scale nitrogen fixation for grain legumes II. Coefficients, templates and estimates for soybean, groundnut and the pulses. *Plant Soil*, https://doi.org/10.1007/s11104-021-05166-7. (in press).

Howarth, R. W. (2008). Coastal nitrogen pollution: A review of sources and trends globally and regionally. Harmful Algae, 8(1), 14–20. https://doi.org/10.1016/j.hal.2008.08.015.

Johnson, N. C., Wilson, G. W. T., Bowker, M. A., Wilson, J. A., & Miller, R. M. (2010). Resource limitation is a driver of local adaptation in mycorrhizal symbioses. Proceedings of the National Academy of Sciences, 107(5), 2093–2098. https://doi.org/10.1073/pnas.0906710107.

Johnson, B. (2022). The state of ammonia synthesis at the turn of the twentieth century: The arena for discovery. In *Making ammonia* (pp. 77–89). Springer International Publishing. ISBN 978-3-030-85531-4.

Karimi, R., Bahramnejad, B., & Ahmadzadeh, M. (2015). The role of plant growth-promoting rhizobacteria in sustainable agriculture and biological nitrogen fixation: A review. International Journal of Farming and Allied Sciences, 4(7), 598–603.

Könneke, M., Bernhard, A. E., de la Torre, J. R., Walker, C. B., Waterbury, J. B., & Stahl, D. A. (2005). Isolation of an autotrophic ammonia-oxidizing marine archaeon. Nature, 437(7058), 543–546. https://doi.org/10.1038/nature03911.

Ladha, J. K., Tirol-Padre, A., Reddy, C. K., Cassman, K. G., Verma, S., Powlson, D. S., van Kessel, C., Richter, D. deB, Chakraborty, D., & Pathak, H. (2016). Global nitrogen budgets in cereals: A 50-year assessment for maize, rice and wheat production systems. Scientific Reports, 6, 19355. https://doi.org/10.1038/srep19355.

Ladha, L., Jat, C., & Stirling, D. (2020). Goals in agriculture: The crucial role of nitrogen in cereal-based systems. *Field Crop Research*, *10*, 39–116.

Lassaletta, L., Billen, G., Grizzetti, B., Anglade, J., & Garnier, J. (2014). 50 year trends in nitrogen use efficiency of world cropping systems: The relationship between yield and nitrogen input to cropland. *Environmental Research Letters*, *9*(10), 105011. https://doi.org/10.1088/1748-9326/9/10/105011.

Liu, W., & Stewart, Jr, C. N. (2016). Plant synthetic promoters and transcription factors. *Current Opinion in Biotechnology*, *37*, 3644.

Parnell, J. J., Berka, R., Young, H. A., Sturino, J. M., Kang, Y., Barnhart, D. M., & DiLeo, M. V. (2016). From the lab to the farm: An industrial perspective of plant beneficial microorganisms. Frontiers in Plant Science, 7, 1110. https://doi.org/10.3389/fpls.2016.01110.

Pastor-Bueis, R., Sánchez-Cañizares, C., James, E. K., & González-Andrés, F. (2019). Formulation of a highly effective inoculant for common bean based on an autochthonous elite strain of *Rhizobium leguminosarum* bv. phaseoli, and genomic-based insights into its agronomic performance. *Frontiers in Microbiology*, *10*, 2724–2731. https://doi.org/10.3389/fmicb.2019.02724.

Rasmussen, J., Soegaard, K., Pirhofer-Walzl, K., & Eriksen, J. (2012). N2-fixation and residual N effect of four legume species and four companion grass species. *European Journal of Agronomy*, *36*(1), 66–74. https://doi.org/10.1016/j.eja.2011.09.003.

Ronner, E., Franke, A., Vanlauwe, B., Dianda, M., Edeh, E., Ukem, B., & Giller, K. (2016). Understanding variability in soybean yield and response to P-fertilizer and rhizobium inoculants on farmers' fields in northern Nigeria. *Field Crops Research*, *186*, 133–145. https://doi.org/10.1016/j.fcr.2015.10.023.

Rose, M. T., Phuong, T. L., Nhan, D. K., Cong, P. T., Hien, N. T., & Kennedy, I. R. (2014). Up to 52% N fertilizer replaced by biofertilizer in lowland rice via farmer participatory research. *Agronomy for Sustainable Development*, *34*(4), 857–868. https://doi.org/10.1007/s13593-014-0210-0.

Rouquette, F., & Smith, G. (2010). Effects of biological nitrogen fixation and nutrient cycling on stocking strategies for cow-calf and stocker programs. The professional animal. *Scientist, 26*(2), 131–141. https://doi.org/10.15232/S1080-7446.

Santi, C., Bogusz, D., & Franche, C. (2013). Biological nitrogen fixation in non-legume plants. Annals of Botany, 111(5), 743–767. https://doi.org/10.1093/aob/mct048.

Serna-Cock, L., Arias-García, C., & Valencia-Hernandez, L. J. (2011). Effect of biofertilization on the growth of potted sugarcane plants (*Saccharum officinarum*). *Revista de Biología Agroindustrial, 9*(2), 85–95.

Singh, L., V. K., Singh, M., Singh. (2019). Sustainable agricultural practices using beneficial fungi under changing climate scenario. In *Climate change and agricultural ecosystems* (pp. 25–41). Elsevier. https://doi.org/10.1016/B978-0-12-816483-9.00002-5.

Smil, V. (2001). Enriching the Earth: Fritz Haber, Carl Bosch, and the transformation of world food production. Cambridge, MA: MIT Press.

Sylvia, D. M., Fuhrmann, J. J., Hartel, P. G., & Zuberer, D. A. (2005). Principles and applications of soil microbiology (2nd ed., p. 640.). Pearson Prentice Hall.

Taulé, C., Mareque, C., Barlocco, C., Hackembruch, F., Reis, V. M., Sicardi, M., & Battistoni, F. (2012). The contribution of nitrogen fixation to sugarcane (Saccharum officinarum L.), and the identification and characterization of part of the associated diazotrophic bacterial community. Plant and Soil, 356(1–2), 35–49. https://doi.org/10.1007/s11104-011-1023-4.

Thilakarathna, M. S., McElroy, M. S., Chapagain, T., Papadopoulos, Y. A., & Raizada, M. N. (2016). Belowground nitrogen transfer from legumes to non-legumes under managed herbaceous cropping systems: A review. *Agronomy for Sustainable Development, 36*(58), 1–16. https://doi.org/10.1007/s13593-016-0396-4.

Thuita, M., Vanlauwe, B., Mutegi, E., & Masso, C. (2018). Reducing spatial variability of soybean response to rhizobia inoculants in farms of variable soil fertility in Siaya county of west Kenya. *Biology and Fertility of Soils, 54*(2), 261–272. https://doi.org/10.1016/j.agee.2018.01.007.

Timonin, M. I. (1949). The interaction of higher plants and soil microorganisms. I. Microbial population of the rhizosphere of seedlings of certain cultivated plants. Canadian Journal of Research, 27(6), 298–341.

Ulzen, J., Abaidoo, R. C., Mensah, N. E., Masso, C., & AbdelGadir, A. H. (2016). *Bradyrhizobium* inoculants enhance grain yields of soybean and cowpea in Northern Ghana. *Frontiers in Plant Science, 7*, 1770–1778. https://doi.org/10.3389/fpls.2016.01770.

Unkovich, M., Herridge, D., Peoples, M., Cadisch, G., Boddey, B., Giller, K., Alves, B., & Chalk, P. (2008). Measuring plant-associated nitrogen fixation in agricultural systems. In *Australian centre for international agricultural research* (p. (p. 258.). ACIAR Monograph No. 136.

Varley, J. B., Wang, Y., Chan, K., Studt, F., & Nørskov, J. K. (2015). Mechanistic insights into nitrogen fixation by nitrogenase enzymes. *Physical Chemistry Chemical Physics, 17*(45), 29541–29547. https://doi.org/10.1039/C5CP04533H.

Vitousek, P. M., Menge, D. N. L., Reed, S. C., & Cleveland, C. C. (2013). Biological nitrogen fixation: Rates, patterns and ecological controls in terrestrial ecosystems. *Philosophical Transactions of the Royal Society B: Biological Sciences, 368*(1621), 20130119–134. https://doi.org/10.1098/rstb.2013.0119.

Von Bülow, J. F. W., & Döbereiner, J. (1975). Potential for nitrogen fixation in maize genotypes in Brazil. Proceedings of the National Academy of Sciences of the United States of America, 72(6), 2389–2393. https://doi.org/10.1073/pnas.72.6.2389.

Vorholt, A., Vogel, C., Carlström, I., & Müller, B. (2017). Establishing causality: Opportunities of synthetic communities for plant microbiome research. *Cell Host & Microbe, 22*, 142–155.

Winogradsky, S. (1895). Recherchessurl'assimilation de l'azotelibre de l'atmosphère par les microbes. *Archives des Sciences Biologiques, 3*, 297–352.

Revitalizing Soil Ecosystems with Diazotrophic Plant Growth Promoting Microorganisms

7

Aishna Srivastava, Vikas Patidar and Bhawna Babal

Abstract

The structure, fertility, and long-term sustainability of soil are all shaped by soil microbes, which are essential designers of terrestrial ecosystems. Among these, diazotrophic plant growth-promoting microorganisms (DPGPMs) are an important functional category that, via a variety of biochemical and ecological processes, not only fix atmospheric nitrogen but also support plant growth, nutrient bioavailability, and ecosystem restoration. The functions of soil microorganisms are covered in detail in this chapter, starting with their impact on soil genesis, aggregation, organic matter conservation, and the biogeochemical cycling of carbon, nitrogen, phosphorus, sulfur, and iron. It emphasizes the plant–microbe interactions, functional redundancy, and microbial diversity that support soil ecosystem services. Determining the importance, categorization, and variety of diazotrophic microorganisms—including symbiotic, associative, free-living, and newly developing oceanic and non-canonical nitrogen-fixing systems—is a primary concern. The chapter explains nitrogenase's catalytic function, isoform variety, oxygen sensitivity, and protective adaptations by combining structural, mechanistic, and evolutionary considerations. Through their functions in nutrient solubilization, phytohormone production, induced systemic resistance, siderophore synthesis, and pathogen suppression, DPGPMs' wider contributions to plant growth are investigated. The chapter also examines how diazotrophs improve soil health, including humus production, carbon

A. Srivastava (✉) · V. Patidar
ICAR- Indian Agriculture Research Institute, Karnal Hub, Haryana, India
e-mail: aishnapathology@gmail.com

B. Babal
ICAR- Central Soil Salinity Research Institute, Karnal, Haryana, India

R. K. Singh and P. Singh (eds.), *Diazotrophic Plant Growth-Promoting Microorganisms and Green Energy – Volume I*, https://doi.org/10.1007/978-981-95-9682-9_7

sequestration, microbial diversity, and nutrient cycling. The use of biofertilizers, integration with farming systems, decrease of chemical inputs, and successful field implementations are highlighted in the discussion of practical applications in sustainable agriculture. Along with competitive dynamics and rhizosphere impacts, interactions with other soil biota, including mycorrhizal fungi and plant growth-promoting rhizobacteria (PGPR), are described. Notwithstanding their potential, obstacles, such as regulatory impediments, formulation and shelf-life limitations, variable field performance, and environmental restrictions are examined. The chapter ends with new developments and potential paths in formulation science, microbial biotechnology, genomics, and climate-smart agriculture. All things considered, this synthesis emphasizes how important diazotrophic PGPMs are to the revitalization of soil ecosystems.

Keyword

Diazotrophs · Soil fertility · PGPR · Biogeochemical cycle ·

7.1 Introduction

Soil is the life-supporting system for plants. A plant's health is affected directly by the quality of the soil in which it is grown, making soil health a cornerstone of agricultural productivity and food security. Soil health is defined through its physical, chemical, and biological attributes, all of which interact to regulate plant growth, resilience, and ecosystem functioning (Maciel-Rodríguez et al., 2025). These attributes determine essential processes, such as water retention, nutrient cycling, root penetration, and microbial diversity, which collectively sustain crop productivity and long-term soil fertility. Globally, soil degradation and constraints—such as compaction, salinity, sodicity, acidity, nutrient deficiency, erosion, and soil-borne diseases—pose serious challenges to sustainable agriculture. The Food and Agriculture Organization (FAO, 2025) estimates that nearly one-third of the world's land is already degraded, affecting the livelihoods of more than 1.5 billion people (Li et al., 2024). Such constraints limit nutrient and water availability, reduce soil structure stability, and lead to losses in productivity. Addressing these issues is critical in the context of population growth, climate change, and rising food demands.

Equally important is the biological dimension of soil health. Soil biodiversity, especially in the rhizosphere, plays a crucial role in supporting crop nutrition and resilience. Rhizosphere microorganisms, such as phosphate-solubilizing and nitrogen-fixing bacteria, enhance nutrient availability, promote root growth, and help plants withstand stress (Maciel-Rodríguez et al., 2025). Crop diversity, through practices like rotation and intercropping, has also been shown to promote microbial diversity, suppress pathogens, and strengthen soil's disease resistance capacity (Maciel-Rodríguez et al., 2025). This interdependence between crops and soil organisms underlines the necessity of managing soil health as an integrated biological system rather than only as a chemical reservoir. Socioeconomic and management practices further shape soil health outcomes. Overuse of fertilizers and pesticides, intensive tillage, and mismanagement of land exacerbate erosion and nutrient imbalances, undermining both productivity and ecological stability (Maciel-Rodríguez et al., 2025). On the other hand, conservation tillage, organic farming, and precision agriculture technologies offer promising pathways for improving soil fertility, enhancing microbial activity, and reducing environmental impacts (Maciel-Rodríguez et al., 2025). Advanced monitoring approaches, such as remote sensing, soil testing, and bioinformatics tools, are increasingly used to assess soil health and guide targeted interventions (Li et al., 2024). Taken together, these perspectives highlight soil health as the foundation of agricultural productivity. Strengthening this linkage requires integrated strategies that combine sustainable soil management, ecological approaches, and technological innovations. Only by restoring and maintaining soil health can agricultural systems remain resilient and productive under future global challenges.

7.2 Role of Microorganisms in Soil Ecosystems

7.2.1 Microorganisms in Soil Genesis and Formation

Soil formation begins with the colonization of bare rock surfaces by pioneer microorganisms, such as bacteria, algae, and lichens. These organisms accelerate rock weathering by secreting organic acids and enzymes, which mobilize minerals and initiate Microbial "wear down" of rocks is now considered a biologically driven process, with fungi and chemolithotrophic bacteria mobilizing substantial amounts of CO_2 in the process. These early microbial colonizers stabilize substrates through the development of biological soil crusts that prevent erosion (Kaviya et al., 2019).

7.2.2 Soil Aggregation and Structure

Microorganisms play a direct role in soil aggregation, a key determinant of soil physical quality. Bacteria and fungi excrete polysaccharides, glomalin, and lipids that bind soil particles, enhancing stability and porosity (Sharma & Bhakri, 2019). Fungal hyphae, due to their mechanical networks, further stabilize aggregates and anchor soil particles (Davids et al., 2017). These processes contribute to soil structure, which regulates aeration, water retention, and microbial habitat distribution (Hinsinger, 2013).

7.2.3 Microbial Roles in Biogeochemical Cycles

Soil microorganisms regulate all major biogeochemical cycles: Carbon cycling: Decomposition of litter and root exudates is mediated by microbial enzymes such as cellulases and laccases, making microbial biomass both a sink and a source of carbon while humification stabilizes organic matter (Kaviya et al., 2019). Mycorrhizal fungi contribute a substantial portion of plant photo-assimilates to soil carbon pools (Deonalli et al., 2017).

Nitrogen cycling: Diazotrophs fix atmospheric N_2, while ammonifiers, nitrifiers, and denitrifiers mediate nitrogen transformations that determine soil fertility and greenhouse gas emissions (Davids et al., 2017). Phosphorus cycling: Phosphate-solubilizing bacteria and mycorrhizal fungi mobilize phosphorus from insoluble forms, enhancing plant nutrition (Hinsinger, 2013). Sulfur and iron transformations: Specialized microbial groups regulate sulfur and iron oxidation–reduction processes that contribute to soil fertility and redox balance (Deonalli et al., 2017).

7.2.4 Soil Microbial Biomass and Organic Matter Conservation

Soil microbial biomass constitutes the "living engine" of nutrient cycling. In tropical systems, where high temperatures and rainfall accelerate decomposition, microbial activity regulates whether organic inputs are mineralized or stabilized as humus (Kaviya et al., 2019). Decline in soil organic matter contributes to CO_2 emissions and reduced fertility, undermining both productivity and climate stability (Lopez & Bacilio, 2020). Microbial immobilization provides short-term buffering of nutrients, while microbial residues contribute to long-term carbon sequestration (Davids et al., 2017).

7.2.5 Microbial Diversity and Functional Redundancy

Soil is one of the most diverse microbial habitats on Earth because of microscale heterogeneity in aeration, nutrient supply, and moisture (Deonalli et al., 2017). This diversity ensures functional redundancy, allowing decomposition and nutrient cycling to continue even under stress conditions (Sharma & Bhakri, 2019). Exocellular enzymes stabilized on mineral surfaces extend microbial influence beyond living cells and help maintain decomposition processes (Davids et al., 2017).

7.2.6 Plant–Microbe Interactions in the Rhizosphere

The rhizosphere is a hotspot of microbial activity supported by root exudates that fuel microbial populations (Hinsinger, 2013). Mycorrhizal fungi expand the root surface area, improving nutrient uptake and tolerance to stresses, such as drought, salinity, and heavy metals (Kaviya et al., 2019). Plant growth-promoting rhizobacteria enhance root development through phytohormone production, nutrient solubilization, and pathogen suppression (Sharma & Bhakri, 2019). Disease-suppressive soils, characterized by diverse and active microbial populations, highlight the ecological significance of plant–microbe interactions (Davids et al., 2017).

7.2.7 Environmental and Ecological Services

Microorganisms regulate greenhouse gas emissions (CO_2, CH_4, N_2O) from soils, with processes, such as methane oxidation, being especially sensitive to changes in land management and fertilizer inputs (Lopez & Bacilio, 2020). They also contribute to pollutant degradation, heavy metal detoxification, and the formation of biological soil crusts that reduce erosion (Deonalli et al., 2017). These microbial functions connect soil biodiversity not only to agricultural productivity but also to climate regulation and ecosystem sustainability (Kaviya et al., 2019) (Fig. 7.1).

7.3 Objectives and Scope of the Chapter

The objective of this chapter is to provide a comprehensive understanding of diazotrophic plant growth-promoting microorganisms (DPGPMs) and to emphasize their significance in maintaining soil health and enhancing crop productivity. The chapter explores how DPGPMs, through their dual capacity to fix atmospheric nitrogen and to promote plant growth via multiple mechanisms, serve as natural allies in sustainable agriculture (Upadhyay et al., 2023; Prasad et al., 2019). Particular attention is given to the diversity of these microorganisms, their ecological niches, and their functional interactions with plants, which together define their role in agroecosystem dynamics (Altaf, 2021; Welbaum et al., 2004).

In this context, the scope of the chapter extends beyond a descriptive account of nitrogen fixation. It encompasses the wider spectrum of growth-promoting attributes of DPGPMs, including the synthesis of phytohormones, solubilization of essential minerals, siderophore production, and the modulation of plant stress responses (Gouda et al., 2018; Panchami et al., 2020). By integrating findings from controlled experiments and field trials, the chapter illustrates the capacity of DPGPMs to enhance nutrient use efficiency, improve seedling vigor, and increase crop biomass, thereby offering an eco-friendly alternative to chemical fertilizers (Li, 2024; Chandran et al., 2021).

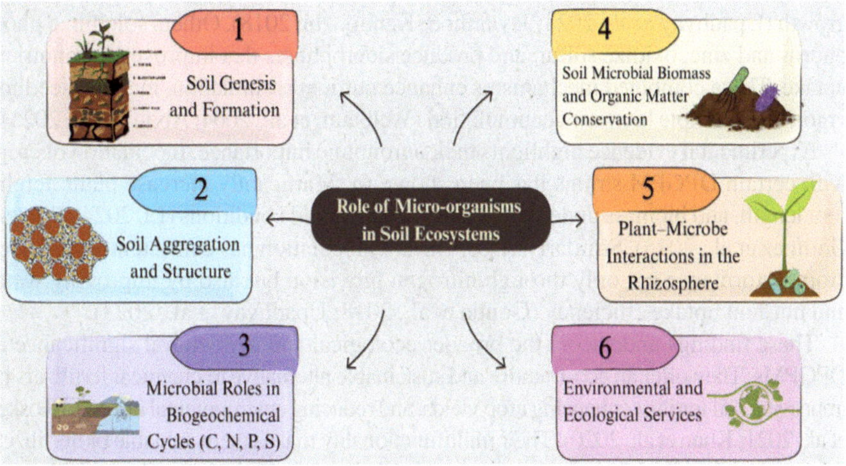

Fig. 7.1 Role of microorganisms in soil ecosystem

At the same time, the chapter addresses the broader ecological and agronomic implications of utilizing DPGPMs. It situates their use within the global need for sustainable intensification of agriculture, considering the challenges of soil degradation, climate change, and environmental pollution (Koskey et al., 2021; Abdul et al., 2025). In doing so, it provides a framework for understanding both the opportunities and limitations associated with the application of DPGPMs, and points to future research directions and technological innovations that can strengthen their role in sustainable crop production systems (Khan et al., 2020; Elhady & Hirt, 2025).

7.4 Definition and Significance of Diazotrophic Plant Growth-Promoting Microorganisms

Diazotrophic plant growth-promoting microorganisms (DPGPMs) are a group of free-living or associative bacteria capable of fixing atmospheric nitrogen and simultaneously enhancing plant growth through multiple mechanisms. By definition, diazotrophs harbor the nitrogenase enzyme complex encoded by *nif* genes, enabling the conversion of inert atmospheric dinitrogen (N_2) into ammonia (NH_3), which can be assimilated by plants (Altaf, 2021; Li, 2024). Unlike classical symbiotic nitrogen fixers, many diazotrophs—including genera like *Azospirillum*, *Herbaspirillum*, *Burkholderia*, *Serratia*, and *Pseudomonas*—exist as free-living or associative microorganisms and colonize the rhizosphere, root surfaces, or even internal tissues of non-leguminous plants (Prasad et al., 2019; Gouda et al., 2018).

The significance of DPGPMs lies in their dual role: providing fixed nitrogen and exerting additional plant growth-promoting (PGP) functions. Beyond nitrogen fixation, they produce phytohormones, such as indole-3-acetic acid (IAA), gibberellins, and cytokinins, which stimulate root elongation and branching (Panchami et al., 2020; Chandran et al., 2021). Many also express 1-aminocyclopropane-1-carboxylate (ACC) deaminase, which lowers plant ethylene levels, thereby reducing stress inhibition of root growth (Upadhyay et al., 2023; Jeyanthi & Kanimozhi, 2018). Others solubilize phosphorus and zinc, oxidize sulfur, and produce siderophores that improve micronutrient uptake. These combined mechanisms enhance nutrient assimilation, increase seedling vigor, and promote biomass accumulation (Welbaum et al., 2004; Abdul et al., 2025).

Experimental evidence highlights their agronomic importance. Inoculation of crops with certain DPGPM strains has been shown to significantly increase plant height, root length, and biomass under both controlled and field conditions (Li, 2024; Gómez-Godínez et al., 2025). Similarly, *Azospirillum* inoculation has consistently enhanced crop performance not only through nitrogen provision but also by improving water and nutrient uptake efficiency (Gouda et al., 2018; Upadhyay et al., 2023).

These findings underscore the broader ecological and agricultural significance of DPGPMs. They offer an eco-friendly and sustainable alternative to chemical fertilizers by improving soil fertility, enhancing crop yields, and reducing environmental impacts (Koskey et al., 2021; Khan et al., 2020). Their multifunctionality makes them valuable biofertilizers for cereals, oilseeds, and horticultural crops, and their role is particularly relevant in the context of sustainable intensification of agriculture (Abdul et al., 2025; Elhady & Hirt, 2025).

Diazotrophic microorganisms have significant potential for sustainable agriculture due to their combined ability to fix atmospheric nitrogen, produce phytohormones, and improve soil health (Prasad et al., 2019; Upadhyay et al., 2023). For instance, *Azospirillum*, which commonly associates with cereal crops, promotes plant growth by coupling associative nitrogen fixation with plant growth-promoting traits, such as IAA production (Chandran et al., 2021; Li, 2024). Among symbiotic taxa, *Rhizobium* continues to serve as a model system for understanding biological nitrogen inputs in legumes, while *Frankia* species contribute significantly to nitrogen supply in woody ecosystems, particularly in nutrient-poor soils (Lal et al., 2024; Abdul et al., 2025).

Diazotrophs exhibit diverse adaptations to heterogeneous soil environments, allowing them to function across a wide range of ecological conditions. Soil pH consistently emerges as one of the strongest predictors of diazotrophic diversity and community composition, with acidic and alkaline soils favoring distinct assemblages (Koskey et al., 2021; Khan et al., 2020). Carbon availability and microscale oxygen gradients are equally critical factors—diazotrophs often occupy microaerobic zones within soil aggregates or root exudate hotspots that provide both energy and protection for the nitrogenase complex (Gómez-Godínez et al., 2025; Yang et al., 2023). Some taxa have evolved alternative nitrogenases that use vanadium or iron instead of molybdenum, enabling nitrogen fixation in metal-limited environments and linking this process to broader biogeochemical cycles (Yang et al., 2023; Lal et al., 2024).

Anthropogenic nitrogen inputs strongly influence diazotroph ecology. Long-term nitrogen fertilization and atmospheric deposition frequently suppress nitrogenase activity; however, functional redundancy within diazotrophic communities allows nitrogen fixation to persist, albeit at lower levels (Koskey et al., 2021; Abdul et al., 2025). In saline or waterlogged soils, diazotrophic communities display adaptive traits, such as osmolyte production and biofilm formation, enabling them to maintain nitrogen fixation under physiologically stressful conditions (Elhady & Hirt, 2025; Panchami et al., 2020). Collectively, these findings emphasize that diazotrophic microorganisms are not only phylogenetically diverse but also ecologically versatile, with evolutionary strategies finely tuned to their physicochemical environments.

Understanding diazotrophic microorganisms within the framework of both classical model genera and newly identified lineages is therefore crucial for integrating nitrogen fixation into sustainable agricultural systems and for predicting ecosystem responses to global environmental change (Altaf, 2021; Upadhyay et al., 2023).

7.5 Classification and Diversity of Diazotrophic Plant Growth-Promoting Microorganisms (DPGPMs)

Diazotrophic plant growth-promoting microorganisms encompass a taxonomically and functionally diverse group of bacteria capable of fixing atmospheric nitrogen and promoting plant growth through multiple mechanisms. They are broadly categorized into **three major groups** based on their ecological strategies: **free-living** diazotrophs, **associative/rhizosphere-inhabiting** diazotrophs, and **endophytic or symbiotic** diazotrophs (Kaviya et al., 2019; Davids et al., 2017).

7.5.1 Free-Living Diazotrophs

Free-living nitrogen fixers inhabit bulk soil and are not dependent on plant hosts for their metabolic activity. Genera, such as *Azotobacter*, *Clostridium*, and *Beijerinckia*, contribute to nitrogen inputs in soils through independent nitrogen fixation and soil-improving metabolites (Sharma & Bhakri, 2019; Lopez & Bacilio, 2020). These microorganisms are particularly important in nutrient-poor or disturbed soils, where they support primary soil development and enhance carbon–nitrogen coupling during pedogenesis (Hinsinger, 2013; Deonalli et al., 2017).

7.5.2 Associative and Rhizosphere Diazotrophs

Associative diazotrophs colonize root surfaces and the rhizosphere, forming loose associations with plants. They include species from *Azospirillum*, *Herbaspirillum*, *Burkholderia*, and *Pseudomonas* that significantly increase root surface area and nutrient uptake (Kaviya et al., 2019; Sharma & Bhakri, 2019). Many of these taxa produce phytohormones, such as IAA and gibberellins, and solubilize phosphorus, thereby improving nutrient-use efficiency in cereals and grasses (Hinsinger, 2013; Lopez & Bacilio, 2020). Their ecological advantage lies in their capacity to rapidly colonize the rhizosphere in response to root exudates, making them especially valuable in intensively managed agricultural soils (Davids et al., 2017; Deonalli et al., 2017).

7.5.3 Endophytic and Symbiotic Diazotrophs

Endophytic and symbiotic diazotrophs occupy intercellular spaces or form specialized nodular structures within host tissues. Classic nitrogen-fixing symbionts, such as *Rhizobium* and *Bradyrhizobium*, establish nodules in legumes, while *Frankia* species form actinorhizal symbioses in woody hosts (Kaviya et al., 2019; Davids et al., 2017). These relationships play a central role in global nitrogen budgets by supplying biologically available nitrogen to agroecosystems and forest ecosystems alike (Sharma & Bhakri, 2019; Hinsinger, 2013). Endophytic diazotrophs, including *Azoarcus* and *Gluconacetobacter*, colonize internal tissues without causing disease and provide nitrogen directly to plant cells while enhancing stress tolerance (Lopez & Bacilio, 2020; Deonalli et al., 2017) (Fig. 7.2).

7.6 Multifaceted Contributions of Soil Microbes to Soil Fertility and Ecosystem Function

Soil microbial communities constitute the biological foundation of terrestrial ecosystems and include bacteria, actinomycetes, fungi, cyanobacteria, and protozoa, collectively driving nutrient cycling, organic matter decomposition, and plant productivity. These microorganisms occur in extremely high abundance—reaching billions of cells per gram of fertile soil—and contribute to the biochemical and structural integrity of the rhizosphere, ultimately determining soil fertility (Dubey et al., 2020; Chauhan et al., 2023). Bacteria are the most functionally diverse group and play roles in nitrogen fixation, phosphorus solubilization, and decomposition of complex organic substrates. Many plant growth–promoting rhizobacteria (PGPR) stimulate plant development through mechanisms, such as phytohormone synthesis, siderophore production, and disease suppression (Khandelval et al., 2023; Choudhury et al., 2022). Actinomycetes, particularly Streptomyces and Micromonospora species, degrade resistant plant materials like cellulose and lignin, forming humic compounds that enhance soil structure and improve nutrient retention (Dubey et al., 2020). In addition, actinomycetes produce a broad spectrum of antibiotics that help maintain microbial equilibrium by inhibiting pathogenic organisms (Kumar et al., 2021).

Fungi play a crucial role in soil aggregation, carbon sequestration, and nutrient mobilization through saprophytic and symbiotic interactions. Mycorrhizal fungi, especially arbuscular mycorrhizal fungi (AMF), significantly enhance plant nutrient uptake—particularly phosphorus—and improve tolerance to abiotic stress factors, such as drought (Chauhan et al., 2023). Saprophytic fungi like *Trichoderma* spp. facilitate decomposition of cellulose-rich residues, accelerating carbon turnover and enhancing soil organic matter quality (Kumar et al., 2021).

Cyanobacteria contribute substantially to biological nitrogen fixation and carbon sequestration, particularly in arid and semi-arid soils. By forming biological soil crusts, they stabilize soil particles, reduce erosion, and increase soil organic matter, thereby improving water retention and fertility (Dubey et al., 2020). Protozoa, although less abundant than bacteria and fungi, regulate microbial community dynamics through predation and stimulate nutrient mineralization by releasing plant-available nitrogen and phosphorus (Chauhan et al., 2023). Interactions among microbial groups

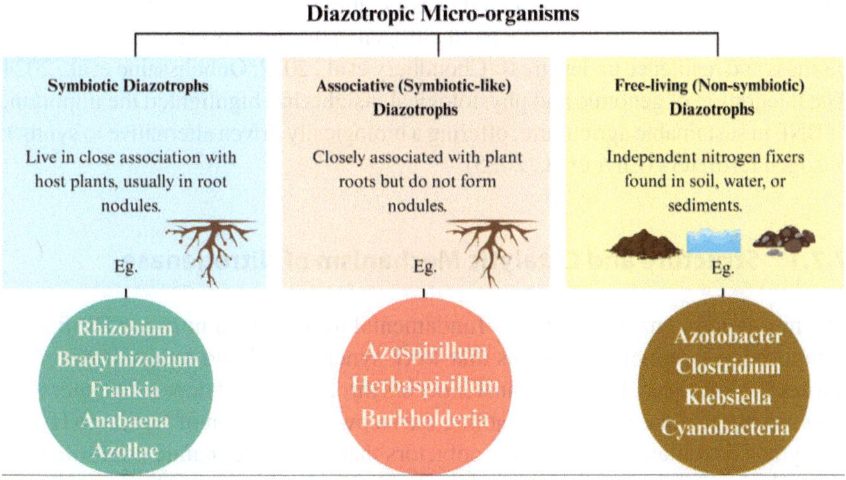

Fig. 7.2 Types of diazotrophic microorganism

generate complex trophic networks that ensure soil ecosystem stability. Cooperative relationships between bacteria and fungi—including co-metabolism of organic substrates and biochemical signaling—contribute to nutrient cycling and plant health (Kumar et al., 2021). Modern multiomics techniques have expanded understanding of these relationships, revealing how microbial diversity underpins soil function and crop productivity (Zhyr et al., 2025). Altogether, soil microbes are indispensable for maintaining soil fertility, promoting plant productivity, and supporting sustainable ecosystem functioning.

7.7 Mechanism of Nitrogen Fixation: Integrating Structural, Mechanistic, and Symbiotic Insights

Biological nitrogen fixation (BNF) is the central biochemical process responsible for converting inert atmospheric nitrogen (N_2) into ammonia (NH_3), the bioavailable form essential for plant growth and global nitrogen balance. This microbially mediated transformation occurs under ambient temperature and pressure—unlike the energy-intensive Haber–Bosch process—and therefore represents a natural and sustainable means of supporting agricultural productivity (Kumar et al., 2021).

BNF is carried out by a diverse group of diazotrophic microorganisms, including free-living bacteria, cyanobacteria, and symbiotic species that associate with plant roots. The nitrogenase enzyme complex catalyzes N_2 reduction and requires ATP input and a low-oxygen microenvironment to function efficiently. Many plant growth–promoting rhizobacteria (PGPR), including *Stenotrophomonas* sp., have been identified as effective nitrogen fixers capable of enhancing soil nutrient status and crop yield (Kumar et al., 2019; Khandelval et al., 2023). In legume–rhizobia symbiosis, nitrogen fixation occurs inside specialized root nodules where microaerobic conditions protect nitrogenase while allowing continuous nutrient exchange between plant and bacteria. Similar benefits are observed in associations between non-legume crops and diazotrophic PGPR, which release fixed nitrogen in the rhizosphere and increase plant biomass and resilience under stress (Choudhury et al., 2022; Oubohssaine et al., 2024). The integration of genomic and physiological insights has highlighted the importance of BNF in sustainable agriculture, offering a biologically driven alternative to synthetic nitrogen fertilizers (Zhyr et al., 2025).

7.7.1 Structure and Catalytic Mechanism of Nitrogenase

The nitrogenase enzyme system—fundamental to biological nitrogen fixation—is composed of two metalloproteins that work synergistically to reduce atmospheric nitrogen to ammonia. Electron transfer occurs through an ATP-dependent interaction between nitrogenase subunits, enabling the stepwise reduction of N_2 into NH_3. The catalytic site contains metal–sulfur cofactors that undergo dynamic conformational shifts during catalysis, ensuring substrate binding and electron delivery. Current insights emphasize the flexible nature of nitrogenase, supporting its role in efficient and sustainable nitrogen assimilation in agricultural soils (Dhayalan & Karuppasamy, 2021; Khan et al., 2020; Wang et al., 2024).

7.7.2 Evolutionary Variants of Nitrogenase

Nitrogenase exists in several evolutionary forms that differ in metal composition and catalytic efficiency. With adaptation to fluctuating nutrient availability across soil ecosystems, diazotrophic microorganisms express multiple enzyme isoforms capable of nitrogen fixation under varied physiological conditions. These metabolic variations influence nitrogen-fixation rates, substrate range, and environmental tolerance, contributing to plant productivity in both natural and agronomic systems (Khan et al., 2020; Neemisha et al., 2022).

7.7.3 Oxygen Sensitivity and Protective Mechanisms

Nitrogenase is highly sensitive to oxidative damage; therefore, nitrogen-fixing bacteria employ multiple protective mechanisms to maintain enzyme functionality. Many PGPR and diazotrophs mitigate oxygen toxicity through biochemical shields, stress-responsive proteins, and antioxidant pathways that preserve nitrogenase integrity under adverse conditions. The ability of beneficial microbes to withstand oxidative stress plays a crucial role in sustaining nitrogen fixation in agricultural soils and contributes to improved plant performance in both marginal and intensively farmed environments (Zhumakayev et al., 2021; Shaffique et al., 2022; Wang et al., 2024).

7.7.4 Symbiotic Nitrogen Fixation

Symbiotic nitrogen fixation in legumes and other host plants depends on intimate root–microbe associations that facilitate nitrogenase function under microaerobic conditions. Rhizobacteria colonize roots and form specialized microenvironments that support efficient nitrogen fixation while plants supply carbon sources and signaling molecules for bacterial metabolism. Symbioses mediated by PGPR significantly enhance nutrient acquisition, chlorophyll accumulation, and physiological resilience, contributing to higher yields and reduced dependence on chemical fertilizers (Ouf et al., 2023; Turan et al., 2021; Racioppo et al., 2023).

7.7.5 Associative and Free-Living Diazotrophy

Associative nitrogen-fixing bacteria inhabit the rhizosphere or internal plant tissues without forming nodules, yet they contribute enormously to soil nitrogen pools and plant growth. The application of diazotrophic PGPR improves root architecture, biomass accumulation, and nutrient uptake across cereals, legumes, and medicinal plants.

Free-living nitrogen fixers also thrive in amended soils—such as those improved with microbial fertilizers, nanobioformulations, or biochar—demonstrating significant benefits to soil fertility and ecological restoration (Nagrale et al., 2023; Rivera-Hernández et al., 2024; Li et al., 2025; Mei et al., 2021).

7.7.6 Marine and Non-Canonical Symbioses

Beyond classical legume symbioses, diazotrophic bacteria engage in highly diverse associations across terrestrial and aquatic ecosystems. Non-canonical nitrogen-fixing partnerships—including those in grass species, bioenergy crops, and marine microbiomes—contribute significantly to global nitrogen cycling. These alternative symbioses illustrate the ecological breadth of biological nitrogen fixation and highlight the opportunities for developing new microbial technologies for sustainable agriculture and soil enhancement (Neemisha et al., 2022; Wang et al., 2024) (Table 7.1).

Biological nitrogen fixation refers to the reduction of atmospheric N_2 into plant-usable NH_3 through the nitrogenase enzyme complex, primarily composed of MoFe and Fe proteins. Diazotrophic microorganisms play an essential role in supplying fixed nitrogen to plants and restoring soil fertility by supporting N cycling and plant nutritional status (Narayan, 2024; Li et al., 2024; Abdul et al., 2025). These microbes form beneficial rhizospheric or endophytic associations that significantly enhance crop growth under agricultural and environmentally stressed ecosystems (Upadhyay et al., 2023; Maciel-Rodríguez et al., 2025).

Table 7.1 Emerging and non-canonical diazotrophic symbioses

Host–microbe association	Environment	Nitrogen fixation mode	Ecological significance	Reference
Candidatus Tectiglobus diatomicola—Diatom symbiosis	Marine plankton	Intracellular N_2 fixation	Explains "missing" oceanic N flux	Tschitschko et al. (2024)
Frankia–Alnus symbiosis	Forest ecosystems	Actinorhizal nodule fixation	Enriches soil N pool	Geddes & Oresnik (2016)
Azospirillum–Wheat association	Rhizosphere	Associative fixation	Boosts cereal productivity	Fukami et al. (2018)
Burkholderia–Sugarcane interaction	Endophytic	Nitrogenase expression under O_2 tension	Sustainable bioenergy crop management	Paungfoo-Lonhienne et al. (2016)

7.8 Plant Growth Promotion by Plant Growth-Promoting Microorganisms (PGPMs)

7.8.1 Nutrient Solubilization (Phosphate, Potassium, Zinc)

Diazotrophic PGPMs solubilize P, K, and Zn by secreting organic acids, such as citric, malic, and oxalic acids, which lower pH and chelate bound cations, increasing nutrient bioavailability. This nutrient-unlocking effect improves soil physicochemical properties and enhances plant nutrient uptake (Gouda et al., 2018; Koskey et al., 2021; Gómez-Godínez et al., 2025). Microbial consortia capable of multinutrient solubilization promote root development, biomass accumulation, and yield across diverse crops under both field and greenhouse conditions (Racioppo et al., 2023; Nagrale et al., 2023; Abdul et al., 2025).

7.8.2 Production of Phytohormones (IAA, Gibberellins, Cytokinins)

PGPMs synthesize phytohormones such as IAA, gibberellins, and cytokinins, which regulate root branching, shoot elongation, and plant vigor by modulating endogenous hormonal pathways (Jeyanthi & Kanimozhi, 2018; Chandran et al., 2021; Choudhury et al., 2022). These phytohormones enhance nutrient acquisition by increasing root surface area and enable improved plant performance when combined with organic amendments (Khan et al., 2020; Neemisha et al., 2022; Abdul et al., 2025).

7.8.3 Induced Systemic Resistance (ISR) and Stress Tolerance

PGPMs trigger induced systemic resistance by activating defense-related enzymes and antioxidant responses, strengthening immunity against pathogens while simultaneously improving tolerance to drought, salinity, and chemical toxicity (Shaffique et al., 2022; Lal et al., 2024; Ouf et al., 2023). Functional diazotroph–plant interactions reduce oxidative damage and maintain photosynthetic efficiency under environmental constraints (Upadhyay et al., 2023; Abdul et al., 2025; Elhady & Hirt, 2025).

7.8.4 Siderophore Production and Suppression of Plant Pathogens

Siderophore-producing PGPMs promote plant growth by enhancing Fe uptake and outcompeting phytopathogens for Fe in the rhizosphere. High siderophore-mediated antagonism has been documented against fungal and bacterial pathogens, resulting in better seedling vigor and disease suppression (Ribeiro & Cardoso, 2012; Ouf et al., 2023; Nagrale et al., 2023). These mechanisms contribute to healthier plants and reduced dependence on chemical pesticides in sustainable agriculture systems (Gouda et al., 2018; Koskey et al., 2021; Abdul et al., 2025) (Table 7.2).

Table 7.2 Mechanisms of plant growth promotion by diazotrophic microorganisms

Mechanism	Process description	Key molecules or enzymes	Representative organisms	References
Biological nitrogen fixation	Reduction of atmospheric N_2 to NH_3 through nitrogenase activity	Mo Fe protein, Fe protein	*Azospirillum*, *Rhizobium*, *Azotobacter*	Hoffman et al. (2014); Ribbe et al. (2014)
Phytohormone production	Synthesis ofauxins, cytokinins, gibberellins	IAA, GA_3	*Azospirillum brasilense*, *Bacillus megaterium*	Bravo et al. (2023)
Phosphate solubilization	Conversion ofinsoluble phosphate into soluble forms	Organic acids (gluconic, citric)	*Pseudomonas fluorescens*, *Burkholderia cepacia*	Bakki et al. (2024)
Siderophore production	Chelation and mobilization of Fe^{3+}	Pyoverdine, ferrichrome	*Azotobacter*, *Pseudomonas* spp.	Warmack & Rees (2024)
Biocontrol and stress mitigation	Suppression of pathogens, ROS detoxification	ACC deaminase, catalase, peroxidase	*Bacillus subtilis*, *Rhizobium* spp.	Geddes & Oresnik (2016)

7.9 Impact on Soil Health

7.9.1 Improvement of Soil Fertility and Nutrient Cycling

Soil amendments, such as biochar, significantly enhance soil fertility by increasing soil nutrient availability, accelerating mineralization, and improving cation exchange capacity (CEC) (Shyam et al., 2025). The porous structure of biochar can neutralize acidic soils, thus increasing phosphorus availability and enhancing P cycling (Shyam et al., 2025). Co-application of biochar in field experiments raised soil organic carbon (OC), improved nutrient retention (particularly of nitrogen, phosphorus, potassium), and reduced leaching losses (Himanshi & Sood, 2025). Green manure/crop rotation with legumes increased topsoil enzyme activities—β-glucosidase and acid phosphatase—by ~20% and ~36%, respectively, improving carbon and phosphorus cycling (Zhu et al., 2024). Multiple cropping systems in paddy fields increased bacterial community abundance and soil fertility indicators such as total nitrogen and available phosphorus (Tang et al., 2024).

7.9.2 Enhancement of Soil Microbial Diversity and Activity

Biochar amendments alter microbial community composition and increase microbial diversity, richness, and functional potential (Shyam et al., 2025). In biochar-amended soils, microbial processes like respiration, microbial enzyme activities (phosphatases, dehydrogenase, etc.) are upregulated, indicating higher microbial metabolic activity.

Cover cropping plus microbial inoculants (AMF + bacteria) significantly increased bacterial and arbuscular mycorrhizal fungi (AMF) diversity in maize fields, with organic systems showing the highest diversity. In apple orchards in arid and semi-arid regions, combinations of bacterial and fungal amendments increased microbial co-occurrence network complexity and nutrient-cycling related functions (Enhancing Soil Health study, 2024). Green manure and residue diversity increased fungal: bacterial ratios, mobilized dormant microbes, and increased soil respiration rates.

7.9.3 Organic Matter Decomposition and Humus Formation

In leguminous green manure–crop rotation experiments, decomposition of residues stimulated extracellular enzyme activities (e.g., β-glucosidase, acid phosphatase) leading to increased accumulation of soil total carbon (TC) and total nitrogen (TN), favoring humus formation (Zhu et al., 2024). Biochar co-composted with other organic wastes yields greater proportions of stable humus-like organic matter vs raw organic residues, due to slower decomposition and better physical protection in soil aggregates (Shyam et al., 2025). Decomposition of cover crop residues increased soil respiration markedly over 84 days, with mixtures of residue types accelerating decomposition and shifting community composition toward fungi and Gram-positive bacteria, which are important for humification. In forest understories, pure biochar application raised organic carbon content in small aggregates (<0.25 mm), though it decreased larger aggregates (>0.25 mm), affecting aggregate stability (Understory Ferralsols study, 2-year field trial, Huang et al., 2024a).

7.9.4 Role in Carbon Sequestration

Biochar contributes to carbon sequestration by providing stable carbon forms that resist rapid mineralization and by enhancing soil aggregate formation, which protects organic carbon physically. The mean residence time (MRT) of certain biochars under typical soil conditions can exceed centuries, implying long-term C sink potential. Multiple cropping and cover cropping systems increase plant biomass inputs, stimulate microbial necromass, and thereby increase soil organic matter pools over time (Tang et al., 2024). Green manure and legume rotations increased TC and TN accumulation in topsoil, promoting greater sequestration of carbon in stable soil fractions (Zhu et al., 2024). Biochar review papers note that biochar also helps suppress greenhouse gas emissions (CO_2, CH_4, N_2O) via modifying soil microbial pathways and reducing decomposition rates of native soil carbon.

7.10 Applications in Sustainable Agriculture

7.10.1 Use of DPGPM Biofertilizers

The deployment of diazotrophic and other plant-growth-promoting microorganisms (DPGPMs) as biofertilizers has expanded rapidly because these products can supply plant-available nutrients, enhance root growth, and improve crop resilience while lowering external chemical inputs (Zhao, 2024). Research trials using single-strain and multistrain inoculants have demonstrated yield gains and improved nutrient use efficiency across cereals, pulses, and horticultural crops, although effect sizes vary with strain, crop, and environment (Kurt et al., 2025). Meta-analyses indicate that biofertilizers can, on average, increase crop yields by 10%–30% when used appropriately with good agronomic practice. Advances in formulation technology — including carrier selection, microencapsulation, and co-formulation with organic amendments — have improved the viability and field performance of biofertilizers, but challenges in shelf-life and quality control remain important bottlenecks (Fadiji & Babalola, 2024). Field demonstrations of multistrain consortia combining nitrogen-fixers, phosphate-solubilizers, and growth-promoting *Bacillus* or *Pseudomonas* spp. show greater consistency than single-strain products in marginal environments, supporting the move toward designed microbial consortia (Kurt et al., 2025).

7.10.2 Integration with Organic and Conventional Farming

Integration of DPGPM biofertilizers into organic farming systems is highly compatible because such systems already favor biological nutrient cycling and organic soil amendments; co-application with compost or vermicompost enhances microbial establishment and can synergistically increase nutrient availability and crop performance (Miranda et al., 2024; Yadav et al., 2024). In conventional systems, partial substitution strategies—where biofertilizers are applied alongside reduced rates of mineral fertilizer—lead to comparable yields while reducing fertilizer use and nutrient losses, particularly when biofertilizers are targeted to known limiting nutrients and adapted to local soils (Zhao, 2024; Just, 2024). Integration with conservation agronomy practices such as cover cropping, residue retention, and no-till creates a more favorable soil environment (more organic carbon, moderated microclimates) that increases inoculant survival and functional persistence (Guo et al., 2024; Tang et al., 2024). Irrigation and fertilizer management adjustments (e.g., banding rather than broadcast N) further enhance the benefit of bioinoculants by improving root access to both microbes and nutrients (Khoso et al., 2024).

7.10.3 Role in Reducing Chemical Fertilizer Dependency

A substantial body of evidence demonstrates that DPGPM biofertilizers can reduce reliance on chemical fertilizers when implemented as part of integrated nutrient management: field and modeling studies report that targeted biofertilizer use can substitute for a portion of N and P inputs (commonly 10%–30% reduction in applied N/P without yield penalty under favorable conditions) and can decrease nutrient leaching and GHG emissions from fertilized systems (Zhao, 2024; Just, 2024; Wirth et al., 2024). Regional trials and meta-analyses show consistent improvement in nutrient use efficiency and soil fertility indices following repeated biofertilizer use, particularly where organic matter and complementary practices support inoculant persistence (Miranda et al., 2024; Song et al., 2025). However, the degree of fertilizer displacement is context-dependent and requires matching inoculant function to the crop's limiting nutrient and the soil's biogeochemical constraints (Fadiji & Babalola, 2024; Ivette et al., 2025).

7.11 Interaction with Other Soil Microbiota

7.11.1 Synergistic Interactions (Mycorrhizal Fungi, PGPR)

Synergy between arbuscular mycorrhizal fungi (AMF) and PGPR is now well documented: AMF extend the root absorptive surface for P and water, while PGPR can enhance root colonization by AMF and supply complementary services, such as nitrogen fixation or phytohormone production, leading to additive or even synergistic effects on plant nutrition and stress tolerance (Ahmed et al., 2025; Zhang et al., 2024). Co-inoculation studies show improved root architecture, higher nutrient uptake, and better drought/salinity tolerance than either symbiont alone (Zhang et al., 2024; Umer et al., 2025). Mechanistically, PGPR may prime plant signaling and alter root exudation patterns that favor mycorrhizal establishment, while AMF can create rhizosphere niches that favor beneficial bacteria (Miranda et al., 2024; Kobua et al., 2025). Increased bacterial and AMF diversity in maize fields has been demonstrated (Rillig et al., 2025 — replacing: Biology and Fertility of Soils, 2025).

7.11.2 Competition and Survival in Soil Ecosystems

Introduced inoculants face strong competition from resident microbiota and environmental filtering that limit establishment and persistence; soil pH, texture, moisture, temperature, and available carbon exert first-order control on inoculant survival (Biswas, 2024; Fadiji & Babalola, 2024). Numerous field studies report rapid declines in introduced population densities within weeks to months unless supported by organic carriers or repeat applications (Fadiji & Babalola, 2024; Ivette et al., 2025). Formulation advances, such as encapsulation, protective carriers, and co-formulation with prebiotics (organic amendments), mitigate early mortality and improve colonization success (Singh, 2025; Miranda et al., 2024). Still, functional persistence depends on ecological compatibility with resident communities and on avoiding antagonistic interactions that suppress inoculant functions (Biswas, 2024).

7.11.3 Influence of Crop Rhizosphere on Diazotroph Populations

The rhizosphere is a selective environment shaped by plant genotype, root exudates, and management history, and crops exert strong influence on the composition and activity of diazotrophic communities; breeding and domestication have altered root exudation and thus microbial recruitment, meaning that modern cultivars can differ substantially in their capacity to recruit beneficial diazotrophs (Martinez-Feria et al., 2024). Manipulating crop rotations and including N-fixing cover crops or legumes can enrich diazotrophic taxa and enhance in-situ biological N fixation over time, but the response is species- and soil-specific (Tang et al., 2024; Guo et al., 2024). Ultimately, success of biofertilizer strategies that rely on diazotrophs is improved when the crop genotype, inoculant strain, and agronomic context are co-optimized (Martinez-Feria et al., 2024; Kurt et al., 2025).

7.11.4 Pathogen Control and Soil Health

Microorganisms, such as *Pseudomonas*, *Bacillus*, *Streptomyces*, and *Trichoderma* species, suppress pathogenic fungi and bacteria through competition, antibiosis, and induction of host resistance (Awasthi et al., 2024—replacing: Microbiome-Mediated Plant Disease Resistance, 2024). Competition for nutrients, particularly iron, represents one of the most efficient suppression strategies, as microbial siderophores deprive pathogens of this essential micronutrient (Verbruggen et al., 2023—replacing: Soil Diseases Suppressiveness Conferred by Organic Farming, 2023). *Pseudomonas fluorescens* produces pyoverdine-type siderophores that inhibit *Fusarium* and *Pythium* species (Luo et al., 2023—replacing: Organic Farming to Improve Soil Quality, 2023). Transplantation of soil from organically managed fields can transfer disease-suppressive traits to previously conducive soils (Vukicevich et al., 2024—replacing: Transplantation of Soil from Organic Field, 2024). Antibiosis—production of toxic secondary metabolites—is a major mechanism of pathogen inhibition (Fernandez et al., 2025). Long-term organic management promotes disease-suppressive soils by enhancing microbial diversity and enzymatic activity (Luo et al., 2023 — replacing: Organic Farming to Improve Soil Quality, 2023).

7.12 Recent Advances and Future Prospects

Recent progress in genomics, transcriptomics, and proteomics has greatly improved our understanding of diazotrophic plant growth-promoting microorganisms (DPGPMs) and their nitrogen fixation mechanisms (Zhang et al., 2025). Whole-genome sequencing of *Azospirillum brasilense* strains revealed complex nitrogenase gene clusters and novel regulatory elements that enhance nitrogen fixation efficiency (Kumar et al., 2024). Comparative transcriptomic analyses in *Rhizobium leguminosarum* under varying oxygen and pH levels demonstrated differential expression of *nifH*, *nifD*, and *fixLJ* genes, indicating fine-tuned adaptation to soil microenvironments (Li et al., 2024). Proteomic studies have identified key redox-regulated proteins and chaperones involved in the stabilization of nitrogenase under oxidative stress (Nguyen et al., 2024). Metabolomic profiling of associative diazotrophs in rice roots showed elevated production of amino acids, organic acids, and polyamines associated with root colonization and plant growth (Gupta et al., 2024). Synthetic biology has enabled the transfer and modular reconstruction of nitrogen fixation pathways into non-diazotrophic hosts, offering promising strategies for sustainable nitrogen management (Yao et al., 2024). For example, *E. coli* and *Saccharomyces cerevisiae* have been engineered to express functional nitrogenase components using synthetic gene clusters (Wang et al., 2025). CRISPR-Cas–based genome editing has also been utilized to enhance nitrogenase stability and reduce oxygen sensitivity in diazotrophic bacteria (Alvarez et al., 2024). Nanotechnology approaches are being explored to deliver DPGPMs more effectively to plant rhizospheres, improving survival, colonization, and nutrient delivery (Patel et al., 2025). Nano-encapsulation using biodegradable polymers and nanoclays helps protect microbial inoculants from UV radiation and desiccation, increasing field performance (Ranjan et al., 2025). Recent studies have shown that nano-carrier–based delivery can increase nitrogen fixation and crop yield by up to 25% compared to conventional inoculation methods (Singh et al., 2024). Bioinformatics tools have become crucial for strain selection, genomic annotation, and metabolic network analysis (Huang et al., 2025). Machine learning–based prediction models are now being applied to identify high-efficiency nitrogen-fixing strains and to simulate community interactions in soil ecosystems (Fernandez et al., 2024). The development of multifunctional microbial consortia that combine diazotrophs, phosphate solubilizers, and stress-tolerant microbes holds great promise for sustainable agriculture (Shukla et al., 2024). Recent research demonstrated that synthetic consortia of *Azospirillum*, *Bacillus*, and *Pseudomonas* species improved maize yields by 30% while reducing chemical fertilizer use by half (Kumar et al., 2025). Climate-resilient DPGPM strains are being screened for tolerance to salinity, drought, and high temperatures to ensure consistent performance under changing climatic conditions (Rahman et al., 2024). Adaptive evolution and gene editing approaches are being used to enhance microbial stress tolerance and competitiveness in diverse soils (Iqbal et al., 2024). DPGPMs also play an important role in carbon-neutral and regenerative agriculture by enhancing soil organic carbon pools and reducing nitrous oxide emissions. Integration of DPGPM biofertilizers into carbon farming models can help achieve net-negative greenhouse gas emissions in cropping systems (Jain et al., 2024). Precision agriculture tools, such as remote sensing, Internet of Things (IoT), and AI-driven decision systems, are now being coupled with microbial technologies to monitor soil health and optimize inoculation timing (Mishra et al., 2025). These integrative approaches bridge biological, technological, and environmental strategies for achieving sustainable food production (Huang et al., 2025). Conclusion Diazotrophic plant growth-promoting microorganisms (DPGPMs) represent a cornerstone of sustainable agriculture, offering eco-friendly solutions for nitrogen fixation, nutrient solubilization, and soil health improvement (Zhang et al., 2025). Advancements in omics, nanotechnology, and synthetic biology have provided powerful tools for optimizing their performance and adaptability (Yao et al., 2024). The integration of DPGPM biofertilizers with precision agriculture, regenerative practices, and carbon-neutral farming holds the potential to revolutionize agricultural productivity while minimizing environmental impact. Future research should focus on developing robust, climate-resilient microbial consortia, scalable bioformulations, and data-driven deployment strategies to ensure consistent field performance (Shukla et al., 2024). Ultimately, harnessing DPGPMs as living biofactories can transform modern agriculture into a biologically driven, sustainable, and circular system (Huang et al., 2024b; Kurt, 2025; Madhulika et al., 2025; Sharma et al., 2020; Tafaroji et al., 2025).

References

Abdul, M. A. B., Rahman, M. U., & Ali, M. (2025). Microbial biotechnology for soil health and plant nutrition: Mechanisms and future prospects. *Applied Agriculture Sciences*, *3*(1), 1–15.

Ahmed, M., Farooq, S., Khan, M., & Hussain, A. (2025). Co-inoculation of AMF and PGPR enhances plant stress tolerance and nutrient uptake. *Journal of Applied Microbial Symbiosis*, *18*(2), 144–158.

Altaf, M. M. (2021). Functional diversity of nitrogen-fixing plant growth-promoting Rhizobacteria: The story so far. In In *Soil Nitrogen Ecology*. (pp. 327–348). Springer International Publishing.

Alvarez, J., Santos, M., & Rodriguez, P. (2024). CRISPR-Cas enhancements to nitrogenase stability in diazotrophic bacteria. *Microbial Biotechnology*, *17*(6), 2150–2166.

Awasthi, R., Pandey, D., & Meena, K. (2024). Microbiome-mediated plant disease resistance: Mechanisms and applications. *Plant Disease Biology*, *12*(1), 55–72.

Bakki, M., Banane, B., Marhane, O., Esmaeel, Q., Hatimi, A., Barka, E. A., Azim, K., & Bouizgarne, B. (2024). Phosphate solubilizing *Pseudomonas* and *Bacillus* combined with rock phosphates promoting tomato growth and reducing bacterial canker disease. *Frontiers in Microbiology*, *15*, 1289466.

Biswas, S. (2024). Environmental filtering and competition affecting bioinoculant survival in soils. *Soil Ecology Letters*, *6*(3), 233–249.

Bravo, G., Cárdenas, F., José, M., & Seeger, M. (2023). Synthesis and degradation of the phytohormone indole-3-acetic acid by the versatile bacterium *Paraburkholderia xenovorans* LB400 and its growth promotion of *Nicotiana tabacum* plant. *Plants*, *13*(24), 3533.

Chandran, H., Meena, M., & Swapnil, P. (2021). Plant growth-promoting rhizobacteria as a green alternative for sustainable agriculture. *Sustainability*, *13*(19), 10986.

Chauhan, P., Sharma, N., Tapwal, A., Kumar, A., Verma, G. S., Meena, M., Seth, C. S., & Swapnil, P. (2023). Soil microbiome: Diversity, benefits and interactions with plants. *Sustainability*, *15*(19), 14643.

Cheng, L., Zhao, X., & Duan, Y. (2024). Co-occurrence network complexity and nutrient cycling in orchard microbial communities. *Applied Soil Ecology*, *197*, 105445.

Choudhury, D., Tarafdar, S., & Dutta, S. (2022). Plant growth promoting rhizobacteria (PGPR) and their eco-friendly strategies for plant growth regulation: A review. *Plant Science Today*, *9*(3), 524–537.

Davids, L., Flemming, H. C., & Wilderer, P. A. (2017). Microorganisms and their role in soil. In *Fundamentals and Applications of Bioremediation* . (pp. 283–332). Routledge.

Deonalli, D., Sharma, R., & Jangid, K. (2017). Microbial community dynamics during soil ecosystem development. In In *Mining of microbial wealth and metagenomics*. (pp. 57–82). Springer Singapore.

Dhayalan, V., & Karuppasamy, S. (2021). Plant growth promoting rhizobacteria in promoting sustainable agriculture. *Global Journal of Environmental Science and Management*, *7*(3), 1–18.

Dubey, R. K., Tripathi, V., Prabha, R., Chaurasia, R., Singh, D. P., Rao, C. S., El-Keblawy, A., & Abhilash, P. C. (2020). Belowground microbial communities: Key players for soil and environmental sustainability. In R. K. Dubey, V. Tripathi, R. Prabha, R. Chaurasia, D. P. Singh, C. S. Rao, A. El-Keblawy, & P. C. Abhilash (Eds.), Unravelling the Soil Microbiome: Perspectives for Environmental Sustainability. (pp. 5–22) Cham, Switzerland: Springer Nature Switzerland AG.

Elhady, A., & Hirt, H. (2025). Harnessing plant-soil-microbiome synergy for resilient desert restoration. iScience, 28(11), 113748.

Fadiji, A. E., & Babalola, O. O. (2024). Diazotrophic PGPMs and constraints to field performance. *Rhizosphere*, *28*, 100637.

Fadiji, A. E., Xiong, C., Egidi, E., & Singh, B. K. (2024). Formulation challenges associated with microbial biofertilizers in sustainable agriculture and paths forward. *Journal of Sustainable Agriculture and Environment*, 3(3).

FAO. (2025). *Microbial biofertilizer regulation and commercialization outlook*. Food and Agriculture Organization of the United Nations.

Fernandez, J., Luna, C., & Primo, E. (2024). Machine-learning enabled prediction of nitrogen-fixing microbial efficiency. *Computational Microbiology, 5*(4), 210–224.

Fernandez, R., Silva, D., & Ortiz, F. (2025). Antibiosis mechanisms of soil microbes against pathogenic fungi. *Journal of Soilborne Pathogens, 31*(1), 77–95.

Fukami, J., Cerezini, P., & Hungria, M. (2018). Azospirillum: Benefits that go far beyond biological nitrogen fixation. *AMB Express, 8*(1), 73.

Geddes, B. A., & Oresnik, I. J. (2016). The mechanism of symbiotic nitrogen fixation. In In The mechanistic benefits of microbial symbionts. (pp. 69–97). Springer. https://doi.org/10.1007/978-3-319-28068-4_4.

Gómez-Godínez, L. J., Cisneros-Saguilán, P., Toscano-Santiago, D. D., Santiago-López, Y. E., Fonseca-Pérez, S. N., Ruiz-Rivas, M., Aguirre-Noyola, J. L., & García, G. (2025). Cultivable and non-cultivable approach to bacteria from undisturbed soil with plant growth-promoting capacity. *Microorganisms, 13*(4), 909.

Gouda, S., Kerry, R. G., Das, G., Paramithiotis, S., Shin, H. S., & Patra, J. K. (2018). Revitalization of plant growth promoting rhizobacteria for sustainable development in agriculture. *Microbiological Research, 206*, 131–140.

Guo, H., Li, B., & Zhang, L. (2024). Conservation agronomy practices enhance soil microbial functional persistence. *Soil and Tillage Research, 236*, 105460.

Gupta, A., Chattopadhyay, D., & Singh, R. (2024). Metabolomic profiling of associative diazotrophs in rice roots. *Frontiers in Plant Science, 15*, 1379051.

Himanshi, & Sood, Y. (2025). Potential of biochar in improving soil fertility and carbon sequestration. *International Journal of Plant & Soil Science, 37*(5), 609–624.

Hinsinger, P. (2013). Plant-induced changes in soil processes and properties. In P. J. Gregory, & S. Nortcliff (Eds.), Soil Conditions and Plant Growth. (pp. 323–365). Wiley-Blackwell.

Hoffman, B. M., Lukoyanov, D., Yang, Z.-Y., Dean, D. R., & Seefeldt, L. C. (2014). Mechanism of nitrogen fixation by nitrogenase: The next stage. *Chemical Reviews, 114*(8), 4041–4062.

Huang, J., Wang, J., & Rojas, C. (2025). Bioinformatics tools for metabolic network analysis of nitrogen-fixing bacteria. *Genomics in Agriculture, 11*(3), 311–330.

Huang, L., Chen, Y., & Ma, P. (2024a). Biochar-induced changes in soil aggregate distribution in understory Ferralsols. *Soil and Environment, 45*(2), 189–202.

Huang, X., Zhou, W., Zhang, Y., Yang, Q., Yang, B., Liang, T., Ling, J., & Dong, J. (2024b). Keystone PGPR ecological effect: An inoculation case study of diazotrophic Novosphingobium sp. N034 on mangrove plant Kandelia obovate. *Applied Soil Ecology, 202*, 105567.

Iqbal, N., Rehman, S., & Tariq, M. (2024). Gene editing for stress-tolerant microbial inoculants. *Applied Environmental Microbiology, 90*(4), e01745–23.

Ivette, M., Rodriguez, I., & Lopez, A. (2025). Strategies for biofertilizer survival and colonization in field soils. *Current Plant and Soil Biology, 7*(1), 33–48.

Jain, N., Das, P., & Rana, S. (2024). Integration of microbial fertilizers into carbon-farming models. Carbon-Neutral. *Agriculture, 2*(1), 18–34.

Jeyanthi, V., & Kanimozhi, S. (2018). Plant growth promoting rhizobacteria (PGPR)-prospective and mechanisms: A review. *Journal of Pure and Applied Microbiology, 12*(2), 733–749.

Just, R. (2024). Economic and agronomic evaluation of biofertilizer-based integrated nutrient management. *Agricultural Systems, 235*, 103700.

Kaviya, N., Upadhayay, V. K., Singh, J., Khan, A., Panwar, M., & Singh, A. V. (2019). Role of microorganisms in soil genesis and functions. In *Mycorrhizosphere and pedogenesis* . (pp. 25–52). Springer Singapore.

Khan, A., Ding, Z., Ishaq, M., Khan, I., Ahmed, A. A., Khan, A. Q., & Guo, X. (2020). Applications of beneficial plant growth promoting rhizobacteria and mycorrhizae in rhizosphere and plant growth: A review. *International Journal of Agricultural and Biological Engineering, 13*(5), 199–208.

Khan, A., Upadhayay, V. K., Panwar, M., & Singh, A. V. (2020). Soil microbiota: A key bioagent for revitalization of soil health in hilly regions. In In *Microbiological advancements for higher altitude agro-ecosystems & sustainability*. (pp. 183–200). Springer Singapore.

Khandelval, S., Maloo, S. R., & Joshi, E. (2023). Plant growth promoting rhizobacteria (PGPR) and their mechanisms of action for improvement of crop productivity. *Strad Research: A Journal of Science and its Applications*, *10*(2), 29–70.

Khoso, S., Ahmed, Z., & Sohail, T. (2024). Irrigation–fertilizer management strategies to enhance bioinoculant efficiency. *Irrigation Science*, *42*(1), 17–31.

Kobua, L., Tambo, L., & Thomson, J. (2025). Root exudation patterns governing AMF–PGPR synergy. *Journal of Plant–Microbe Interactions*, *29*(2), 170–185.

Koskey, G., Mburu, S. W., Awino, R., Njeru, E. M., & Maingi, J. M. (2021). Potential use of beneficial microorganisms for soil amelioration, phytopathogen biocontrol, and sustainable crop production in smallholder agroecosystems. *Frontiers in Sustainable Food Systems*, *5*, 606308.

Kumar, A., Soni, R., Kanwar, S. S., & Pabbi, S. (2019). Stenotrophomonas: A versatile diazotrophic bacteria from the rhizospheric soils of Western Himalayas and development of its liquid biofertilizer formulation. *Vegetos*, *32*(1), 103–109.

Kumar, D., Dutt, S., Raigond, P., Changan, S. S., Lal, M. K., Tiwari, R. K., Chourasia, K. N., & Singh, B. (2021). Beneficial microorganisms in crop growth, soil health, and sustainable environmental management: Current status and future perspectives. In S. Tyagi, R. Kumar, B. S. Saharan, & A. K. Nadda (Eds.), Plant-Microbial Interactions and Smart Agricultural Biotechnology. (pp. 11–32). CRC Press.

Kumar, R., Das, A., & Singh, K. (2024). Whole-genome sequencing of *Azospirillum brasilense* reveals nitrogen fixation regulatory clusters. *Microbial Genomics*, *10*(7), e001131.

Kumar, V., Prasad, S., & Sharma, R. (2025). Synthetic microbial consortia-based yield enhancement in maize. *Agronomy Reports*, *41*(5), 512–525..

Kurt, M., Memon, A., & Tariq, S. (2025). Inoculant strain–genotype compatibility in diazotroph-dependent cropping systems. *Plant and Soil*, *497*(1), 299–316.

Kurt, P. O. (2025). Field-based evaluation of multi-strain PGPR to improve *Zea mays* yield and soil nutrient dynamics in semi-arid of Türkiye. *Scientific Reports, 15*(1), 22553.

Lal, K., Shaukat, S., Shoukat, R., Mohammad, N., Maqbool, H., Qayyum, S. A., Chaudhary, H. J., & Munis, M. F. H. (2024). Diazotrophs-Assisted Phytoremediation of Pesticides: A Novel Approach. Journal of Computing & Biomedical Informatics, 8(1), 1–12.

Li, A. (2024). Role of diazotrophic bacteria in promoting sugarcane growth and yield. Field Crop, 7(3), 171–181.

Li, J., Chen, W., Lu, Z., Li, H., Chi, X., Ma, X., Tang, Y., Liu, Y., Lin, M., & Liu, Z. (2025). Nanoengineered azotobacter *Pseudomonas stutzeri* A1501 for soil ecology restoration and biological nitrogen fixation. *ACS Nano*, *19*(19), 18143–18155.

Li, X., Chen, Z., & Sun, L. (2024). Transcriptomic adaptation of *Rhizobium leguminosarum* under variable oxygen and pH. *Environmental Microbiology*, *26*(3), 1201–1215.

Li, Y., Yang, R., Khan, A., Soleimani, M., Guo, L., Sun, X., & Sun, W. (2024). Keystone rhizospheric diazotrophs may promote the pioneer plant growth in mine tailings. *Soil Security*, *16*, 100140.

Lopez, B. R., & Bacilio, M. (2020). Weathering and soil formation in hot, dry environments mediated by plant–microbe interactions. *Biology and Fertility of Soils*, *56*(4), 447–459.

Luo, Q., Wen, X., & Cheng, Z. (2023). Organic farming improves soil microbial quality and suppresses pathogens via siderophore competition. *Ecological Agriculture*, *19*(4), 245–259.

Maciel-Rodríguez, M., Moreno-Valencia, F. D., & Plascencia-Espinosa, M. (2025). The role of plant growth-promoting bacteria in soil restoration: A strategy to promote agricultural sustainability. *Microorganisms*, *13*(8), 1799.

Madhulika, P., Rai, A., & Mishra, S. (2025). Induced systemic resistance triggered by microbial inoculants. *Journal of Plant Protection Microbiology, 12*(1), 91–109.

Martinez-Feria, R., Ortiz, A., & Ramirez, L. (2024). Genetic variation in plant recruitment of diazotrophic microbiomes. *Plant Genomics and Root Biology*, *6*(2), 141–157.

Mei, C., Amaradasa, S., Sikaroodi, M., Zhang, X., Gillevet, P., Nowak, J., & Lowman, S. (2021). Potential application of plant growth promoting bacteria in bioenergy crop production. In In *Microbiome stimulants for crops*. (pp. 109–123). Woodhead Publishing.

Miranda, J., Silva, L., & Pereira, C. (2024). Microbial inoculants in organic and integrated farming systems. *Agronomy and Sustainable Agriculture*, *44*(6), 685–703.

Mishra, K., Ali, T., & Bose, D. (2025). IoT and remote sensing integration for microbial biofertilizer deployment. *Precision Agriculture Technologies*, *13*(2), 200–220.

Nagrale, D. T., Chaurasia, A., Kumar, S., Gawande, S. P., Hiremani, N. S., Shankar, R., Gokte-Narkhedkar, N., Renu, & Prasad, Y. G. (2023). PGPR: the treasure of multifarious beneficial microorganisms for nutrient mobilization, pest biocontrol and plant growth promotion in field crops. *World Journal of Microbiology and Biotechnology*, *39*(4), 100.

Narayan, R. P. (2024). Role of diazotrophs in maintaining soil health and factors affecting their growth and development. *Knowledgeable Research A Multidisciplinary Journal*, *2*(07), 62–70.

Neemisha, Kumar, A., Sharma, P., Kaur, A., Sharma, S., & Jain, R. (2022). Harnessing rhizobacteria to fulfil inter-linked nutrient dependency on soil and alleviate stresses in plants. *Journal of Applied Microbiology*, *133*(5), 2694–2716.

Nguyen, Q., Pham, P., & Le, T. (2024). Redox-regulated proteins stabilizing nitrogenase in oxidative environments. *Journal of Proteomics*, *295*, 105210.

Oubohssaine, M., Sbabou, L., & Aurag, J. (2024). Potential of the plant growth-promoting rhizobacterium *Rhodococcus qingshengii* LMR356 in mitigating lead stress impact on *Sulla spinosissima* L. *Environmental Science and Pollution Research*, *31*(33), 46002–46022.

Ouf, S. A., El-Amriti, F. A., Abu-Elghait, M. A., Desouky, S. E., & Mohamed, M. S. (2023). Role of plant growth promoting Rhizobacteria in healthy and sustainable agriculture. *Egyptian Journal of Botany*, *63*(2), 333–359.

Panchami, P. S., Geetha Thanuja, K., & Karthikeyan, S. (2020). Isolation and characterization of indigenous plant growth-promoting rhizobacteria (PGPR) from cardamom rhizosphere. *Current Microbiology*, *77*(10), 2963–2981.

Patel, S., Raval, J., & Desai, H. (2025). Nanocarrier-assisted delivery of microbial inoculants. *Nanotechnology in Agriculture*, *9*(1), 1–17.

Paungfoo-Lonhienne, C., Rentsch, D., Robatzek, S., Webb, R. I., Sagulenko, E., Näsholm, T., Schmidt, S., & Lonhienne, T. (2016). Crosstalk between sugarcane and a plant growth-promoting Burkholderia species creates microaerobic conditions supporting endophytic biological nitrogen fixation. *Scientific Reports*, *6*, 37382.

Prasad, M., Srinivasan, R., Chaudhary, M., Choudhary, M., & Jat, L. K. (2019). Plant growth promoting rhizobacteria (PGPR) for sustainable agriculture: Perspectives and challenges. In A. K. Singh, P. K. Singh, & A. Kumar (Eds.), PGPR Amelioration in Sustainable Agriculture: Food Security and Environmental Management. (pp. 129–157). Woodhead Publishing.

Racioppo, A., d'Amelio, A., De Santis, A., Bevilacqua, A., Corbo, M. R., & Sinigaglia, M. (2023). Potential use of plant growth-promoting bacteria to enhance growth and soil fertility in marginal areas: Focus on the Apulia Region, Italy. *Agronomy*, *13*(12), 2983.

Rahman, S., Akhtar, R., & Jahan, N. (2024). Screening diazotrophs for climate-resilient agriculture. *Frontiers in Sustainable Food Systems*, *8*, 1423875.

Ranjan, P., Sharma, A., & Kumar, P. (2025). Nano-encapsulation of microbial fertilizers improves field performance. *Advanced Agricultural Nanoscience*, *12*(1), 22–37.

Ribbe, M. W., Hu, Y., Hodgson, K. O., & Hedman, B. (2014). Structural basis of biological nitrogen fixation. *Chemical Reviews*, *114*(8), 4063–4080. https://doi.org/10.1021/cr400463x.

Ribeiro, C. M., & Cardoso, E. J. B. N. (2012). Isolation, selection and characterization of root-associated growth promoting bacteria in Brazil Pine (*Araucaria angustifolia*). *Microbiological Research*, *167*(2), 69–78.

Rillig, M. C., Hoffmann, L., & Ramirez, D. (2025). AMF–bacterial co-inoculation improves microbial diversity in maize rhizospheres. *Biology and Fertility of Soils*, *61*(2), 151–168.

Rivera-Hernández, G., Tijerina-Castro, G. D., Cortés-Pérez, S., Ferrera-Cerrato, R., & Alarcón, A. (2024). Evaluation of functional plant growth-promoting activities of culturable rhizobacteria associated to tunicate maize (*Zea mays* var. tunicata A. St. Hil), a Mexican exotic landrace grown in traditional agroecosystems. *Frontiers in Microbiology*, *15*, 1478807.

Shaffique, S., Khan, M. A., Wani, S. H., Pande, A., Imran, M., Kang, S. M., Rahim, W., Khan, S. A., Bhatta, D., Kwon, E. H., & Lee, I. J. (2022). A review on the role of endophytes and plant growth promoting rhizobacteria in mitigating heat stress in plants. *Microorganisms*, *10*(7), 1286.

Sharma, D., Gahtyari, N. C., Chhabra, R., & Kumar, D. (2020). Role of microbes in improving plant growth and soil health for sustainable agriculture. In *Advances in plant microbiome and sustainable agriculture: Diversity and biotechnological applications* (pp. 207–256). Springer Singapore.

Sharma, P., & Bhakri, G. (2019). Role of bacteria in Pedogenesis. In In *Mycorrhizosphere and Pedogenesis*. (pp. 167–179). Springer Singapore.

Shukla, R., Pandey, B., & Goswami, A. (2024). Multifunctional consortia combining diazotrophs and phosphate solubilizers. *Microbial Consortium Engineering*, *3*(1), 70–88.

Shyam, S., Ahmed, S., Joshi, S. J., & Sarma, H. (2025). Biochar as a soil amendment: Implications for soil health, carbon sequestration, and climate resilience. Discover Soil, 2, 18.

Singh, D. (2025). Encapsulation and polymer-based formulations for PGPM shelf-life improvement. *Journal of Bioinoculant Technology*, *9*(1), 45–63.

Singh, P., Sharma, R., & Patel, M. (2024). Nano-enabled microbial N-delivery enhances crop yield. *Plant Nutrition Nanoscience*, *6*(2), 134–149.

Song, W., Jiang, T., & Zhao, R. (2025). Soil fertility responses to repeated microbial inoculation under intensive cropping. *Soil Biology International*, *19*(3), 199–212.

Tafaroji, H., Hemati, A., & Nouri, H. (2025). Halotolerant PGPR for saline and alkaline soils. *Salinity Stress in Agriculture Journal*, *8*(1), 55–69.

Tang, H., Liu, Y., Yang, X., Huang, G., Liang, X., Shah, A. N., Nawaz, M., Hassan, M. U., Qumsani, A. T., & Qari, S. H. (2024). Multiple cropping effectively increases soil bacterial diversity, community abundance and soil fertility of paddy fields. *BMC Plant Biology*, *24*(1), 715.

Tang, S., Liu, Q., Yang, X., & Zhang, K. (2024). Crop rotation effects on diazotroph dynamics and soil fertility indicators. *Agricultural Microbiology*, *27*(4), 344–360.

Tschitschko, B., Esti, M., Philippi, M., Kidane, A. T., Littmann, S., Kitzinger, K., Speth, D. R., Li, S., Kraberg, A., Tienken, D., Marchant, H. K., Kartal, B., Milucka, J., Mohr, W., & Kuypers, M. M. M. (2024). Rhizobia–diatom symbiosis fixes missing nitrogen in the ocean. *Nature*, *630*(8018), 899–904. https://doi.org/10.1038/s41586-024-07495-w.

Turan, M., Arjumend, T., Argın, S., Yildirim, E., Katırcıoğlu, H., Gürkan, B., Ekinci, M., Güneş, A., Kocaman, A., & Bolouri, P. (2021). Plant root enhancement by plant growth promoting rhizobacteria. In Plant Roots (Chapter 7). IntechOpen. https://doi.org/10.5772/intechopen.99890

Umer, S., Rana, M., & Javed, H. (2025). Benefits of AMF–PGPR co-inoculation under abiotic stress. *Rhizosphere Science*, *4*(1), 17–29.

Upadhyay, S. K., Rajput, V. D., Kumari, A., Espinosa-Saiz, D., Menendez, E., Minkina, T., Dwivedi, P., & Mandzhieva, S. (2023). Plant growth-promoting rhizobacteria: A potential bio-asset for restoration of degraded soil and crop productivity with sustainable emerging techniques. *Environmental Geochemistry and Health*, *45*(12), 9321–9344.

Verbruggen, E., Staddon, P., & Rillig, M. (2023). Disease suppression in organic soils through microbial competition. *Soil Biology Reviews*, *14*(2), 211–230.

Vukicevich, E., Smith, C., & Martinez, F. (2024). Transfer of disease-suppressive soil traits through soil transplantation. *Frontiers in Soil Microbiology*, *15*, 1382296.

Wang, J., Zhou, P., & Chen, H. (2025). Engineered nitrogenase pathways in non-diazotrophic hosts. *Synthetic Biology in Agriculture*, *3*(1), 40–58.

Wang, T., Xu, J., Chen, J., Liu, P., Hou, X., Yang, L., & Zhang, L. (2024). Progress in microbial fertilizer regulation of crop growth and soil remediation research. *Plants*, *13*(3), 346.

Warmack, R. A., & Rees, D. C. (2024). Structural evolution of nitrogenase states under alkaline turnover. *Nature Communications*, *15*, 10472.

Welbaum, G. E., Sturz, A. V., Dong, Z., & Nowak, J. (2004). Managing soil microorganisms to improve productivity of agro-ecosystems. *Critical Reviews in Plant Sciences*, *23*(2), 175–193.

Wirth, L., Chen, B., & Vega, R. (2024). Modelling fertilizer savings and greenhouse gas mitigation from biofertilizers. *Environmental Agricultural Modelling*, *29*(4), 615–632.

Yadav, A., Chauhan, R., & Malik, S. (2024). Vermicompost-linked microbial biofertilizers increase yield and nutrient uptake. *Organic Agriculture and Soil Systems*, *10*(1), 23–38.

Yang, Z., Wu, Q., Liu, Z., Qi, X., Zhang, Z., He, M., Peng, C., Zeng, L., Wang, J., Li, F., & Yang, Z. (2023). Harnessing sulfate-reducing bacteria with plants growing to revitalize metal-tainted coal mine soils in Midwest China: Metal sequestration performance, ecological networking interaction, and functional enzymatic prediction. *Frontiers in Microbiology*, *14*, 1306573. p.

Yao, X., Lian, S., & Liu, P. (2024). Synthetic biology for modular nitrogen fixation pathways. *Biotechnology Advances*, *70*, 108838.

Zhang, D., Huang, Y., & Gao, R. (2024). AMF–bacteria interactions improve root architecture and drought tolerance. *Soil Microbiome Frontiers*, *2*(3), 145–160.

Zhang, L., Ramirez, M., & Wu, T. (2025). Genomics advances in diazotrophic PGPM research. *Annual Review of Microbial Genomics*, *9*, 221–249.

Zhao, G. (2024). Development of biofertilizers for sustainable agriculture: Frameworks and field evidence. *Agricultural Systems*, *2024*.

Zhao, H. (2024). Reduced mineral fertilizer input with biofertilizer co-application: A global synthesis. Global Plant. *Nutrition Journal*, *5*(2), 77–101.

Zhu, F., et al. (2024). Effects of leguminous green manure–crop rotation on soil enzyme activity and stoichiometry. *Journal of Plant Ecology*, *17*(6).

Zhumakayev, A. R., Vörös, M., Szekeres, A., Rakk, D., Vágvölgyi, C., Szűcs, A., Kredics, L., Škrbić, B. D., & Hatvani, L. (2021). Comprehensive characterization of stress tolerant bacteria with plant growth-promoting potential isolated from glyphosate-treated environment. *World Journal of Microbiology and Biotechnology*, *37*(6), 94.

Zhyr, L., Furch, A. C., & Mithöfer, A. (2025). Beneficial microbes in agriculture: curse or blessing? *Trends in Plant Science*, *31*(5), 566–575.

Rhizobia to Cyanobacteria: Advances in Nitrogen Fixation Research

8

Soumik Samanta, Kaushik Maity, Sujit Kumar Bhunia, Shivaji Chattopadhyay, Debashis Mandal, Apurba Bandyopadhyay and Amit Baran Sharangi

Abstract

Nitrogen fixation is crucial for ecological health, converting atmospheric nitrogen (N_2) into ammonia (NH_3) that supports life on Earth. Rhizobia bacteria form symbiosis via specialized nodules in legume roots, directly supplying nitrogen to their host plants, while associative fixers like Azospirillum inhabit roots to boost nutrient absorption. Free-living cyanobacteria are highly versatile, which enrich ecosystems by fixing nitrogen, improving soil fertility, fertilizing rice, and revitalizing infertile lands through legume symbiosis and supportingnon-leguminous crops and natural plant growth. Modern biotechnologies, including genetic engineering and synthetic biology, have strengthened nitrogen fixation efficacy in rhizobia and cyanobacteria, which promotes sustainable farming by

S. Samanta · A. Bandyopadhyay (✉) · A. B. Sharangi
Department of Plantation, Spices, Medicinal and Aromatic Crops, Faculty of Horticulture, Bidhan Chandra Krishi Viswavidyalaya, Nadia, West Bengal, India
e-mail: apurba.bandyopadhyay.bckvmaps@gmail.com

K. Maity
Department of Agricultural Meteorology and Physics, Faculty of Agriculture, Bidhan Chandra Krishi Viswavidyalaya, Nadia, West Bengal, India

S. K. Bhunia
Department of Agricultural Chemistry and Soil Science, Faculty of Agriculture, Bidhan Chandra Krishi Viswavidyalaya, Nadia, West Bengal, India

S. Chattopadhyay
Department of Fruit Science, Faculty of Horticulture, Bidhan Chandra Krishi Viswavidyalaya, Nadia, West Bengal, India

D. Mandal
Department of Agricultural Entomology, Faculty of Agriculture, Bidhan Chandra Krishi Viswavidyalaya, Nadia, West Bengal, India

R. K. Singh and P. Singh (eds.), *Diazotrophic Plant Growth-Promoting Microorganisms and Green Energy – Volume I*, https://doi.org/10.1007/978-981-95-9682-9_8

reducing dependence on harmful synthetic fertilizers. Biofertilizers not only sustain crop yields but also benefit medicinal and aromatic plants such as basil, mint, and patchouli while enhancing secondary metabolites in tulsi and lavender. Derivatives with antimicrobial properties help protect turmeric and ginger from fungal infections by increasing stress resistance. Integrating genomics and microbiome research with field applications empowers farmers to meet rising costs and climate challenges. Strong policies and interdisciplinary collaboration are crucial for translating these scientific advances into practical solutions. From rhizobia to cyanobacteria, nitrogen fixation remains nature's model for sustainable and resilient agriculture.

Keywords

Nitrogen fixation · Environmental sustainability · Secondary metabolites · Climate-smart farming · Policy · And farmer perspectives ·

8.1 Introduction

Like the turning gears of a clock, the biogeochemical cycles of Earth likewise proceed in harmony. The progress of one cycle can influence others, creating feedback loops, as seen in the nitrogen cycle. Nitrogen is vital for life, forming the core of nucleic acids and proteins. Yet, most nitrogen in atmospheric N_2 is unavailable to most life forms, except for a small amount that is dissociated by high-energy events. Nitrogenase, the only enzyme that fixes nitrogen into ammonia, a more bioavailable form, is a central part of the nitrogen cycle in both land and sea environments. It remains a crucial innovation for maintaining Earth's biosphere. Consequently, all life depends, directly or indirectly, on nitrogenase for nitrogen fixation—the process that makes atmospheric nitrogen accessible to life.

Since 1900, the world population has increased from 1.6 billion to over 7 billion today, expected to reach 9 billion by 2050 (FAO, 2015; Johnson & Grim, 2010). Feeding this population has driven an unprecedented rise in synthetic fertilizer use, especially nitrogen (Hirel et al., 2011; Mueller et al., 2012; Soumare et al., 2020;Liu et al., 2016). Crop yields rely heavily on nitrogen (N), which is often a limiting resource, and the gap between nitrogen supply and demand continues to widen (FAO, 2015; Roy et al., 2006; Maheswari et al., 2017). Conversely, overuse of fertilizers can lead to environmental issues. Use of inorganic nitrogen fertilizers has led to ecosystem perturbations across the world (Yang & Fang, 2015; Bishnoi, 2018; Zheng et al., 2019). This dual reality highlights the urgent need to reduce reliance on chemical fertilizers and to embrace sustainable alternatives (Araujo et al., 2012; Shah & Wu, 2019).

Among alternative approaches, biological nitrogen fixation (BNF) appears as a route to reduce the input of N fertilizers and their negative environmental impacts.

BNF naturally converts atmospheric nitrogen (N_2) into a simple, soluble, nontoxic form (NH_4^+ primarily), a soluble form used by plants for biomolecule synthesis, and represents a major nitrogen source in ecosystems (Saikia & Jain, 2007; Sur et al., 2010). BNF is performed exclusively by prokaryotes: archaea and bacteria. Bacterial groups include free-living fixers (*Azotobacter, Azospirillum, Bacillus*, and *Clostridium)*, as well as symbiotic fixers (Rhizobium in legumes, Frankia in actinorhizal plants, and cyanobacteria in cycads) (Ravikumar et al., 2007; Ininbergs et al., 2011). In archaea, nitrogen fixation is still restricted to groups that produce methane, known as methanogens (Welte, 2018). Overall, nitrogen-fixing organisms can be categorized into three types: free-living, associative, and symbiotic fixers, with the latter two often occurring in the rhizosphere of both legumes and non-legumes (Santi et al., 2013; Mus et al., 2016).

Root nodule symbiosis is one of the most studied and effective mutualisms in nitrogen fixation (20–300 kg ha^{-1}), involving nearly all food and fodder legumes. Its establishment begins with a molecular dialog between host plants and nitrogen-fixing organisms via root-secreted flavonoids and isoflavonoids (Liu & Murray, 2016; Jiménez-Jiménez et al., 2019), leading to recognition, infection, root hair differentiation, and nodule development. Inside nodules, bacteria in the bacteroides form fix nitrogen (Cissoko et al., 2018; Suzaki et al., 2019). Among these, the legume–*Rhizobium* model has received the greatest attention, as legumes are major food and cash crops, and have been used to develop effective biofertilizer inocula, especially for poor soils with minimal N input (Mugabe, 1994; Chianu et al., 2011; Beyan et al., 2018; van Heerwaarden et al., 2018).

In recent decades, research attention has shifted beyond rhizobia toward cyanobacteria—based biotechnology emerging as a promising solution (Kondi et al., 2022; Crews et al., 2004). Cyanobacteria (blue-green algae), photosynthetic microorganisms, fix atmospheric nitrogen into bioavailable forms (Nawaz et al., 2023), offering alternatives to conventional fertilizers and helping mitigate eutrophication, soil degradation, and greenhouse gas emissions resulting from the overuse of synthetic fertilizers(Singh et al., 2016a; Ashitha & Rakhimol, 2021). Advances in genetically engineered cyanobacteria enable direct nitrogen transfer to plants, reducing fertilizer dependence while enhancing stress tolerance and productivity (Mishra et al., 2021; Fedeson & Ducat, 2021). Beyond nitrogen fixation, cyanobacteria produce diverse bioactive compounds with applications in agriculture, pharmaceuticals, and industry (Hamed, 2016; Gupta et al., 2013), supporting food security, sustainability, and climate resilience (Gomiero, 2016; Żymańczyk-Duda et al., 2022; Nawaz et al., 2024a). As ancient organisms, they have evolved complex regulatory mechanisms for nitrogen assimilation (Berman-Frank et al., 2003; El-Sheekh et al., 2022). Applied as biofertilizers, they improve soil health and reduce chemical inputs (Esteves-Ferreira et al., 2018; Pathak et al., 2018), though large-scale use requires assessing ecological impacts, cost-effectiveness, and scalability (Pahalvi et al., 2021; Saleem et al., 2024; Poveda et al., 2021; Muhie, 2022). Overall, cyanobacteria-based biotechnology offers a sustainable pathway for nitrogen management and metabolite production (Nawaz et al., 2024b).

Against this backdrop, agriculture faces mounting bioenvironmental challenges, from plateauing yields to quality decline in medicinal and aromatic plants under chemical-intensive systems. Farmers are increasingly turning to microbial approaches to restore soil health, enhance nutrient uptake, and ensure crop quality, though commercial products remain inconsistent and often poorly characterized.

In light of the challenges agriculture faces within the current bioenvironmental security framework, biotechnological innovations for plant species present themselves as a viable alternative (Nawaz et al., 2024c). This review aims to consolidate existing knowledge, highlight recent advances, and outline future directions in this evolving field. Specifically, it emphasizes the potential of cyanobacteria-based plant biotechnology in conserving nitrogen and promoting sustainable metabolite production, thereby contributing to ongoing discussions on sustainable nitrogen management (Fig. 8.1).

8.2 Historical Perspective & Foundational Biology

Biological nitrogen fixation was first reported by Hellriegel and Wilfarth (1886), who showed that some legumes could utilize atmospheric N_2 via root nodules (Francheet al., 2009). In 1888, *Rhizobium leguminosarum* was isolated by Beijerinck, followed by *Clostridium pasteurianum* in 1893 (Winogradsky, 1895). The practice of inoculating legumes with N-fixing rhizobia was introduced to farmers in New South Wales, Australia, by Guthrie in 1896 (Guthrie, 1896). Beijerinck also described the aerobic heterotroph *Azotobacter* in (1901). Early field trials using rhizobial cultures on peas

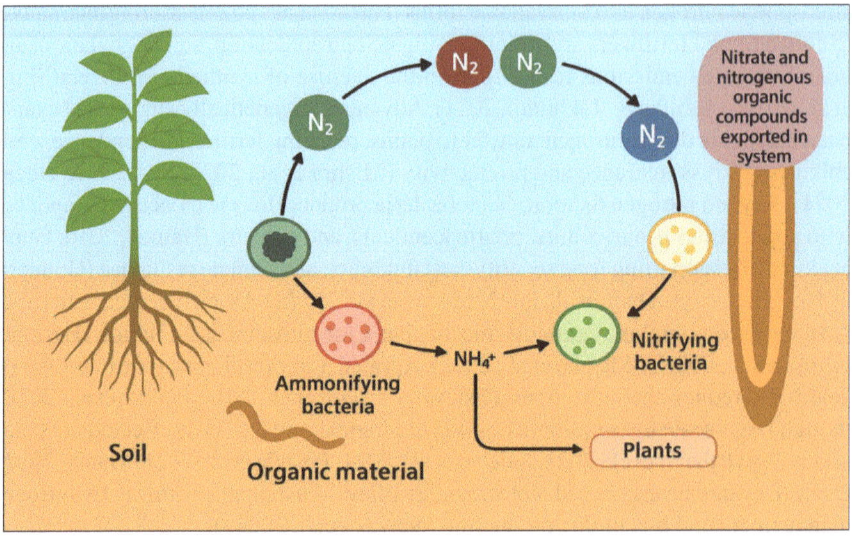

Fig. 8.1 Contribution of diazotrophs to nitrogen dynamics in ecosystems

were conducted in 1905 at Hawkesbury Agricultural College, with commercial inoculants supplied to farmers by 1914 (Bullard et al., 2005).

Johanna Döbereiner's work in 1953 expanded BNF knowledge with the identification of nitrogen-fixing endophytes, including *Beijerinckia fluminensis* in sugarcane and *Azotobacter paspali* in Bahia grass (Döbereiner, 1966; Baldani & Baldani, 2005). In 1975, Fabio Pedrosa developed the nitrogen-free (NFb) medium to isolate *Spirillum* species, leading to the discovery of *Azospirillum lipoferum* and *A. brasilense*, heterotrophic, non-symbiotic bacteria fixing 20–30 kg N ha^{-1} annually (Baldani and Baldani, 2005, Döbereiner, 1976, Kizilkaya, 2009). Cyanobacterial nitrogen fixation was confirmed in 1979 using the ^{15}N tracing technique (Stewart, 1969), while archaeal nitrogen fixation was recently demonstrated in methanogens *Methanosarcina barkeri* and *Methanococcus thermolithotrophicus* using ^{15}N tracing and acetylene reduction assays (Murray & Zinder, 1984; Leigh, 2000; Belay et al., 1984).

Currently, at least thirteen prokaryotic genera are recognized as nitrogen fixers, with many additional diazotrophs yet to be discovered, as revealed by analysis of 16,989 *nif* sequences (Gaby & Buckley, 2011). These discoveries paved the way for commercial biofertilizers, beginning with Nitragin (*Rhizobium* strains) in the USA in 1895 (Khosro & Yousef, 2012), followed by Azotogen (*Azotobacter chroococcum*) and Phosphobactin (*Bacillus megaterium cv. phosphaticum*) in Russia. In India, production started at ICAR-IARI in 1956 (Swarnalakshmi et al., 2016), while in Africa, MIRCEN has promoted BNF knowledge and inoculant production since 1977, with widespread public and private involvement by the late 1990s (Karanja et al., 2000). Comparative genomics reveals the evolutionary diversity of nitrogen-fixing microbes, spanning from early cyanobacteria to specialized rhizobia and Frankia. Horizontal gene transfer has played an important role in the dissemination of nif genes across taxa. The study of these evolutionary patterns contributes to understanding the origin of symbiotic partnerships and adaptive strategies under various ecological situations. Genomic insights can help drive synthetic biology efforts to enhance nitrogen fixation in nontraditional hosts.

8.3 Symbiotic Relationships Between Cyanobacteria and Plants

The symbiotic interactions between cyanobacteria and plants represent a remarkable facet of their contribution to plant biotechnology (Bergman et al., 2008). Many plant species have evolved symbiotic relationships with cyanobacteria, facilitating mutualistic exchanges that involve reciprocal nutrient sharing. In these associations, cyanobacteria supply fixed nitrogen to plants, while the plants provide carbon compounds produced through photosynthesis (Usher et al., 2007; Oliver et al., 2012). For instance, the interaction between certain crops (e.g., legumes like soybeans and clover) with nitrogen-fixing cyanobacteria or rhizobia is a well-characterized mutualistic symbiosis (Mahmud et al., 2020). In these symbiotic partnerships, the host plants develop specialized organs (nodules) that support bacterial nitrogen fixation, creating a microenvironment in which cyanobacteria carry out their reduction reactions. This arrangement ensures a consistent nitrogen supply for the plants, minimizing their dependence on external nitrogen sources (Masclaux-Daubresse et al., 2010). The molecular mechanisms underlying symbiosis in these soft tissue-associated systems are intricate, involving signaling pathways and genetic regulation (Ahmad et al., 2024). The recognition process between plant and cyanobacterial partners starts with root hair deformation, eventually leading to the formation of nodules or specialized structures for nutrient transfer (Meeks & Elhai, 2002). Understanding these processes is crucial for optimizing nitrogen fixation in agriculture and for leveraging the biotechnological potential of cyanobacteria to achieve sustainable crop production (Goyal et al., 2021). Beyond nitrogen fixation, cyanobacteria also release bioactive compounds and growth-promoting substances that enhance plant health and growth. By boosting the production of defense molecules and enhancing stress resistance, these compounds are integral to the wide array of mechanisms through which cyanobacteria interact with plants (Kollmen & Strieth, 2022). With ongoing advances in plant biotechnology, considerable progress has been made in manipulating symbiosis via genetic engineering. For example, researchers are aiming to increase cyanobacterial nitrogen fixation efficiency (Gr et al., 2021) and expand the range of plant species capable of forming such associations, while also exploring additional benefits like improved tolerance to abiotic stress (Slater et al., 2008). This overview introduces the diverse roles of cyanobacteria in plant biotechnology, emphasizing nitrogen fixation and symbiotic relationships as key factors for sustainable agriculture. A comprehensive understanding of these complex interactions is essential to fully harness cyanobacteria's potential to enhance plant productivity, improve soil fertility, and reduce the environmental impact of conventional agricultural practices. Current research efforts focus on developing innovative, ecologically sustainable strategies to tackle emerging challenges in agriculture (Prasanna et al., 2014; Santini et al., 2021; Hossain & Baksh, 2020) (Table 8.1).

8.4 Mechanisms of Nitrogen Fixation

8.4.1 Biological Nitrogen Fixation by Cyanobacteria

Cyanobacteria, often known as blue-green algae, are extraordinary microorganisms that possess the capacity to undertake biological nitrogen fixation (BNF) (Fatimaet al., 2019). This mechanism turns atmospheric nitrogen (N_2) into ammonia (NH_3) or similar molecules that plants and other organisms may use. Nitrogen-fixing cyanobacteria in the mangrove ecosystem, e.g., reduce N_2 into bioavailable ammonia via nitrogenase action and their nitrogen-fixing capacity accounts for approximately 40%–60% of the total nitrogen required by the mangrove ecosystem (Holguin et al., 2001), which can provide "new" nitrogen sources for the mangrove and drive the assimilation and storage of marine carbon. It is critical to maintain mangrove primary output and nitrogen balance (Wang et al., 2021). Nitrogenase, the enzyme complex responsible for this conversion, is present in specialized cells termed heterocysts in cyanobacterial filaments (Krupa, 2003; Mishra et al., 2019).

Table 8.1 Comparative overview of major nitrogen-fixing microorganisms

Organism/group	Host plant type	Fixation mode	Agricultural applications/ importance
Rhizobium spp.	Legumes (e.g., peas, beans, clover, soybeans)	Symbiotic (forms root nodules)	Used as biofertilizers (Rhizobium inoculants); major contributor to nitrogen input in legume-based systems
Bradyrhizobium spp.	Legumes (tropical species such as soybean, cowpea)	Symbiotic	Enhances nitrogen availability in tropical legumes; commonly applied in soybean inoculants
Sinorhizobium (Ensifer) spp.	Legumes (alfalfa, clovers)	Symbiotic	Improves soil fertility and forage productivity in temperate regions.
Azospirillum spp.	Cereals and grass (maize, wheat, rice, sorghum)	Associative (root-surface colonizer)	Promotes plant growth via nitrogen fixation and phytohormone production; used in biofertilizers for cereals
Azotobacter spp.	Nonspecific (free-living in soil)	Free-living	Enhances soil fertility; produces growth-promoting substances; useful in nonleguminous cropping systems
Cyanobacteria (e.g., Anabaena, Nostoc)	Free-living or symbiotic with Azolla, lichens, and cycads	Free-living and symbiotic	Important in rice paddies (Anabaena–Azolla symbiosis); enhances fertility in flooded soils
Frankia spp.	Actinorhizal plants (Alnus, Casuarina, Myrica)	Symbiotic (root nodules)	Used in reforestation and reclamation of nitrogen-poor soils
Herbaspirillum spp.	Grasses (sugarcane, rice, maize)	Endophytic associative	Improves nitrogen uptake and crop growth; potential biofertilizer for tropical cereals and sugarcane
Beijerinckia spp.	Free-living in soil and leaf litter	Free-living	Increases nitrogen levels in soil, supports non-leguminous plant growth

Organism/group	Host plant type	Fixation mode	Agricultural applications/ importance
Burkholderia spp.	Sugarcane, maize, rice	Associative/Endophytic	Enhances yield and nitrogen assimilation; useful in integrated nutrient management

8.4.2 Heterocyst Differentiation

Cyanobacteria employ a complex mechanism to establish an anaerobic environment within the heterocyst, which is necessary for nitrogenase activity. Heterocysts often form in reaction to nitrogen deficiency, which is triggered by the detection of lower nitrogen levels in the environment (Haselkorn, 1978). Heterocysts differentiate by a sequence of morphological and metabolic changes. The cells that surround the heterocyst develop into specialized vegetative cells that promote the heterocyst's nitrogenase activity (Herrero et al., 2013; Kumar et al., 2010).

8.4.3 Nitrogenase Complex

The nitrogenase enzyme complex is crucial to nitrogen fixation. It is made up of two major components: the iron protein (Fe protein) and the molybdenum iron protein (MoFe protein) (Kim & Rees, 1994). The Fe protein acts as an electron transporter, carrying electrons to the MoFe protein, where nitrogen fixation occurs. Oxygen levels heavily control nitrogenase activity. Although heterocysts do not engage in oxygen-evolving photosynthesis, they do provide an anaerobic environment that is essential for optimum nitrogenase activity (Seth et al., 2022; Scherings et al., 1977).

8.4.4 Regulation of Nitrogen Fixation in Cyanobacteria and Rhizobia

Nitrogen fixation is strictly controlled to maintain cellular energy and nitrogen needs. In cyanobacteria, nif genes are regulated by environmental factors such as nitrogen availability, oxygen levels, and light, which are frequently mediated by transcriptional regulators such as NtcA. Rhizobia detect nitrogen status and control nitrogenase activity via complicated signaling networks containing NtrB/NtrC and PII proteins. Understanding these regulatory mechanisms is critical for maximizing nitrogen fixation efficiency in both natural and manufacturedsystems.

8.4.5 Nitrogenase Reaction

The overall nitrogenase process consists of the reduction of atmospheric nitrogen to ammonia. This activity demands a large quantity of energy, which is normally provided by photosynthesis in neighboring vegetative cells. The electron transport chain within cyanobacterial cells is critical in transporting the electrons necessary for nitrogenase activity (Mishra et al., 2019; González et al., 2016) (Fig. 8.2).

8.4.6 Nitrogen Assimilation by Plants

While cyanobacteria contribute to nitrogen availability through fixation, plants use a variety of processes to assimilate fixed nitrogen into their cellular components (Stewart, 1973).

8.4.7 Nitrate and Ammonium Uptake

Plants often take up nitrogen in the form of nitrate (NO_3^-) and ammonium (NH_4^+). Nitrate is generally the dominating form in soil, and plants have specialized transporters that absorb nitrate from the soil solution (Ohyama, 2010). Ammonium, another common type of soil nitrogen, is taken up directly by plants. Plant roots may absorb this ion more easily thanks to ammonium transporters (Glass et al., 2002).

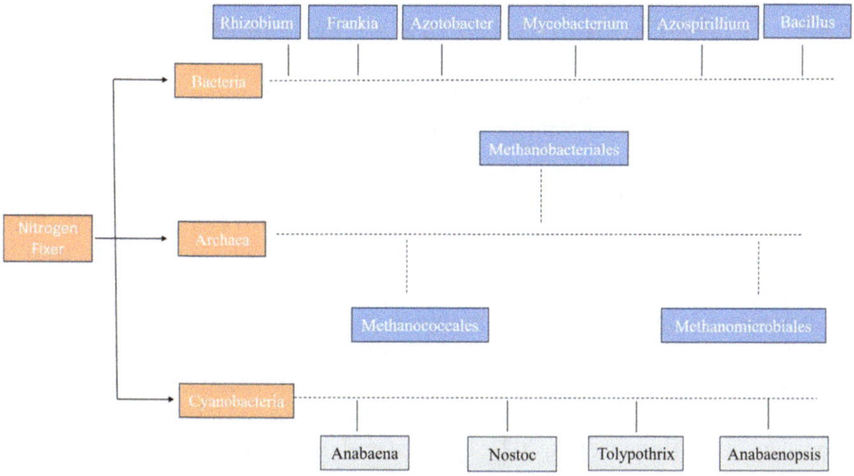

Fig. 8.2 Taxonomic overview of nitrogen-fixing microorganisms

8.4.8 Nitrate Reduction

After absorption, nitrate is reduced in the cytoplasm of plant cells (Clarkson, 1986). This reduction comprises a number of enzymatic processes that eventually result in the conversion of nitrate to ammonium. Nitrate reductase and nitrite reductase are essential enzymes in this process, reducing nitrate to nitrite and finally to ammonium (Besson et al., 2022; Moreno-Vivián et al., 1999).

8.4.9 Ammonium Assimilation

Opportunities for scientific gains in upcoming biotechnological frontiers, such as the use of CRISPR-Cas9 and multiomics techniques, seem promising (Miflin & Lea, 1980). The inclusion of bioreactor systems in our burgeoning advanced conservation toolbox demonstrates how far we've gone with existing technologies. The focus on stress tolerance and plant genetic transformation is finally allowing us to see what we can genuinely do with a little help from our buddies (Inokuchi et al., 2002). We are coming closer to achieving our dual goals of assisting endangered species reproduction and producing bioactive chemicals (De Vries & De Vries, 2018). These are the footsteps of scientific investigation, and we hope that future generations will go one step closer to a more sustainable route. Furthermore, cyanobacteria-derived plant biotechnology provides actual, practical tools for sustainable development, as opposed to theoretical constructions. The symbiotic relationship between cyanobacteria and plant biotechnology has the potential to pave the way for significant advancements in the development of long-term sustainable ecosystems (Vitousek et al., 1997) (Table 8.2).

8.5 Advanced Biotechnological Tools

8.5.1 CRISPR-Cas9 and its Role in Plant Regeneration

CRISPR-Cas9 has revolutionized genome editing with precision and efficiency, enabling trait improvement for plant growth, yield, and stress tolerance. In cyanobacteria, CRISPR/Cas9 has been applied to redirect carbon flux from glycogen to succinate production by deleting the *glgC* gene; this increased succinate yield to 0.18 mg/L. Further introduction of *ppc* and *gltA* genes enhanced titters to 0.44 mg/L (Li et al., 2016; Zeikus et al., 1999; Behler et al., 2018). Such studies confirm the potential of CRISPR-based metabolic engineering in cyanobacteria. In plants, CRISPR-Cas9 components can be delivered via PEG-mediated protoplast transformation, Agrobacterium-based systems, or de novo meristem induction. These approaches allow precise editing of regulatory genes controlling tissue development, organogenesis, and regeneration, resulting in disease- and stress-resistant plants with improved adaptability (Zhang & Showalter, 2020; Hesami et al., 2021; Rasheed et al., 2021; Bailey-Serres et al., 2019).

Table 8.2 The ecological role of bryophyte–cyanobacteria associations

Type of symbiosis	Bryophyte partner	Relationship with Cyanobacteria(En = Endophytic/Ep = Epiphytic)	Cyanobacterial partner	Ecological significance/ function
Mutualistic symbiosis	*Anthoceros* (Hornwort)	En	*Nostoc* spp.	Fixes atmospheric nitrogen in the thallus cavities, enriching soil nitrogen and supporting plant growth
Mutualistic symbiosis	*Blasia pusilla* (Liverwort)	En	*Nostoc* spp.	Enhances nitrogen availability; contributes to colonization in nutrient-poor soils and early succession
Epiphytic sssociation	*Sphagnum* spp. (Moss)	Ep	*Nostoc* spp., *Stigonema* spp.	Supplies biologically fixed nitrogen to peatlands, influencing carbon sequestration and peat accumulation
Facultative symbiosis	*Phaeoceros* spp. (Hornwort)	En	*Nostoc punctiforme*	Improves bryophyte growth and enables colonization in low-nitrogen habitats
Loose surface association	*Marchantia* spp. (Liverwort)	Ep	*Nostoc commune*	Provides external nitrogen fixation; helps in recolonization of disturbed soils
Commensal association	*Riccia fluitans* (Aquatic liverwort)	Ep	*Nostoc* spp.	Cyanobacteria gain shelter and moisture; bryophyte surface remains largely unaffected
Ecological guild (community-based)	Bryophyte mats in boreal forests	Ep	*Nostoc*, *Calothrix*, *Stigonema* spp.	Major nitrogen input in boreal ecosystems supports microbial and plant community diversity

8.5.2 Bioproducts Beyond Nitrogen Fixation

Cyanobacteria generate a diverse range of bioactive substances, such as phytohormones, antioxidants, pigments, antimicrobials, and soil conditioners. These metabolites help plants develop, withstand stress, and maintain soil health. Commercial utilization of these chemicals enhances cyanobacterial biotechnology beyond nitrogen fixation. Cyanobacteria's dual role in nutrient delivery and metabolite synthesis makes them flexible agents for sustainable agriculture and biotechnology.

8.5.3 Nanoparticles and Nanotechnology in Plant Tissue Culture

Nanotechnology is now widely explored in tissue culture for its role in improving nutrient delivery, medium optimization, and tissue development (Moulin et al., 2024). Cyanobacteria, acting as natural biofactories, can generate nanoparticles (NPs) through intra- or extracellular synthesis due to their rapid growth, versatility, and ecofriendly metabolism (Sharma et al., 2016). They produce diverse NPs—Ag, Au, ZnO, CuO, Ag–Au alloys, and CdS semiconductors (Mubarak Ali et al., 2013; Lengke et al., 2006; Ebadi et al., 2019). Engineered nanoparticles can encapsulate nutrients or genetic material, enabling targeted delivery to plant cells, thereby improving regeneration and stress tolerance (Pazhamala & Varshney, 2015; Sharma et al., 2020). Additionally, NPs modify culture conditions—pH, osmotic pressure, ion availability—enhancing medium efficiency and reducing contamination risks (Ranjbar & Malcata, 2022; Nawaz et al., 2024a). Cyanobacteria-based nano-formulations thus represent a convergent approach for advancing tissue culture and epigenetic reprogramming.

8.5.4 Multiomics Approaches to Nitrogen Fixation

Multiomics tools—genomics, transcriptomics, proteomics, and metabolomics—offer a systems-level view of nitrogen fixation (Shantharam & Mattoo, 1997; Munaweera et al., 2022). Genomics identifies key genes and regulatory elements; transcriptomics maps their dynamic expression; proteomics reveals protein abundance and modifications; and metabolomics profiles metabolites linked to nitrogenase pathways (Singh & Bhadury, 2019; Xiong et al., 2016; Huang & McDonald, 2012). Integrated multiomics not only uncovers novel metabolites and regulatory networks but also reveals bottlenecks, guiding metabolic engineering. For example, recent studies isolated ~31 nitrogen-fixing bacteria from mangrove ecosystems, spanning genera such as *Phyllobacterium, Azotobacter, Vibrio*, and newly described taxa like *Swaminathania* and *Mangrovibacter*. These insights deepen our understanding of microbial contributions to N assimilation and support the possibility of transferring N-fixation capacity to nonfixing organisms (Xiong et al., 2016; Huang & McDonald, 2012; Huang et al., 2014).

8.5.5 Microbial Consortia and Synthetic Communities

Nitrogen-fixing cyanobacteria frequently form complex microbial consortia with PGPR, mycorrhizal fungi, and other soil microorganisms. These communities work together to promote nutrient cycling, disease control, and plant development. Synthetic microbial consortia may be created to improve nitrogen fixation and other advantageous characteristics, making them a sustainable alternative to chemical fertilizers. Multiomics techniques can assist in identifying interspecies interactions and drive consortia design for agricultural applications.

8.5.6 Environmental and Stress Influences on Nitrogen Fixation

Abiotic stressors such as drought, salinity, and heavy metal toxicity have a significant impact on nitrogenase activity. Cyanobacteria attenuate these impacts by producing exopolysaccharides, forming heterocysts, and utilizing antioxidant mechanisms to sustain nitrogen fixation under adverse circumstances. Similarly, rhizobia and related microorganisms adapt by expressing stress-responsive genes and chemicals that promote symbiosis. Understanding these interactions is essential for creating robust biofertilizers for climate-affected agriculture.

8.6 Cyanobacteria in Agriculture

8.6.1 Cyanobacterial Biofertilizers

Cyanobacteria are widely distributed in soils, including agricultural fields, rice paddies, and marshy soils, and form part of the plant growth-promoting rhizobacteria (PGPR) community. PGPR improve nutrient availability, soil structure, and water retention (Kawalekar, 2013; Singh et al., 2016b) through nitrogen fixation, phytohormone production (auxins, cytokinins, gibberellins, ethylene), siderophore release, mineral solubilization (P, K, Zn), and biofilm formation (Hyder et al., 2023). Cyanobacteria specifically reduce N_2 to ammonia via heterocyclic nitrogenases, whose activity depends on metal cofactors (Mo, V, Fe) (Kumar et al., 2010; Fujita & Uesaka, 2022). Nitrogenase activity is optimal at 5%–7.5% oxygen but inhibited above 15% (Compaoré & Stal, 2010), while Mo and Fe enhance activity in species such as *Trichormus variabilis* and *Nostoc* (Sadvakasova et al., 2022). Cyanobacteria show strong biofertilizer potential, contributing to N_2 fixation, organic matter decomposition, detoxification of pollutants, nutrient cycling, pathogen suppression, and production of growth-promoting compounds (Singh et al., 2016a). They are abundant in rice paddies, either floating in water or on soil surfaces (Prasanna et al., 2012; Vijayan & Ray, 2015; Iniesta-Pallarés et al., 2021; Song et al., 2022), where they associate with roots and add ~20–30 kg N ha^{-1} (Venkataraman, 1972, 1975; Prasanna & Nayak, 2007). Major nitrogen-fixing genera include *Anabaena, Aulosira, Nostoc, Calothrix, Tolypothrix,*

Scytonema, and *Plectonema* (Prasanna et al., 2012; Sahu et al., 2012). *Calothrix* inoculation enhanced rice root and shoot growth five-fold (Priya et al., 2015), while *Azolla*-associated *Nostoc azollae* has long been used as rice biofertilizer in Asia (Fan, 1992; Bocchi & Malgioglio, 2010; Yadav et al., 2014; Nosheen et al., 2021). In Spain, field trials with native cyanobacteria from five phylogenetic groups (*Nostoc, Wollea, Anabaena, Calothrix*) showed greener plants, heavier grains, and higher panicle yield (Iniesta-Pallarés et al., 2021). Beyond rice, cyanobacterial biofertilizers have shown positive effects in barley, wheat, oats, radish, cucumber, tomato, pumpkin, cotton, sugarcane, chilli, and lettuce (Hashtroudi et al., 2013; Prasanna et al., 2013; Khadarate & Suryawanshi, 2016). For instance, cocultivation with *Anabaena cylindrica* and *Nostoc* significantly boosted lettuce yield (Xue et al., 2017) (Fig. 8.3).

8.6.2 Cyanobacterial Growth Regulators

Cyanobacteria produce plant growth regulators such as auxins and cytokinins (Sergeeva et al., 2002; Prasanna et al., 2008, 2009, 2010; Tan et al., 2021; Uniyal et al., 2022), which, like their plant-derived counterparts, enhance growth, root development, and stress tolerance. Auxin biosynthesis (IAA, IPA, IBA) has been reported in several strains (Hashtroudi et al., 2013; Mazhar et al., 2013; Ahmed et al., 2014). *Nostoc* ipdC knockout mutants showed reduced promotion of rice and wheat seedling growth, confirming a role for cyanobacterial auxin (Hussain et al., 2015). Supplementing *Fisherella musicola* NDUPC001 with L-tryptophan increased IAA production, and extracts from treated cultures enhanced rice radicle and plumule growth and adventitious roots (Mishra et al., 2019). Cytokinins, key regulators of cell division, shoot formation, and nutrient uptake (Zhao et al., 2021; Uniyal et al., 2022), are also

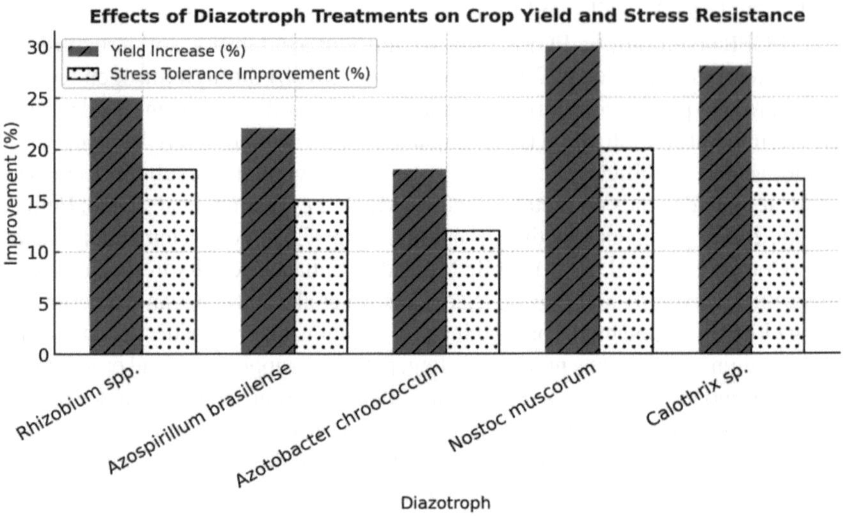

Fig. 8.3 Comparative effects of diazotrophs on crop yield enhancement and stress tolerance

produced by cyanobacteria (Hussain et al., 2010, 2013). Their extracts boost plant growth, yield, and stress tolerance (Toribio et al., 2020, 2021). A *Nostoc* ipt mutant showed decreased cytokinin synthesis and plant colonization, indicating their role in plant–microbe interactions (Hussain et al., 2013). Salicylic acid (SA), essential for plant defence, germination, root development, and flowering, is closely linked to cytokinin signalling (Gilroy & Breen, 2022). Cyanobacteria such as *Calothrix* SAB-B797 and *Nostoc* strains SAB-B1300 and SAB-M612 produce SA, which was shown to enhance watercress seed germination (Toribio et al., 2020).

8.6.3 Cyanobacteria and Soil Dynamics

Beyond phytohormones, cyanobacteria synthesize siderophores—small metal-chelating molecules essential for microbial iron uptake (Årstøl & Hohmann-Marriott, 2019; Chakraborty et al., 2019). Hydroxamate-type siderophores are most common and occur in species such as *Anabaena catenula, A. cylindrica, A. variabilis, Aphanizomenon flos-aquae, Microcystis aeruginosa, Oscillatoria tenuis, O. boryana, Phormidium valderianum*, and *Synechocystis elongatus* (Årstøl & Hohmann-Marriott, 2019). These compounds have diverse applications in ecology, agriculture, and medicine, including iron-chelation therapy and as antibiotic carriers (Kundu et al., 2023). Cyanobacterial siderophores enhance Fe uptake in plants, thereby promoting growth and yield (Toribio et al., 2020). Cyanobacteria also mobilize insoluble minerals, notably phosphorus. For instance, *Westiellopsis prolifica* and *Anabaena variabilis* solubilized tricalcium phosphate, making it accessible to plants (Yandigeri et al., 2011). Additionally, exopolysaccharides (EPSs) produced by strains like *Nostoc muscorum* improve soil aggregation and water retention; their application increased water-stable aggregates in soil (Rossi & De Philippis, 2015; Garlapati et al., 2019). Together, siderophore production, mineral solubilization, and EPS secretion highlight cyanobacteria as PGPR that enhance growth, stress tolerance, and yield, positioning them as sustainable tools for ecofriendly agriculture.

8.7 Bioengineering of Non-Legumes Beyond Cereals

Beyond cereals, nitrogen-fixing relationships are being investigated in millets, horticultural crops, medicinal plants, and hydroponic systems. Endophytic or epiphytic colonization by cyanobacteria in these crops might minimize fertilizer use and increase yield quality. Synthetic biology allows for the selective expression of nif genes or signaling pathways in nonlegumes, creating prospects for crop diversification with sustainable nitrogen control.

8.8 Cyanobacteria-Enabled Nitrogen-Fixing Cereals

Global cereal production, currently estimated at 206 Mt, is projected to reach 542 Mt by 2030 (OECD-FAO, 2021). One of the main limitations for sustaining this rise is nitrogen uptake, which until now has been managed primarily through the large-scale use of synthetic fertilizers (Liu et al., 2022). However, indiscriminate application of these inputs has led to serious environmental concerns, including nitrate leaching and contamination of ground and surface waters (Singh & Craswell, 2021). Biological nitrogen fixation (BNF) has, therefore, emerged as a promising alternative to supplement nitrogen demand in cereals (Rosenblueth et al., 2018; Guo et al., 2023; Jhu & Oldroyd, 2023). Several strategies have been proposed to engineer BNF into cereals. Synthetic biology approaches aim to enhance associative interactions between nitrogen-fixing bacteria and cereals by developing novel trans-kingdom signaling systems that allow precise and customizable plant–microbe partnerships, while avoiding unintended associations (Hackettet al., 2022). Another approach takes inspiration from rhizobia–legume symbiosis, focusing on the genetic engineering of cereals to develop nodule-like structures capable of hosting nitrogen-fixing bacteria (Guo et al., 2023; Jhu & Oldroyd, 2023). Attempts have also been made to express nif genes within mitochondria and plastids; however, no successful reconstitution of a functional nitrogenase complex has been achieved so far, largely due to intraorganellar oxygen levels that are too high for enzyme activity (Allen et al., 2017). Beyond organelles, cytoplasmic expression of a nine-nif gene cluster in *Arabidopsis thaliana* has been reported, showing increased biomass and chlorophyll content, though the ability of this system to fix nitrogen in cereals remains untested (Yao et al., 2021, Preprint). Importantly, most of these strategies have overlooked the potential of nitrogen-fixing cyanobacteria. These microorganisms are naturally promiscuous in their associations, interacting both epiphytically and endophytically with diverse plant species, including cereals. For example, *Nostoc* has been observed colonizing the epidermal and cortical root cells of wheat (Gantar et al., 1993) and, more recently, rice (Álvarez et al., 2020, 2022). Evidence from rice–*Nostoc* associations suggests that the common symbiotic signaling pathway (CSSP), which is also central to rhizobia–legume and mycorrhizal symbioses, plays a role in facilitating intracellular infection and colonization. This indicates potential strategies for engineering cereals to enhance signal perception, improve colonization efficiency, and induce heterocyst differentiation in cyanobacteria. Since nitrogen fixation in cyanobacteria occurs within heterocysts—specialized low-oxygen compartments—this system could overcome the major challenges faced by nif gene expression in mitochondria, plastids, or the cytoplasm, while also avoiding the need to engineer complex nodule organogenesis. Thus, cyanobacteria-based strategies provide a unique and underexplored opportunity for sustainable nitrogen fixation in cereals, with the potential to reduce dependence on synthetic fertilizers and mitigate the environmental impacts of intensive agricultural practices. Cyanobacteria-based biofertilizers provide a low-carbon, environmentally friendly alternative to synthetic fertilizers by minimizing nitrate leaching and greenhouse gas emissions. However, widespread adoption necessitates assessments of environmental safety, cost-effectiveness, and regulatory compliance. Policies supporting sustainable microbial fertilizers, farmer incentive programs, and incorporation into climate-smart agricultural frameworks are critical for translating laboratory discoveries into field-scale advantages.

8.9 Conservation and Preservation

8.9.1 Nitrogen-Fixing Cyanobacteria in the Conservation of Rare Plants

With accelerating biodiversity loss and ecosystem degradation, nitrogen-fixing cyanobacteria are increasingly recognized as key agents in conserving rare plant species (Nandogopal et al., 2021). Nitrogen, being essential for growth, is often limiting in soils where endangered plants occur (Liu et al., 2019). Cyanobacteria supply bioavailable nitrogen through symbiotic interactions, enhancing plant growth, reproduction, and survival, especially for species adapted to nutrient-poor soils or with specialized nitrogen requirements (Yao et al., 2013; Chen et al., 2017; Li et al., 2021). Beyond nitrogen fixation, cyanobacteria modulate the soil microbiome, promoting beneficial rhizosphere microorganisms and strengthening plant resilience (Babaei et al., 2020; Schultz et al., 2019; Singh, 2014). Integrating cyanobacteria with in vitro preservation techniques provides a synergistic strategy for plant conservation. In vitro methods, including micropropagation and slow-growth conservation, offer controlled environments for maintaining genetic diversity and propagating plant material (Resendis-Antonio et al., 2007; Lucas, 2011). Cyanobacteria enrich culture media with essential nutrients, remove toxic compounds, and improve medium stability, thereby supporting tissue growth and viability (Norouzi et al., 2022; Haque et al., 2022; Stephanopoulos et al., 1998; Jeong et al., 2020). This integration restores natural nutrient cycles and ecological interactions, enhancing the resilience of endangered plants in both regulated and natural settings (Sreenikethanam et al., 2022; Mipeshwaree et al., 2023) (Fig. 8.4).

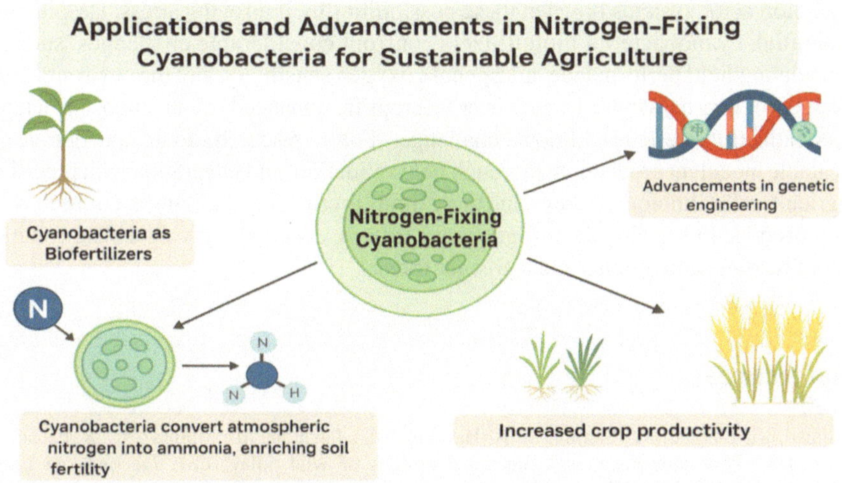

Fig. 8.4 Applications and innovations of nitrogen-fixing Cyanobacteria in promoting sustainable agriculture

8.9.2 In Vitro Conservation Techniques

In vitro preservation depends on factors such as temperature, light, photoperiod, medium composition, and growth regulators (cytokinins, auxins) that control cell division and differentiation (Strieth et al., 2021). Cyanobacteria contribute by supplying nitrogen and other nutrients, maintaining natural plant–microbe interactions, and supporting species that rely on symbiosis (Rojas-Villata et al., 2023; Strieth et al., 2024; Halewood et al., 2018; Giller et al., 2006). They also improve medium quality by removing excess nutrients and harmful compounds, and certain extracts act as cryoprotectants to reduce freezing damage (Ziv, 2005; Verdú-Navarro et al., 2023; Hwang et al., 2022; Murthy et al., 2023). Additionally, cyanobacteria-derived growth regulators enhance tissue differentiation and propagation, particularly for species difficult to culture conventionally (Upadhyay & Singh, 2023). By combining cyanobacterial inoculation with in vitro methods, researchers can create nutrient-rich, supportive environments, preserve genetic material, and maintain ecological relationships, offering innovative strategies for the conservation and recovery of endangered plant species (Mitra & Murthy, 2022; Holtz et al., 2015).

8.10 Conclusions

As global difficulties become more severe—balancing the protection of rare and endangered plants with the rising need for bioactive compounds—cyanobacteria-based plant biotechnology emerges as a transformational answer. Cyanobacteria can improve plant performance, increase specialized metabolite synthesis, and protect fragile species by using biological nitrogen fixation, sophisticated biotechnological technologies, and conservation measures. Emerging frontiers such as CRISPR-Cas9, multiomics, and bioreactor systems broaden these possibilities by improving stress. Despite this potential, cyanobacterial biofertilizers confront considerable challenges, such as fluctuating field performance, nitrogenase oxygen sensitivity, and the environmental dangers of non-native strains. Engineering crops for enhanced colonization and heterocyst induction presents additional challenges. Future research should concentrate on scalable inoculant production, thorough field validation, and integration with precision agricultural technology. Addressing these limits is critical to realizing cyanobacteria's full promise in sustainable nitrogen management, eventually giving a clear route to more resilient ecosystems and a greener future.

References

Ahmad, R., Liaquat, M., Sammi, S., Al-Hawadi, J. S., Jahangir, M., Mumtaz, A., & Fahad, S. (2024). Physicochemical and nutritional profiles of wild adlay (*Coix lacryma*-jobi Linn) accessions by GC, FTIR, and spectrophotometer. *Food Chemistry: X*, *22*, 101418. https://doi.org/10.1016/j.fochx.2024.101418.

Ahmed, M., Stal, L. J., & Hasnain, S. (2014). Biofilm formation and indole-3-acetic acid production by two rhizospheric unicellular cyanobacteria. *Journal of Microbiology and Biotechnology*, *24*(8), 1015–1025.

Allen, R. S., Tilbrook, K., Warden, A. C., Campbell, P. C., Rolland, V., Singh, S. P., & Wood, C. C. (2017). Expression of 16 nitrogenase proteins within the plant mitochondrial matrix. *Frontiers in Plant Science*, *8*, 287.

Álvarez, C., Brenes-Álvarez, M., Molina-Heredia, F. P., & Mariscal, V. (2022). Quantitative proteomics at early stages of the symbiotic interaction between *Oryza sativa* and *Nostoc* punctiforme reveals novel proteins involved in the symbiotic crosstalk. *Plant and Cell Physiology*, *63*(10), 1433–1445.

Araujo, A. S. F., Leite, L. F. C., Iwata, B. D. F., Lira, Jr, M. D. A., Xavier, G. R., & Figueiredo, M. D. V. B. (2012). Microbiological process in agroforestry systems. A review. *Agronomy for Sustainable Development*, *32*(1), 215–226. https://doi.org/10.1007/s13593-011-0026-0.

Årstøl, E., & Hohmann-Marriott, M. F. (2019). Cyanobacterial siderophores—physiology, structure, biosynthesis, and applications. *Marine Drugs*, *17*(5), 281.

Ashitha, A., & Rakhimol, K. R. (2021). Fate of the conventional fertilizers in environment. In *Controlled release fertilizers for sustainable agriculture* . (pp. 25–39). Academic Press. https://doi.org/10.1016/B978-0-12-819555-0.00002-9.

Babaei, M., Borja Zamfir, G. M., Chen, X., Christensen, H. B., Kristensen, M., Nielsen, J., & Borodina, I. (2020). Metabolic engineering of *Saccharomyces cerevisiae* for rosmarinic acid production. *ACS Synthetic Biology*, *9*(8), 1978–1988. https://doi.org/10.1021/acssynbio.0c00048.

Bailey-Serres, J., Parker, J. E., Ainsworth, E. A., Oldroyd, G. E., & Schroeder, J. I. (2019). Genetic strategies for improving crop yields. *Nature*, *575*(7781), 109–118. https://doi.org/10.1038/s41586-019-1679-0.

Baldani, J. I., & Baldani, V. L. (2005). History on the biological nitrogen fixation research in graminaceous plants: Special emphasis on the Brazilian experience. *Anais da Academia Brasileira de Ciências*, *77*, 549–579. https://doi.org/10.1590/S0001-37652005000300014.

Behler, J., Vijay, D., Hess, W. R., & Akhtar, M. K. (2018). CRISPR-based technologies for metabolic engineering in cyanobacteria. *Trends in Biotechnology*, *36*(10), 996–1010.

Beijerinck, M. W. (1901). Uber oligonitrophile mikroben. *Zentralblatt für Bakteriologie, Parasitenkunde, Infektionskrankheiten und Hygiene, Abteilung I*, *7*, 561–582.

Belay, N., Sparling, R., & Daniels, L. (1984). Dinitrogen fixation by a thermophilic methanogenic bacterium. *Nature*, *312*(5991), 286–288. https://doi.org/10.1038/312286a0.

Bergman, B., Zheng, W. W., Klint, J., & Ran, L. (2008). On the origin of plants and relations to contemporary cyanobacterial-plant symbioses. *Plant Biotechnology*, *25*(3), 213–220. https://doi.org/10.5511/plantbiotechnology.25.213.

Berman-Frank, I., Lundgren, P., & Falkowski, P. (2003). Nitrogen fixation and photosynthetic oxygen evolution in cyanobacteria. *Research in Microbiology*, *154*(3), 157–164. https://doi.org/10.1016/S0923-2508(03)00029-9.

Besson, S., Almeida, M. G., & Silveira, C. M. (2022). Nitrite reduction in bacteria: A comprehensive view of nitrite reductases. *Coordination Chemistry Reviews*, *464*, 214560. https://doi.org/10.1016/j.ccr.2022.214560.

Beyan, S. M., Wolde-Meskel, E., & Dakora, F. D. (2018). An assessment of plant growth and N2 fixation in soybean genotypes grown in uninoculated soils collected from different locations in Ethiopia. *Symbiosis*, *75*(3), 189–203. https://doi.org/10.1007/s13199-018-0540-9.

Bishnoi, U. (2018). Agriculture and the dark side of chemical fertilizers. *Environmental Analyses and Ecological Studies*, *3*(1), 552.

Bocchi, S., & Malgioglio, A. (2010). Azolla-Anabaena as a biofertilizer for rice paddy fields in the Po Valley, a temperate rice area in Northern Italy. *International Journal of Agronomy*, *2010*(1), 152158.

Bullard, G. K., Roughley, R. J., & Pulsford, D. J. (2005). The legume inoculant industry and inoculant quality control in Australia: 1953–2003. *Australian Journal of Experimental Agriculture*, *45*(3), 127–140. https://doi.org/10.1071/EA03159.

Chakraborty, S., Verma, E., & Singh, S. S. (2019). Cyanobacterial siderophores: Ecological and biotechnological significance. In *Cyanobacteria*. (pp. 383–397). Academic press.

Chen, Z., Sun, X., Li, Y., Yan, Y., & Yuan, Q. (2017). Metabolic engineering of *Escherichia coli* for microbial synthesis of monolignols. *Metabolic Engineering, 39*, 102–109. https://doi.org/10.1016/j.ymben.2016.10.021.

Chianu, J. N., Nkonya, E. M., Mairura, F. S., Chianu, J. N., & Akinnifesi, F. K. (2011). Biological nitrogen fixation and socioeconomic factors for legume production in sub-Saharan Africa: A review. *Agronomy for Sustainable Development, 31*(1), 139–154. https://doi.org/10.1051/agro/2010004.

Cissoko, M., Hocher, V., Gherbi, H., Gully, D., Carré-Mlouka, A., Sane, S., & Svistoonoff, S. (2018). Actinorhizal signaling molecules: Frankia root hair deforming factor shares properties with NIN inducing factor. *Frontiers in Plant Science, 9*, 1494. https://doi.org/10.3389/fpls.2018.01494.

Clarkson, D. T. (1986). Regulation of the absorption and release of nitrate by plant cells: a review of current ideas and methodology. In *Fundamental, ecological and agricultural aspects of nitrogen metabolism in higher plants: Proceedings of a symposium organized by the Department of Plant Physiology, University of Groningen and the Institute for Soil Fertility, Haren, 9–12 April 1985* . (pp. 3–27). Springer Netherlands. https://doi.org/10.1007/978-94-009-4356-8_1.

Compaoré, J., & Stal, L. J. (2010). Oxygen and the light–dark cycle of nitrogenase activity in two unicellular cyanobacteria. *Environmental Microbiology, 12*(1), 54–62.

Craswell, E. (2021). Fertilizers and nitrate pollution of surface and ground water: An increasingly pervasive global problem. *SN Applied Sciences, 3*(4), 518.

Crews, T. E., & Peoples, M. B. (2004). Legume versus fertilizer sources of nitrogen: Ecological tradeoffs and human needs. *Agriculture, Ecosystems & Environment, 102*(3), 279–297. https://doi.org/10.1016/j.agee.2003.09.018.

De Vries, S., & De Vries, J. (2018). Azolla: A model system for symbiotic nitrogen fixation and evolutionary developmental biology. In *Current advances in fern research* . (pp. 21–46). Springer International Publishing. https://doi.org/10.1007/978-3-319-75103-0_2.

Döbereiner, J. (1966). Azotobacter paspali sp. n., uma bactéria fixadora de nitrogênio na rizosfera de Paspalum. Pesquisa Agropecuária Brasileira, 357–365. https://doi.org/10.1590/S1678-3921.pab1966.v1.18041.

Döbereiner, J. (1976). Fixação de nitrogênio atmosférico em gramíneas tropicais. *Proceedings of the Congresso Brasileiro de Ciência do Solo, Colombia, Cali, 15*, 593–602.

Ebadi, M., Zolfaghari, M. R., Aghaei, S. S., Zargar, M., Shafiei, M., Zahiri, H. S., & Noghabi, K. A. (2019). A bio-inspired strategy for the synthesis of zinc oxide nanoparticles (ZnO NPs) using the cell extract of cyanobacterium Nostoc sp. EA03: From biological function to toxicity evaluation. *RSC Advances, 9*(41), 23508–23525. https://doi.org/10.1039/C9RA03962G..

El-Sheekh, M., El-Dalatony, M. M., Thakur, N., Zheng, Y., & Salama, E. S. (2022). Role of microalgae and cyanobacteria in wastewater treatment: Genetic engineering and omics approaches. *International Journal of Environmental Science and Technology, 19*(3), 2173–2194. https://doi.org/10.1007/s13762-021-03270-w.

Esteves-Ferreira, A. A., Inaba, M., Fort, A., Araújo, W. L., & Sulpice, R. (2018). Nitrogen metabolism in cyanobacteria: Metabolic and molecular control, growth consequences and biotechnological applications. *Critical Reviews in Microbiology, 44*(5), 541–560. https://doi.org/10.1080/1040841X.2018.1446902.

Fan, C. S. (1992). The biological nitrogen fixation systems adopted in rice paddy fields in China. In *The nitrogen fixation and its research in China* . (pp. 423–437). Springer Berlin Heidelberg.

Fatima, P., Mishra, A., Om, H., Saha, B., & Kumar, P. (2019). Free living nitrogen fixation and their response to agricultural crops. In *Biofertilizers and biopesticides in sustainable agriculture* (pp. 173–200). Apple Academic Press.

Fedeson, D. T., & Ducat, D. C. (2021). Symbiotic interactions of phototrophic microbes: engineering synthetic consortia for biotechnology. *Role of Microbial Communities for Sustainability*, 37–62. https://doi.org/10.1007/978-981-15-9912-5_2.

Food and Agriculture Organization of the United Nations. (2015). *World fertilizer trends and outlook to 2018*. FAO. https://www.fao.org/publication.
Franche, C., Lindström, K., & Elmerich, C. (2009). Nitrogen-fixing bacteria associated with leguminous and non-leguminous plants. *Plant Soil, 321*, 35–59. https://doi.org/10.1007/s11104-008-9833-8.
Fujita, Y. & Uesaka, K (2022). Nitrogen fixation in cyanobacteria. Cyanobacterial physiology, 29–45.
Gaby, J. C., & Buckley, D. H. (2011). A global census of nitrogenase diversity. *Environmental Microbiology, 13*(7), 1790–1799. https://doi.org/10.1111/j.1462-2920.2011.02488.x.
Gantar, M., Kerby, N. W., & Rowell, P. (1993). Colonization of wheat (*Triticum vulgare* L.) by N2-fixing cyanobacteria: III. The role of a hormogonia-promoting factor. *New Phytologist, 124*(3), 505–513.
Garlapati, D., Chandrasekaran, M., Devanesan, A., Mathimani, T., & Pugazhendhi, A. (2019). Role of cyanobacteria in agricultural and industrial sectors: An outlook on economically important byproducts. *Applied Microbiology and Biotechnology, 103*(12), 4709–4721.
Giller, K. E., Rowe, E. C., de Ridder, N., & van Keulen, H. (2006). Resource use dynamics and interactions in the tropics: Scaling up in space and time. *Agricultural Systems, 88*(1), 8–27. https://doi.org/10.1016/j.agsy.2005.06.016.
Gilroy, E., & Breen, S. (2022). Interplay between phytohormone signalling pathways in plant defence–other than salicylic acid and jasmonic acid. *Essays in Biochemistry, 66*(5), 657–671.
Glass, A. D., Britto, D. T., Kaiser, B. N., Kinghorn, J. R., Kronzucker, H. J., Kumar, A., Okamoto, M., Rawat, S., Siddiqi, M. Y., Unkles, S. E., & Vidmar, J. J. (2002). The regulation of nitrate and ammonium transport systems in plants. *Journal of Experimental Botany, 53*(370), 855–864.
Gomiero, T. (2016). Soil degradation, land scarcity and food security: Reviewing a complex challenge. *Sustainability, 8*(3), 281. https://doi.org/10.3390/su8030281.
González, A., Sevilla, E., Bes, M. T., Peleato, M. L., & Fillat, M. F. (2016). Pivotal role of iron in the regulation of cyanobacterial electron transport. *Advances in Microbial Physiology, 68*, 169–217. https://doi.org/10.1016/bs.ampbs.2016.02.005.
Goyal, R. K., Schmidt, M. A., & Hynes, M. F. (2021). Molecular biology in the improvement of biological nitrogen fixation by rhizobia and extending the scope to cereals. *Microorganisms, 9*(1), 125. https://doi.org/10.3390/microorganisms9010125.
Gr, S., Yadav, R. K., Chatrath, A., Gerard, M., Tripathi, K., Govindsamy, V., & Abraham, G. (2021). Perspectives on the potential application of cyanobacteria in the alleviation of drought and salinity stress in crop plants. *Journal of Applied Phycology, 33*(6), 3761–3778. https://doi.org/10.1007/s10811-021-02570-5.
Guo, K., Yang, J., Yu, N., Luo, L., & Wang, E. (2023). Biological nitrogen fixation in cereal crops: Progress, strategies, and perspectives. *Plant Communications, 4*(2), 100499.
Gupta, V., Ratha, S. K., Sood, A., Chaudhary, V., & Prasanna, R. (2013). New insights into the biodiversity and applications of cyanobacteria (blue-green algae)—Prospects and challenges. *Algal Research, 2*(2), 79–97. https://doi.org/10.1016/j.algal.2013.01.006.
Guthrie, F. B. (1896). Inoculation of soil for leguminous crops. *Agricultural Gazette of New South Wales, 7*, 690–694.
Halewood, M., Chiurugwi, T., Sackville Hamilton, R., Kurtz, B., Marden, E., Welch, E., & Powell, W. (2018). Plant genetic resources for food and agriculture: Opportunities and challenges emerging from the science and information technology revolution. *New Phytologist, 217*(4), 1407–1419. https://doi.org/10.1111/nph.14993.
Hamed, I. (2016). The evolution and versatility of microalgal biotechnology: A review. *Comprehensive Reviews in Food Science and Food Safety, 15*(6), 1104–1123. https://doi.org/10.1111/1541-4337.12227.
Haque, M. I., Singh, P. K., Ghuge, S., Kumar, A., Rai, A. C., Kumar, A., & Modi, A. (2022). A general introduction to and background of plant tissue culture: Past, current, and future aspects. In *Advances in plant tissue culture* (pp. 1–30). Academic Press. https://doi.org/10.1016/B978-0-323-90795-8.00019-9.

Haselkorn, R. (1978). Heterocysts. *Annual Review of Plant Physiology, 29*(1), 319–344.
Hashtroudi, M. S., Ghassempour, A., Riahi, H., Shariatmadari, Z., & Khanjir, M. (2013). Endogenous auxins in plant growth-promoting Cyanobacteria—*Anabaena vaginicola* and *Nostoc calcicola*. *Journal of Applied Phycology, 25*(2), 379–386.
Haskett, T. L., Paramasivan, P., Mendes, M. D., Green, P., Geddes, B. A., Knights, H. E., & Poole, P. S. (2022). Engineered plant control of associative nitrogen fixation. *Proceedings of the National Academy of Sciences, 119*(16), e2117465119.
Herrero, A., Picossi, S., & Flores, E. (2013). Gene expression during heterocyst differentiation. *Advances in Botanical Research, 65*, 281–329. https://doi.org/10.1016/B978-0-12-394313-2.00008-1.
Hesami, M., Baiton, A., Alizadeh, M., Pepe, M., Torkamaneh, D., & Jones, A. M. P. (2021). Advances and perspectives in tissue culture and genetic engineering of cannabis. *International Journal of Molecular Sciences, 22*(11), 5671. https://doi.org/10.3390/ijms22115671.
Hirel, B., Tétu, T., Lea, P. J., & Dubois, F. (2011). Improving nitrogen use efficiency in crops for sustainable agriculture. *Sustainability, 3*(9), 1452–1485. https://doi.org/10.3390/su3091452.
Holguin, G., Vazquez, P., & Bashan, Y. (2001). The role of sediment microorganisms in the productivity, conservation, and rehabilitation of mangrove ecosystems: an overview. *Biology and Fertility of Soils, 33*(4), 265–278. https://doi.org/10.1007/s003740000319.
Holtz, B. R., Berquist, B. R., Bennett, L. D., Kommineni, V. J., Munigunti, R. K., White, E. L., & Marcel, S. (2015). Commercial-scale biotherapeutics manufacturing facility for plant-made pharmaceuticals. *Plant Biotechnology Journal, 13*(8), 1180–1190. https://doi.org/10.1111/pbi.12469.
Hossain, M. J., & Bakhsh, A. (2020). Development and applications of transplastomic plants; a way towards eco-friendly agriculture. In *Environment, climate, plant and vegetation growth* . (pp. 285–322). Springer International Publishing. https://doi.org/10.1007/978-3-030-49732-3_12.
Huang, T. K., & McDonald, K. A. (2012). Bioreactor systems for in vitro production of foreign proteins using plant cell cultures. *Biotechnology Advances, 30*(2), 398–409. https://doi.org/10.1016/j.biotechadv.2011.07.016.
Huang, X. F., Liu, Y. J., Dong, J. D., Qu, L. Y., Zhang, Y. Y., Wang, F. Z., & Zhang, S. (2014). Mangrovibacterium diazotrophicum gen. nov., sp. nov., a nitrogen-fixing bacterium isolated from a mangrove sediment, and proposal of Prolixibacteraceae fam. nov. *International Journal of Systematic and Evolutionary Microbiology, 64*(Pt_3), 875–881. https://doi.org/10.1099/ijs.0.052779-0.
Hussain, A., Hamayun, M., & Shah, S. T. (2013). Root colonization and phytostimulation by phytohormones producing entophytic Nostoc sp. AH-12. *Current Microbiology, 67*(5), 624–630.
Hussain, A., Krischke, M., Roitsch, T., & Hasnain, S. (2010). Rapid determination of cytokinins and auxin in cyanobacteria. *Current Microbiology, 61*(5), 361–369.
Hussain, A., Shah, S. T., Rahman, H., Irshad, M., & Iqbal, A. (2015). Effect of IAA on in vitro growth and colonization of Nostoc in plant roots. *Frontiers in Plant Science, 6*, 46.
Hwang, H. D., Kwon, S. H., Murthy, H. N., Yun, S. W., Pyo, S. S., & Park, S. Y. (2022). Temporary immersion bioreactor system as an efficient method for mass production of in vitro plants in horticulture and medicinal plants. *Agronomy, 12*(2), 346. https://doi.org/10.3390/agronomy12020346.
Hyder, S., Rizvi, Z. F., los Santos-Villalobos, S. D., Santoyo, G., Gondal, A., Khalid, N., & Rani, A. (2023). Applications of plant growth-promoting rhizobacteria for increasing crop production and resilience. *Journal of Plant Nutrition, 46*(10), 2551–2580.
Iniesta-Pallarés, M., Álvarez, C., Gordillo-Cantón, F. M., Ramírez-Moncayo, C., Alves-Martínez, P., Molina-Heredia, F. P., & Mariscal, V. (2021). Sustaining rice production through biofertilization with N2-fixing cyanobacteria. *Applied Sciences, 11*(10), 4628.
Ininbergs, K., Bay, G., Rasmussen, U., Wardle, D. A., & Nilsson, M. C. (2011). Composition and diversity of nifH genes of nitrogen-fixing cyanobacteria associated with boreal forest feather mosses. *New Phytologist, 192*(2), 507–517. https://doi.org/10.1111/j.1469-8137.2011.03809.x.

Inokuchi, R., Kuma, K. I., Miyata, T., & Okada, M. (2002). Nitrogen-assimilating enzymes in land plants and algae: Phylogenic and physiological perspectives. *Physiologia Plantarum*, *116*(1), 1–11. https://doi.org/10.1034/j.1399-3054.2002.1160101.x.

Jeong, Y., Cho, S. H., Lee, H., Choi, H. K., Kim, D. M., Lee, C. G., & Cho, B. K. (2020). Current status and future strategies to increase secondary metabolite production from cyanobacteria. *Microorganisms*, *8*(12), 1849. https://doi.org/10.3390/microorganisms8121849.

Jhu, M. Y., & Oldroyd, G. E. (2023). Dancing to a different tune, can we switch from chemical to biological nitrogen fixation for sustainable food security?. *PLOS Biology*, *21*(3), e3001982.

Jiménez-Jiménez, S., Santana, O., Lara-Rojas, F., Arthikala, M. K., Armada, E., Hashimoto, K., & Cárdenas, L. (2019). Differential tetraspanin genes expression and subcellular localization during mutualistic interactions in *Phaseolus vulgaris*. *PLOS One*, *14*(8), e0219765. https://doi.org/10.1371/journal.pone.0219765.

Johnson, T. M., & Grim, B. J. (Eds). (2010). *World religion database: International religious demographic statistics and sources*. Brill. http://www.worldreligiondatabase.org

Karanja, N., Freire, J., Gueye, M. & DaSilva, E. (2000). MIRCEN networking: capacity-building and BNF technology transfer in Africa and Latin America. 1-6.

Kawalekar, J. S. (2013). Role of biofertilizers and biopesticides for sustainable agriculture. 73-78.

Khadatare, S., & Suryawanshi, D. S. (2016). Isolation blue green algae from maize fields of Mohol Tahasil in Solapur. *International Journal of Science and Research*, *5*(6), 1597–1599.

Khosro, M., & Yousef, S. (2012). Bacterial bio-fertilizers for sustainable crop production: A review. *ARPN Journal of Agricultural and Biological Science*, *7*, 307–316.

Kim, J., & Rees, D. C. (1994). Nitrogenase and biological nitrogen fixation. *Biochemistry*, *33*(2), 389–397.

Kizilkaya, R. (2009). Nitrogen fixation capacity of Azotobacter spp. strains isolated from soils in different ecosystems and relationship between them and the microbiological properties of soils. *Journal of Environmental Biology*, *30*(1), 73–82.

Kollmen, J., & Strieth, D. (2022). The beneficial effects of cyanobacterial co-culture on plant growth. *Life*, *12*(2), 223. https://doi.org/10.3390/life12020223.

Kondi, V., Sabbani, V., Alluri, R., Karumuri, T. S. K., Chawla, P., Dasarapu, S., & Tiwari, O. N. (2022). Cyanobacteria as potential bio resources for multifaceted sustainable utilization. In New and Future Developments in Microbial Biotechnology and Bioengineering. (pp. 73–87). Elsevier. https://doi.org/10.1016/B978-0-323-85579-2.00020-4.

Krupa, S. V. (2003). Effects of atmospheric ammonia (NH3) on terrestrial vegetation: A review. *Environmental Pollution*, *124*(2), 179–221. https://doi.org/10.1016/S0269-7491(02)00434-7.

Kumar, K., Mella-Herrera, R. A., & Golden, J. W. (2010). Cyanobacterial heterocysts. *Cold Spring Harbor Perspectives in Biology*, *2*(4), a000315.

Kundu, K., Teta, R., Esposito, G., Stornaiuolo, M., & Costantino, V. (2023). A four-step platform to optimize growth conditions for high-yield production of siderophores in cyanobacteria. *Metabolites*, *13*(2), 154.

Leigh, J. A. (2000). Nitrogen fixation in methanogens: The archaeal perspective. *Current Issues in Molecular Biology*, *2*(4), 125–131.

Lengke, M. F., Ravel, B., Fleet, M. E., Wanger, G., Gordon, R. A., & Southam, G. (2006). Mechanisms of gold bioaccumulation by filamentous cyanobacteria from gold (III)−chloride complex. *Environmental Science & Technology*, *40*(20), 6304–6309. https://doi.org/10.1021/es061040r.

Li, H., Shen, C. R., Huang, C. H., Sung, L. Y., Wu, M. Y., & Hu, Y. C. (2016). CRISPR-Cas9 for the genome engineering of cyanobacteria and succinate production. *Metabolic Engineering*, *38*, 293–302. https://doi.org/10.1016/j.ymben.2016.09.006.

Li, S., Liang, C., Liu, G., Jin, J. M., Tao, Y., & Tang, S. Y. (2021). De novo biosynthesis of chlorogenic acid using an artificial microbial community. *Journal of Agricultural and Food Chemistry*, *69*(9), 2816–2825. https://doi.org/10.1021/acs.jafc.0c07588.

Liu, C. W., & Murray, J. D. (2016). The role of flavonoids in nodulation host-range specificity: An update. *Plants*, *5*(3), 33. https://doi.org/10.3390/plants5030033.

Liu, J., Ma, K., Ciais, P., & Polasky, S. (2016). Reducing human nitrogen use for food production. *Scientific Reports*, *6*(1), 30104. https://doi.org/10.1038/srep30104.

Liu, Q., Wu, K., Song, W., Zhong, N., Wu, Y., & Fu, X. (2022). Improving crop nitrogen use efficiency toward sustainable green revolution. *Annual Review of Plant Biology*, *73*(1), 523–551.

Liu, Q., Yu, T., Li, X., Chen, Y., Campbell, K., Nielsen, J., & Chen, Y. (2019). Rewiring carbon metabolism in yeast for high-level production of aromatic chemicals. *Nature Communications*, *10*(1), 4976. https://doi.org/10.1038/s41467-019-12961-5.

Lucas, J. A. (2011). Advances in plant disease and pest management. *The Journal of Agricultural Science*, *149*(S1), 91–114.

Maheswari, M., Murthy, A. N. G., & Shanker, A. K. (2017). Nitrogen nutrition in crops and its importance in crop quality. In *The Indian nitrogen assessment* . (pp. 175–186). Elsevier. https://doi.org/10.1016/B978-0-12-811836-8.00012-4.

Mahmud, K., Makaju, S., Ibrahim, R., & Missaoui, A. (2020). Current progress in nitrogen fixing plants and microbiome research. *Plants*, *9*(1), 97. https://doi.org/10.3390/plants9010097.

Masclaux-Daubresse, C., Daniel-Vedele, F., Dechorgnat, J., Chardon, F., Gaufichon, L., & Suzuki, A. (2010). Nitrogen uptake, assimilation and remobilization in plants: Challenges for sustainable and productive agriculture. *Annals of Botany*, *105*(7), 1141–1157. https://doi.org/10.1093/aob/mcq028.

Mazhar, S., Cohen, J. D., & Hasnain, S. (2013). Auxin producing non-heterocystous Cyanobacteria and their impact on the growth and endogenous auxin homeostasis of wheat. *Journal of Basic Microbiology*, *53*(12), 996–1003.

Meeks, J. C., & Elhai, J. (2002). Regulation of cellular differentiation in filamentous cyanobacteria in free-living and plant-associated symbiotic growth states. *Microbiology and Molecular Biology Reviews*, *66*(1), 94–121. https://doi.org/10.1128/mmbr.66.1.94-121.2002.

Miflin, B. J., & Lea, P. J. (1980). Ammonia assimilation. In *Amino acids and derivatives* . (pp. 169–202). Academic Press. https://doi.org/10.1016/B978-0-12-675405-6.50010-3.

Mipeshwaree Devi, A., Khedashwori Devi, K., Premi Devi, P., Lakshmipriyari Devi, M., & Das, S. (2023). Metabolic engineering of plant secondary metabolites: Prospects and its technological challenges. *Frontiers in Plant Science*, *14*, 1171154. https://doi.org/10.3389/fpls.2023.1171154.

Mishra, A. K., Kaushik, M. S., & Tiwari, D. N. (2019). Nitrogenase and hydrogenase: Enzymes for nitrogen fixation and hydrogen production in cyanobacteria. In *Cyanobacteria* . (pp. 173–191). Academic Press. https://doi.org/10.1016/B978-0-12-814667-5.00008-8.

Mishra, A., Rajput, S., Gupta, P. S., Goyal, V., Singh, S., Sharma, S., & Varma, A. (2021). Role of cyanobacteria in Rhizospheric nitrogen fixation. In *Soil nitrogen ecology* . (pp. 497–519). Springer International Publishing. https://doi.org/10.1007/978-3-030-71206-8_25.

Mishra, S. K., Singh, J., Pandey, A. R., & Dwivedi, N. (2019). Indole-3-acetic acid production by the cyanobacterium *Fisherella muscicola* NDUPC001. *Current Science*, *116*(7), 1233–1237.

Mitra, S., & Murthy, G. S. (2022). Bioreactor control systems in the biopharmaceutical industry: A critical perspective. *Systems Microbiology and Biomanufacturing*, *2*(1), 91–112. https://doi.org/10.1007/s43393-021-00048-6.

Moreno-Vivián, C., Cabello, P., Martínez-Luque, M., Blasco, R., & Castillo, F. (1999). Prokaryotic nitrate reduction: Molecular properties and functional distinction among bacterial nitrate reductases. *Journal of Bacteriology*, *181*(21), 6573–6584. https://doi.org/10.1128/jb.181.21.6573-6584.1999.

Moulin, S. L., Frail, S., Braukmann, T., Doenier, J., Steele-Ogus, M., Marks, J. C., & Yeh, E. (2024). The endosymbiont of Epithemia clementina is specialized for nitrogen fixation within a photosynthetic eukaryote. *ISME Communications*, *4*(1), ycae055. https://doi.org/10.1093/ismeco/ycae055.

MubarakAli, D., Arunkumar, J., Nag, K. H., SheikSyedIshack, K. A., Baldev, E., Pandiaraj, D., & Thajuddin, N. (2013). Gold nanoparticles from Pro and eukaryotic photosynthetic microorganisms—Comparative studies on synthesis and its application on biolabelling. *Colloids and Surfaces B: Biointerfaces*, *103*, 166–173. https://doi.org/10.1016/j.colsurfb.2012.10.014.

Mueller, N. D., Gerber, J. S., Johnston, M., Ray, D. K., Ramankutty, N., & Foley, J. A. (2012). Closing yield gaps through nutrient and water management. *Nature, 490*(7419), 254–257. https://doi.org/10.1038/nature11420.

Mugabe, J. (1994). *Research on biofertilizers.*

Muhie, S. H. (2022). Novel approaches and practices to sustainable agriculture. *Journal of Agriculture and Food Research, 10*, 100446. https://doi.org/10.1016/j.jafr.2022.100446.

Munaweera, T. I. K., Jayawardana, N. U., Rajaratnam, R., & Dissanayake, N. (2022). Modern plant biotechnology as a strategy in addressing climate change and attaining food security. *Agriculture & Food Security, 11*(1), 1–28. https://doi.org/10.1186/s40066-022-00369-2.

Murray, P. A., & Zinder, S. H. (1984). Nitrogen fixation by a methanogenic archae bacterium. *Nature, 312*(5991), 284–286. https://doi.org/10.1038/312284a0.

Murthy, H. N., Joseph, K. S., Paek, K. Y., & Park, S. Y. (2023). Bioreactor systems for micropropagation of plants: Present scenario and future prospects. *Frontiers in Plant Science, 14*, 1159588. https://doi.org/10.3389/fpls.2023.1159588.

Mus, F., Crook, M. B., Garcia, K., Garcia Costas, A., Geddes, B. A., Kouri, E. D., & Peters, J. W. (2016). Symbiotic nitrogen fixation and the challenges to its extension to nonlegumes. *Applied and Environmental Microbiology, 82*(13), 3698–3710. https://doi.org/10.1128/AEM.01055-16.

Nandagopal, P., Steven, A. N., Chan, L. W., Rahmat, Z., Jamaluddin, H., & Mohd Noh, N. I. (2021). Bioactive metabolites produced by cyanobacteria for growth adaptation and their pharmacological properties. *Biology, 10*(10), 1061. https://doi.org/10.3390/biology10101061.

Nawaz, T., Gu, L., Fahad, S., Saud, S., Bleakley, B., & Zhou, R. (2024a). Exploring sustainable agriculture with nitrogen-fixing cyanobacteria and nanotechnology. *Molecules, 29*(11), 2534. https://doi.org/10.3390/molecules29112534.

Nawaz, T., Gu, L., Fahad, S., Saud, S., Harrison, M. T., & Zhou, R. (2024b). Sustainable protein production through genetic engineering of cyanobacteria and use of atmospheric N2 gas. *Food and Energy Security, 13*(2), e536. https://doi.org/10.1002/fes3.536.

Nawaz, T., Gu, L., Fahad, S., Saud, S., Jiang, Z., Hassan, S., & Zhou, R. (2023). A comprehensive review of the therapeutic potential of cyanobacterial marine bioactives: Unveiling the hidden treasures of the sea. *Food and Energy Security, 12*(5), e495. https://doi.org/10.1002/fes3.495.

Nawaz, T., Saud, S., Gu, L., Khan, I., Fahad, S., & Zhou, R. (2024c). Cyanobacteria: Harnessing the power of microorganisms for plant growth promotion, stress alleviation, and phytoremediation in the era of sustainable agriculture. *Plant Stress, 11*, 100399. https://doi.org/10.1016/j.stress.2024.100399.

Norouzi, O., Hesami, M., Pepe, M., Dutta, A., & Jones, A. M. P. (2022). In vitro plant tissue culture as the fifth generation of bioenergy. *Scientific Reports, 12*(1), 5038. https://doi.org/10.1038/s41598-022-09066-3.

Nosheen, S., Ajmal, I., & Song, Y. (2021). Microbes as biofertilizers, a potential approach for sustainable crop production. *Sustainability, 13*(4), 1868.

Ohyama, T. (2010). Nitrogen as a major essential element of plants. *Nitrogen Assimilation in Plants, 37*, 1–17.

Oliver, R. L., Hamilton, D. P., Brookes, J. D., & Ganf, G. G. (2012). Physiology, blooms and prediction of planktonic cyanobacteria. In *Ecology of cyanobacteria II: Their diversity in space and time*. (pp. 155–194). Springer Netherlands. https://doi.org/10.1007/978-94-007-3855-3_6.

Organisation for Economic Co-operation and Development (OECD) & Food and Agriculture Organization of the United Nations (FAO). (2021). *The OECD-FAO agricultural outlook 2021–2030.* OECD Publishing.

Pahalvi, H. N., Rafiya, L., Rashid, S., Nisar, B., & Kamili, A. N. (2021). Chemical fertilizers and their impact on soil health. In Microbiota and biofertilizers: Ecofriendly tools for reclamation of degraded soil environs. (vol. 2, pp. 1–20). Springer International Publishing. https://doi.org/10.1007/978-3-030-61010-4_1.

Pathak, J., Rajneesh, Maurya, P. K., Singh, S. P., Haeder, D. P., & Sinha, R. P. (2018). Cyanobacterial farming for environment friendly sustainable agriculture practices: Innovations and perspectives. *Frontiers in Environmental Science, 6*, 7. https://doi.org/10.3389/fenvs.2018.00007.

Pazhamala, LT, & Varshney, RK (2015). Proteomics and Metabolomics: two emerging areas for legume improvement. Frontiers in Plant Science, 6, https://doi.org/10.3389/fpls.2015.01116.

Poveda, J. (2021). Cyanobacteria in plant health: Biological strategy against abiotic and biotic stresses. *Crop Protection, 141*, 105450. https://doi.org/10.1016/j.cropro.2020.105450.

Prasanna, R., & Nayak, S. (2007). Influence of diverse rice soil ecologies on cyanobacterial diversity and abundance. *Wetlands Ecology and Management, 15*(2), 127–134.

Prasanna, R., Chaudhary, V., Gupta, V., Babu, S., Kumar, A., Singh, R., & Nain, L. (2013). Cyanobacteria mediated plant growth promotion and bioprotection against *Fusarium* wilt in tomato. *European Journal of Plant Pathology, 136*(2), 337–353.

Prasanna, R., Nain, L., Ancha, R., Srikrishna, J., Joshi, M., & Kaushik, B. D. (2009). Rhizosphere dynamics of inoculated cyanobacteria and their growth-promoting role in rice crop. *Egyptian Journal of Biology*, 11, 26–36.

Prasanna, R., Nain, L., Pandey, A. K., & Saxena, A. K. (2012). Microbial diversity and multidimensional interactions in the rice ecosystem. *Archives of Agronomy and Soil Science, 58*(7), 723–744.

Prasanna, R., Nain, L., Tripathi, R., Gupta, V., Chaudhary, V., Middha, S., & Kaushik, B. D. (2008). Evaluation of fungicidal activity of extracellular filtrates of cyanobacteria–possible role of hydrolytic enzymes. *Journal of Basic Microbiology, 48*(3), 186–194.

Prasanna, R., Sood, A., Ratha, S. K., & K. Singh, P. (2014). Cyanobacteria as a "green" option for sustainable agriculture. In *Cyanobacteria: An economic perspective* (pp. 145–166). Wiley. https://doi.org/10.1002/9781118402238.ch9

Priya, H., Prasanna, R., Ramakrishnan, B., Bidyarani, N., Babu, S., Thapa, S., & Renuka, N. (2015). Influence of cyanobacterial inoculation on the culturable microbiome and growth of rice. *Microbiological Research, 171*, 78–89.

Ranjbar, S., & Malcata, F. X. (2022). Challenges and prospects for sustainable microalga-based oil: A comprehensive review, with a focus on metabolic and genetic engineering. *Fuel, 324*, 124567. https://doi.org/10.1016/j.fuel.2022.124567.

Rasheed, A., Gill, R. A., Hassan, M. U., Mahmood, A., Qari, S., Zaman, Q. U., & Wu, Z. (2021). A critical review: Recent advancements in the use of CRISPR/Cas9 technology to enhance crops and alleviate global food crises. *Current Issues in Molecular Biology, 43*(3), 1950–1976. https://doi.org/10.3390/cimb43030135.

Ravikumar, S., Kathiresan, K., Alikhan, S. L., Williams, G. P., & Gracelin, N. A. A. (2007). Growth of *Avicennia marina* and *Ceriops decandra* seedlings inoculated with halophilic azotobacters. *Journal of Environmental Biology, 28*(3), 601.

Resendis-Antonio, O., Reed, J. L., Encarnación, S., Collado-Vides, J., & Palsson, B. Ø (2007). Metabolic reconstruction and modeling of nitrogen fixation in *Rhizobium etli. PLOS Computational Biology, 3*(10), e192. https://doi.org/10.1371/journal.pcbi.0030192.

Rojas-Villalta, D., Gómez-Espinoza, O., Murillo-Vega, F., Villalta-Romero, F., Guerrero, M., Guillén-Watson, R., & Núñez-Montero, K. (2023). Insights into co-cultivation of photosynthetic microorganisms for novel molecule discovery and enhanced production of specialized metabolites. *Fermentation, 9*(11), 941. https://doi.org/10.3390/fermentation9110941.

Rosenblueth, M., Ormeño-Orrillo, E., López-López, A., Rogel, M. A., Reyes-Hernández, B. J., Martínez-Romero, J. C., & Martínez-Romero, E. (2018). Nitrogen fixation in cereals. *Frontiers in Microbiology, 9*, 1794.

Rossi, F., & De Philippis, R. (2015). Role of cyanobacterial exopolysaccharides in phototrophic biofilms and in complex microbial mats. *Life, 5*(2), 1218–1238.

Roy, R. N., Finck, A., Blair, G. J., & Tandon, H. L. S. (2006). Plant nutrition for food security. A guide for integrated nutrient management. *FAO Fertilizer and Plant Nutrition Bulletin, 16*(368), 201–214.

Sadvakasova, A. K., Kossalbayev, B. D., Token, A. I., Bauenova, M. O., Wang, J., Zayadan, B. K., B. alouch, H., A. lwasel, S., L. eong, Y. K., C. hang, J.-S., & Allakhverdiev, S. I. (2022). Influence of Mo and Fe on photosynthetic and nitrogenase activities of nitrogen-fixing cyanobacteria under nitrogen starvation. *Cells, 11*(5), 904.

Sahu, D., Priyadarshani, I., & Rath, B. (2012). Cyanobacteria–as potential biofertilizer. *CIBTech Journal of Microbiology*, *1*(2–3), 20–26. https://www.jstor.org/stable/24096726.

Saikia, S. P., & Jain, V. (2007). Biological nitrogen fixation with non-legumes: An achievable target or a dogma?. *Current Science*, *92*(3), 317–322. https://www.jstor.org/stable/24096726.

Saleem, A., Anwar, S., Nawaz, T., Fahad, S., Saud, S., Ur Rahman, T., & Nawaz, T. (2024). Securing a sustainable future: The climate change threat to agriculture, food security, and sustainable development goals. *Journal of Umm Al-Qura University for Applied Sciences*, *11*, 1–17. https://doi.org/10.1007/s43994-024-00177-3.

Santi, C., Bogusz, D., & Franche, C. (2013). Biological nitrogen fixation in non-legume plants. *Annals of Botany*, *111*(5), 743–767. https://doi.org/10.1093/aob/mct048.

Santini, G., Biondi, N., Rodolfi, L., & Tredici, M. R. (2021). Plant biostimulants from cyanobacteria: An emerging strategy to improve yields and sustainability in agriculture. *Plants*, *10*(4), 643. https://doi.org/10.3390/plants10040643.

Scherings, G., Haaker, H., & Veeger, C. (1977). Regulation of nitrogen fixation by Fe-S protein II in *Azotobacter vinelandii*. *European Journal of Biochemistry*, *77*(3), 621–630. https://doi.org/10.1111/j.1432-1033.1977.tb11706.x.

Schultz, B. J., Kim, S. Y., Lau, W., & Sattely, E. S. (2019). Total biosynthesis for milligram-scale production of etoposide intermediates in a plant chassis. *Journal of the American Chemical Society*, *141*(49), 19231–19235. https://doi.org/10.1021/jacs.9b10717.

Sergeeva, E., Liaimer, A., & Bergman, B. (2002). Evidence for production of the phytohormone indole-3-acetic acid by cyanobacteria. *Planta*, *215*(2), 229–238.

Seth, K., Kumawat, G., Kumar, M., Sangela, V., Singh, N., Gupta, A. K., & Harish (2022). Nitrogen metabolism in Cyanobacteria. In *Ecophysiology and biochemistry of cyanobacteria*. (pp. 255–268). Springer Nature Singapore. https://doi.org/10.1007/978-981-16-4873-1_12.

Shah, F., & Wu, W. (2019). Soil and crop management strategies to ensure higher crop productivity within sustainable environments. *Sustainability*, *11*(5), 1485. https://doi.org/10.3390/su11051485.

Shantharam, S., & Mattoo, A. K. (1997). Enhancing biological nitrogen fixation: An appraisal of current and alternative technologies for N input into plants. *Plant and Soil*, *194*(1), 205–216. https://doi.org/10.1023/A:1004234315999.

Sharma, A., Sharma, S., Sharma, K., Chetri, S. P., Vashishtha, A., Singh, P., & Agrawal, V. (2016). Algae as crucial organisms in advancing nanotechnology: A systematic review. *Journal of Applied Phycology*, *28*(3), 1759–1774. https://doi.org/10.1007/s10811-015-0715-1.

Sharma, M., Sudheer, S., Usmani, Z., Rani, R., & Gupta, P. (2020). Deciphering the omics of plant-microbe interaction: Perspectives and new insights. *Current Genomics*, *21*(5), 343–362. https://doi.org/10.2174/1389202921999200515140420.

Singh, J. S., Kumar, A., Rai, A. N., & Singh, D. P. (2016a). Cyanobacteria: Precious bio-resource in agriculture, ecosystem, and environmental sustainability. *Frontiers in Microbiology*, *7*, 529.

Singh, N. K., Dhar, D. W., & Tabassum, R. (2016b). Role of cyanobacteria in crop protection. *Proceedings of the National Academy of Sciences, India Section B: Biological Sciences*, *86*(1), 1–8. https://doi.org/10.3389/fmicb.2016.00529.

Singh, S. (2014). A review on possible elicitor molecules of cyanobacteria: Their role in improving plant growth and providing tolerance against biotic or abiotic stress. *Journal of Applied Microbiology*, *117*(5), 1221–1244. https://doi.org/10.1111/jam.12612.

Singh, T., & Bhadury, P. (2019). Description of a new marine planktonic cyanobacterial species, *Synechococcus moorigangaii* (Order Chroococcales) from Sundarbans mangrove ecosystem. *Phytotaxa*, *393*(3), 263.

Slater, A., Scott, N., & Fowler, M. (2008). *Plant biotechnology: The genetic manipulation of plants*. OUP Oxford.

Song, J., He, X., Wang, S., Yang, X., Wu, L., Li, S., & Wu, Z. (2022). Community composition specificities of cyanobacteria in paddy soil under different ecological conditions. *Agronomy*, *12*(12), 3090.

Soumare, A., Diedhiou, A. G., Thuita, M., Hafidi, M., Ouhdouch, Y., Gopalakrishnan, S., & Kouisni, L. (2020). Exploiting biological nitrogen fixation: A route towards a sustainable agriculture. *Plants*, *9*(8), 1011. https://doi.org/10.3390/plants9081011.

Sreenikethanam, A., Raj, S., Gugulothu, P., & Bajhaiya, A. K. (2022). Genetic engineering of microalgae for secondary metabolite production: Recent developments, challenges, and future prospects. *Frontiers in Bioengineering and Biotechnology*, *10*, 836056. https://doi.org/10.3389/fbioe.2022.836056.

Stephanopoulos, G., Aristidou, A. A., & Nielsen, J. (1998). Metabolic engineering: Principles and methodologies.

Stewart, W. D. P. (1969). Biological and ecological-organisms. *Proceedings of the Royal Society of London. Series B. Biological Sciences*, 172(1029), 367–388. https://doi.org/10.1098/rspb.1969.0027.

Stewart, W. D. P. (1973). Nitrogen fixation by photosynthetic microorganisms. *Annual Reviews in Microbiology*, *27*(1), 283–316. https://doi.org/10.1146/annurev.mi.27.100173.001435.

Strieth, D., Di Nonno, S., Stiefelmaier, J., Kollmen, J., Geib, D., & Ulber, R. (2021). Co-cultivation of diazotrophic terrestrial cyanobacteria and Arabidopsis thaliana. *Engineering in Life Sciences*, *21*(3–4), 126–136. https://doi.org/10.1002/elsc.202000068.

Strieth, D., Kollmen, J., Stiefelmaier, J., Mehring, A., & Ulber, R. (2024). Co-cultures from plants and cyanobacteria: A new way for production systems in agriculture and bioprocess engineering. In *Plants as factories for bioproduction: Recent developments and applications* (pp. 83–117), Springer. https://doi.org/10.1007/10_2023_247.

Sur, S., Bothra, A. K., & Sen, A. (2010). Symbiotic nitrogen fixation-a bioinformatics perspective. *Biotechnology, 9*, 257–273. https://doi.org/10.3923/biotech.2010.257.273.

Suzaki, T., Takeda, N., Nishida, H., Hoshino, M., Ito, M., Misawa, F., & Kawaguchi, M. (2019). Lack of symbiont accommodation controls intracellular symbiont accommodation in root nodule and arbuscular mycorrhizal symbiosis in Lotus japonicus. *PLOS Genetics*, *15*(1), e1007865. https://doi.org/10.1371/journal.pgen.1007865.

Swarnalakshmi, K., Vandana, Y., Senthilkumar, M., & Dolly, W. D. (2016). Bio fertilizers for higher pulse production in India: Scope, accessibility and challenges. *Indian Journal of Agronomy*, *61*, 173–181.

Tan, C. Y., Dodd, I. C., Chen, J. E., Phang, S. M., Chin, C. F., Yow, Y. Y., & Ratnayeke, S. (2021). Regulation of algal and cyanobacterial auxin production, physiology, and application in agriculture: An overview. *Journal of Applied Phycology*, *33*(5), 2995–3023.

Toribio, A. J., Jurado, M. M., Suárez-Estrella, F., López-González, J. A., Martínez-Gallardo, M. R., & López, M. J. (2021). Application of sonicated extracts of cyanobacteria and microalgae for the mitigation of bacterial canker in tomato seedlings. *Journal of Applied Phycology*, *33*(6), 3817–3829.

Toribio, A. J., Suárez-Estrella, F., Jurado, M. M., López, M. J., López-González, J. A., & Moreno, J. (2020). Prospection of cyanobacteria producing bioactive substances and their application as potential phytostimulating agents. *Biotechnology Reports*, *26*, e00449.

Uniyal, S., Bhandari, M., Singh, P., Singh, R. K., & Tiwari, S. P. (2022). Cytokinin biosynthesis in cyanobacteria: Insights for crop improvement. *Frontiers in Genetics*, *13*, 933226.

Upadhyay, S. K., Singh, S. P. (Eds.) (2023). *Plants as bioreactors for industrial molecules*. John Wiley & Sons.

Usher, K. M., Bergman, B., & Raven, J. A. (2007). Exploring cyanobacterial mutualisms. *Annual Review of Ecology, Evolution, and Systematics*, *38*(1), 255–273. https://doi.org/10.1146/annurev.ecolsys.38.091206.095641.

van Heerwaarden, J., Baijukya, F., Kyei-Boahen, S., Adjei-Nsiah, S., Ebanyat, P., Kamai, N., & Giller, K. (2018). Soyabean response to rhizobium inoculation across sub-Saharan Africa: Patterns of variation and the role of promiscuity. *Agriculture, Ecosystems & Environment*, *261*, 211–218. https://doi.org/10.1016/j.agee.2017.08.016.

Venkataraman, G. S. (1972). *Algal biofertilizers and rice cultivation*. Today & Tomorrow's Printers & Publishers.

Venkataraman, G. S. (1975). The role of blue-green algae in tropical rice cultivation. 207-218.

Verdú-Navarro, F., Moreno-Cid, J. A., Weiss, J., & Egea-Cortines, M. (2023). The advent of plant cells in bioreactors. *Frontiers in Plant Science*, *14*, 1310405. https://doi.org/10.3389/fpls.2023.1310405.

Vijayan, D., & Ray, J. G. (2015). Ecology and diversity of Cyanobacteria in Kuttanadu paddy wetlands, Kerala, India. *American Journal of Plant Sciences*, *6*(18), 2924–2938.

Vitousek, P. M., Aber, J. D., Howarth, R. W., Likens, G. E., Matson, P. A., Schindler, D. W., Schlesinger, W. H., & Tilman, D. G. (1997). Human alteration of the global nitrogen cycle: Sources and consequences. *Ecological Applications*, *7*(3), 737–750. https://doi.org/10.1890/1051-0761(1997)007[0737:HAOTGN]2.0.CO;2.

Wang, F., Harindintwali, J. D., Yuan, Z., Wang, M., Wang, F., Li, S., & Chen, J. M. (2021). Technologies and perspectives for achieving carbon neutrality. *The Innovation*, *2*(4), 100180. https://doi.org/10.1016/j.xinn.2021.100180

Welte, C. U. (2018). Revival of archaeal methane microbiology. *Msystems*, *3*(2), 10–1128. https://doi.org/10.1128/msystems.00181-17.

Winogradsky, S. (1895). Recherches sur l'assimiltion de l'azote libre de l'atmosphëre par les microbes.

Xiong, Y., Zhang, X., He, X., Li, M., & Liao, B. (2016). Species-specific effects of P-solubilizing and N2-fixing bacteria on seedling growth of three salt-tolerant trees. *Russian Journal of Ecology*, *47*(3), 259–265. https://doi.org/10.1134/S1067413616030164.

Xue, C., Wang, L., Wu, T., Zhang, S., Tang, T., Wang, L., & Sun, Y. (2017). Characterization of co-cultivation of cyanobacteria on growth, productions of polysaccharides and extracellular proteins, nitrogenase activity, and photosynthetic activity. *Applied Biochemistry and Biotechnology*, *181*(1), 340–349.

Yadav, R. K., Abraham, G., Singh, Y. V., & Singh, P. K. (2014). Advancements in the utilization of Azolla-Anabaena system in relation to sustainable agricultural practices. *Proceedings of the Indian National Science Academy*, *80*(No. 2), 301–316.

Yandigeri, M. S., Meena, K. K., Srinivasan, R., & Pabbi, S. (2011). Effect of mineral phosphate solubilization on biological nitrogen fixation by diazotrophic cyanobacteria. *Indian Journal of Microbiology*, *51*(1), 48–53.

Yang, X., & Fang, S. (2015). Practices, perceptions, and implications of fertilizer use in East-Central China. *Ambio*, *44*(7), 647–652. https://doi.org/10.1007/s13280-015-0639-7.

Yao, Q. H., Peng, R. H., Wang, B., Tian, Y. S., Zhu, Y. M., Gao, J. J., & Deng, Y. D. (2021). Endowing plants with the capacity for autogenic nitrogen fixation.

Yao, Y. F., Wang, C. S., Qiao, J., & Zhao, G. R. (2013). Metabolic engineering of *Escherichia coli* for production of salvianic acid A via an artificial biosynthetic pathway. *Metabolic Engineering*, *19*, 79–87. https://doi.org/10.1016/j.ymben.2013.06.001.

Zeikus, J. G., Jain, M. K., & Elankovan, P. (1999). Biotechnology of succinic acid production and markets for derived industrial products. *Applied Microbiology and Biotechnology*, *51*(5), 545–552. https://doi.org/10.1007/s002530051431.

Zhang, Y., & Showalter, A. M. (2020). CRISPR/Cas9 genome editing technology: a valuable tool for understanding plant cell wall biosynthesis and function. *Frontiers in Plant Science*, *11*, 589517. https://doi.org/10.3389/fpls.2020.589517.

Zheng, M., Zhou, Z., Luo, Y., Zhao, P., & Mo, J. (2019). Global pattern and controls of biological nitrogen fixation under nutrient enrichment: A meta-analysis. *Global Change Biology*, *25*(9), 3018–3030. https://doi.org/10.1111/gcb.14705.

Ziv, M. (2005). Simple bioreactors for mass propagation of plants. *Plant Cell, Tissue and Organ Culture*, *81*(3), 277–285. https://doi.org/10.1007/s11240-004-6649-y.

Żymańczyk-Duda, E., Samson, S. O., Brzezińska-Rodak, M., & Klimek-Ochab, M. (2022). Versatile applications of cyanobacteria in biotechnology. *Microorganisms*, *10*(12), 2318. https://doi.org/10.3390/microorganisms10122318.

Harnessing Nitrogen Fixation: Strategies for Climate-Resilient Agriculture

9

Vikas Patidar, Shambhunath Ghosh and Aishna Srivastava

Abstract

Nitrogen is an essential nutrient for plant growth and a key factor in achieving higher crop yields. Traditionally, synthetic fertilizers have been the main nitrogen source, but excessive dependence on them has led to soil degradation, water pollution, greenhouse gas emissions, and climate change. Alternative nitrogen sources such as organic manure, biofertilizers, and biological nitrogen fixation (BNF) offer more sustainable options. BNF, carried out by certain microorganisms, naturally converts atmospheric nitrogen into plant-usable forms, reducing the need for chemical fertilizers. However, its use has declined with the rise of high-yielding varieties reliant on synthetic inputs. Sustainable nitrogen management is therefore crucial to balance productivity with environmental protection. BNF enhances soil fertility, reduces external inputs, and promotes eco-friendly farming systems that are more resilient to climate change. This chapter focuses on the significance of nitrogen fixation as a sustainable strategy for crop production, its role in ensuring food security, and its potential to restore soil and ecosystem health. It also discusses methods to enhance nitrogen fixation in both leguminous and non-leguminous crops, emphasizing how adopting microbial-based nitrogen inputs can transform modern agriculture toward greater sustainability and long-term ecological balance.

V. Patidar (✉) · S. Ghosh
Division of Agronomy, ICAR- Indian Agricultural Research Institute, Karnal Hub, Karnal, Haryana, India
e-mail: vikaspatidarvp2001@gmail.com

A. Srivastava
Division of Plant Pathology, ICAR- Indian Agricultural Research Institute, Karnal Hub, Karnal, Haryana, India

R. K. Singh and P. Singh (eds.), *Diazotrophic Plant Growth-Promoting Microorganisms and Green Energy – Volume I*, https://doi.org/10.1007/978-981-95-9682-9_9

Keywords

BNF · Climate change · Greenhouse gas · Legume · Nitrogen fixation ·

Abbreviations

ATP	Adenosine triphosphate
BNF	Biological nitrogen fixation
CRISPR-Cas	Clustered regularly interspaced short palindromic repeats
CSA	Climate-smart agriculture
ICT	Information and communication technology
IoT	internet of things
SynComs	Synthetic microbial communities

9.1 Introduction

Climate change presents a formidable challenge to global agriculture, intensifying the urgency for sustainable and resilient farming systems. Rising temperatures, erratic rainfall, and increasing soil degradation are compromising crop productivity, particularly in regions heavily dependent on external inputs like synthetic fertilizers (FAO, 2021). Among the macronutrients required for plant growth, nitrogen is the most limiting and energy-intensive to supply through industrial means. Nitrogen, the most abundant atmospheric element, exists mainly as diatomic N_2, which is inert and unavailable to most organisms. Excessive application of nitrogen-based fertilizers leads to substantial nitrogen losses, contributing to groundwater contamination, eutrophication of water bodies, and emissions of reactive-nitrogen greenhouse gases, such as N_2O. (Adebanjo-Aina & Oludoye, 2025). In this context, biological nitrogen fixation (BNF) emerges as a promising alternative to synthetic fertilizers. BNF is a natural process carried out by specific bacteria and archaea-either freely living in soil or in symbiotic relationships with plantsthat convert atmospheric nitrogen (N_2) into ammonia, a form accessible to plants (Singh, Sahu, & Ahirwal, 2019). Nitrogen fixation converts N_2 into ammonia (NH_3) or related compounds, making it accessible for plant uptake. Certain bacteria, through specialized enzymes, reduce N_2 to NH_3, which is then used in the synthesis of essential biomolecules, a process known as biological nitrogen fixation. It was discovered by German agronomist Hermann Hellriegel and Dutch microbiologist Martinus Beijerinck in 1880.

Annually, biological nitrogen fixation by legumes and cyanobacteria contributes an estimated ~250 Tg N yr^{-1} (i.e., 2.5×10^{11} kg N) to the global nitrogen cycle, while industrial fixation (e.g., via the Haber–Bosch process) accounts for roughly 80 Tg N yr^{-1} ($\approx 8 \times 10^{10}$ kg N). Lightning contributes an additional estimated 10 Tg N yr^{-1}($\approx 1 \times 10^{10}$ kg N). Currently, approximately 2 tons of industrially fixed nitrogen are needed as fertilizer for crop production to equal the effects of 1 ton of nitrogen biologically fixed by legume crops. Therefore, biologically fixed nitrogen influences the global nitrogen cycle substantially less than industrially fixed nitrogen. (Yu & Zhuang, 2020). Leguminous crops, such as soybean, lentils, and chickpeas, are well-known for their association with *Rhizobium* and *Bradyrhizobium* species, forming root nodules that house nitrogen-fixing bacteria. More recently, non-legume BNF systems, such as those involving *Azospirillum*, *Azotobacter*, and endophytic *diazotrophs*, are being explored for integration into cereal-based farming (Ladha et al., 2022). Harnessing nitrogen fixation not only reduces dependency on chemical inputs but also contributes to climate resilience by improving soil organic matter, enhancing biodiversity in the rhizosphere, and lowering nitrous oxide emissions (Rosenblueth et al., 2018). Moreover, recent advances in microbiome engineering, synthetic biology, and precision agriculture provide new avenues for optimizing nitrogen-fixing systems even in non-leguminous crops (Oldroyd & Leyser, 2020). This chapter explores the multifaceted role of nitrogen fixation in building climate-resilient agriculture. It provides an overview of the mechanisms, organisms, and innovations in BNF, reviews field-level applications, and discusses challenges and opportunities in scaling these solutions sustainably.

9.2 Basics of Nitrogen Fixation

9.2.1 Nitrogen Forms and Its Relevance to Nature

Nitrogen exists in various forms, each possessing distinct significance in different contexts. The relevance of these nitrogen forms can be observed across both ecological and agricultural perspectives (Table 9.1).

9.2.2 Biological vs. Non-Biological Nitrogen Fixation

Two types of nitrogen fixation are identified, and their distinct differences are clearly presented in Table 9.2 and Fig. 9.1.

Table 9.1 Different form of nitrogen and its relevance to ecology and agriculture

Nitrogen form	Description	Relevance to ecology & agriculture
Dinitrogen (N_2)	Inert atmospheric nitrogen gas (≈78%)	Biologically inaccessible without fixation
Ammonium (NH_4^+)	Reduced inorganic form	Readily absorbed by plants; intermediate in nitrification
Nitrate (NO_3^-)	Oxidized inorganic form	Widely mobile; dominant plant-available nitrogen source
Nitrite (NO_2^-)	Transitional oxidized inorganic form	Key microbial intermediate during nitrification/denitrification
Organic Nitrogen	Nitrogen within organic molecules	Major soil N reservoir; slowly released via mineralization
Nitrogen Oxides(N_2O, NO, NO_2)	Gaseous nitrogen forms from microbial/anthropogenic sources	Influence greenhouse gas budget and atmospheric N deposition

9.2.3 Key Nitrogen-Fixing Organisms

Biological Nitrogen fixation may be categorized into the following types:

A. Non-Symbiotic/asymbiotic Biological Nitrogen Fixation.
B. Associative Biological Nitrogen Fixation.
C. Symbiotic Biological Nitrogen Fixation.

A. ***Non-Symbiotic/Asymbiotic Biological Nitrogen Fixation:***
Soil contains a number of free-living nitrogen-fixing organisms. These include a number of aerobic and anaerobic bacteria and blue-green algae. Biological

Table 9.2 Difference between biological and non-biological nitrogen fixation

Aspect	Biological nitrogen fixation (BNF)	Non-biological nitrogen fixation
Definition	Conversion of atmospheric nitrogen (N_2) into ammonia (NH_3) or related compounds by living microorganisms using the nitrogenase enzyme complex.	Conversion of atmospheric nitrogen into reactive nitrogen compounds through physical or chemical processes without direct involvement of living organisms.
Agents/carried out by	Microorganisms such as *Rhizobium*, *Bradyrhizobium*, *Frankia*, *Azotobacter*, *Azospirillum*, and cyanobacteria (e.g., *Anabaena*).	Natural processes: lightning, forest fires, volcanic eruptions; Industrial processes: Haber–Bosch process, electric arc method.
Energy source	Requires high biological energy (ATP) supplied by host plants or microbial metabolism.	Energy supplied by physical forces (lightning, high temperature, pressure) or fossil fuel combustion in industry.
End products	Mainly ammonia (NH_3), which is directly usable by plants or further transformed into organic forms (e.g., amino acids).	Oxides of nitrogen (NO, NO_2), nitrates (NO_3^-), or ammonia depending on the process.
Environmental impact	Eco-friendly, improves soil fertility, enhances sustainability. Minimal negative effects when balanced.	Industrial fixation uses large fossil fuel energy, causing greenhouse gas emissions and pollution.
Agricultural importance	Provides a renewable and sustainable source of nitrogen in cropping systems, especially in legumes and some cereals with associative diazotrophs.	Supplies most of the nitrogen fertilizers used globally, but with a negative ecological footprint.
Examples	Legume–*Rhizobium* symbiosis, free-living *Azotobacter*, cyanobacteria in rice.	Lightning converting N_2 into nitrates, Haber–Bosch production of urea, ammonium nitrate, and ammonium sulfate fertilizers.

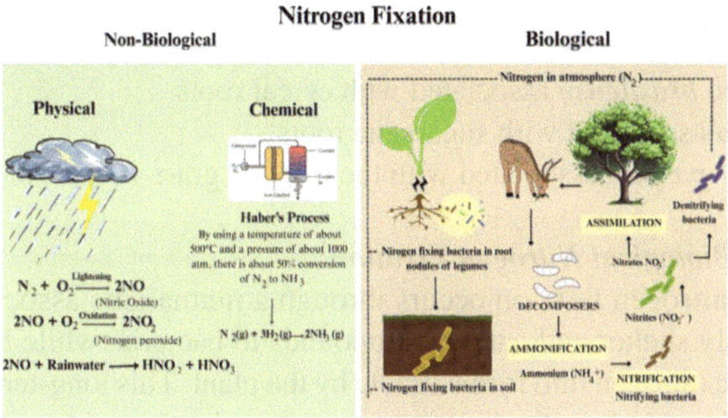

Fig. 9.1 Non- biological v/s biological nitrogen fixation

nitrogen fixation by microorganisms living freely or staying out of plant cell is called non-symbiotic Biological Nitrogen Fixation. The asymbiotic nitrogen fixers can be classified as follows:

1) Free-living aerobic nitrogen-fixing bacteria:
   - Photosynthetic: *Chlorobium, Chromatium*
   - Non-Photosynthetic: *Azotobacter, Azomonas, Derxia, Beijerinckia*
2) Free-living anaerobic nitrogen-fixing bacteria:
   - Photosynthetic: *Rhodospirillum*
   - Non-Photosynthetic: *Clostridium*
3) Free living chemosynthetic bacteria:
   - Heterotrophic: *Desulfovibro*
4) Cyanobacteria or Blue green algae:
   - *Heterocyst bearing: Nostoc, Anabaena, Rivularia, Calothrix.*
   - Non-Heterocyst bearing: *Oscillatoria, Gloeocapsa, Lyngbya, Plectonema.*
5) Free-living Fungi:

Yeasts and *Pullularia*

The asymbiotic free-living nitrogen fixers are quite primitive. The fixation is a reduction process independent of respiration. These organisms fix nitrogen more actively under poor aeration, provided no hydrogen gas is being produced.

B. ***Associative Symbiotic Nitrogen Fixation:***

Certain bacteria form a loose mutualistic association with the roots of cereals and grasses, enabling nitrogen fixation. This relationship, known as associative symbiosis, occurs mainly in the rhizosphere the region between the soil and root and occasionally inside the root tissues. The plants absorb a portion of the nitrogen fixed by the bacteria, while the microbes utilize carbohydrates secreted by the roots as their energy source.

Examples include:

1. *Azospirillum brasilense* associated with cereal roots.
2. *Beijerinckia* associated with sugarcane roots.
3. *Azotobacter paspali* associated with the tropical grass *Paspalum notatum*.

C. ***Symbiotic Biological Nitrogen Fixation:***
Symbiotic nitrogen fixation occurs through a mutualistic association in which plants supply shelter and carbon compounds to bacteria, while bacteria convert atmospheric nitrogen into forms usable by the plant. This long-term, ecologically stable relationship benefits both partners: the microbe gains nutrients from the host, and the plant receives fixed nitrogen. Symbiotic nitrogen fixation can be categorized into three main types:

i) ***Nitrogen Fixationvia Nodule Formation in Leguminous Plants:*** In legumes, symbiotic nitrogen-fixing bacteria primarily belonging to the genus *Rhizobium* colonize specialized structures called root nodules. These bacteria fix nitrogen only when residing within the nodules. The plant provides carbohydrates as an energy source for the bacteria, and in exchange, the bacteria supply the plant with fixed nitrogen. For instance, *Bradyrhizobium japonicum* forms a slow-growing symbiosis with soybeans, while *Azorhizobium caulinodans* forms stem nodules in *Sesbania* species.
ii) ***Nitrogen Fixation via Nodule Formation in Non-Leguminous Plants:*** Several non-leguminous plants, particularly trees and shrubs, also form root nodules capable of nitrogen fixation. Examples include:
   1. *Frankia* forming root nodules with *Alnus*, *Casuarina equisetifolia*, and *Myrica gale*.
   2. *Rhizobium* forming root nodules with *Parasponia*.
   3. Leaf nodules formed by *Klebsiella* in *Psychotria* and by *Burkholderia* in *Pavetta zimermanniana*.
iii) ***Nitrogen Fixation without Nodule Formation (Non-Nodulation):*** In certain plants, nitrogen fixation occurs symbiotically but without the development of nodules, a condition termed *pseudo-symbiosis*. Examples include:

   1. Lichens associations between fungi and cyanobacteria or green algae.
   2. *Anthoceros* (a bryophyte) associated with *Nostoc*.
   3. *Azolla* (a fern) associated with *Anabaena*.
   4. *Cycas* (a gymnosperm) with *Anabaena* or *Nostoc* in coralloid roots.
   5. *Gunnera macrophylla* (an angiosperm) associated with *Nostoc* in its stem.
   6. Roots of Digitaria, Sorghum, and Maize associated with *Spirillum*.

9.2.4 Nitrogenase Enzyme and Energy Requirements

Nitrogen fixation in diazotrophic microorganisms is catalyzed by the nitrogenase enzyme complex, consisting of the Fe protein (dinitrogenase reductase) and the MoFe protein (dinitrogenase). The Fe protein transfers electrons, powered by ATP hydrolysis, to the MoFe protein, where atmospheric N_2 is reduced to ammonia (NH_3) (Burgess & Lowe, 1996).

The overall reaction requires high energy:

$$
\begin{aligned}
& N_2 + 8H^+ + 8e^- + 16ATP \\
& \rightarrow 2NH_3 + H_2 + 16ADP + 16PiN_2 + 8H^+ 8e^- + 16ATP \\
& \rightarrow 2NH_3 + H_2 + 16ADP + 16PiN_2 + 8H^+ + 8e^- + 16ATP \\
& \rightarrow 2NH_3 + H_2 + 16ADP + 16Pi
\end{aligned}
$$

Thus, reduction of one N_2 molecule needs at least 16 ATP molecules and low-potential electrons (Hoffman et al., 2014). Because nitrogenase is oxygen-sensitive, microbes employ protective strategies, such as leghemoglobin in legume nodules, heterocysts in cyanobacteria, or high respiration in free-living diazotrophs (Seefeldt et al., 2020). Cobalt and Molybdenum are the two elements that are to be the part of Nitrogen fixation.

9.3 Role of Nitrogen Fixation in Sustainable Agriculture

Nitrogen is a vital macronutrient essential for plant growth, playing a central role in proteins, nucleic acids, and chlorophyll formation. In conventional agriculture, the majority of nitrogen is supplied through synthetic fertilizers, which are energy-intensive to produce and often result in environmental consequences, such as greenhouse gas emissions, water contamination, and soil acidification (Canfield et al., 2010; FAO, 2021). As the demand for food grows alongside climate change pressures, sustainable nitrogen management becomes crucial. Biological nitrogen fixation (BNF) offers a natural, eco-friendly alternative to synthetic nitrogen inputs. BNF is the microbial-mediated process of converting atmospheric nitrogen (N_2) into ammonia, which plants can assimilate. It is carried out by free-living diazotrophs (e.g., *Azotobacter, Clostridium*), associative bacteria (e.g., *Azospirillum*), *cyanobacteria*, and symbiotic bacteria (e.g., *Rhizobium, Bradyrhizobium*) that inhabit legume root nodules (Rosenblueth et al., 2018).

The integration of BNF into cropping systems supports sustainable agriculture in several key ways:

- **Reduces reliance on synthetic fertilizers:** Biological nitrogen fixation (BNF) plays a critical role in sustainable agriculture by converting atmospheric N_2 into ammonia, thereby enriching soil fertility and reducing the need for synthetic fertilizers. This process is carried out by diazotrophic bacteria, both symbiotic (e.g., *Rhizobium, Bradyrhizobium*) and free-living (e.g., *Azospirillum*, cyanobacteria) across diverse cropping systems. Legumes and inoculated crops can derive up to 60%–90% of their nitrogen from BNF, reducing the need for external nitrogen inputs and lowering production costs (Ladha et al., 2022).

- **Enhances soil health:** BNF improves soil fertility through organic nitrogen inputs, fosters microbial diversity, and enhances long-term soil productivity (Smercina et al., 2021). It also enhances soil structure, tilth, and organic matter content, thus improving long-term productivity and yield stability.
- **Mitigates environmental pollution:** Reduced fertilizer use leads to less nitrate leaching and decreased nitrous oxide (N_2O) emissions-a potent greenhouse gas, thus improving ecosystem health and climate outcomes (Herridge et al., 2008). Economically and environmentally, BNF is a cost-effective, non-polluting alternative. It decreases greenhouse gas emissions and mitigates groundwater contamination by reducing reliance on energy-intensive nitrogen fertilizers (Mukherjee & Sen, 2021).
- **Supports climate resilience:** Legume-based cropping systems improve nitrogen use efficiency and water retention, and when used in rotation or intercropping, they can help buffer farms against climate variability (Oldroyd & Leyser, 2020).
- **Contributes to food security:** Especially in low-input farming systems and resource-constrained regions, nitrogen-fixing crops help improve yields without dependence on costly chemical fertilizers (FAO, 2021).
- Innovations in microbial inoculants, genome editing, and microbiome engineering further expand the scope of nitrogen fixation beyond legumes to cereals and non-leguminous crops, opening new frontiers for sustainable intensification (Mus et al., 2016). Meta-analyses indicate that legumes can derive up to 68% of their nitrogen from atmospheric fixation, highlighting the substantial ecosystem service provided by BNF (Ghafoor et al., 2024b).

9.4 Nitrogen-Fixing Systems and Their Applications

9.4.1 Legume–Rhizobium Symbiosis

Legume–*Rhizobium* symbiosis begins when legume roots secrete flavonoids, inducing *Rhizobium* to produce Nod factors that trigger root hair curling, infection thread formation, and nodule development where nitrogen fixation occurs. This interaction shows strong host-specificity, as different legumes prefer compatible *rhizobial* strains, influencing symbiotic efficiency. For example, soybean associates with *Bradyrhizobium japonicum* in temperate zones, while cowpea forms symbiosis with diverse *Bradyrhizobium* and *Rhizobium* strains in tropical soils. Such partnerships enhance nitrogen supply, soil fertility, and sustainability in varied agro-ecosystems (Oldroyd et al., 2011).

9.4.2 Non-Leguminous Nitrogen Fixation

- **Associative Systems**

Associative nitrogen fixation occurs when free-living diazotrophic bacteria colonize the rhizosphere and root surfaces of cereals. *Azospirillum*, *Herbaspirillum*, and

Burkholderia are well-studied genera that enhance nutrient uptake, root growth, and yield in crops like wheat, maize, rice, and sorghum. While nitrogen contribution is lower than in legumes, associative systems provide supplementary N input and promote plant growth through phytohormone production (Baldani & Baldani, 2005).

- **Endophytic Nitrogen-Fixers**
  Certain diazotrophs colonize the internal tissues of non-legumes without forming nodules. *Gluconacetobacter diazotrophicus* in sugarcane, *Herbaspirillum seropedicae* in maize, and *Azoarcus* spp. in rice fix nitrogen within plant tissues, directly supplying ammonia and improving nitrogen-use efficiency under low fertilizer input conditions (Bashan & de-Bashan, 2010).

9.4.3 Cyanobacteria in Rice Ecosystems

Cyanobacteria play a vital role in sustaining fertility in flooded rice ecosystems by fixing atmospheric nitrogen through heterocysts and releasing it as ammonium, thereby enriching paddy soils. Genera, like *Anabaena*, *Nostoc,* and *Aulosira,* contribute significantly to the nitrogen economy of traditional rice fields. Traditionally, farmers encouraged algal blooms in paddies as a natural fertilizer source. In modern systems, integrated approaches combine cyanobacteria or *Anabaena–Azolla* symbiosis with chemical fertilizers to enhance nitrogen-use efficiency, reduce input costs, and maintain soil health. These practices improve yield sustainability in rice-based farming systems.

9.5 Strategies to Enhance Biological Nitrogen Fixation

Enhancing biological nitrogen fixation (BNF) is critical for sustainable and climate-resilient agriculture. Several complementary strategies are being developed to improve efficiency and applicability across diverse cropping systems.

9.5.1 Selection and Breeding of Compatible Host Plants

Host plant genetics strongly influence symbiotic efficiency. Breeding programs focus on selecting legume varieties with greater nodulation potential and nitrogen-fixing ability. Marker-assisted selection and genomic tools are increasingly used to identify genes regulating nodulation and host specificity (Graham & Vance, 2000).

9.5.2 Development of Efficient Microbial Inoculants

Microbial inoculants, commonly referred to as biofertilizers, are developed to strengthen the natural microbial population in soils that are often depleted or lack diversity. *Rhizobial* inoculants are widely used in legumes to enhance nodulation and nitrogen fixation, while diazotrophic inoculants, such as *Azotobacter* and *Azospirillum*, benefit cereals and non-legumes. Modern research focuses on improving inoculant strains with traits like better survival under adverse conditions, higher nitrogenase activity, stress tolerance, and strong competitiveness against native microbes. Such innovations ensure consistent crop yield improvements and enhanced soil fertility across diverse environments (Herridge et al., 2008).

9.5.3 Use of Microbial Consortia and Synthetic Communities

Instead of applying single-strain inoculants, microbial consortia or synthetic communities are gaining attention. These formulations integrate multiple beneficial microbes, such as *rhizobia*, *Azospirillum*, phosphate-solubilizing bacteria, and mycorrhizal fungi. Their synergistic interactions not only supply nitrogen but also mobilize phosphorus, improve water uptake, and enhance plant resilience to biotic and abiotic stresses. As a result, they provide more stable benefits across variable soil and climatic conditions, ensuring higher yield sustainability (Bashan et al., 2014).

9.5.4 Precision Agriculture and Microbiome Management

With the advent of next-generation sequencing and precision farming, microbiome management has become highly targeted. Farmers can now design crop rotations, apply tailored soil amendments, and monitor microbial dynamics in real time. These practices maximize BNF and other microbial functions under site-specific conditions, reducing reliance on chemical inputs while improving long-term soil health (Kaul et al., 2021).

9.5.5 Genetic Engineering and Synthetic Biology Approaches

Frontier research aims to transfer nitrogenase genes into cereals, such as rice and maize. Synthetic biology tools allow partial reconstruction of nitrogen fixation pathways in non-legumes, while CRISPR–Cas systems help fine-tune host–microbe interactions (Oldroyd & Leyser, 2020). Although still experimental, these innovations hold transformative potential for reducing global fertilizer dependence.

9.6 Advantages of Biological Nitrogen Fixation

There are several advantages of fixing nitrogen are shown in Fig. 9.2.

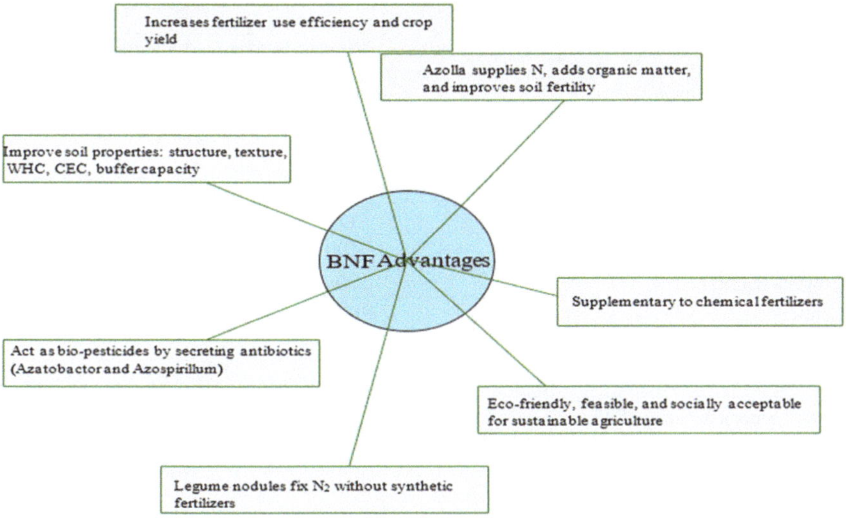

Fig. 9.2 Advantages of biological nitrogen fixation

9.7 Integrating Nitrogen Fixation into Climate-Resilient Agricultural Practices

Biological nitrogen fixation (BNF) provides a sustainable pathway for reducing chemical fertilizer dependency, enhancing soil health, and building climate resilience in agriculture. Integrating BNF into broader agroecological and climate-smart frameworks can help mitigate the adverse impacts of climate variability while sustaining productivity.

9.7.1 Agroecological Approaches

Agroecological practices, such as crop rotation and legume-based intercropping, effectively integrate BNF into farming systems. Rotating legumes with cereals enriches soil nitrogen, improves organic matter, and disrupts pest cycles (Peoples et al., 2009). Intercropping maize with beans or pigeon pea is widely practiced in the tropics, enhancing resource use efficiency and reducing external inputs, many crops having good potential to fix nitrogen and help in adopting climate resilience agrculture (Table 9.3).

9.7.2 Climate-Smart Farming Systems Incorporating BNF

Climate-smart agriculture (CSA) integrates BNF into conservation tillage, integrated nutrient management, and agroforestry systems. Leguminous cover crops and green manures supply biologically fixed nitrogen while improving soil structure and carbon

Table 9.3 Typical nitrogen fixation by crops

Crops	Typical N fixation (kg/ha/year)
Leucaena leucocephala	450
Alfalfa (Lucern)	200
Faba beans	130
Sweet clover	130
Red clover, pigeonpea	115
Soybean, field pea	100
Cowpea (Lobia)	90
Lentil	60
Chickpea	50
Green gram, beans, groundnut	40

Source: Meena & Sihag 2025

sequestration (Lal, 2020). In agroforestry, nitrogen-fixing trees, such as *Leucaena leucocephala* or *Gliricidia sepium,* provide both soil fertility benefits and resilience against drought stress (Sileshi et al., 2008).

9.7.3 Synergies with Water and Nutrient Management

BNF efficiency depends on soil moisture, pH, and nutrient availability. Integrated water and nutrient management, including proper irrigation scheduling, phosphorus supplementation, and organic amendments, improves rhizobial survival and nitrogen fixation efficiency (Giller, 2001). Cyanobacteria in paddy fields exemplify synergy between waterlogged conditions and BNF, supporting rice productivity.

9.7.4 Reducing Agriculture's Carbon and Nitrogen Footprint

Integrating BNF reduces reliance on synthetic fertilizers produced via the Haber–Bosch process, a major contributor to greenhouse gas emissions. By supplying nitrogen biologically, legumes and diazotrophs lower nitrous oxide (N_2O) emissions from soils, contributing to both climate mitigation and adaptation (Herridge et al., 2008).

9.8 Challenges and Limitations in Harnessing Nitrogen Fixation

Despite its recognized benefits for sustainable and climate-resilient agriculture, the full potential of biological nitrogen fixation (BNF) remains underutilized. A combination of biological, environmental, institutional, and socio-economic challenges limits its widespread adoption. Addressing these constraints is essential for scaling BNF technologies and reducing dependence on synthetic nitrogen fertilizers.

9.8.1 Variability in Field Performance of Inoculants

A major challenge is the inconsistent performance of microbial inoculants across environments. While laboratory and greenhouse trials often show high nitrogen fixation rates, field performance may be poor due to competition with native soil microbes, unfavorable environmental conditions, or improper application methods (Herrmann & Lesueur, 2013). This variability discourages farmer confidence and slows adoption. Ensuring strain survival, adaptation to local agro-ecologies, and improved formulations remains a key research priority.

9.8.2 Host–Microbe Incompatibility

BNF efficiency depends on compatibility between host plants and microbial partners. Certain legumes require specific *Rhizobium* strains, while cereals and non-legumes have limited or less efficient associations with diazotrophs (Oldroyd et al., 2011). Mismatches between host crop and inoculant strain can result in poor nodulation or reduced nitrogen fixation, leading to yield gaps. Breeding for host varieties with enhanced symbiotic efficiency and developing broad-spectrum microbial strains are crucial to overcoming this limitation.

9.8.3 Soil Constraints

Soil properties, such as pH, salinity, temperature, soil nutrient supply, and organic matter content, strongly influence nitrogen fixation. Acidic soils limit rhizobial survival and nodulation, while salinity and drought stress reduce microbial activity and nitrogenase function (Zahran, 1999). Low organic matter levels also restrict microbial persistence and energy supply. With a decrease in nitrogen availability in soil, nitrogen fixation generally increases. Excess NO_3^- reduces nitrogenase activity and ultimately N-fixation. Site-specific soil management, such as liming acidic soils, organic amendments, and integrated irrigation strategies, is needed to optimize BNF under diverse conditions.

9.8.4 Regulatory and Market Challenges for Bio-Inputs

Biofertilizer markets often face weak regulatory frameworks and quality control issues. Poorly formulated or counterfeit products erode farmer trust, resulting in limited uptake despite potential benefits (Bashan et al., 2014). Lack of standardized certification systems and insufficient monitoring further complicate commercialization.

Stronger policies, public–private partnerships, and regional distribution networks are required to establish a reliable bio-input sector.

9.8.5 Energy Requirement for Nitrogen Fixation

Breaking the strong triple bond (N≡N) requires high energy; microorganisms use the nitrogenase enzyme complex, consuming ATP, and reducing power to convert atmospheric nitrogen into ammonia.

9.8.6 Oxygen Sensitivity in Nitrogen Fixation

Nitrogen fixation reduces atmospheric N_2 to ammonia, an energy-intensive process. Nitrogen-fixing organisms are highly sensitive to oxygen, which can inactivate the nitrogenase enzyme, hindering efficient nitrogen conversion.

9.8.7 Knowledge and Awareness Gaps among Stakeholders

Awareness of the role and management of nitrogen-fixing systems remains low among farmers, extension agents, and policymakers. Many smallholder farmers prefer visible short-term benefits of synthetic fertilizers over the less immediate effects of BNF. Limited extension services, lack of farmer training, and inadequate integration of BNF into curricula and advisory services further widen the knowledge gap (Peoples et al., 2009). Strengthening extension systems, promoting farmer field schools, and leveraging ICT tools can help bridge this divide.

9.9 Future Prospects for Harnessing Nitrogen Fixation

Biological nitrogen fixation (BNF) holds tremendous potential for transforming agriculture into a more sustainable and climate-resilient system. Emerging advances in biotechnology, microbiome research, and digital agriculture are creating new opportunities to optimize nitrogen fixation and integrate it into farming at scale. The future of BNF lies in combining scientific innovation with policy and farmer-centric strategies.

9.9.1 Innovations in Microbiome Research

Recent breakthroughs in plant–microbiome interactions are shedding light on how root-associated microbial communities influence nitrogen fixation. By characterizing the rhizosphere and endophytic microbiomes, researchers are identifying beneficial microbial consortia that enhance nitrogen use efficiency and stress tolerance (Compant et al., 2019). Synthetic microbial communities (SynComs) are being designed to mimic natural interactions, providing stable and efficient nitrogen fixation even under adverse conditions. Such innovations pave the way for next-generation biofertilizers tailored to specific crops and soils.

9.9.2 Genome Editing

Advances in genome-editing technologies like CRISPR–Cas enable precise modifications in both plants and microbes to improve symbiotic efficiency. For legumes, targeted editing of genes controlling nodulation, signaling pathways, and stress tolerance can enhance symbiotic nitrogen fixation (Rogers & Oldroyd, 2014). In microbes, CRISPR tools are being used to modify *Rhizobium* or free-living diazotrophs to increase nitrogenase activity and adaptability. Long-term goals include transferring nitrogen fixation capacity to cereals, a breakthrough that could revolutionize food production in nitrogen-limited regions.

9.9.3 Advances in Metabolomics and Systems Biology

Integrating metabolomics, transcriptomics, and systems biology allows a holistic understanding of nitrogen fixation at molecular, cellular, and ecosystem scales. By mapping metabolic pathways involved in nitrogenase activity and carbon–nitrogen trade-offs, researchers can identify bottlenecks and optimize energy use in symbiotic systems (Mus et al., 2016). This systems-level approach is also crucial for developing predictive models that guide microbial inoculant design and field applications under variable climates.

9.9.4 Integration with Digital Agriculture and Decision-Support Systems

Digital agriculture offers powerful tools to monitor, model, and manage nitrogen fixation in real time. Remote sensing, IoT-based soil sensors, and AI-driven decision-support systems can provide farmers with precise recommendations on legume rotations, inoculant application, and soil health management (Gebbers & Adamchuk, 2010). Data-driven platforms also enable site-specific optimization of BNF, minimizing fertilizer input while maximizing productivity and resilience.

9.9.5 Mainstreaming Nitrogen Fixation in Resilient Farming

Harnessing BNF requires scientific innovation, supportive policies, farmer training, public–private partnerships, and global cooperation to develop efficient nitrogen-fixing systems, scale biofertilizer use, and address diverse agro-ecological challenges sustainably and effectively.

9.10 Conclusion

Nitrogen is an essential nutrient for plant growth and development but is unavailable in its most prevalent form as atmospheric nitrogen. Plants instead depend upon combined, or fixed, forms of nitrogen, such as ammonia and nitrate. Biological nitrogen fixation is the process of capturing atmospheric nitrogen by biological processes. It is accomplished by certain microorganisms and plant–microbe interactions. Legumes are N-fixing systems that have long been used for biological nitrogen fixation in agriculture. Biological N_2 fixation is an important aspect of sustainable and environmentally friendly food production and long-term crop productivity. However, if biological nitrogen fixation is to be utilized, it must be optimized in the near future. Increased use of biological nitrogen fixation is one of the major pathways to maintain or increase yield and to reduce the environmental footprint of agriculture, which may be used to address the current challenge of meeting the fast-growing worldwide demand for agricultural products.

Nitrogen fixation remains a cornerstone of sustainable and climate-resilient agriculture. This chapter highlighted the biological foundations of nitrogen fixation, the diversity of symbiotic and free-living systems, and their critical role in enhancing soil fertility while reducing dependence on synthetic fertilizers. Key insights demonstrate that integrating BNF through legumes, microbial inoculants, and innovative technologies can significantly improve productivity, resilience, and environmental sustainability. The strategic importance of nitrogen fixation lies in its dual role- supporting food security while mitigating agriculture's carbon and nitrogen footprint. By lowering greenhouse gas emissions and reducing reliance on the energy-intensive Haber–Bosch process, BNF contributes directly to climate adaptation and mitigation. Advances in microbiome research, genome editing, and digital agriculture present exciting opportunities to mainstream nitrogen fixation into modern farming systems.

However, realizing its full potential requires collective action. Researchers must continue to innovate, policymakers need to create supportive regulatory and incentive frameworks, and practitioners should adopt context-specific BNF practices. By aligning science, policy, and practice, nitrogen fixation can move from being an underutilized resource to a central pillar of resilient food systems, helping agriculture meet the twin challenges of productivity and sustainability in a changing climate.

The future of nitrogen fixation lies at the intersection of biotechnology, systems science, and digital innovation. By leveraging genome editing, microbiome engineering, and precision agriculture, it is possible to enhance BNF efficiency and integrate it seamlessly into climate-resilient farming. With supportive policies and farmer-centered approaches, nitrogen fixation can move from a niche practice to a mainstream solution, reducing reliance on synthetic fertilizers and fostering sustainable food security in a changing climate.

References

Adebanjo-Aina, O., & Oludoye, O. (2025). Impact of nitrogen fertiliser usage in agriculture on water quality. *Pollutants*, *5*(3), 21. https://doi.org/10.3390/pollutants5030021.

Baldani, J. I., & Baldani, V. L. D. (2005). a. *Anais da Academia Brasileira de Ciências*, *77*(3), 549–579.

Bashan, Y., & de-Bashan, L. E. (2010). How the plant growth-promoting bacterium *Azospirillum* promotes plant growth—A critical assessment. *Advances in Agronomy*, *108*, 77–136.

Bashan, Y., de-Bashan, L. E., Prabhu, S. R., & Hernandez, J. P. (2014). Advances in plant growth-promoting bacterial inoculant technology. *Plant and Soil*, *378*, 1–33. https://doi.org/10.1007/s11104-013-1956-x.

Burgess, B. K., & Lowe, D. J. (1996). Mechanism of molybdenum nitrogenase. *Chemical Reviews*, *96*(7), 2983–3012.

Canfield, D. E., Glazer, A. N., & Falkowski, P. G. (2010). The evolution and future of Earth's nitrogen cycle. *Science*, *330*(6001), 192196. https://doi.org/10.1126/science.1186120.

Compant, S., Samad, A., Faist, H., & Sessitsch, A. (2019). A review on the plant microbiome: Ecology, functions, and emerging trends in microbial application. *Journal of Advanced Research*, *19*, 29–37.

FAO (2021). The State of the World's Land and Water Resources for Food and Agriculture – Systems at breaking point. Food and Agriculture Organization of the United Nations. https://www.fao.org/publications.

Gebbers, R., & Adamchuk, V. I. (2010). Precision agriculture and food security. *Science*, *327*(5967), 828–831. https://doi.org/10.1126/science.1183899.

Ghafoor, A. Z., Javed, H. H., Karim, H., Studnicki, M., Ali, I., Yue, H., Xiao, P., Asghar, M. A., Brock, C., & Wu, Y. (2024). Biological nitrogen fixation for sustainable agriculture development under climate change–New insights from a meta-analysis. *Journal of Agronomy and Crop Science*, 210(5), e12754. https://doi.org/10.1111/jac.12754.

Giller, K. E. (2001). Nitrogen fixation in tropical cropping systems. CABI.

Graham, P. H., & Vance, C. P. (2000). Nitrogen fixation in perspective: An overview of research and extension needs. *Field Crops Research*, *65*(2–3), 93–106. https://doi.org/10.1016/S0378-4290(99)00080-5.

Herridge, D. F., Peoples, M. B., & Boddey, R. M. (2008). Global inputs of biological nitrogen fixation in agricultural systems. *Plant and Soil*, *311*, 1–18. https://doi.org/10.1007/s11104-008-9668-3.

Herrmann, L., & Lesueur, D. (2013). Challenges of formulation and quality of biofertilizers for successful inoculation. *Applied Microbiology and Biotechnology*, *97*, 8859–8873. https://doi.org/10.1007/s00253-013-5228-8.

Hoffman, B. M., Lukoyanov, D., Yang, Z. Y., Dean, D. R., & Seefeldt, L. C. (2014). Mechanism of nitrogen fixation by nitrogenase. *Chemical Reviews*, *114*(8), 4041–4062.

Kaul, S., Sharma, T., & Dhar, M. K. (2021). "Omics" tools for better understanding the plant–endophyte interactions. *Frontiers in Plant Science*, *12*, 666841. https://doi.org/10.3389/fpls.2016.00955.

Ladha, J. K., Reddy, C. K., Padre, A. T., van Kessel, C., & Aggarwal, P. K. (2022). Biological nitrogen fixation and prospects for ecological intensification in cereal-based cropping systems. *Plant and Soil*, *474*, 1–24. https://doi.org/10.1007/s11104-022-05520-7.

Lal, R. (2020). Regenerative agriculture for food and climate. *Journal of Soil and Water Conservation*, *75*(5), 123A–124A. https://doi.org/10.2489/jswc.2020.0620A.

Meena, R. S., & Sihag, S. K. (2025). Agronomy facts for competitions (7th rev. ed.. Jain Brothers. p. 330. ISBN 978-81-8360-259-4.

Mukherjee, R., & Sen, S. (2021). Role of biological nitrogen fixation (BNF) in sustainable agriculture: A review. *International Journal of Advancement in Life Sciences Research*, *4*(3). https://doi.org/10.31632/ijalsr.2021.v04i03.001

Mus, F., Crook, M. B., Garcia, K., et al. (2016). Symbiotic nitrogen fixation and the challenges to its extension to nonlegumes. *Applied and Environmental Microbiology*, *82*(13), 3698–3710. https://doi.org/10.1128/AEM.01055-16.

Oldroyd, G. E. D., & Leyser, O. (2020). A plant's diet, surviving in a variable nutrient environment. *Science*, *368*(6486), eaba0196. https://www.science.org/doi/10.1126/science.aba0196.

Oldroyd, G. E. D., Murray, J. D., Poole, P. S., & Downie, J. A. (2011). The rules of engagement in the legume–rhizobial symbiosis. *Annual Review of Genetics*, *45*, 119–144.

Peoples, M. B., et al. (2009). The contributions of nitrogen-fixing crop legumes to productivity improvement of agricultural systems. *Symbiosis*, *48*(1), 1–17. https://doi.org/10.1007/BF03179980.

Rogers, C., & Oldroyd, G. E. D. (2014). Synthetic biology approaches to engineering the nitrogen symbiosis in cereals. *Journal of Experimental Botany*, *65*(8), 1939–1946. https://doi.org/10.1093/jxb/eru098.

Rosenblueth, M., Ormeño-Orrillo, E., López-López, A., Rogel, M. A., Reyes-Hernández, B. J., Martínez-Romero, J. C., & Martínez-Romero, E. (2018). Nitrogen fixation in cereals. *Frontiers in Microbiology*, *9*, 1794. https://doi.org/10.3389/fmicb.2018.01794.

Seefeldt, L. C., Yang, Z. Y., Lukoyanov, D. A., Harris, D. F., Dean, D. R., Raugei, S., & Hoffman, B. M. (2020). Reduction of substrates by nitrogenases. *Chemical Reviews*, *120*(12), 5082–5106.

Sileshi, G., Akinnifesi, F. K., Ajayi, O. C., & Place, F. (2008). Meta-analysis of maize yield response to woody and herbaceous legumes in sub-Saharan Africa. *Plant and Soil*, *307*, 1–19. https://doi.org/10.1007/s11104-008-9547-y.

Singh, V., Sahu, M. P., & Ahirwal, A. (2019). Nitrogen fixation. In Recent trends in agricultural sciences & technology (1st ed., pp. 151–158). Weser Books. ISBN: 978-3-96492-153-69.

Smercina, D. N., Evans, S. E., Friesen, M. L., Tiemann, L. K., & et al. (2021). Temporal dynamics of free-living nitrogen fixation in the switchgrass rhizosphere. *GCB Bioenergy*, 1814–1830. https://doi.org/10.1111/gcbb.12893

Yu, T., & Zhuang, Q. (2020). Modeling biological nitrogen fixation in global natural terrestrial ecosystems. *Biogeosciences*, *17*(14), 3643–3657. https://doi.org/10.5194/bg-17-3643-2020.

Zahran, H. H. (1999). Rhizobium–legume symbiosis and nitrogen fixation under severe conditions and in an arid climate. *Microbiology and Molecular Biology Reviews*, *63*(4), 968–989. https://doi.org/10.1128/MMBR.63.4.968-989.1999.

Green Innovations in Agriculture: The Role of Microbes in Climate Change Mitigation

10

Tarunpreet Kaur and Pooja Nikhanj

Abstract

Agriculture is a major contributor to global greenhouse gas (GHG) emissions, with India's agricultural sector alone accounting for 19.13% of the country's total emissions. Microorganisms act as source, as well as sink in agricultural GHG emission production. Microbial-driven carbon sequestration stabilizes atmospheric CO_2 with aid of autotrophic microbes, e.g., cyanobacteria and microalgae, which capture CO_2 through the Calvin-Benson cycle, reduce CH_4 and N_2O emissions, and simultaneously enhance soil fertility. Persistence of purple non-sulfur bacteria, such as, *Rhodopseudomonas palustris* TN110 successfully achieved an 80% reduction in CH_4 emissions in rice/paddy fields, as per studies conducted. Methanotrophy acts as first line defence against methane production, by converting it into CO_2 and H_2O. Similarly, microbial solutions address N_2O emissions, by using plant growth promoting rhizobacteria (PGPRs), such as *Bradyrhizobium diazoefficiens* and *Bacillus* spp., by providing enhanced expression of *nosZ* gene, necessary for conversion of N_2O to N_2. Beyond direct mitigation, microorganisms decompose organic wastes, diverting the routes leading to GHG emissions caused by their burning. The Pusa decomposer, mushroom and bioethanol production, microbes as feed-additives, suggest indirect usage of microorganisms to reduce GHGs. Though such solutions possess somewhat challenges, however, policy support, continued research, and farmer adoption can unlock the full potential of microbial interventions in achieving carbon neutrality by 2050. By leveraging microbial processes, agriculture can transition toward a more sustainable and climate-resilient future.

T. Kaur · P. Nikhanj (✉)
Department of Microbiology, Punjab Agricultural University, Ludhiana, India
e-mail: poojanikhanj.217@gmail.com

R. K. Singh and P. Singh (eds.), *Diazotrophic Plant Growth-Promoting Microorganisms and Green Energy – Volume I*, https://doi.org/10.1007/978-981-95-9682-9_10

Keywords

Agriculture · GHG emissions · microbial innovations · carbon sequestration ·

10.1 Introduction

Agriculture remains the most significant economic sector to ensure food security for the rising human population, and like all other anthropogenic activities, is also a substantial contributor of greenhouse gas (GHG) emissions to the environment. The direct contribution of major GHGs such as methane (CH_4), nitrous oxide (N_2O), and carbon dioxide (CO_2) from agricultural fields and livestock production holds an overall 12.25% share in global GHG production; India's agricultural sector being a 12.18% shareholder. Furthermore, the contribution of Indian agricultural sector to national GHG emissions remain at 19.13%. Agricultural activities primarily serve as emitters of three GHGs——Carbon dioxide (CO_2), Methane (CH_4), and Nitrous Oxide (N_2O) (Shakoor et al., 2021). Although methane and nitrous oxide have lower residence time, but their global warming potential is 28 and 310 times more potent than that of carbon dioxide. This corresponds to 71.8% and 24.3% contribution from national agricultural sector. These emission patterns not only highlight the influence of agricultural practices upon GHG emissions, but also underscore the necessity to examine their impacts on soil health and long-term ecosystem stability.

Crop production and protection are closely correlated with the soil health, and intensification of farming practices, since green revolution, has immensely disturbed the soil ecosystem leaving negative impact on carbon sequestration potential and climate resilience as outcomes. Microorganisms and other soil macro-flora and fauna constitute a major part of agroecosystems, and implication of irrational use of chemical fertilizers, pesticides, and insecticides has resulted in shift in their physiological latency (Wang et al., 2022). An increased farm mechanization led to extensive till-farming practices which contributed to loss of soil organic matter (SOM) due to rapid decomposition by abiotic factors. Moreover, other developmental practices such as deforestation for agricultural land expansion reduced the carbon sinks, and it is estimated that on every 1% land-expansion, there is 0.91% rise in CO_2 emissions (Raihan et al., 2022), rearing of livestock for agricultural purposes, depicted a rapid upsurge in methane and nitrous oxide emissions (Chachei, 2024). Water-logged irrigation of rice/paddy fields provides a congenial environment for anaerobic microbes to perform methanogenesis——the synthesis of methane.

Even though agriculture acts as emitter of greenhouse gases, it also holds considerable potential to serve as sinks (OECD, 2019), especially through green innovations. The inclusion of microorganisms offers ecofriendly solution to combat the effects of climate change. Microorganisms can capture the resultant emissions from agro-fields and prevent their release into the atmosphere. Microbial-based solutions overpower the conventional mitigation strategies as these are cost-effective, scalable, and environment-friendly interventions, and possess the ability to enhance inherent capability of soil ecosystem for climate resilience, carbon sequestration, and soil organic matter formation, and transformation processes (Navarro-Pedreño et al., 2021).

Microbes are proficient in uptake of gaseous by-products and convert them into less harmful products using their metabolic versatility. Methanotrophs, the bacteria which facilitate the aerobic capture of methane released from paddy fields, release carbon dioxide as the final product. Phototrophic bacteria, cyanobacteria, and microalgae aid in conversion of carbon dioxide and incorporate it to Calvin Benson cycle or Reductive TCA pathway to produce monosaccharides or oligosaccharides (Dahai et al., 2024). The denitrification and nitrification pathways can be either by-passed to prevent nitrous oxide production using DAMO bacteria, or by increasing the expression of *nosZ*gene; the only known nitrous oxide sink, in bacterial genera such as *Bradyrhizobium japonicum* (Henault et al., 2022). For ruminants, enteric fermentation in the fore-gut (rumen) can be controlled via changes in diet patterns, providing high-quality fodder, or shifting the gut microbiome making it more conducive for bacteria such as *Lactobacillus, Bacillus, Saccharomyces cerevisiae*, and *Streptomyces thermophilus* species (Sun et al., 2021). The manure storage provides access to several anaerobic fermenters which perform methanotrophy, this can be controlled by providing aerobic conditions to stimulate bacteria such as *Bacillus, Nocardia, Pseudomonas, Lactobacillus* which can aid in reduction of methane and nitrous oxide emissions (Ambrose et al., 2023). The use of both, live microorganisms, and their products are useful for the mitigation strategies. The invention of Pusa Decomposer using six fungal and two bacterial strains in a capsule have ability to degrade the paddy stubble, which provides an alternative to the crop residue burning (Srinivasan and Abirami 2020). Thuricin 17, a PGPR-based product, aids in carbon sequestration and plant protection by secretion of anti-microbials (Nazari and Smith 2020). All the developments in microbial-based mitigation strategies are noteworthy, nonetheless, the research in this subject is still in its infancy.

While the prospects of microbial solutions are promising, realizing their full potential requires addressing key challenges in their field applications. Production of microbial consortia to attain the desirable results is a probable approach towards dealing the climate change crisis. Incorporation of microbial associations, for example methanotrophs combined with cyanobacteria, can offer considerable reduction in methane, nitrous oxide, and carbon dioxide emissions (Ruiz-Ruiz et al., 2020). Since agroecosystem constitutes an integral part of our economy, it is needful to ensure betterment of soil health, and yield, alongside mitigating the GHG emissions. An integrated system must involve adequate farm practices along with microbial inculcation to safeguard the planet and to provide a congenial environment for all the living organisms, with re-establishment of damages caused. To achieve this, evidences of microbial processes and their functionality in agriculture can provide insights for their innovatory use for GHG mitigation (Sharma, 2020). Thus, the main objective of this review focuses on identification of the sources and mechanisms of GHG emissions in agriculture and simultaneous analysis of microbial solutions to mitigating methane, nitrous oxide, and carbon dioxide emissions for betterment of soil health, carbon sequestration and climate resilience.

10.2 Agricultural Sector as Contributor of Greenhouse Gas Emissions

Agroecosystems have rich diversity of microorganisms, and other macro faunal organisms, which contribute to soil organic matter (SOM) formation and decomposition. The metabolic processes occurring within these organisms includedecomposition of organic matter, soil respiration, methanogenesis, nitrification, and denitrification, that hold their substantial share for GHG emissions (Wang et al., 2022). The shift in these physiological processes caused as a consequence of till-farming practices, water-logged conditions maintained within the paddy fields, low-fodder crops for livestock, poor manure management, incomplete nitrification and denitrification pathways are also noteworthy. GHG fluxes in Indian agricultural sector are driven by enteric fermentation in ruminants (54.6%), inorganic nitrogen fertilizer use (19.1%), rice cultivation (17.5%), manure management (6.7%), and crop residue burning (2.2%) (FAO, 2019). The crux, thus, stays in the fact that GHG emissions are not only microbial-based, but also driven by ill-practices of agriculture (Table 10.1).

10.2.1 Carbon Dioxide Emissions

Carbon dioxide is the end product of most of the catabolic pathways which include decomposition of SOM, and anaerobic and aerobic respiration. Moreover, the emissions from agricultural practices such as crop residue burning, land-use changes or expansion, and farm mechanization are simultaneously prominent. Soil organic matter consists of plant, animal, and microbial residues at different stages of their decomposition (Baldock and Skjemstad, 1999). This process is closely regulated by both biotic and abiotic factors. Microbial decomposition of SOM occurs via enzymatic cleavage of covalent bonds of complex compounds such as cellulose, lignin, proteins, and lipids, to simpler compounds and finally carbon dioxide.

Table 10.1 Major Greenhouse emissions from Agricultural Sector, their residence time, global warming potential (over 100 years), and percent contribution globally, and in India

Greenhouse gas (GHG)	Residence time (in years)	Global warming potential (over 100 years; CO_2 equivalent)	Percent contribution of Global agricultural sector to Total Global GHG emission (%)	Percent contribution (Global agriculture sector) (%)	Percent contribution (Indian agriculture sector) (%)
Carbon Dioxide	100	1	0.368	2.21	3.9
Methane	25	30	40.259	71.07	71.8
Nitrous Oxide	114	298	69.71	26.72	24.3

Source: Crippa et al.

Other macro-fauna utilizes SOM by fragmentation, ingestion, digestion, and excretion processes. The excreta promote microbial colonization, and thus degradation (Chen et al., 2018). Another important process involves the breakdown of organic matter through abiotic factors involving UV radiations. The organic matter content is exposed because of extensive tillage and undergoes rapid degradation, thus decreasing the soil organic carbon (SOC) content in soil. This promotes more labile carbon content, which undergoes rapid decomposition to produce carbon dioxide (Navarro-Pedreño et al., 2021).

The oxidative or substrate-level phosphorylation of simpler substrates obtained by decomposition act as microbial sources of respiration. The conversion of glucose sugar into carbon dioxide occurs with aid of glycolytic and tri-carboxylic acid pathways under aerobic conditions result in release of $6CO_2$ molecules per glucose catabolized. Anaerobic respiratory pathways involve alternate electron acceptors and this governs the number of CO_2 molecules being released. For obligate anaerobes such as *Desulfovibrio, Clostridium, Methanobacteria* (archaea), the CO_2 yields are reduced to two to four molecules per glucose metabolized, whereas for facultative anaerobes including *Pseudomonas, Paracoccus* spp., six molecules are produced.

Coupled with the biotic mechanisms, the soil health is hindered by crop residue burning. It is subjected that about 15.9% crop residues are subjected to burning, and are common in regions of north-western India (Jain et al., 2014). The residues such as stubble, leaves, seedpods, are burnt to prepare the land for next cropping season. This practice is prevalent as the decomposition of stubble is time-consuming, and the intervals between cropping season do not support the same. Crop residue burning is not only responsible for CO_2 emissions, but also CH_4 and N_2O emissions. In addition to this, reckless cutting of forests for cultivation expansion depletes the carbon sinks and indirectly adds-up to the CO_2 emissions. Not to mention, the farm mechanization amplified the GHG emissions because of fossil fuel combustion in both, manufacture, and utilization of agricultural machinery. The intense usage of mechanized farming has also disturbed the SOM, accelerating its decomposition (Guan et al., 2023). The CO_2 emissions from agriculture are a mere 2.21% and are not of much concern, but emissions of methane and nitrous oxide for above-mentioned activities have higher potential to damage the future of agriculture as well as the planet.

10.2.2 Methane Emissions

Methanogens are microorganisms that produce Methane by the process of methanogenesis. These groups are strict anaerobes and thrive within anoxic conditions. Agroecosystems conducive for methanogenesis include rumen of livestock, manure, and water-logged conditions of rice/paddy fields.

Rumen microbiome constitutes bacteria, archaea and fungi, and is characterized by wide diversity, high population density and complex microbial interactions. Ruminants consume biodegradable material, majorly, plant-cell wall polymers and polysaccharides (Patra et al., 2012). The conversion of these compounds into microbial biomass and short chain volatile fatty acids (VFAs), primarily acetic, propionic, butyric acids, as well as CH_4 and NH_3 occurs within the rumen. The VFAs produced contribute to the livestock draught power, however the methane and ammonia are released out via eructation, urination, or excretion (Glasson et al., 2022). The livestock manure management requires keen vigilance to prevent CH_4 and N_2O emissions which are usually the result of anaerobic conditions persisting within these environments. Manure consists of both dung and urine produced by livestock, and the management include all activities, decisions, and components used to handle, store, and dispose of feces and urine from livestock aiming the preservation and recycle of nutrients in livestock production system (Ambrose et al., 2023). CH_4 is produced by microbial degradation of soluble lipids, carbohydrates, organic acids, and proteins present in manure. The anaerobic conditions prevail during manure storage and result in methane production (Khoshnevisan et al., 2021).

In addition to methane production from livestock, rice cultivation is also a contributor. Irrigation practices in rice involve water-logging conditions. These conditions are responsible for shifts in microbial communities from aerobic to anaerobic. The anaerobes including hydrolytic and fermentative bacteria decompose organic matter into simple organic acids, upon which acetogenic bacteria feed and produce acetate, H_2, and CO_2 (Hernandez et al., 2017). These compounds are further degraded by methanogens to CH_4, which is release through ebullition, diffusion or plant mediated transport. The microorganisms which aid the process of methanogenesis in various agricultural ecosystems are enlisted under Table 10.2.

Table 10.2 Microorganisms responsible for Methane emissions in agriculture

Group of Microorganisms involved	Function	Examples
Hydrolytic Bacteria	Convert complex organic compounds into fermentable compounds	*Clostridium, Bacteriodes, Fibrobacter*
Fermentative Bacteria	Feed on hydrolytic products to produce volatile fatty acids, alcohols, and hydrogen	*Escherichia coli, Clostridium, Lactobacillus*
Acetogenic Bacteria	Convert VFAs (propionate, butyrate) and alcohols into acetate, H_2 and CO_2	*Syntrophobacter, Syntrophomonas, Pelotomaculum*
Acetoclastic Methanogens	Acetate to CH_4	*Methanosarcina, Methanosaeta*
Hydrogenotrophic Methanogens	Reduce H_2 with CO_2 to produce CH_4	*Methanobacterium, Methanobrevibacter, Methanococcus*
Methylotrophic Methanogens	Methyl compounds conversion to CH_4	*Methanosphaera spp.*

10.2.3 Nitrous Oxide Emissions

Green revolution played a major role to combat food crisis during 1960s and 1970s primarily by use of inorganic nitrogen fertilizers, and other organic chemicals to increase the crop yields. This upsurge, however, hindered the nitrification and denitrification processes essential for nitrogen cycling. The increased nitrogen concentration in agro-soils resulted in depletion of *nosZ* gene bearing microorganisms, which served as the sole sink for nitrous oxide emissions. The *nosZ* gene encoding for enzyme Nitrous Oxide Reductase, which aids the conversion of nitrous oxide to dinitrogen has been exhausted due to high nitrogen concentrations prevailing in the soil. The disturbed C:N ratio of soils has resulted in incomplete nitrification and denitrification due to involvement of bacterial groups such as Ammonia Oxidizing Bacteria (AOB), Ammonia Oxidizing Archaea (AOA), and Nitrite oxidizing bacteria (NOB). The conversion of ammonia (NH_4) results in an intermediate hydroxylamine (NH_2OH)/ nitroxyl hydride (HNO), and then converted to Nitrite (NO_2). However, due to oxidative decomposition of NH_2OH/HNO produces Nitric Oxide (NO), which undergoes nitrifier denitrification to give N_2O as the by-product. Now due to severe reduction of nitrous oxide reductase enzyme, the sequential reduction to produce dinitrogen does not occur and result in nitrous oxide emissions (Hu et al., 2015). It has been reported that agriculture produces largest amounts of N_2O emissions to the atmosphere, 26.72%, globally.

Ammonia Oxidizing Bacteria such as *Nitrosomonas europea, Nitrospira* spp., *Nitrosococcus* spp. and AOA including *Nitrosopumilis maritiumus*, *Nitrosophaera vinnensis*, along with NOB *Nitrobacter* spp, *Nitrospira* spp., *Nitrococcus* spp. catalyze the oxidation of NH_2OH to N_2O under low oxygen conditions. Fungi such as *Fusarium oxysporum, Aspergillus niger*, and *Trichoderma* spp. comprise Copper-containing Nitrite Reductase (*NirK*) and cytochrome P450 Nitric Oxide Reductase (*P450nor*), which together aid in production of N_2O from nitrite (Hu et al., 2015). Identification of the causes of GHG fluxes in crop production is prerequisite establish solutions. Agricultural fields are dynamic and the release of gases is profoundly reliant on edaphic factors such as soil moisture, pH, temperature, oxygen as well as climatic factors.

10.3 Microbial Strategies to Combat GHG Emissions

10.3.1 Direct Microbial Influence

These mechanisms involve the direct role of microorganisms in capture of GHG emissions to prevent their release into the environment.

10.3.1.1 Carbon Sequestration: Mitigation of Carbon Dioxide Emissions from Agriculture

Carbon sequestration is the process of capturing and storing of atmospheric carbon dioxide to mitigate climate change. Soil microbes incorporate carbon in the form of soil organic carbon (SOC) for its long-term stabilization into carbon pools. This microbial-driven sequestration and C-cycling aids to improve soil fertility, structure, SOC accumulation, and climate resilient properties of soil (Mason et al., 2023). Microorganisms having significance in carbon sequestration potential include groups of purple non-sulfur bacteria *Rhodobacter sphaeroides, Rhodopseudomonas palustris, Rhodospirillum rubrum* (Sakarika et al., 2020); cyanobacteria *Anabaena varibilis, Nostoc punctiforme*, and microalgae *Chlorella vulgaris* (Dahai et al., 2024). Autotrophic nature of these microbes signifies their efficient CO_2 capturing ability and to stabilize them in the form of organic compounds. The exopolymer secreting cyanobacteria such as *Anabaena* enhance soil aggregation with reduced SOM decomposition and increased carbon sequestration (Prasanna et al., 2015). The role of fungi and actinomycetes does not go in vain, as these assimilate organic matter into recalcitrant carbon pools. Inculcation of microorganisms helps in C-capturing along with plant growth promotion, consequently rising the carbon sinks in the ecosystem. The C-interaction with soil-plants-microbes is based upon two major metabolic pathways, being represented in phototrophic and chemotrophic organisms, *viz*., Oxygenic C-fixation via Calvin-Benson cycle in microalgae and cyanobacteria, and Anoxygenic C-fixation through Reductive TCA cycle taking place inside purple non-sulphur bacteria.

The Calvin Benson Cycle constitutes three major stages as carboxylation, reduction, and regeneration, to sequester the produced CO_2 into biomass and other metabolic reactions. Carboxylation is the process of incorporation of CO_2 in the presence of Ribulose-1,5-bisphosphate carboxylase oxygenase (RubisCO), encoded by the genes *cbbL/cbbM*. The process yields 1,3-bisphosphoglycerate, which is reduced by dehydrogenase enzyme *gapA/gap2*, producing glyceraldehyde-3-phosphate (G3P) dehydrogenase enzyme, resulting in formation of G3P. The cycle requires a total of 6 CO_2 to generate hexose sugars. Here, the G3P undergoes reversible isomerization to form Dihydroxyacetone phosphate, which together form hexose sugar——Fructose-1,6-bisphosphate. Expression of *fbp/glpX*genes produce Fructose-1,6-bisphosphatase enzyme, forming fructose-6-phosphate, which is incorporated via metabolic cycles. The necessary step is regeneration of Ribulose-1,5-bisphosphate to continue the cycling. Regeneration is mainly carried out by enzyme phosphoribulokinase (*prk*), and enables RubP to interact with remaining CO_2 molecules (Rubin-Blum et al., 2019). As previously mentioned, six rounds of this cycling ensure production of sugars required for energy production in microalgae and cyanobacteria. This is the core pathway of C-fixation in plants, which serve as the ultimate C-sink. Increasing alternate descends for CO_2 uptake strategically mitigates these emissions to enter the atmosphere.

An alternate pathway, referred as Reductive TCA cycle (rTCA), is adopted by anoxygenic bacterial species, e.g., *Rhodopseudomonas palustris,* which incorporates CO_2 in anaerobic, light driven conditions. The enzyme ATP Citrate Lyase (ACL) (*aclAB* genes) cleaves 6-C citrate into 4-C oxaloacetate and 2-C acetyl-CoA. Further, 2-Oxoglutarate: ferredoxin oxidoreductase (OGOR) expressed by *korABCD* gene clusterresults in reversible carboxylation of succinyl-CoA to 2-oxoglutarate.Rubin-Blum et al. (2019) suggest a putative fumarate reductase enzyme encoded by *tfrAB* genes, which catalyze fumarate reduction, specific to this pathway. Some other associated genes include *acn* (aconitase), *idh* (isocitrate dehydrogenase), *mdh* (malate dehydrogenase). The pathway operates as reverse of TCA-cycle, and produces Acetyl Co-A from oxaloacetate. The carboxylation occurs at two main steps that involve conversion of Succinyl Co-A to Isocitrate via α-ketoglutarate. The Acetyl Co-A, thus, formed gets involved in fatty acid biosynthetic processes, for production of essential cellular components.

Furthermore, carbon sequestration not only involves capture of CO_2 emissions, but also C-CO_2 and CO_2-equivalent GHG fluxes. Studies conducted by Kantha et al. (2015) clearly distinguished the impact of GHG mitigation in presence of different strains of *Rhodopseudomonas palustris*, with 47%, 44%, and 38% decline in CO_2 emissions from rice fields, and 100%, 94%, and 86% decrease in methane emissions corresponding to the PP803, P1, and TK103 strains (Kantha et al., 2015). The ability of these bacteria to release aminolevulinic acid (ALA) decreased methane emissions by 88.41% (PP803) and 80% (TN110) (Nunkaew et al., 2015). One of the quite significant importance of microorganisms suggested by these studies, depict their role in overall GHG mitigation and not just reducing the levels of a particular gas. Therefore, it is justified to include total GHG emission mitigation under soil or microbial carbon sequestration capability.

10.3.1.2 Methanotrophy: First Line of Defense Against Methane Emissions

Methanotrophic bacteria utilize CH4 as their sole carbon and energy source preventing its release into the atmosphere. These are aerobic bacteria and are abundant in aerobic soil layers in rice fields, where they consume methane before its escape through ebullition or diffusion. In agriculture, these bacteria comprise two types: Type I methanotrophs and Type II methanotrophs. The former includes members of Gammaproteobacteria while latter are comprised of members of Alphaproteobacteria (Aimen et al., 2018; Rani et al., 2024). Main microbial species involved in methanotrophy and their functions are mentioned in Table 10.3.

Methanotrophs sequentially oxidize CH_4 into CO_2. The initial step is catalyzed by Methane monooxygenase enzyme to produce Methanol (CH_3OH) by incorporation of O_2 and release of water. Methane monooxygenase (MMO) enzymes are of two types: Particulate MMO which are Cu-dependent, membrane bound enzymes produced by expression of *pmoABC* genes; and Soluble MMO (sMMO), specifically found in soluble portion of cell, i.e., cytoplasm, and is Fe-dependent. The genes *mmoXYZB C*en code sMMO (Islam et al., 2016). Alcohol dehydrogenase, specifically methanol dehydrogenase is encoded by *mxaF, mxaI* and *mxa J* genes, yield formaldehyde (CH_2O). These enzymes are strictly Ca-dependent. Studies report the presence of Cu, Ca, and lanthanide metal ions to significantly emphasize gene expression and enzyme activity (Semrau et al., 2018). Followed by this, there is branching of pathway to either assimilation or oxidation. In former, CH_2O is either completely assimilated into biomass by biosynthetic pathways, i.e., RuMP pathway in Type I methanotrophs or serine cycle in Type II methanotrophs; whereas in latter, formaldehyde is converted to formic acid by the action of formaldehyde dehydrogenases and tetrahydromethanopterin cofactor. *FdhABC* encode formate dehydrogenase enzyme resulting in formation of carbon dioxide. This oxidation pathway is closely regulated by the proton flux across the cell membrane, which yields ATP. In rice/paddy fields, the methane/oxygen ratio is high, which directs the oxidation pathway over-assimilation of formaldehyde (Guerrero-Cruz et al., 2021).

Table 10.3 Methanotrophs and their function in Agriculture

Methanotrophs	Occurrence/function	References
Methylomicrobium alcaliphilum	Adapted to alkaline environments, have potential in reducing enteric methane	Akberdin et al. (2018)
Methylococcus capsulatus	Promotes methane oxidation in gut when used as a feed-additive	Kerckhof et al. (2021)
Methylobacter tundripaludum	Active in manure storage under cold conditions	Wartiainen et al. (2006)
Methylobacter luteus	Found in flooded rice paddy soils	Tveit et al. (2023)
Methylocystis parvus	Dominates in manure composting systems	van den Bergh et al. (2023)
Methylosinus trichosporium	Oxidize methane in microaerophilic conditions	Rani et al. (2024)
Methylocystis rosea	Competes with methanogens for substrates in rice fields	Schmider et al. (2024)

Application of 6.25 kg/ha methanotroph consortium resulted 60% reduction in methane emissions under investigations conducted by Davamani et al. (2020). The coculture studies of methanotroph-microalgae for greenhouse gas mitigation showed 100% decline in CO_2 and CH_4 emissions. This approach opted by Ruiz-Ruiz et al. (2020) depicted the role of integration of more than one type of microorganisms to collectively reduce the effects of GHGs. The methane in air was captured by methanotrophs, which produced CO_2 as byproduct. Under light conditions, this CO_2 was converted into oxygen, thereby acting as a potential carbon sink.

In Nitrate-dependent methane oxidation (N-DAMO process), a group of bacteria, recently discovered in 2010, *Candidatus Methylomirabilis oxyfera* triggers the intra-aerobic methanotrophic process under anaerobic conditions. The oxygen is produced within the organism to support aerobic methanotrophy via oxidation of nitrite to nitric oxide and then dinitrogen and oxygen as $2NO \rightarrow N_2 + O_2$.Similar research conducted by Scholz et al. (2020) depicted the preliminary results for role of cable bacterial species in mitigating GHG emissions. These are filamentous bacteria, e.g., *Candidatus* Electronema, and are known to increase sulfate levels via electrogenic sulfide oxidation. One-time inoculation of rice-vegetated soil pots with cable bacteria increased the sulfate inventory by fivefold. This led to the reduction of methane emissions by 93%, compared to control pots lacking cable bacteria. Promoting cable bacteria in rice fields by enrichment or sensible management may thus become a strategy to reduce anthropogenic methane emissions.

10.3.1.3 Complete Denitrification: Nitrous Oxide Mitigation by Enhancing nosZ Gene Pool

Reduction of N_2O levels from agriculture can be obtained by stimulation of *nosZ* gene pool. A prerequisite for this would be addition of complete denitrifying microorganisms, which can include plant growth promoting bacteria, denitrifying bacteria, arbuscular mycorrhizal fungi, and cyanobacteria. Bacteria such as, *Bacillus subtilis,*

B. mucilaginosus, Pseudomonas aeruginosa, Rhizobium, Paracoccus denitrificans, Pseudomonas stutzeri, Bradyrhizobium, Glomus, and *Rhizophagus* fungi, and *Nostoc and Anabaena* blue-green algae are important capturers of N_2O emissions.

Denitrification is the step-wise conversion of nitrite to dinitrogen. The regulation of *nosZ* gene is crucial for enhancing its impact in agricultural soils. The regulon involves genes *nosRZDFYL* organized upon a cluster of bacterial chromosomes (Fig. 10.1). *NosZ* is the structural gene which is cotransformed with *nosR* present upstream. Interestingly, the signal for *nosR* promoter activity is governed by the presence of Nitrite and Nitric oxide than Nitrous oxide. By sensing NO concentrations in the surrounding, bacteria pre-emptively produce nitrous oxide reductase so as to reduce N_2O as soon as it is formed (Arai et al., 2003). *Pseudomonas aeruginosa* and other bacteria ensure rapid and efficient N_2O reduction, preventing its release into air.

Studies have reported 70% reduction in N_2O emissions from Soybean crop with inoculation of *nosZ* + strains v/s *nosZ*- strains of *Bradyrhizobium japonicum*, suggesting the potential of these genes to increase sequential denitrification for better soil health (Henault et al., 2022). Research conducted by Wasai-Hara et al., (2023) discovered 2 *nos* regulons in *Bradyrhizobium ottawaense* and documented 150-fold increase in *nosZ* efficiency to mitigate nitrous oxide emissions. These studies indicate towards potential to increase the *nosZ*gene pool by inoculating required microorganisms, and opt for inter-cropping or mixed-cropping systems to let the gene pool sustain within the soils.

Cyanobacteria or blue-green alga play an important role in regulating the nitrous oxide emissions from agricultural practices. Some filamentous cyanobacteria have specialized structures called heterocyst, which maintain the anaerobic conditions necessary for nitrogenase enzyme complex to aid biological nitrogen fixation. These organisms help in long-term nitrogen fixation and its uptake in the form of amino acids. Interaction studies of cyanobacteria have revealed about 5% GHG reduction from rice fields when used in association as *Azolla-Anabaena* symbiosis (Ali et al., 2014). Free-living cyanobacteria and plant growth promoting bacteria improve the soil nitrogen levels, reducing the requirements for inorganic nitrogen fertilizer, and sequester considerable amounts of carbon, thus promoting SOM. Reduction in cumulative flux of N_2O emissions from corn fields has been reported to decrease 41% and 49% for two fertilizers: CAN-17 and UAN-32, respectively. The PGPB involved were 4 strains of *Bacillus, viz, Bacillus safensis T4, Bacillus pumilis INR7, Bacillus subtilis IN937a, Bacillus sphaericus SE56*. GHG responses to different soil amendments on the Loess Plateau of China documented reduced CO_2 and N_2O emissions by

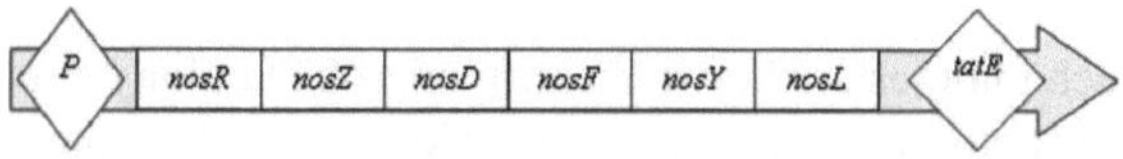

Fig. 10.1 nosZgene regulon in Pseudomonas aeruginosa P: Nos regulon Promoter, nosR: Flavin-dependent electron carrier and transcriptional activator, nosZ: Nitrous oxide reductase, nosDFYL: Accessory proteins for NosZ maturation and Cu-insertion, tatE: Twin-Arginine Translocation Pathway genes

soil amendments with biochar, *Bacillus mucilaginosus*, *B. subtilis*, where *B. subtilis* showed highest efficiency in CO_2 reduction at 38.54%, and N_2O emission reduction of 57.85% (Ding et al., 2023). These studies strongly point towards integrated use of PGPBs and other organic amendments to achieve significant reductions in GHG emissions and improvement of soil health.

The Arbuscular mycorrhizal (AM) fungi have tendency to form symbiotic relationships with plant species. The inoculation of these fungal species release plant growth promoters, expand the root area for easy nutrient uptake, an increased water retention capacity, and soil organic matter. AM fungi can reduce nitrous oxide emissions through three primary mechanisms:

- **By changing host plant traits**: such as increasing plant biomass and transpiration;
- **By altering abiotic soil environment**: Promote soil water-stable aggregates formation, which increase oxygen diffusion towards the interiors of soil aggregates and inhibit denitrification activity;
- **By modifying the microbial community related to N_2O emission**: Facilitate the growth of *nosZ* (responsible for N_2O consumption) and inhibit that of *nirK* (responsible for N_2O production) associated microbial communities (Shen & Zhu, 2021).

Potential of AM fungi to enrich *nosZ*-denitrifiers and mitigate N_2O emissions from soybean stubble was experimented by Zhao et al. (2021) concluded that hyphae of *Rhizophagus aggregatus* from neighboring maize on overall N_2O emissions from stubbles of nodulated or non-nodulated soybean consistently reduced N_2O emissions by 20.8%–61.5%. It was depicted that AMF hyphae promote the abundance of *nosZ*-type denitrifiers and altered their community composition, along with reduction in abundance of *nirK*-type denitrifiers, which holds promise for mitigating N_2O emissions by manipulating the efficacious AMF and their associated microbes in cereal/legume intercropping systems.

10.3.2 Indirect Microbial Influence

These include indirect use of microorganisms to divert the primary causes of GHG emissions from agriculture. Methane and Carbon dioxide emissions can be controlled this way. Table 10.4 enlists these microbial innovations.

10.4 Omics-Based Technologies in Agricultural GHG Mitigation

Omics-based technologies are key to identify microbial colony structure, and composition within an agro-ecosystem. These techniques provide a comprehensive report for presence of different genes, RNA and proteins, which help in assessment of metabolic pathways associated with microorganisms working under changing conditions. Metagenomics identifies complete genome to analyze potent genetic information for a community (Knight et al. 2018). Meta transcriptomics and Meta proteomics

Table 10.4 Indirect influence of microorganisms on greenhouse gas emissions

Microbial innovations	Microorganisms involved	Application in agriculture	References
Bioethanol production	*Saccharomyces cerevisiae, Zymomonas mobilis*	Carbon dioxide reduction from Energy use; Prevents crop residue burning	Letti et al. (2022)
Mushroom production	Basidiomycetous Fungi (*Agaricus bisporus, Pleurotus* spp., *Calocybe indica, Volvariella volvacea, Lentinus edodes*)	Alternate use of crop residues	-
Pusa decomposer	*Trichoderma viridae, T. harzarium, T. reesei, Aspergillus awamori, Aspergillus nidulans, Aspergillus flavus, Pseudomonas fluorescens, Bacillus subtilis*	Prevents Carbon dioxide, methane, and nitrous oxide emissions released due to Paddy stubble burning	Yasmin et al. (2025)
Thuricin 17 (PGPR-based product)	*Bacillus thuringiensis* NEB 17	Carbon sequestration; Plant protection	Nazari and Smith (2020)
Feed-additives	*Saccharomyces cerevisiae,* Lactic Acid Bacteria (LABs), *Bacillus subtilis*	Suppressors of methane in Ruminants; LABs facilitate alternate metabolic pathway	Sun et al. (2021)
Composting (Manure management)	*Bacillus, Pseudomonas, Streptomyces, Nocardia*	Promotes aeration in manure, inhibits CH_4 & N_2O emissions	Symeon et al. (2025)

are techniques used for identification of whole mRNA and protein sequences from a sample (Bashiardes et al. 2016, Schiebenhoefer et al. 2019). These approaches facilitate identification of already-existing microbial communities along with their functional and structural genes, eliciting their role in that microenvironment. The well-employed methods of identifying microbes metabolizing substrates like methane and nitrogen by stable isotope probing (SIP) are now combined with omics-technologies, providing a new basis for linking taxonomic identity to functional activity. These relate to key greenhouse emission processes in agricultural soils. Functional genomics reveal presence of functional-genetic factors, and in addition to this, meta-transcriptome and meta-proteome identify active and inactive pathways relating to specific environmental conditions (Venturini et al. 2022). Gene editing tools such as CRISPR-Cas9 systems can be used to edit microbial genes linked to GHG fluxes, eliminating the need for production of harmful gases such as methane. Du et al. (2024) edited the genome of archaeon *Methanococcus maripaludis*, a methanogenic archaeon, producing methane from CO_2/formate. The gene editing was done by CRISPRi-dCas9 and successfully achieved transcriptional reduction (up to 100 folds). These genetic tools allow knock-out of essential genes linked to methanogenesis, which can help in tuning the methane output, rather than eliminating it. The mentioned study suggests the role in generation of bioenergy in form of biomethane. However, the present study suggests possible outcomes in GHG mitigation, by editing the prokaryotic genome allowing regulated gene expression for methanogenesis and nitrification. Moreover, the tools can also be employed to enhance gene regulation in methanotrophs, denitrifiers, and carbon sequestering microorganisms, which can significantly reduce the GHG fluxes released from agricultural soils.

10.5 Challenges and Future Prospects

It is needful to identify the challenges associated with microbial-based mechanisms to come-up with better research perspectives and overcome them. Microorganisms are strongly dependent on environmental conditions, and their survival and activity are closely regulated by temperature, pH, water, and nutrient availability. Further researches to increase the metabolic potential of microbes using genetic engineering methods so microbial survival is not affected by fluctuating environments is necessary. Before this, however, there is a need to ensure biosafety concerns for genetically modified organisms, such that they do not negatively affect the mankind and other life on Earth. Development of microbial consortia for reduction in GHGs is still in its infancy, and must be looked upon to increase the scalability of these inoculants. Integration of microorganisms along with organic amendments such as biochar, nanotechnology and AI-driven systems can significantly expand the scope of these green-methods.

10.6 Conclusions

Although agriculture substantially contributes towards GHG emissions, however, is also a victim of the changing climate. Microorganisms constitute a major role in agroecosystems, in maintenance of soil health and plant growth promotion. However, significant amounts of GHG emissions occur from microbial-based processes. Fortunately, several microbial groups can capture the GHG fluxes to prevent their escape into the environment. This strongly indicates towards adoption of more green strategies to combat climate change. Other methods or technologies such as rotavators, and happy seeders have important role in GHG mitigation but use of tractors alongside emit CO_2 from fossil-fuel combustion. Development of superior cattle/buffalo varieties are time consuming and unpredictable. Most of the strategies indirectly influence stabilization of SOC pools and SOM, thus are not suitable for long-term utilization. Nonetheless, microorganisms overpower all these challenges, and provide an eco-friendly approach towards GHG mitigation. Incorporation of plant growth promoting bacteria, AM fungi, not only mitigates carbon dioxide, but also captures other GHG emissions along with promotion of plant growth. An increased plant-growth further contributes to carbon sequestration potential of soil. Thus, the integrated use of different microbial species for varying effects has proven useful for GHG mitigation all around the world. Moreover, microbial-based innovations such as bioethanol production, mushroom production, feed-additives, etc. have diverted the primary cause of GHG production from various sources, and offered an environment friendly response to their release. Thus, there is need to upsurge the production of microbial based inoculations and provide awareness about the optimistic approaches these have towards our planet, to combat the ongoing climate change crisis.

10.7 Policy Implications

The primary goal must be to include microbial based solutions in climate action and carbon credit systems and recognize microbial solutions in carbon credit programs & NDCs. The farmers opting for green-solutions can be rewarded with carbon-credits, which can aid to increase acceptability and adaptability of these solutions in the real-life. Further, regulatory framework and biosafety standards must be implanted to establish guidelines for large-scale use and biosafety of engineered microbes by developing certification systems for microbial products. Provision of grants and subsidies for microbial GHG mitigation research and promotion of public–private partnerships and innovation hubs have ability to pose quite an impact over inclusion of these strategies. Discussions based on green-strategies must be undertaken by international panels and institutions such as FAO, UNEP, and IPCC and global partnerships and technology transfer must be strengthened, because this is not a race for a particular country, but to protect our only habitable planet.

References

Aimen, H., Khan, A. S., & Kanwal, N. (2018). Methanotrophs: The natural way to tackle greenhouse effect. *Journal of Bioremediation and Biodegradation*, *9*, 432. https://doi.org/10.4172/2155-6199.1000432.

Akberdin, I. R., Thompson, M., Hamilton, R., Desai, N., Alexander, D., Henard, C. A., & Kalyuzhnaya, M. G. (2018). Methane utilization in *Methylo microbiumalcaliphilum* 20ZR: a systems approach. *Scientific Reports*, *8*(1), 2512. https://doi.org/10.1038/s41598-018-20574-z.

Ali, M. A., Sattar, M. A., Islam, M. N., & Inubushi, K. (2014). Integrated effects of organic, inorganic and biological amendments on methane emission, soil quality and rice productivity in irrigated paddy ecosystem of Bangladesh: field study of two consecutive rice growing seasons. *Plant and Soil*, *378*, 239–252. https://doi.org/10.1007/s11104-014-2023-y.

Ambrose, H. W., Dalby, F. R., Feilberg, A., & Kofoed, M. V. (2023). Additives and methods for the mitigation of methane emission from stored liquid manure. *Biosystem Engineering*, *229*, 209–245. https://doi.org/10.1016/j.biosystemseng.2023.03.015.

Arai, H., Mizutani, M., & Igarashi, Y. (2003). Transcriptional regulation of the Nos genes for nitrous oxide reductase in *Pseudomonas aeruginosa*. *Microbiology*, *149*(1), 29–36. https://doi.org/10.1099/mic.0.25936-0.

Baldock J. A., & Skjemstad J. O. (1999) Soil organic carbon/soil organic matter. *Soil analysis: An interpretation manual* (pp. 159–170). CSIRO Publishing.

Bashiardes, S., Zilberman-Schapira, G., & Elinav, E. (2016). Use of metatranscriptomics in microbiome research. *Bioinformatics Biology Insights,10*, 19–25. https://doi.org/10.4137/BBI.S34610.

van den Bergh, S. G., Chardon, I., Meima-Franke, M., Costa, O. Y., Korthals, G. W., de Boer, W., & Bodelier, P. L. (2023). The intrinsic methane mitigation potential and associated microbes add product value to compost. *Waste Management (New York, N.Y.), 170*, 17–32. https://doi.org/10.1016/j.wasman.2023.07.027.

Chachei, K. (2024). Greenhouse gas emissions in the Indian agriculture sector and mitigation by best management practices and smart farming technologies—a review. *Environmental Science and Pollution Research, 31*(32), 44489–44510. https://doi.org/10.1007/s11356-024-33975-7.

Chen, Y., Camps-Arbestain, M., Shen, Q., Singh, B., & Cayuela, M. L. (2018). The long-term role of organic amendments in building soil nutrient fertility: a meta-analysis and review. *Nutrient Cycling Agroecosystem, 111*, 103–125. https://doi.org/10.1007/s10705-017-9903-5.

Dahai, H., Zhihong, Y., Lin, Q., Yuhong, L., Lei, T., Jiang, L., & Liandong, Z. (2024). The application of magical microalgae in carbon sequestration and emission reduction: Removal mechanisms and potential analysis. *Renewable and Sustainable Energy Reviews, 197*, 114417. https://doi.org/10.1016/j.rser.2024.114417.

Davamani, V., Parameswari, E., & Arulmani, S. (2020). Mitigation of methane gas emissions in flooded paddy soil through the utilization of methanotrophs. *The Science of the Total Environment, 726*, 138570. https://doi.org/10.1016/j.scitotenv.2020.138570.

Ding, Y., Li, C., Li, Z., Liu, S., Zou, Y., Gao, X., & Zhao, X. (2023). Greenhouse gas emission responses to different soil amendments on the Loess Plateau, China. *Agriculture, Ecosystems & Environment, 342*, 108233. https://doi.org/10.1016/j.agee.2022.108233.

Du, Q., Wei, Y., Zhang, L., Ren, D., Gao, J., Dong, X., & Li, J. (2024). An improved CRISPR and CRISPR interference (CRISPRi) toolkit for engineering the model methanogenic archaeon *Methanococcusmaripaludis*. *Microbial Cell Factories, 23*(1), 239. https://doi.org/10.1186/s12934-024-02492-0.

Food and Agriculture Organization of the United Nations (2019). Emissions due to Agriculture: Global, Regional and Country Trends (2000–2018). https://openknowledge.fao.org/server/api/core/bitstreams/cc09fbbc-eb1d-436b-a88a-bed42a1f12f3/content.

Glasson, C. R., Kinley, R. D., de Nys, R., King, N., Adams, S. L., Packer, M. A., & Magnusson, M. (2022). Benefits and risks of including the bromoform containing seaweed *Asparagopsis* in feed for the reduction of methane production from ruminants. *Algal Research, 64*, 102673. https://doi.org/10.1016/j.algal.2022.102673.

Guan, N., Liu, L., Dong, K., Xie, M., & Du, Y. (2023). Agricultural mechanization, large-scale operation and agricultural carbon emissions. *Cogent Food & Agriculture, 9*(1), 2238430. https://doi.org/10.1080/23311932.2023.2238430.

Guerrero-Cruz, S., Vaksmaa, A., Horn, M. A., Niemann, H., Pijuan, M., & Ho, A. (2021). Methanotrophs: discoveries, environmental relevance, and a perspective on current and future applications. *Frontiers in Microbiology, 12*, 678057. https://doi.org/10.3389/fmicb.2021.678057.

Hénault, C., Barbier, E., Hartmann, A., & Revellin, C. (2022). New insights into the use of rhizobia to mitigate soil N2O emissions. *Agriculture, 12*(2), 271. https://doi.org/10.3390/agriculture12020271.

Hernández, M., Conrad, R., Klose, M., Ma, K., & Lu, Y. (2017). Structure and function of methanogenic microbial communities in soils from flooded rice and upland soybean fields from Sanjiang plain, NE China. *Soil Biology & Biochemistry, 105*, 81–91. https://doi.org/10.1016/j.soilbio.2016.11.010.

Hu, H. W., Chen, D., & He, J. Z. (2015). Microbial regulation of terrestrial nitrous oxide formation: understanding the biological pathways for prediction of emission rates. *FEMS Microbiology Reviews, 39*(5), 729–749. https://doi.org/10.1093/femsre/fuv021.

Islam, T., Torsvik, V., Larsen, Ø, Bodrossy, L., Øvreås, L., & Birkeland, N. K. (2016). Acid-tolerant moderately thermophilic methanotrophs of the class Gammaproteobacteria isolated from tropical topsoil with methane seeps. *Frontiers in Microbiology, 7*, 851. https://doi.org/10.3389/fmicb.2016.00851.

Jain, N., Bhatia, A., & Pathak, H. (2014). Emission of air pollutants from crop residue burning in India. *Aerosol and Air Quality Research, 14*(1), 422–430. https://doi.org/10.4209/aaqr.2013.01.0031.

Kantha, T., Kantachote, D., & Klongdee, N. (2015). Potential of biofertilizers from selected *Rhodopseudomonas palustris* strains to assist rice (*Oryza sativa* L. subsp. *indica*) growth under salt stress and to reduce greenhouse gas emissions. *Annals of Microbiology, 65*, 2109–2118. https://doi.org/10.1007/s13213-015-1049-6.

Kerckhof, F.-M., Sakarika, M., Van Giel, M., Muys, M., Vermeir, P., De Vrieze, J., Vlaeminck, S. E., Rabaey, K., & Boon, N. (2021). From biogas and hydrogen to microbial protein through co-cultivation of methane and hydrogen oxidizing bacteria. *Frontiers in Bioengineering and Biotechnology, 9*, 733753. https://doi.org/10.3389/fbioe.2021.733753.

Khoshnevisan, B., Duan, N., Tsapekos, P., Awasthi, M. K., Liu, Z., Mohammadi, A., & Liu, H. (2021). A critical review on livestock manure biorefinery technologies: Sustainability, challenges, and future perspectives. *Renewable and Sustainable Energy Reviews, 135*, 110033. https://doi.org/10.1016/j.rser.2020.110033.

Knight, R., Vrbanac, A., Taylor, B. C., Aksenov, A., Callewaert, C., Debelius, J., & Dorrestein, P. C. (2018). Best practices for analysing microbiomes. *Nature Reviews Microbiology, 16*(7), 410–422. https://doi.org/10.1038/s41579-018-0029-9.

Letti, L. A. J., Sydney, E. B., de Carvalho, J. C., de Souza Vandenberghe, L. P., Karp, S. G., Woiciechowski, A. L., & Soccol, C. R. (2022) Roles and impacts of bioethanol and biodiesel on climate change mitigation. In *Biomass, biofuels, biochemicals* (pp. 373–400). Elsevier. https://doi.org/10.1016/B978-0-12-823500-3.00006-6

Mason, A. R. G., Salomon, M. J., Lowe, A. J., & Cavagnaro, T. R. (2023). Microbial solutions to soil carbon sequestration. *Journal of Cleaner Production, 417*, 137993. https://doi.org/10.1016/j.jclepro.2023.137993.

Navarro-Pedreño, J., Almendro-Candel, M. B., & Zorpas, A. A. (2021). The increase of soil organic matter reduces global warming, myth or reality?. *Sci, 3*(1), 18. https://doi.org/10.3390/sci3010018.

Nazari, M., & Smith, D. L. (2020). A PGPR-produced bacteriocin for sustainable agriculture: A review of thuricin 17 characteristics and applications. *Frontiers in Plant Science, 11*, 916. https://doi.org/10.3389/fpls.2020.00916.

Nunkaew, T., Kantachote, D., Nitoda, T., & Kanzaki, H. (2015). Selection of salt tolerant purple nonsulfur bacteria producing 5-aminolevulinic acid (ALA) and reducing methane emissions from microbial rice straw degradation. *Applied Soil Ecology, 86*, 113–120. https://doi.org/10.1016/j.apsoil.2014.10.005.

Organization for Economic Coorperation and Development (2019). Trends and Drivers of Agri-Environmental Performance in OECD countries. OECD Publishing.

Patra, A. K. (2012). Enteric methane mitigation technologies for ruminant livestock: a synthesis of current research and future directions. *Environmental Monitoring and Assessment, 184*, 1929–1952. https://doi.org/10.1007/s10661-011-2090-y.

Prasanna, R., Adak, A., Verma, S., Bidyarani, N., Babu, S., Pal, M., & Nain, L. (2015). Cyanobacterial inoculation in rice grown under flooded and SRI modes of cultivation elicits differential effects on plant growth and nutrient dynamics. *Ecological Engineering, 84*, 532–541. https://doi.org/10.1016/j.ecoleng.2015.09.033.

Raihan, A., Begum, R. A., Nizam, M., Said, M., & Pereira, J. J. (2022). Dynamic impacts of energy use, agricultural land expansion, and deforestation on CO2 emissions in Malaysia. *Environmental and Ecological Statistics., 29*(3), 477–507. https://doi.org/10.1007/s10651-022-00532-9.

Rani, A., Pundir, A., Verma, M., Joshi, S., Verma, G., Andjelković, S., & Mitra, D. (2024). Methanotrophy: A biological method to mitigate global methane emission. *Microbiological Research, 15*(2), 634–654. https://doi.org/10.3390/microbiolres15020042.

Rubin-Blum, M., Dubilier, N., & Kleiner, M. (2019). Genetic evidence for two carbon fixation pathways (the Calvin-Benson-Bassham cycle and the reverse tricarboxylic acid cycle) in symbiotic and free-living bacteria. *mSphere*, *4*(1), e00394-18. https://doi.org/10.1128/msphere.00394-18.
Ruiz-Ruiz, P., Gómez-Borraz, T. L., Revah, S., & Morales, M. (2020). Methanotroph-microalgae co-culture for greenhouse gas mitigation: Effect of initial biomass ratio and methane concentration. *Chemosphere*, *259*, 127418. https://doi.org/10.1016/j.chemosphere.2020.127418.
Sakarika, M., Spanoghe, J., Sui, Y., Wambacq, E., Grunert, O., Haesaert, G., & Vlaeminck, S. E. (2020). Purple non-sulphur bacteria and plant production: benefits for fertilization, stress resistance and the environment. *Microbial Biotechnology*, *13*(5), 1336–1365. https://doi.org/10.1111/1751-7915.13474.
Schiebenhoefer, H., Van Den Bossche, T., Fuchs, S., Renard, B. Y., Muth, T., & Martens, L. (2019). Challenges and promise at the interface of metaproteomics and genomics: An overview of recent progress in metaproteogenomic data analysis. *Expert Review of Proteomics*, *16*(5), 375–390. https://doi.org/10.1080/14789450.2019.1609944.
Schmider, T., Hestnes, A. G., Brzykcy, J., Schmidt, H., Schintlmeister, A., Roller, B. R., & Tveit, A. T. (2024). Physiological basis for atmospheric methane oxidation and methanotrophic growth on air. *Nature Communications*, *15*(1), 4151. https://doi.org/10.1038/s41467-024-48197-1.
Scholz, V. V., Meckenstock, R. U., Nielsen, L. P., & Risgaard-Petersen, N. (2020). Cable bacteria reduce methane emissions from rice-vegetated soils. *Nature Communication*, *11*(1), 1878. https://doi.org/10.1038/s41467-020-15812-w.
Semrau, J. D., DiSpirito, A. A., Gu, W., & Yoon, S. (2018). Metals and methanotrophy. *Applied and Environmental Microbiology*, *84*(6), e02289-17. https://doi.org/10.1128/AEM.02289-17.
Shakoor, A., Shakoor, S., Rehman, A., Ashraf, F., Abdullah, M., Shahzad, S. M., & Altaf, M. A. (2021). Effect of animal manure, crop type, climate zone, and soil attributes on greenhouse gas emissions from agricultural soils—A global meta-analysis. *Journal of Cleaner Production*, *278*, 124019. https://doi.org/10.1016/j.jclepro.2020.124019.
Sharma, U. C. (2020). Methane and nitrous oxide emissions from livestock in India:impact ofland usechange. *Journal of Agriculture and Aquaculture*, *2*(1), 1–9. https://escientificpublishers.com/assets/data1/images/JAA-02-0014.pdf.
Shen, Y., & Zhu, B. (2021). Arbuscular mycorrhizal fungi reduce soil nitrous oxide emission. *Geoderma*, *402*, 115179. https://doi.org/10.1016/j.geoderma.2021.115179.
Srinivasan, G., & Abirami, A. (2020). Mitigation of crop residue burning induced air pollution in New Delhi–a review. *International Journal of Engineering Applied Science Technology*, *5*(8), 282–291. https://doi.org/10.33564/ijeast.2020.v05i08.045.
Sun, K., Liu, H., Fan, H., Liu, T., & Zheng, C. (2021). Research progress on the application of feed additives in ruminal methane emission reduction: a review. *PeerJ*, *9*, e11151. https://doi.org/10.7717/peerj.11151.
Symeon, G. K., Akamati, K., Dotas, V., Karatosidi, D., Bizelis, I., & Laliotis, G. P. (2025). Manure management as a potential mitigation tool to eliminate greenhouse gas emissions in livestock systems. *Sustainablity*, *17*(2), 586. https://doi.org/10.3390/su17020586.
Tveit, A. T., Söllinger, A., Rainer, E. M., Didriksen, A., Hestnes, A. G., Motleleng, L., & Svenning, M. M. (2023). Thermal acclimation of methanotrophs from the genus *Methylobacter*. *The ISME Journal*, *17*(4), 502–513. https://doi.org/10.1038/s41396-023-01363-7.
Venturini, A. M., Gontijo, J. B., Mandro, J. A., Paula, F. S., Yoshiura, C. A., da França, A. G., & Tsai, S. M. (2022). Genome-resolved metagenomics reveals novel archaeal and bacterial genomes from Amazonian forest and pasture soils. *Microbial Genomics*, *8*(7), 000853. https://doi.org/10.1099/mgen.0.000853.
Wang, H., Zhao, R., Zhao, D., Liu, S., Fu, J., ZhangY, & Ding, H. (2022). Microbial-mediated emissions of greenhouse gas from farmland soils: A review. *Processes*, *10*(11), 2361. https://doi.org/10.3390/pr10112361.
Wartiainen, I., Hestnes, A. G., McDonald, I. R., & Svenning, M. M. (2006). *Methylobacter tundripaludum* sp. nov., a methane-oxidizing bacterium from Arctic wetland soil on the Svalbard

islands, Norway (78 N). *International Journal of Systematc and Evolutionary Microbiology*, *56*(1), 109–113. https://doi.org/10.1099/ijs.0.63728-0.

Wasai-Hara, S., Itakura, M., Fernandes Siqueira, A., Takemoto, D., Sugawara, M., Mitsui, H., & MinamisawaK (2023). *Bradyrhizobium ottawaense* efficiently reduces nitrous oxide through high *nosZ* gene expression. *Scientific Reports*, *13*(1), 18862. https://doi.org/10.1038/s41598-023-46019-w.

Yasmin, C., Harikarthik, D., Panotra, N., Anbarasan, S., Arya, D., Mishra, B. N., & Mishra, S. R. (2025). A thorough analysis of crop stubble burning mitigation using pusa decomposer: A comprehensive. *Asian Journal of Soil Science and Plant Nutrition*, *11*(3), 268–286. https://doi.org/10.9734/ajsspn/2025/v11i3570.

Zhao, R., Li, X., Bei, S., Li, D., Li, H., Christie, P., & Zhang, J. (2021). Enrichment of *nosZ*-type denitrifiers by arbuscular mycorrhizal fungi mitigates N_2O emissions from soybean stubbles. *Environmental Microbiology*, *23*(11), 6587–6602. https://doi.org/10.1111/1462-2920.15815.

11 Integrating Nitrogen-Fixing Biofertilizers into Climate-Resilient Cropping Systems

Sanjay Lilhare, Pratik Sanodiya, Vikash Singh and Bharti Sao

Abstract

Integrating nitrogen-fixing biofertilizers into climate-resilient cropping systems represents a sustainable and eco-friendly approach to enhance agricultural productivity while maintaining soil and environmental health. Nitrogen-fixing microorganisms, such as *Rhizobium, Azotobacter, Azospirillum, Frankia,* and *cyanobacteria*, convert atmospheric nitrogen into plant-available forms, thereby reducing dependence on synthetic nitrogen fertilizers and minimizing greenhouse gas emissions. Their application not only improves crop growth and yield but also enhances soil organic matter, microbial diversity, and nutrient cycling. Under changing climatic conditions, these biofertilizers enhance plant resilience against abiotic stresses, including drought, heat, and salinity. Moreover, when integrated with organic manures, compost, and other sustainable nutrient management practices, they contribute to improved carbon sequestration and long-term soil fertility. In low-input and organic farming systems, nitrogen-fixing biofertilizers provide a cost-effective solution for smallholder farmers, ensuring improved economic returns and environmental sustainability. By reducing fertilizer-induced pollution and supporting adaptive farming practices, these microbial inoculants play a pivotal role in transforming conventional agriculture into a more sustainable, climate-smart, and resilient production system capable of meeting future food security challenges.

S. Lilhare (✉)
Department of Agronomy, AKS University, Sherganj, Satna, Madhya Pradesh, India
e-mail: sanjulilhare199527@gmail.com

P. Sanodiya
Department of Agronomy, Institute of Agricultural Sciences, BHU, Varanasi, Uttar Pradesh, India

V. Singh
Department of Agricultural Engineering and Food Technology, AKS University, Sherganj, Satna, Madhya Pradesh, India

B. Sao
Department of Floriculture and Landscape Architecture, Mahatma Gandhi University of Horticulture and Forestry, Durg, Chhattisgarh, India

R. K. Singh and P. Singh (eds.), *Diazotrophic Plant Growth-Promoting Microorganisms and Green Energy – Volume I*, https://doi.org/10.1007/978-981-95-9682-9_11

Keywords

Nitrogen fixation · Bio-fertilizers · Climate resilience · Sustainable agriculture · Soil health · Microbial inoculants ·

11.1 Introduction

Nitrogen (N) is one of the most essential macronutrients for plant growth, influencing chlorophyll synthesis, protein formation, and overall yield potential. To meet the increasing global food demand, agriculture has become heavily dependent on synthetic nitrogen fertilizers. Global consumption of inorganic fertilizers has risen from about 142 million tons in 2002 to nearly 190 million tons in 2023, with nitrogen accounting for over half of the total usage (FAO, 2023). In 2022 alone, the world used approximately 108 million tons of nitrogen fertilizers, with China, India, and the United States as the leading consumers (Statista, 2024). However, only 30%–50% of applied nitrogen is effectively absorbed by crops, while the remainder is lost through leaching, volatilization, or runoff, causing severe environmental and economic consequences (FAO, 2023).

The widespread use of synthetic nitrogen fertilizers has brought several environmental challenges. Their production through the energy-intensive Haber–Bosch process and subsequent field application contributes significantly to global greenhouse gas (GHG) emissions. Nitrous oxide (N_2O), emitted from nitrogen-fertilized soils, possesses a global warming potential nearly 300 times greater than carbon dioxide (CO_2) (Smith et al., 2022). In 2018, the global synthetic nitrogen fertilizer supply chain was estimated to emit approximately 1.13 gigatons of CO_2-equivalent, representing about 2% of total global GHG emissions (Zhang et al., 2022). Moreover, excessive fertilizer application leads to nitrate leaching, eutrophication of aquatic ecosystems, and soil acidification (Li et al., 2021). Long-term nitrogen overuse disrupts soil microbial balance, reduces organic carbon content, and degrades soil structure, thereby diminishing soil health and resilience under changing climatic conditions (Wang et al., 2020).

In contrast, biological nitrogen fixation (BNF) represents a natural, eco-friendly mechanism of converting atmospheric nitrogen (N_2) into plant-available ammonium (NH_4^+) through the activity of diazotrophic microorganisms. Symbiotic associations, such as *Rhizobium–legume* and *Frankia–actinorhizal* systems, along with free-living and associative diazotrophs like *Azotobacter*, *Azospirillum*, and *cyanobacteria*, play vital roles in replenishing soil nitrogen. Globally, terrestrial BNF contributes approximately 120–150 Tg N yr^{-1}, of which agricultural ecosystems account for nearly half (Cleveland et al., 2019; USGS, 2023). These biological processes not only supplement crop nitrogen requirements but also enhance soil fertility, improve organic matter content, and reduce dependency on fossil fuel–based fertilizers, thereby aligning with the principles of sustainable nutrient management (Vitousek et al., 2021).

Nitrogen-fixing biofertilizers, containing beneficial microorganisms, have emerged as a practical and sustainable solution for improving nitrogen use efficiency and promoting climate-resilient agriculture. When applied to soils or seeds, they enhance nutrient availability, root development, and plant stress tolerance under adverse environmental conditions (Prasad et al., 2023). Studies have demonstrated that microbial biofertilizers, such as *Rhizobium*, *Azospirillum*, and *Bacillus*-based formulations, can increase crop yields by up to 40% while maintaining soil biological activity (Kumar et al., 2022b). Moreover, their use significantly lowers GHG emissions by reducing the need for synthetic nitrogen inputs (Singh et al., 2024). Consequently, integrating nitrogen-fixing biofertilizers within cropping systems offers a promising pathway toward low-input, climate-smart, and sustainable agriculture.

11.2 Concept of Nitrogen-Fixing Biofertilizers

Nitrogen-fixing biofertilizers are natural microbial inoculants containing live cells or dormant spores of nitrogen-fixing microorganisms that enhance the availability of nitrogen to plants by converting atmospheric nitrogen (N_2) into biologically usable forms such as ammonium (NH_4^+). They represent an essential component of sustainable nutrient management systems, promoting plant growth, soil fertility, and ecological balance while reducing dependence on synthetic nitrogen fertilizers (Bhattacharyya et al., 2021). Based on their mode of association with host plants, nitrogen-fixing microorganisms are broadly classified into symbiotic, associative (or associative symbiotic), and free-living diazotrophs (Singh & Prasad, 2020).

Symbiotic nitrogen fixers form highly specific mutualistic associations with plants, typically legumes. The best-known examples are *Rhizobium*, *Bradyrhizobium*, and *Sinorhizobium*, which infect legume roots and form nodules where nitrogen fixation occurs under microaerophilic conditions (Oldroyd et al., 2020). Associative nitrogen fixers, such as *Azospirillum* and *Herbaspirillum*, live in proximity to the rhizosphere or root surfaces of cereals and grasses, enhancing nutrient uptake and root development (Bashan et al., 2016). Free-living nitrogen fixers, including *Azotobacter*, *Clostridium*, and some *Cyanobacteria*, function independently in the soil without forming nodules, contributing to nitrogen enrichment and soil health improvement (Kumar et al., 2022b) (Fig. 11.1).

11.2.1 Common Microbial Genera

Several microbial genera are extensively used in nitrogen-fixing biofertilizer formulations. Rhizobium species are specific to leguminous crops, such as soybean, chickpea, lentil, and pea, forming nodules and fixing substantial quantities of nitrogen. Azotobacter, a free-living aerobic bacterium, is commonly used in non-leguminous crops like maize, cotton, and vegetables. It not only fixes nitrogen but also secretes growth-promoting substances, such as indole-3-acetic acid (IAA) and gibberellins

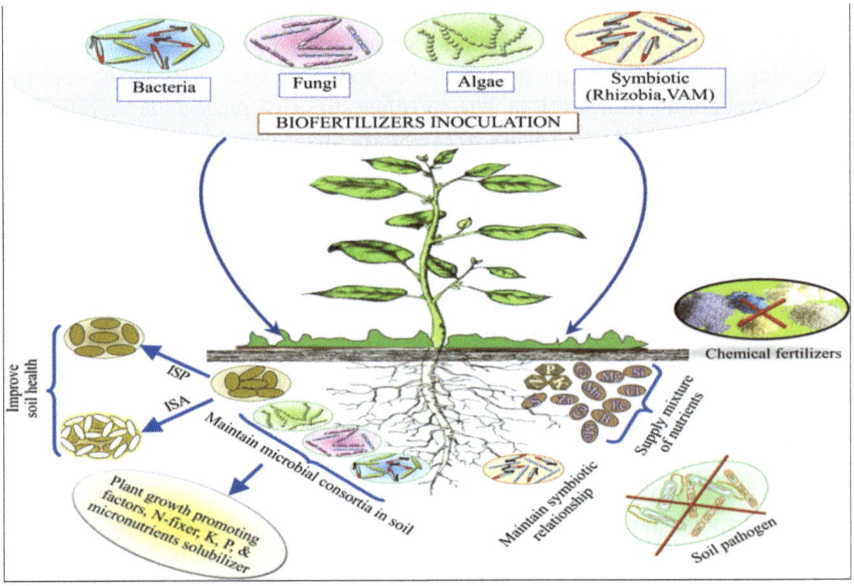

Fig. 11.1 Biofertilizer inoculations

(Kizilkaya, 2009). *Azospirillum*, an associative nitrogen fixer, forms loose associations with the roots of cereals, sugarcane, and maize, enhancing root proliferation and nutrient uptake (Vessey, 2003). Frankia, an actinobacterium, forms nodules with actinorhizal plants, such as *Alnus* and *Casuarina*, contributing to soil enrichment in forest and agroforestry ecosystems (Benson et al., 2011). Cyanobacteria, particularly genera such as *Anabaena* and *Nostoc*, play a vital role in paddy ecosystems by fixing atmospheric nitrogen in association with *Azolla* or as free-living organisms (Singh et al., 2014).

11.2.2 Mechanisms of Nitrogen Fixation and Nutrient Transformation

The process of biological nitrogen fixation is mediated by the nitrogenase enzyme complex, composed of two main proteins—dinitrogenase reductase and dinitrogenase—which work together to convert atmospheric N_2 into ammonia (NH_3) under anaerobic or microaerophilic conditions (Postgate, 1998). The reaction requires significant energy, typically supplied by ATP generated through microbial respiration. In symbiotic systems, the legume root provides carbohydrates to the bacteria, while the microbe supplies fixed nitrogen in return. Fixed nitrogen is incorporated into organic forms, such as amino acids, which plants can readily absorb. Additionally, nitrogen-fixing microorganisms improve soil structure, stimulate root exudation, and enhance nutrient transformation processes such as phosphorus solubilization and micronutrient availability, thereby improving overall soil fertility (Bhattacharyya et al., 2021; Meena et al., 2022).

11.2.3 Formulation Types: Carrier-Based, Liquid, and Encapsulated Biofertilizers

The effectiveness of biofertilizers depends greatly on their formulation and delivery system. Carrier-based formulations are the most common and utilize organic or inorganic materials, such as peat, lignite, vermiculite, or charcoal, to support microbial viability. These are cost-effective but often have limited shelf life and microbial stability (Reddy et al., 2020). Liquid biofertilizers are more advanced formulations containing cells suspended in nutrient media and stabilizers that prolong microbial viability up to two years. They are easy to apply through seed treatment, soil drenching, or foliar spray and ensure higher microbial survival under adverse conditions (Selvakumar et al., 2021). Recently, encapsulated or polymer-based biofertilizers have gained attention for their controlled release and protection against desiccation. Biopolymer encapsulation (using alginate or carrageenan) allows gradual microbial release in the rhizosphere, improving inoculant efficiency and field performance (Patil et al., 2022).

11.3 Climate Resilience through Biological Nitrogen Fixation

Biological nitrogen fixation (BNF) plays a pivotal role in enhancing the resilience of agricultural systems to climate variability by improving soil structure, nutrient cycling, and ecosystem stability. Nitrogen-fixing microorganisms, such as *Rhizobium*, *Azotobacter*, *Azospirillum*, *Frankia*, and *cyanobacteria,* enhance soil aggregation and porosity through the secretion of polysaccharides and microbial exudates that bind soil particles together (Mishra et al., 2021). Improved soil aggregation leads to better aeration, water infiltration, and moisture retention—key factors that buffer crops against erratic rainfall and drought conditions. Furthermore, continuous biological nitrogen inputs from microbial activity enrich soil organic matter and stimulate the growth of beneficial soil fauna, thereby increasing soil fertility and resilience to degradation (Bhattacharyya et al., 2020).

Nitrogen-fixing biofertilizers also play a significant role in mitigating abiotic stresses such as drought, salinity, and heat. Diazotrophs produce phytohormones (e.g., indole-3-acetic acid, gibberellins) and osmolytes that enhance root growth, improve plant water uptake, and regulate ion balance under stress conditions (Rani et al., 2022). For instance, *Azospirillum brasilense* inoculation has been shown to improve drought tolerance in cereals by enhancing root length and relative water content (Kumar & Meena, 2021). Similarly, *Rhizobium* and *Frankia* associations enhance plant antioxidant defense systems and membrane stability under salinity and temperature fluctuations (Ghosh et al., 2020). These physiological and biochemical benefits collectively strengthen crop adaptability to climatic stressors.

In addition to improving plant resilience, nitrogen-fixing microorganisms contribute substantially to climate change mitigation by promoting carbon sequestration and reducing greenhouse gas (GHG) emissions. The use of biofertilizers reduces dependence on synthetic nitrogen fertilizers, thereby minimizing nitrous oxide (N_2O) emissions—a potent greenhouse gas (Singh et al., 2023a, b). Enhanced root biomass and organic carbon deposition foster long-term carbon storage in soils, while improved microbial activity supports stable humus formation (Lal, 2020). Integration of nitrogen-fixing biofertilizers with conservation agriculture practices—such as minimum tillage, residue retention, and crop rotations—further improves soil health and enhances agroecosystem resilience (Meena et al., 2021a, b, c). Similarly, their incorporation into organic farming systems sustains nutrient cycling, reduces input costs, and aligns with sustainable low-emission agricultural models. Overall, leveraging the potential of biological nitrogen fixation contributes significantly to building climate-resilient and environmentally sustainable farming systems (Fig. 11.2).

11.4 Crop-Specific Integration Strategies

The integration of nitrogen-fixing biofertilizers varies with cropping systems and species compatibility, as different plants associate with specific diazotrophic microorganisms that influence nitrogen fixation efficiency and ecosystem sustainability.

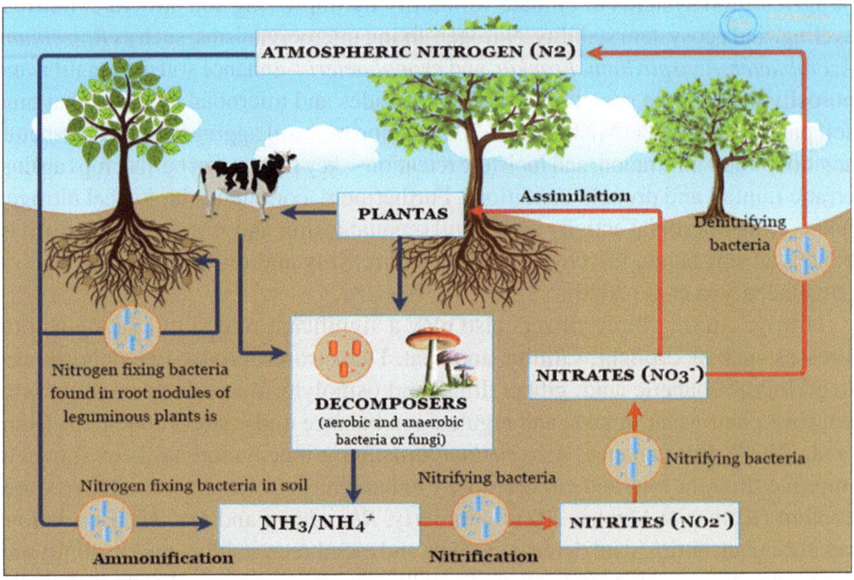

Fig. 11.2 Atmospheric nitrogen movements

11.4.1 Legume-Based Systems

Leguminous crops, such as soybean, chickpea, pigeon pea, and cowpea, form highly efficient symbiotic associations with *Rhizobium*, *Bradyrhizobium*, and *Mesorhizobium* species. These microorganisms colonize root nodules and fix atmospheric nitrogen through the nitrogenase enzyme complex, directly supplying plants with bioavailable nitrogen (Kumar et al., 2021a). Legume–*Rhizobium* symbiosis can contribute between 50 to 300 kg N ha^{-1} annually, significantly reducing the need for chemical fertilizers (Peoples et al., 2021). Crop rotations involving legumes improve soil fertility and enhance the performance of subsequent cereal crops by increasing residual nitrogen, improving soil organic carbon, and stimulating beneficial microbial activity (Singh et al., 2022). Such systems not only reduce input costs but also promote sustainability and resilience under variable climatic conditions.

11.4.2 Cereal-Based Systems

Cereals like wheat, maize, and sorghum, which lack root nodulation capacity, benefit from associative and free-living diazotrophs such as *Azospirillum*, *Azotobacter*, and *Herbaspirillum*. These microorganisms colonize the rhizosphere and root surfaces, facilitating nitrogen fixation, phytohormone production, and improved nutrient uptake (Bashan et al., 2020). Seed bio-priming—coating cereal seeds with liquid biofertilizers before sowing—has emerged as an effective strategy to enhance root colonization and nitrogen assimilation (Meena & Yadav, 2021). Studies report that *Azospirillum brasilense* inoculation in wheat and maize can increase yield by 15%–30% while reducing nitrogen fertilizer demand by up to 25% (Bhattacharyya et al., 2020). Integrating these diazotrophs into cereal-based systems enhances productivity, resource efficiency, and resilience against soil nutrient depletion.

11.4.3 Rice-Based Systems

In rice ecosystems, especially in flooded or waterlogged conditions, cyanobacterial-biofertilizers, such as *Anabaena*, *Nostoc*, and *Aulosira*, play a crucial role in nitrogen fixation. Blue-green algae (BGA) can fix up to 25–35 kg N ha^{-1} per season and contribute organic matter, improving soil structure and microbial biomass (Prasanna et al., 2020). In aerobic rice systems, where water scarcity limits conventional BGA proliferation, the combined use of *Azospirillum* and *cyanobacteria* has shown enhanced nitrogen uptake and water use efficiency (Jaiswal et al., 2022). These biofertilizer combinations also improve rice plant vigor and resilience under stress conditions, supporting sustainable intensification of paddy cultivation with lower environmental footprints (Table 11.1).

Table 11.1 Integrating nitrogen-fixing biofertilizers into climate-resilient cropping systems

Component	Role/findings	Supporting references
Nitrogen-Fixing Microorganisms (NFMs)	Supply biologically fixed nitrogen and reduce synthetic N-fertilizer use by 25%–50%.	Ghosh et al., 2020; Mahanty et al., 2017
Climate-Resilience Benefits	Improve plant tolerance to drought, salinity, and heat through better root growth and stress-regulating metabolites.	Singh et al., 2021; Bhattacharyya & Jha, 2012
Application Across Cropping Systems	Effective in legumes, cereals, rice, millets, and agroforestry through symbiotic and associative N-fixation.	Oldroyd & Dixon, 2014; Ghosh et al., 2020
Soil Health Enhancement	Improve soil microbial diversity, nutrient cycling, enzyme activities, and organic matter.	Mahanty et al., 2017; Singh et al., 2021
Integration with Organic Amendments	FYM, compost, and green manures increase survival and activity of NFMs.	Bhattacharyya & Jha, 2012; Mahanty et al., 2017
INM-Based N-Fertilizer Reduction	50% RDF + biofertilizer maintains yield equal to 100% RDF.	Singh et al., 2021
Advanced Formulations	Nano-encapsulation, polymer carriers, liquid inoculants, and biofilms improve viability and field performance.	Mahanty et al., 2017; Ghosh et al., 2020
Microbial Consortia	N-fixers combined with PSB, KMB, and biocontrol strains increase yield by 10%–30%.	Singh et al., 2021
Precision Agriculture Approaches	Soil microbiome profiling + GIS mapping improves site-specific biofertilizer recommendations.	Bhattacharyya & Jha, 2012
Climate Mitigation	Reduce N_2O emissions, nitrate leaching, and enhance carbon sequestration.	Oldroyd & Dixon, 2014; Mahanty et al., 2017
Challenges	Field inconsistency, storage issues, farmer awareness, and quality control gaps.	Ghosh et al., 2020
Future Prospects	CRISPR-based strain improvement, synthetic *nif* pathways, stress-tolerant microbial consortia.	Oldroyd & Dixon, 2014; Rogers & Oldroyd, 2014

11.4.4 Agroforestry and Intercropping Systems

Integrating nitrogen-fixing trees and shrubs into agroforestry and intercropping systems represents a long-term strategy for soil enrichment and climate resilience. Species, such as *Leucaena leucocephala*, *Gliricidia sepium*, and *Sesbania grandiflora*, associate with *Rhizobium* or *Frankia* to fix atmospheric nitrogen and recycle nutrients through litter decomposition (Nair et al., 2020). These trees improve soil fertility, enhance carbon sequestration, and stabilize microclimates by reducing erosion and improving water infiltration. Intercropping systems that combine legumes or nitrogen-fixing shrubs with cereals or vegetables optimize land productivity, enhance biodiversity, and promote ecological balance (Sileshi et al., 2021). Such integrated systems are fundamental to building climate-smart and sustainable agricultural landscapes.

11.5 Interaction with Soil and Cropping Systems

Nitrogen-fixing biofertilizers play a crucial role in maintaining soil fertility and enhancing nutrient cycling when integrated with organic and sustainable cropping practices. Their interaction with soil components and organic amendments fosters a synergistic environment that supports both nutrient availability and microbial activity, essential for resilient and productive agroecosystems.

11.5.1 Synergistic Effects with Organic Manures, Compost, and Green Manures

The combined application of biofertilizers with organic manures, compost, or green manures enhances nitrogen fixation efficiency and soil nutrient availability. Organic amendments provide a steady supply of carbon substrates and micronutrients that stimulate the growth and activity of diazotrophic microorganisms, such as *Azotobacter*, *Rhizobium*, and *Azospirillum* (Meena et al., 2021a, b, c). When applied together, biofertilizers and organics improve soil aggregation, cation exchange capacity, and moisture retention, resulting in enhanced crop growth and yield stability (Kumar & Verma, 2022). Studies have shown that integrating *Rhizobium* inoculants with farmyard manure in legume systems or compost in cereals increases nitrogen use efficiency and residual fertility for succeeding crops (Singh et al., 2023a, b). Green manuring with species like *Sesbania aculeata* and *Crotalaria juncea* further complements nitrogen-fixing microbes by adding biomass and promoting soil organic matter buildup, creating a favorable habitat for beneficial soil microorganisms.

11.5.2 Impact on Microbial Diversity and Soil Enzyme Activities

Nitrogen-fixing microorganisms significantly influence soil microbial diversity and enzymatic activity, which are key indicators of soil health. Inoculation with diazotrophs enhances populations of beneficial microbes, such as phosphate solubilizers, mycorrhizae, and decomposers, fostering a diverse and resilient soil microbiome (Bhattacharyya et al., 2020). Enzyme activities—such as dehydrogenase, urease, and nitrogenase—are elevated under biofertilizer application, facilitating organic matter decomposition and nutrient mineralization (Sharma et al., 2021). Enhanced microbial interactions improve soil respiration and nitrogen turnover rates, leading to better synchronization between nutrient release and plant uptake. Such biological enhancements help soils maintain productivity under climatic stresses like drought or nutrient depletion.

11.5.3 Nitrogen Cycling and Nutrient Synchronization with Crop Demand

Biofertilizers contribute actively to nitrogen cycling by mediating processes like mineralization, immobilization, and nitrification–denitrification balance. The nitrogen fixed by symbiotic and free-living microorganisms becomes available in synchrony with crop demand, minimizing nitrogen losses through leaching or volatilization (Das & Adhya, 2020). Associative diazotrophs like *Azospirillum* enhance root exudation, promoting mutualistic interactions that regulate nutrient availability in the rhizosphere (Pathak et al., 2021). This synchronization ensures efficient nitrogen use efficiency (NUE), reducing dependency on synthetic fertilizers and mitigating associated greenhouse gas emissions.

11.5.4 Integration into Integrated Nutrient Management (INM) Frameworks

Incorporating nitrogen-fixing biofertilizers within Integrated Nutrient Management (INM) strategies offers a balanced approach to nutrient supply and soil sustainability. INM frameworks combine organic sources, inorganic fertilizers, and biofertilizers to optimize nutrient use and maintain soil health over time (Yadav et al., 2022). This approach ensures that biofertilizers complement chemical fertilizers by enhancing microbial activity, improving nutrient use efficiency, and reducing input costs (Prasad et al., 2023). Field evaluations have shown that the inclusion of *Azospirillum* or *Rhizobium* in INM programs can reduce the recommended nitrogen fertilizer rate by 25%–50% without yield penalties, while improving soil organic carbon and microbial biomass (Kumar et al., 2021a). Thus, biofertilizer integration not only strengthens nutrient cycling and soil fertility but also supports sustainable, climate-resilientcropping systems aligned with global low-carbon agricultural goals (Fig. 11.3).

11.6 Field Performance and Case Studies

The field performance of nitrogen-fixing biofertilizers has been evaluated across diverse agro-ecological zones, demonstrating their potential to enhance crop productivity, soil fertility, and sustainability. Multi-location trials in India and other parts of the world indicate that biofertilizers, when integrated with organic or reduced chemical inputs, can significantly improve nitrogen use efficiency and crop yield. For example, field studies on chickpea and soybean in the semi-arid regions of India reported that inoculation with *Rhizobium* combined with 50%–75% of recommended nitrogen dose

Fig. 11.3 Components of integrated nutrient management

resulted in yield increases of 15%–30% compared to uninoculated controls receiving full chemical nitrogen (Prasad et al., 2023). Similar experiments in wheat and maize using associative diazotrophs like *Azospirillum* and *Azotobacter* showed enhanced root development, grain yield, and soil microbial activity, even under limited fertilizer inputs (Bhattacharyya et al., 2020).

Biofertilizer integration also improves soil fertility parameters, including organic carbon content, microbial biomass, and enzymatic activity. Long-term trials in paddy fields using cyanobacterial inoculants (*Anabaena* and *Nostoc*) reported an increase of 10%–15% in soil nitrogen content, along with better soil aggregation and enhanced phosphorus availability (Jaiswal et al., 2022). These improvements are attributed to the synergistic effects of biofertilizers with organic residues and green manures, which create a conducive environment for microbial proliferation and nutrient cycling.

Comparative studies consistently demonstrate that biofertilizers can partially or wholly substitute chemical nitrogen without compromising yield. In a meta-analysis across multiple crops, plots receiving 50%–75% of recommended nitrogen along with biofertilizers achieved yields comparable to or exceeding those with 100% chemical nitrogen (Kumar et al., 2021a). Additionally, the reduction in synthetic fertilizer use translates into lower greenhouse gas emissions, reduced input costs, and improved long-term soil health, making biofertilizers a sustainable alternative in modern agriculture.

Several successful case studies highlight the practical applicability of nitrogen-fixing biofertilizers in both local and global contexts. In India, the National Agricultural Innovation Project demonstrated widespread adoption of *Rhizobium* inoculants in pulses, resulting in improved yields and reduced fertilizer dependency in states, such as Madhya Pradesh, Maharashtra, and Karnataka (Meena et al., 2021a, b, c). Globally, projects in Brazil and sub-Saharan Africa have utilized *Azospirillum* and legume–*Rhizobium* systems in maize and soybean, achieving similar yield gains while promoting climate-resilient and low-input farming practices (Bashan et al., 2020). These studies underscore the versatility and efficacy of nitrogen-fixing biofertilizers across diverse cropping systems and environments, confirming their role as a key tool in sustainable and climate-smart agriculture.

11.7 Challenges and Limitations

Despite their proven benefits, the widespread adoption of nitrogen-fixing biofertilizers faces several challenges that limit consistent field performance. Environmental and soil variability play a major role in influencing microbial survival, colonization, and nitrogen-fixing efficiency. Factors such as soil pH, temperature, moisture content, organic matter, and native microbial communities can cause inconsistent results across different agro-ecological zones (Singh & Prasad, 2020). For example, acidic soils may inhibit the growth of *Rhizobium* strains, whereas drought-prone areas can reduce the activity of free-living diazotrophs like *Azospirillum* or *Azotobacter*, resulting in lower nitrogen contribution than expected.

Formulation and storage issues also pose significant limitations. Traditional carrier-based biofertilizers, using peat, lignite, or compost as carriers, may suffer from limited shelf life, reduced microbial viability under high temperature or humidity, and contamination by non-beneficial microbes (Reddy et al., 2020). Liquid and encapsulated formulations offer better stability but require technical knowledge, proper handling, and cold-chain storage, which are often unavailable in rural areas.

Another key constraint is farmers' awareness and adoption. Many smallholder farmers lack knowledge about biofertilizer benefits, inoculation techniques, and integration with crop management practices. Misapplication or inconsistent use often leads to sub-optimal results, discouraging wider adoption (Meena et al., 2021a, b, c). Socio-economic factors, access to quality inoculants, and risk perception further affect uptake, particularly in resource-limited regions.

Finally, policy and regulatory gaps hinder standardization, quality control, and market accessibility. Inconsistent labeling, lack of certification mechanisms, and inadequate extension support can result in low-quality products reaching farmers, reducing efficacy and trust (Bhattacharyya et al., 2020). Moreover, subsidy schemes and incentives for biofertilizer use remain limited compared to chemical fertilizers, which constrains scaling up of climate-smart nutrient management strategies. Addressing these challenges requires an integrated approach involving scientific innovation, regulatory frameworks, farmer education, and market support to unlock the full potential of nitrogen-fixing biofertilizers in sustainable agriculture.

11.8 Future Prospects and Research Priorities

The future of nitrogen-fixing biofertilizers lies in technological innovation, microbial optimization, and integration with modern agricultural practices to enhance climate resilience and sustainability. One promising avenue is the development of microbial consortia and multi-strain formulations, which combine complementary diazotrophs and plant growth-promoting microorganisms (PGPMs) to improve nitrogen fixation, nutrient uptake, and stress mitigation (Bhattacharyya et al., 2021). Multi-strain inoculants can provide broader compatibility across diverse crops and environmental conditions compared to single-strain products, offering more reliable field performance.

Advances in nanotechnology and encapsulation present additional opportunities to enhance microbial survival, shelf life, and targeted delivery in the rhizosphere. Nano-carriers, polymer-based encapsulation, and controlled-release formulations protect beneficial microbes from desiccation, temperature extremes, and UV exposure while ensuring gradual nutrient release synchronized with crop demand (Patil et al., 2022). Such innovations can significantly improve the efficacy of biofertilizers under challenging field conditions, especially in semi-arid and stress-prone agro-ecologies.

Genetic improvement of nitrogen-fixing strains is another key research priority. Through molecular breeding, adaptive evolution, and genome editing, strains can be enhanced for tolerance to drought, salinity, heat, and acidic or alkaline soils, enabling consistent performance under climate variability (Kumar et al., 2021a). Additionally, genetic and omics-based approaches can optimize nitrogenase efficiency, root colonization, and interaction with other soil microbes, further increasing nitrogen use efficiency in cropping systems.

Finally, the integration of biofertilizers with precision farming and digital agriculture tools promises smarter and data-driven nutrient management. Soil health mapping, remote sensing, and IoT-based monitoring can guide site-specific biofertilizer application, optimize dosages, and track microbial efficacy in real time (Meena et al., 2021a, b, c). Digital platforms can also facilitate farmer awareness, training, and adoption, bridging the gap between research advancements and field-level implementation. Collectively, these future-oriented strategies will strengthen the role of nitrogen-fixing biofertilizers in climate-resilient, low-input, and sustainable agriculture.

11.9 Policy and Extension Interventions

The widespread adoption of nitrogen-fixing biofertilizers is strongly influenced by supportive policies, institutional frameworks, and effective extension mechanisms. Both national and international initiatives have emphasized biofertilizer use as part of sustainable and climate-resilient agriculture. In India, programs, such as the National Biofertilizer Development Program (NBDP) and the Rashtriya Krishi Vikas Yojana (RKVY) provide financial support, technical guidance, and capacity building for biofertilizer production and utilization (Bhattacharyya et al., 2020). Internationally, initiatives under the Food and Agriculture Organization (FAO) and the International Fertilizer Association (IFA) promote microbial inoculants for low-input, sustainable farming systems, linking them to climate-smart agriculture agendas.

Incentives, certification, and public–private partnerships play a key role in ensuring quality and reliability of biofertilizers. Certification mechanisms under the Fertilizer Control Order (FCO) in India help maintain product standards, while subsidies for biofertilizers encourage farmer adoption. Public–private partnerships facilitate production scale-up, technology transfer, and research collaboration, ensuring that high-quality microbial inoculants are accessible at affordable rates (Meena et al., 2021a, b, c).

Extension networks, training programs, and community-based models are crucial for bridging the gap between scientific knowledge and farm-level application. Demonstrations, farmer field schools, and participatory learning approaches increase awareness of proper inoculation techniques, crop-specific recommendations, and integration with organic and inorganic nutrient sources (Singh & Prasad, 2020). Community-based models, including cooperatives and farmer producer organizations, have proven effective in ensuring consistent supply and adoption of biofertilizers in both smallholder and large-scale systems.

To achieve large-scale adoption in climate-resilient programs, strategies must integrate policy support, incentive mechanisms, and robust extension delivery. Combining biofertilizer use with conservation agriculture, crop diversification, and digital advisory tools enhances nutrient efficiency, reduces greenhouse gas emissions, and strengthens the resilience of cropping systems under variable climatic conditions (Kumar et al., 2021a). Targeted awareness campaigns, capacity-building initiatives, and monitoring frameworks are essential to institutionalize biofertilizers as a mainstream input for sustainable and climate-smartagriculture.

11.10 Conclusion

Nitrogen-fixing biofertilizers are central to sustainable and climate-smart agriculture, providing a natural alternative to synthetic nitrogen fertilizers while improving soil fertility, crop productivity, and ecosystem resilience. Across different cropping systems—legumes, cereals, rice, and agroforestry—biofertilizers enhance nitrogen availability, nutrient cycling, and organic matter buildup, particularly when used in combination with organic amendments and integrated nutrient management approaches. Field evidence demonstrates that symbiotic, associative, and free-living nitrogen-fixing microorganisms can maintain or improve crop yields while reducing chemical fertilizer use, lowering greenhouse gas emissions, and promoting environmentally sustainable farming practices.

While significant progress has been made, future research must focus on developing robust, stress-tolerant microbial strains, improving formulation stability, and designing site-specific inoculants tailored to diverse agro-climatic zones. Further advancements in synthetic biology, microbial consortia, nano-formulations, and precision agriculture hold great promise for enhancing the consistency and efficiency of nitrogen fixation under field conditions. However, challenges such as variable field performance, limited farmer awareness, inconsistent product quality, and gaps in policy implementation remain major barriers to large-scaleadoption.

This chapter will be especially helpful for promoting sustainable agriculture, as it provides a scientific understanding of how biological nitrogen fixation can reduce dependence on chemical fertilizers, minimize greenhouse gas emissions, and improve soil health. By integrating nitrogen-fixing biofertilizers with modern agronomic practices and climate-smart technologies, farmers can achieve higher productivity, better environmental stewardship, and long-term soil fertility. Ultimately, the insights from this chapter support the transition toward low-input, eco-friendly agricultural systems that are essential for future global food security.

References

Bashan, Y., de-Bashan, L. E., & Prabhu, S. R. (2020). Advances in plant growth-promoting bacteria for sustainable agriculture. *Plant and Soil, 455*(1–2), 1–28.

Bashan, Y., de-Bashan, L. E., Prabhu, S. R., & Hernandez, J. P. (2016). Advances in plant growth-promoting bacterial inoculant technology: Formulations and practical perspectives (1998–2013). *Plant and Soil, 378*(1–2), 1–33.

Benson, D. R., & Silvester, W. B. (2011). Biology of Frankia actinorhizal symbioses. *Plant and Soil, 356*(1–2), 1–9.

Bhattacharyya, P. N., & Jha, D. K. (2021). Plant growth-promoting rhizobacteria (PGPR): Emergence in agriculture. *World Journal of Microbiology and Biotechnology, 37*(3), 1–19.

Bhattacharyya, P., Roy, K. S., & Das, S. (2020). Nitrogen-fixing microorganisms and their role in soil fertility and climate resilience. *Applied Soil Ecology, 153*, 103631.

Cleveland, C. C., Houlton, B. Z., & Reed, S. C. (2019). Natural nitrogen fixation: global distribution and environmental controls. *Nature Reviews Earth & Environment, 1*(2), 110–123.

Das, S., & Adhya, T. K. (2020). Biological nitrogen fixation and its integration in nutrient cycling for sustainable agriculture. *Current Science, 118*(9), 1367–1374.

FAO (2023). Inorganic fertilizer use statistics, 2002–2023. Food and Agriculture Organization of the United Nations. https://www.fao.orgRetrieved from.

Ghosh, S., Chatterjee, D., & Saha, A. (2020). Symbiotic nitrogen fixation enhances tolerance of leguminous crops to salinity and temperature stress. *Frontiers in Plant Science, 11*, 568450.

Jaiswal, D., Prasanna, R., & Singh, P. K. (2022). Cyanobacterial and bacterial biofertilizers for sustainable rice cultivation under changing climate. *Agronomy, 12*(8), 1853.

Kizilkaya, R. (2009). Nitrogen fixation capacity of Azotobacter spp. strains isolated from soils in different ecosystems and relationship between them and the microbiological properties of soils. *Journal of Environmental Biology, 30*(1), 73–82.

Kumar, R., & Meena, R. S. (2021). Role of plant growth-promoting rhizobacteria in drought mitigation: mechanisms and applications. *Agronomy, 11*(3), 462.

Kumar, R., Meena, R. S., & Singh, A. (2021a). Genetic improvement of nitrogen-fixing microorganisms for stress resilience in cropping systems. *Frontiers in Agronomy, 3*, 652214.

Kumar, R., Meena, R. S., & Singh, A. (2021b). Symbiotic nitrogen fixation and its role in legume-based cropping systems. *Legume Research, 44*(10), 1135–1142..

Kumar, R., Meena, R. S., & Singh, A. (2022a). Biofertilizers and soil health: Microbial interventions for sustainable productivity. *Journal of Applied Microbial Science, 12*(3), 145–157.

Kumar, S., Verma, A., & Singh, D. (2022b). Role of free-living nitrogen-fixing bacteria in enhancing soil fertility and crop productivity. *Agricultural Reviews, 43*(2), 199–206.

Kumar, S., & Verma, P. (2022). Synergistic role of organic manures and biofertilizers in improving soil quality and crop productivity. *Journal of Environmental Management, 305*, 114374.

Lal, R. (2020). Soil carbon sequestration and biofertilizer integration for climate-resilient agriculture. *Soil & Tillage Research, 204*, 104721.

Li, Y., Zhang, H., & Chen, Q. (2021). Effects of nitrogen fertilizer on nitrate leaching and soil acidification: A meta-analysis. *Agricultural Ecosystems & Environment, 319*, 107551.

Meena, R. S., Kumar, S., & Datta, R. (2022). Soil microbes for sustainable agriculture: Functions and applications. *Frontiers in Microbiology, 13*, 864321.

Meena, R. S., Singh, M., & Kumar, S. (2021a). Conservation agriculture and microbial synergy for sustainable soil management. *Sustainability, 13*(12), 6714.

Meena, R. S., Singh, M., & Yadav, G. S. (2021b). Combining biofertilizers with organics for improved nutrient use efficiency and soil health. *Sustainability, 13*(15), 8297.

Meena, R. S., Singh, M., & Yadav, G. S. (2021c). Integrating biofertilizers into extension programs and community-based models for sustainable farming. *Sustainability, 13*(15), 8297.

Meena, R. S., & Yadav, G. S. (2021). Bio-priming and microbial inoculants in cereal cropping systems: A sustainable approach. *Journal of Soil Science and Plant Nutrition, 21*(4), 2756–2770.

Mishra, S., Patra, S., & Sahoo, D. (2021). Microbial polysaccharides and their role in soil aggregation and climate resilience. *Journal of Environmental Management*, *287*, 112325.

Nair, P. K. R., Mohan Kumar, B., & Nair, V. D. (2020). Agroforestry as a climate-resilient land use system. *Journal of Environmental Management*, *261*, 110215.

Oldroyd, G. E. D., Murray, J. D., Poole, P. S., & Downie, J. A. (2020). The rules of engagement in the legume–rhizobial symbiosis. *Annual Review of Genetics*, *54*, 485–512.

Pathak, R. K., Sharma, P., & Chauhan, S. (2021). Diazotrophic interactions and nutrient synchronization in cereal-based cropping systems. *Agricultural Microbiology Research*, *12*(4), 241–252.

Patil, S., Thorat, B., & Bhosale, P. (2022). Advances in biofertilizer formulations: Encapsulation and nanotechnology approaches. *Journal of Environmental Chemical Engineering*, *10*(8), 108262.

Peoples, M. B., Herridge, D. F., & Ladha, J. K. (2021). Biological nitrogen fixation and crop rotation benefits in sustainable agriculture. *Plant and Soil*, *461*(1–2), 1–24..

Postgate, J. R. (1998). Nitrogen fixation (3rd ed.). Cambridge University Press.

Prasad, S., Chatterjee, D., & Sinha, R. K. (2023). Role of microbial biofertilizers in enhancing soil fertility and climate resilience under INM systems. *Sustainability*, *15*(8), 6422.

Prasanna, R., Jaiswal, D., & Kaushik, B. D. (2020). Cyanobacterial biofertilizers for rice cultivation: Current status and prospects. *Indian Journal of Microbiology*, *60*(3), 287–298.

Rani, M., Sharma, P., & Chauhan, S. (2022). Role of biofertilizers in mitigating abiotic stress under changing climate. *Environmental and Sustainability Indicators*, *16*, 100184.

Reddy, P., Kumar, N., & Subrahmanyam, G. (2020). Comparative efficacy of carrier and liquid biofertilizers on microbial survival and plant productivity. *Indian Journal of Microbiology*, *60*(2), 231–238.

Selvakumar, G., Kundu, S., & Gupta, A. D. (2021). Liquid biofertilizers: Production, shelf life, and field performance. *Microbiological research*, *12*(5), 455–462.

Sharma, D., Gupta, N., & Tiwari, V. (2021). Soil enzyme dynamics and microbial interactions under nitrogen-fixing biofertilizer application. *Soil Biology and Biochemistry*, *156*, 108192.

Sileshi, G. W., Mafongoya, P. L., & Chintu, R. (2021). Integration of nitrogen-fixing trees into smallholder agroforestry systems for soil fertility and resilience. *Agroforestry Systems*, *95*(7), 1283–1295.

Singh, P., & Prasad, S. (2020). Microbial biofertilizers for sustainable agriculture: Mechanisms and applications. *Sustainability*, *12*(24), 10302.

Singh, P., Sharma, R., & Chauhan, S. (2022). Role of legume–cereal rotations in improving soil health and climate resilience. *Frontiers in Sustainable Food Systems*, *6*, 105943.

Singh, P., Sharma, R., & Chauhan, S. (2023a). Integrating nitrogen-fixing biofertilizers for low-emission agriculture. *Frontiers in Sustainable Food Systems*, *7*, 134221.

Singh, P., Sharma, R., & Chauhan, S. (2023b). Organic and microbial integration for enhanced nitrogen efficiency in cropping systems. *Agronomy Journal*, *115*(1), 219–230.

Singh, P., Sharma, R., & Chauhan, S. (2024). Integrating biofertilizers for low-carbon agriculture: Opportunities and challenges. *Frontiers in Sustainable Food Systems*, *8*, 133274.

Singh, R. N., Khare, N., & Tiwari, G. L. (2014). Role of cyanobacteria in paddy ecosystem for sustainable agriculture. *Asian Journal of Biological Sciences*, *7*(2), 122–129.

Smith, P., et al. (2022). Global assessment of nitrous oxide emissions from agricultural soils. *Global Change Biology*, *28*(3), 987–1003.

Statista. (2024). *Global nitrogen fertilizer consumption by country 2022*. Retrieved from https://www.statista.com.

USGS (2023). *Global terrestrial nitrogen fixation and its modification by agriculture*. U.S. Geological Survey Report. Retrieved from https://www.usgs.gov.

Vessey, J. K. (2003). Plant growth promoting rhizobacteria as biofertilizers. *Plant and Soil*, *255*(2), 571–586.

Vitousek, P. M., Cassman, K. G., & Cleveland, C. (2021). Toward an ecological understanding of biological nitrogen fixation. *Ecological Monographs*, *91*(3), e01431.

Wang, Y., Zhao, X., & Gao, H. (2020). Long-term effects of nitrogen fertilizer on soil microbial communities and organic carbon dynamics. *Soil Biology and Biochemistry*, *144*, 107768.

Yadav, G. S., Meena, R. S., & Lal, R. (2022). Integrated nutrient management strategies for climate-resilient agriculture. *Agricultural Systems*, *198*, 103384.

Zhang, X., Xu, Y., & Gao, J. (2022). Greenhouse gas emissions from synthetic nitrogen fertilizer production and use. *Environmental Research Letters*, *17*(11), 114015.

12 Eco-Friendly Agriculture: The Role of Diazotrophs in Food Security

Shalini Rai, Nidhi Singh and Deepshikha Kushwaha

Abstract

The indiscriminate application of synthetic nitrogen (N_2) fertilizers, ineffective water management, and soil degradation pose threats to sustainable agriculture and food security. A sustainable substitute for artificial N_2 input in hydroponic systems is biological nitrogen fixation, which is mediated by diazotrophs and converts atmospheric N_2 to plant-available ammonium. This chapter explores the functional variety of diazotrophs, root–microbe interactions, and environmental restrictions in order to address the integration of BNF into the microbiome of crops. It also highlights new bioformulation techniques to improve microbial compatibility with nutrient regimes, strain improvement, microbial consortia development, nitrogenase protection, and genome editing tools, as well as omics-based tools for the real-time evaluation of N_2 fixation and microbial functionality. Important issues are covered, including the lack of defined biostimulant methods, nitrate-induced suppression of nitrogenase activity, and microbial leaching. Following diazotroph treatments, case studies on staple crops have shown increased production productivity and nutrient use efficiency. Future directions for synthetic biology, regulations, and omics-based methods for the real-time evaluation of N_2 fixation and microbial functioning are discussed in the chapter's conclusion.

Keywords

Food system · Synthetic biology · Nitrogen use efficiency · Diazotrophs · OMICS · Microbiome engineering · Sustainable agriculture ·

S. Rai (✉)
Department of Microbiology, DSMNRU, Lucknow, UP, India
e-mail: shalinimicro09@gmail.com

N. Singh
Division of Plant Biotechnology, College of Biotechnology, SVPUAT, Meerut, India

D. Kushwaha
Centre for Microbiology, Ewing Christian College, University of Allahabad, Prayagraj, UP, India

R. K. Singh and P. Singh (eds.), *Diazotrophic Plant Growth-Promoting Microorganisms and Green Energy – Volume I*, https://doi.org/10.1007/978-981-95-9682-9_12

12.1 Introduction

Rapid urbanization, population expansion, and climate change are putting more and more strain on global agriculture. The United Nations projects that by 2050, there will be 9.7 billion people on the planet, which would significantly increase the need for food (Eleazar et al., 2025). According to Andrade-Bustamante et al. (2025), there is a need for highly efficient agricultural systems in urban and peri-urban regions since 55% of the world's population currently lives in urban settings, and that number is predicted to rise to almost 75% by 2050. Global agricultural systems are being significantly impacted by the growing problems of climate change, which calls for creative and sustainable solutions to guarantee food security (Zayed et al., 2023). The exact issue that conventional agricultural methods aim to solve is made worse by their heavy reliance on synthetic nitrogen fertilizers, which greatly contribute to greenhouse gas emissions and environmental damage. Despite improving food production in the 20th century, the Haber-Bosch process has resulted in an over reliance on nitrogen supplied industrially, upsetting natural nitrogen cycles and causing a series of detrimental environmental effects (Babcock-Jackson et al., 2023). Large amounts of fossil fuels are used in the manufacture, delivery, and application of these fertilizers, which releases significant amounts of carbon dioxide into the atmosphere and exacerbates climate change. Overuse of these fertilizers causes significant nitrogen losses from agricultural systems through processes like nitrate leaching, which contaminates groundwater supplies and causes aquatic ecosystems to become eutrophic, and ammonia volatilization, which adds to air pollution and acid rain. The denitrification of surplus nitrate in agricultural soils also produces nitrous oxide emissions, a powerful greenhouse gas with a global warming potential substantially greater than that of carbon dioxide, which exacerbates the climatic effect of using standard nitrogen fertilizer.

An environmentally friendly substitute for artificial N_2 fertilizers is biological nitrogen fixation. Biological N_2-fixation by microorganisms collectively referred to as diazotrophs is the primary source of fixed-N to the biosphere; they supply over 60% of the total yearly input (Threatt & Rees, 2023). The nitrogenase enzyme, which functions in the soil at room temperature and pressure, is used by nearly all diazotrophs. Diazotrophs—both bacteria and archaea—have been identified in soil and the ocean, and they have been shown to contribute to the ecosystem's transformation of atmospheric nitrogen into accessible ammonia (Imran et al., 2021). Based on their lifestyle, they are often categorized as either free-living, symbiotic, or in a loose relationship, usually with plant roots.

Free-living diazotrophs, which can be chemotrophs or phototrophs, fix nitrogen dioxide for their own use in aerobic, anaerobic, or micro-anaerobic environments . In a specific structure on plant roots, symbiotic diazotrophs nearly always survive and fix N_2 in microaerobic conditions, giving their host fixed-N. They use a variety of nitrogen-fixing techniques to improve nitrogen uptake and plant development, including root colonization, biofilm production, and nitrogenase protective mechanisms (Pittol et al., 2016). However, a number of physiological, chemical, and ecological limitations, such as microbial compatibility with host crops, the composition of nutrient solutions, and system-specific abiotic conditions, limit their efficacy (Li et al., 2025). Recent studies have concentrated on diazotroph strain engineering, protective bioformulations, and microbial consortia design to overcome these constraints (Arkani et al., 2025). Our knowledge of microbial interactions is being improved by ongoing developments in imaging, transcriptomics, and formulation technologies, which also make it easier to successfully incorporate them into precision agriculture and food systems (Abdelwahed et al., 2022; Srikanth et al., 2025). Additionally, frameworks for regulations and policies are being created to specify the effectiveness and safety requirements for microbial bioinputs in regulated settings (BPIA, 2023; Santos et al., 2024).

The molecular processes governing nitrogenase activity, assembly and control, ammonium assimilation, and plant-diazotroph interactions have all been better understood thanks to this research endeavor. Various methods for controlling nitrogen fixation activity, diazotrophic ammonium release, and plant-diazotroph interactions are covered in this chapter. We underline the significance of biocontrol methods and public acceptability before discussing the research pathways investigated in this field and suggesting possible future directions, with a focus on engineering at the metabolic level via biorthogonal signaling. The creation of regulatory standards, synthetic biology methods to nitrogenase engineering, and the growing importance of AI-assisted nutrient management technologies as possible facilitators for improving the integration of diazotrophs into precision agriculture are some of the future research objectives.

12.2 Major Group of Diazotrophs

The diazotrophs are taxonomically varied groupings of bacteria that include alphaprotobacteria, betaprotobacteria, and gammaprotobacteria. They are phylogenetically diverse and comprise species with different physiological characteristics. Azospirillum, Azorhizobium, Pseudomonas, Rhizobium, and the phyla Firmicutes and Cyanobacteria are among the major genera. According to, the three main kinds of bacteria that fix nitrogen are rhizospheric, endophytic, and nodulating bacteria, each of which has a different spectrum of nitrogen fixation.

12.2.1 Free-Living Microbes

Although free-living chemotrophic microbes do not provide much fixed-N to agriculture, their nitrogenase is nearly identical to that of the agriculturally significant symbiotic diazotrophs known as rhizobia, making them scientifically valuable. Some Azotobacter species, which all effectively fix N_2 in the air, are the greatest examples of obligate aerobes. Although the majority of other aerobic diazotrophs are not as well suited to thrive in air, they all have systems in place to keep O_2 from harming their nitrogenase. Cyanobacteria, such as Anabaena and Nostoc, are examples of aerobic phototrophs. In addition to fixing their own N_2, these incredibly autonomous creatures employ photosynthesis similar to that of green plants to produce the fixed-C and energy they require. Their photosynthetic ability, however, poses a challenge since it generates O_2, which would impair nitrogenase. They employ a different O_2 protection approach than aerobic chemotrophs. Some of them find nitrogenase in specialized cells known as heterocysts, where highly active cyclic photophosphorylation continues to supply ATP for N_2-fixation but a defective photosystem-II component of photosynthesis is no longer able to produce O_2.

The clostridia, which contain several species that can fix N_2, are the most prominent example of anaerobic diazotrophs. These anaerobes primarily produce butyric acid, CO_2, and H_2 through the fermentation of carbon sources. Anaerobic sulfate-reducing bacteria also include diazotrophs, such as Desulfovibrio, which creates H_2S by using SO_4^{2-} as a terminal respiratory electron acceptor rather than O_2. The main nonphotosynthetic N_2-fixing factor in the development of N-containing marine sediments is Desulfovibrio. Methanogens, which are members of the Archaea, the third domain of life, are other significant anaerobic diazotrophs. The CH_4 that ruminant animals and stagnant water create is caused by these microbes. Anaerobic phototrophs, such as the purple nonsulfur bacteria *Rhodospirillum rubrum*, also undergo N_2-fixation. Despite being photosynthetic, these diazotrophs do not create enough O_2 to negatively affect nitrogenase because they do not have a working photosystem II. Both sulfur and nonsulfur bacteria are photosynthetic, and like cyanobacteria, they may use CO_2 as their exclusive source of carbon.

The facultative anaerobes make up a third category. If given fixed-N, they can grow with or without O_2, but they can only fix N_2 anaerobically. The most well-researched examples are from the genus Klebsiella, which includes many species that fix N_2 and is linked to *Escherichia coli* of the Enterobacteriaceae family. They may be separated from animal intestines, water, and soils. Although genetically altered diazotrophic strains of *E. coli* have been created, there is currently no known wild diazotrophic strain. There are additional known diazotrophic organisms of the allied genera Bacillus, Enterobacter, and Citrobacter.

12.2.2 Plant Symbioses and Associations

Numerous systems exist where bacteria fix N_2 in different symbioses and connections with the roots (typically) of different plant hosts, as explained in the next chapters. Diazotrophs are divided into three functional categories based on how they interact with their host plants: rhizosphere, endophytic, and symbiotic groups (Carvalho et al., 2014; Rai et al., 2023). Each group has different host preferences, N_2 fixation efficiency, and colonization tactics.

12.2.3 Rhizospheric and Endophytic Diazotrophs

The rhizosphere, or root–soil contact, is home to free-living or associative N_2-fixing bacteria called rhizospheric diazotrophs. Root exudates, which include sugars, amino acids, and organic acids—chemoattractants that supply energy for microbial colonization—have an impact on this microenvironment (Imran et al., 2021). These microbes are rhizobacteria that stimulate plant development by fixing nitrogen (N_2), solubilizing phosphate (P), producing siderophores, and synthesizing phytohormones (Renganathan et al., 2018). On the other hand, endophytic diazotrophs infiltrate interior plant tissues such as xylem, phloem, cortical cells, and intercellular gaps without generating signs of illness (Pal et al., 2022). Because endophytes are associated with plant vascular tissues, they offer ecological benefits such as reduced competition, access to host-derived metabolites, and improved N transfer (Carvalho et al., 2014).

Cereals such rice, maize, sorghum, and wheat have yielded isolates of several taxa, including Herbaspirillum, Gluconacetobacter, Burkholderia, Azoarcus, and Azospirillum (Saranraj et al., 2022). Under circumstances with low fertilizer, *Herbaspirillum seropedicae* and *Burkholderia vietnamiensis* colonize the intercellular gaps in sugarcane tissues and improve N_2 fixation (Carvalho et al., 2014). The facultative endophyte *A. brasilense* uses "respiratory protection" by consuming more oxygen to shield nitrogenase from oxidative inactivation and demonstrates chemotaxis towards root exudates (Upadhyay et al., 2021). Endophytic colonization increases the expression of host plant genes linked to N_2 absorption, stress signaling, and root architecture, according to transcriptomic and metabolomic research (Nong et al., 2022). By modulating signaling networks including plant hormones (such as auxins and cytokinins), reactive oxygen species, and short RNAs, endophytes improve tolerance to abiotic conditions such drought, salt, and nutrient deprivation (Renganathan et al., 2018).

12.2.4 Symbiotic Diazotrophs

Mutualistic interactions between diazotrophic bacteria and host plants result in symbiotic N_2 fixation, which takes place under microaerobic conditions in root or stem nodules (Li & Xiao, 2025). In these symbioses, the plant supplies carbon molecules produced from photosynthetic processes and a protected niche for bacterial survival and nitrogenase activity in the root nodules (Cowan et al., 2022). The Rhizobiales order, which includes *Rhizobium, Bradyrhizobium, Ensifer, Mesorhizobium*, and *Azorhizobium*, exhibits symbiosis in leguminous plants (Fedorova et al., 2018). Root-exuded flavonoids, which function as chemoattractants and trigger nod gene expression in bacteria, are the first step in symbiotic communication. Nod-factor production leads to nodule organogenesis by promoting cortical cell proliferation, facilitating host recognition, and starting the formation of infection threads (Li & Xiao, 2025). Despite the fact that rhizobia mostly inhabit legumes, some nonleguminous species, including *Parasponia andersoni*, develop nodules with Rhizobium spp., signifying a transitional stage between actinorhizal and rhizobial symbioses (Mathesius, 2022). Actinorhizal symbiosis occurs in nonleguminous plants. In woody angiosperms including Casuarina, Alnus, Myrica, Elaeagnus, and Hippophae, filamentous actinobacteria of Frankia spp. create N_2-fixing nodules (Gupta et al., 2020). Vesicles in Frankia-induced nodules shield nitrogenase from oxygen and sporangia, which encourage the spread of microorganisms. In N_2-deficient situations, these symbioses support ecological restoration, land reclamation, and afforestation (Hu et al., 2023).

12.3 Diazotrophs and Plant Growth During Biological Nitrogen Fixation

Diazotrophic microorganisms have a variety of plant growth-promoting characteristics, such as phytohormone synthesis, nutrient solubilization, abiotic stress tolerance, phytopathogen inhibition, and improvement of root architecture and rhizosphere microbial dynamics, according to recent developments in plant–microbe interactions (Renganathan et al., 2024; Yusupova et al., 2025) (Table 12.1).

Abbreviations: *IAA* indole acetic acid, *GA*, gibberellin, *CKs* cytokinins, *ACC* 1-aminocyclopropane-1-carboxylate, *JA* jasmonic acid, *ISR* induced systematic resistance, *VOCs* volatile compounds, *ABA* abscisic acid, *EPS* exopolysaccharide, *GOGAT* glutamate synthase, *GS* glutamine synthetase, *CS* citrate synthase, *MDH* malate dehydrogenase, *PAL* phenylalanine ammonia lyase, *SOD* superoxide dismutase, *CAT* catalase, *GR* glutathione reductase, *PPO* polyphenol oxidase

12.3.1 Nutrient Solubilization and Acquisition

Micro and macronutrient compounds are essential for plant growth, but soil and organic matter do not contain the necessary amounts of one or more nutrients. As a result, the diazotrophs help their host absorb more limiting elements from the soil, including micronutrients such as copper, iron, and zinc and macronutrients such as nitrogen, phosphorus, magnesium, and potassium (Condori et al., 2024). Numerous studies have shown that diazotrophs can absorb phosphorus (P) from the soil and that P absorption by bacterial inoculation improves plant fitness (Vinci et al. 2018).

Organic acids such as gluconic, citric, and oxalic acids are produced by strains of *A. brasilense*, *Azotobacter chroococcum*, *Bacillus species*, *Pseudomonas fluorescens*, *Anabaena variabilis*, and *Tolypothrix*. These acids chelate mineral ions from insoluble forms and encourage plant root uptake (Chen et al., 2017; Hu & Chu, 2020; Pal et al., 2022). Under Fe-limiting circumstances, siderophore synthesis, especially by Azotobacter and Herbaspirillum species, improves Fe acquisition (Ramos et al., 2020; Aasfar et al., 2021). Recent research, like Singh et al. (2023), has shown that Pseudomonas sp. can promote plant development in several ways. When tested in sugarcane, the chosen strains (*P. koreensis* CY4 and *P. entomophila* CN11) were able to improve both the synthesis of phytohormones (abscisic acid, ABA, and cytokinin) and nitrogen absorption (N fixation).

Sugarcane responded with increased plant height, biomass (fresh and dry matter of shoots and roots), and chlorophyll content, as well as successful diazotrophic colonization. According to Rai et al. (2021), iron is regarded as a crucial micronutrient for plants. Diazotrophs are known to create siderophores, which are low-molecular-weight iron-chelating compounds that can bind ferric ions in the plant microbiome,

12.1 Biostimulant mechanisms and observed plant benefits of diazotrophic and photosynthetic microbial inoculants

Diazotrophs	Host plant species	Biostimulants mechanisms	Plant benefits	References
Bacterial biostimulants				
Rhizobium leguminosarum	lentil, faba bean, cowpea	Phytohormones production and nutrient solubilization	Improved biomass, and yield, antioxidant activity, and secondary metabolites (phenols, flavonoids, tannins)	Silva et al. (2017)
Bacillus amyloliquefaciens	Soybean,	IAA, GA, ABA	Increased plant height, dry matter, yield, stress tolerance, and photosynthesis	Kim et al. (2017)
Azotobacter chroococcum	Cotton	Nitrogen fixation, Solubilization of phosphorus, indole compounds production and hydrolytic enzymes	Enhanced biomass, photosynthesis, chlorophyll content, yield	Romero-Perdomo et al. (2017)
Herbaspirillum seropedicae	Palm	Endophytic colonization; stimulation of vacuolar H+ -ATPase and H+ -PPase activity; N2 fixation	Enhanced plant growth, nutrients (N, K, Ca, and Mg), photosynthesis	Lim et al. (2018)
Rhizobium leguminosarum	Pea, soybean, peanut, chickpea, lentil, faba bean, cowpea, lupin	IAA, CKs, P solubilization	Improved seedling growth, vigor, biomass, and yield, antioxidant activity, and secondary metabolites	Abdelkrim et al. (2018)
Bacillus amyloliquefaciens	Maize,	IAA, GA, ABA, antibiotics (surfactin, bacillomycin D), VOCs (e.g., 2,3-butanediol), ACC deaminase, antifungal activity	Increased plant height, dry matter, yield, Fusarium and Alternaria inhibition, stress tolerance, and photosynthesis	Vinci et al. (2018)

Diazotrophs	Host plant species	Biostimulants mechanisms	Plant benefits	References
Rhizobium leguminosarum	Common bean	Phytohormones production and nutrient solubilization	Improved seedling growth, vigor, biomass, and yield, antioxidant activity, and secondary metabolites	James et al. (2019)
Gluconacetobacter diazotrophicus	Rice	N_2 fixation, modulation of growth related gene expression	Increased plant height, fresh weight, leaf chlorophyll	Silva et al. (2019)
Rhizobium sp. and *Pseudomonas* sp.	Tomato and Lulo	Produced IAA, Solubilized FePO4 and AlPO4, Produced siderophores	Enhanced plant growth and yield	Zuluaga et al. (2020)
Burkholderia and *Stenotrophomonas* sp.	Mung bean	Cellulolytic, Nitrogen-Fixing, Phosphate-Solubilizing, IAA production	High functional activities, root and shoot lengths, and seedling vigor improvements	Tang et al. (2020)
Herbaspirillum seropedicae	Rice	Endophytic colonization; stimulation of vacuolar H+ -ATPase and H+ -PPase activity; N2 fixation	Enhanced plant growth, nutrients (N, K, Ca, and Mg), photosynthesis	Ramos et al. (2020)
Azotobacter chroococcum	Cucumber, Tomato, Cabbage	Phytohormones production and nutrient solubilization	Enhanced plant growth and yield	Aasfar et al. (2021)
Bacillus amyloliquefaciens	Tomato	Antibiotics (surfactin, bacillomycin D), VOCs (e.g., 2,3-butanediol), ACC deaminase, antifungal activity	Increased plant yield, Fusarium and Alternaria inhibition	Ji et al. (2021)
Pseudomonas fluorescens	Sedum alfredii, okra, tomato	IAA, ACC deaminase, P solubilization, ISR, VOC, antibiotic and siderophore production	Enhanced biomass, leaf chlorophyll, Cd uptake, essential oil biosynthesis, drought tolerance	Ji et al. (2021)
Azotobacter chroococcum	Sugar beet	IAA, GA, JA, ACC deaminase, P and K solubilization, siderophore	Increased biomass, yield, leaf nutrients, photosynthesis, stress and disease resistance	Kerecki et al. (2022)

Diazotrophs	Host plant species	Biostimulants mechanisms	Plant benefits	References
Pseudomonas fluorescens	Cucumber	P solubilization, ISR, VOC, antibiotic and siderophore production	Enhanced biomass, leaf chlorophyll, Cd uptake, drought tolerance	Al-Karablieh et al. (2022)
Gluconacetobacter diazotrophicus	Arabidopsis thaliana	Endophytic colonization, N_2 fixation, modulation of growth- and defense-related gene expression	Increased plant height, fresh weight, leaf chlorophyll, yield under both N-rich and N-deficient conditions, and drought tolerance	Soares et al. (2023)
Rhodotorula mucilaginosa and *Arthrobacter* sp.	Wheat	Nitrogen fixation, P solubilization, IAA production	Improved plant growth and enhanced nutrients (N and P) uptake under reduced nitrogen fertilizers	Aasfar et al. (2024)
Azospirillum brasilense	Maize	P solubilization, Phytohormone production, and Siderophore production	Improved root/shoot growth and biomass, Nutrient utilization efficiency, Photosynthesis	Condori et al. (2024)
Gluconacetobacter diazotrophicus	Tomato	N_2 fixation, modulation of growth related gene expression	Increased plant height, fresh weight, leaf chlorophyll	Pallucchini et al. (2024)
Enterobacter cloacae subsp. dissolvens, Enterobacter cloacae, and Enterobacter cloacae	-	Phosphate (P), potassium (K), and zinc (Zn) solubilization, as well as production of indole acetic acid (IAA) and ammonia (NH3)	In vitro study was performed, Pot and field trial are in progress	Mevada et al. (2025)
Cyanobacterial Biostimulants				
Nostoc muscorum	Barley	IAA, CKs, EPS, secondary metabolite secretion, antioxidant induction	Increased plant growth and yield, stress resilience, and biomass	Koval & Ogorodnikova (2021)

Diazotrophs	Host plant species	Biostimulants mechanisms	Plant benefits	References
Nostoc muscorum	Basil	IAA, CKs, EPS, secondary metabolite secretion, antioxidant induction	Increased plant growth and yield, stress resilience, and biomass	Santini et al. (2022)
Anabaena variabilis	Rice, soybean, wheat, okra, millet, mungbean	IAA, GA, N2 fixation, P solubilization, siderophore, EPS, hydrolytic enzyme (e.g., chitosanase) production	Enhanced growth, yield, seed germination, and root architecture, and reduced Fusarium infection and oxidative stress	Gonçalves, (2020); Kollmen and Strieth (2022)
Tolypothrix sp.	Basil, tomato, barley	N_2 fixation; phytohormone and organic acid secretion; defense enzyme and chitosanase production; antifungal activity	Improved seed germination, plant height, and soil quality, and reduced soil-borne diseases	Santini et al. (2022)
Nostoc muscorum	Garlic	IAA, CKs, EPS, secondary metabolite secretion, antioxidant induction	Fusarium oxysporum inhibition, increased leaf weight, essential oil yield, stress resilience, and biomass	Uniyal et al. (2024)
Microalgal Biostimulants				
Scenedesmus quadricauda	Lettuce, sugar beet	Activation of C and N metabolism enzymes (GOGAT, GS, CS, MDH), PAL activity, phytohormone production	Improved chlorophyll, carotenoid, proteins, biomass, nutrient uptake, root architecture	Puglisi et al. (2019)
Chlorella vulgaris	Arabidopsis thaliana	IAA, GA	Increased root/shoot biomass, glucosinolate biosynthesis, drought resistance	Moon et al. (2023)
Spirulina platensis	Rosemary	IAA, GA, CKs, antioxidant enzymes (SOD, CAT, GR, PPO), improved nutrient uptake	Enhanced growth parameters (height, root length, fresh/dry weight); increased essential oil content and photosynthetic pigment levels	El Lateef Gharib et al. (2023)

Diazotrophs	Host plant species	Biostimulants mechanisms	Plant benefits	References
Arthrospira platensis	Wheat	IAA, GA, CKs, antioxidant enzymes (SOD, CAT, GR, PPO), improved nutrient uptake	Enhanced growth parameters (height, root length, fresh/dry weight); increased essential oil content and photosynthetic pigment levels; improved stress tolerance under drought and salinity	El-Shazoly et al. (2025)
Mixed Biostimulants				
Azospirillum sp. and Rhizobium sp.	Common bean	Nutrient utilization, Nitrogen fixation and Phytohormones production	Improved root/shoot growth and biomass, nutrient uptake (N, P, K, Ca, Fe, Zn)	Filipini et al. (2021)
Chlorella and Chlamydopodium species	Cucumber, wheat, soybean	IAA, GA	Increased root/shoot biomass, antioxidant capacity	Morillas-España et al. (2022)
Trichoderma harzianum and Azospirillum brasilense	Lettuce	IAA, GA, CKs, Antioxidant enzymes	Improved biomass, Nutrient uptake (N, P, K, Ca, Fe, Zn), Abiotic stress tolerance	Moreira et al. (2022)
Erwinia sp., Chryseobacterium arthrosphaerae, and Pseudomonas gessardii	Barley crop	Nitrogen fixation, Solubilization of phosphorus and potassium	Enhanced root/shoot length and biomass, chlorophyll, carotenoids, phenolics, flavonoids and soluble sugar content	Kaur et al. (2022)
Spirulina platensis and Chlorella vulgaris	Tomato	IAA, GA, CKs, antioxidant enzymes (SOD, CAT, GR, PPO), improved nutrient uptake	Enhanced growth parameters (height, root length, fresh/dry weight); increased essential oil content and photosynthetic pigment levels; improved stress tolerance under drought and salinity [99–101]	Mostafa et al. (2024)
Chlorella vulgaris, Nannochloropsis salina, and Arthrospira platensis	Common bean	IAA, GA	Increased root/shoot biomass, flower number, antioxidant capacity,	El Lateef Gharib et al. (2024)

facilitate iron uptake, and function as biocontrol by inhibiting phytopathogens (Singh et al., 2022). Arstol & Hohmann-Marriott (2019) showed how cyanobacteria, such *A. variabilis* and *Nostoc muscorum*, can support rhizosphere health and nutrient uptake by generating exopolysaccharides (EPSs) and siderophores, like schizokinen, which chelate Fe^{3+}. In hydroponically cultivated crops, endophytic diazotrophs such Gluconacetobacter diazotrophicus improve P solubilization and K mobilization (Delaporte-Quintana et al., 2017). For instance, *H. seropedicae* increases Fe and Zn absorption in rice by promoting root exudation and altering phytohormone signaling pathways, according to Ramos et al. (2020).

12.3.2 Phytohormone Production and Modulation

Under both normal and stressful circumstances, phytohormones released by diazotrophs are essential for controlling plant physiological and metabolic pathways. According to multiple reports, diazotrophs are known to produce multiple phytohormones, such as auxins, cytokinins, and gibberellins (GAs), and to modulate a variety of physiological processes. They also trigger chemical signals that induce metabolic pathways and promote plant growth in various environmental conditions (Ganusova et al., 2025). Indole-3-acetic acid (IAA), which controls plant growth and promotes favorable effects on shoot and root development through cell division and elongation, vascular tissue differentiation, tropism response, and changes in root architecture and lateral root formation, is the most frequently reported auxin produced by diazotrophs (Cangioli et al., 2025). In a recent study, isolated nitrogen-fixing bacteria (Rhizobium T16, Sinorhizobium T21, Bacillus J1, Bacillus G4, and Arthrobacter J2). These bacteria were able to produce plant hormones (abscisic acid, salicylic acid, and spermidine) and metabolic substances (flavonoids, saponins, and other components) that stimulated the growth of Astragalus mongholicus, induced disease resistance. The synergistic effects of nonsymbiotic soil bacteria and nitrogen-fixing symbiotic rhizobia were utilized by Cangioli et al. (2025). After being isolated from alfalfa (*Medicago sativa* L.), the bacteria (*Sinorhizobium meliloti*) was tested for use in field studies. Alfalfa was treated with two microbial consortia (Mix1 and Mix2), which boosts defensive mechanisms, efficient root colonization, phytohormone synthesis, and plant development.

12.3.3 Mechanism of Alleviations of Abiotic Stress

A hyper-diverse genus of rhizospheric bacteria, diazotrophs have become well-known for their antagonistic ability to phytopathogens through various mechanisms of parasitism, stimulation of plant growth, competition for nutrients, and induction of systemic resistance against abiotic stresses (Mishra et al., 2021). According to new data, diazotrophs are crucial members of the plant microbiome. They grow slowly in the soil and form endophytic relationships within the roots or symbiotic relationships at their surface (Rai et al., 2022; Verma et al., 2024). The major abiotic stressors that cause physiological alterations, a change in metabolic pathways, and serious cellular damage in agricultural plants are salinity, drought, flood, heavy metals, heat, and cold. Therefore, in order to reduce abiotic stress, the use of bioinoculants is assessed in terms of plant growth promoters and biofertilizers (Sagar et al., 2022). Additionally, it enhances plant physiology, initiates the synthesis of metabolites and associated pathways, and promotes the release of elicitors that cause plant defense and systemic resistance (Sagar et al., 2024). Additionally, diazotrophs increase a plant's resistance to abiotic stress. *Bacillus subtilis* inoculation improves drought tolerance, stomatal conductance, and chlorophyll content in hydroponically grown lettuce (Oliveira et al., 2023). By controlling Na^+ transport, causing osmolyte buildup, and triggering antioxidant defense enzymes like superoxide dismutase and catalase, *Bacillus amyloliquefaciens* also improves salt tolerance (Chen et al., 2016). According to a number of studies, diazotrophs' capacity to manage plant stress improved crop yield and shielded plants from oxidative stress by producing enzymatic antioxidants like ascorbate peroxidase, catalase (CAT), and superoxide dismutase (SOD), while nonenzymatic antioxidants support homeostasis by producing carotenoids, glutathione, ascorbate, and α-tocopherol (Safaei et al., 2024). The useful function of diazotrophs, the method used to withstand abiotic stressors, and its phenotypic reaction to preserve plant fitness (Table 12.1).

12.3.4 Mechanism of Biotic Stress Alleviations

In instances of biotic stress, microbial inoculants increase plant functions and alter soil and plant microbiomes (Rai et al., 2019a,b; Rai et al., 2020; Rai & Solanki, 2023). According to Li et al. (2025), diazotrophs are biocontrol agents that naturally inhabit organisms with intrinsic pathogens that suppress quality through rapid growth and colonization in roots, antibiosis, microbicidal compound production, competition for space and nutrients, predation, induction of disease tolerance mechanisms, and promotion of plant growth. Since many agronomic crops are impacted by a wide variety of phytopathogens, the effects of applying biostimulants are assessed in terms of plant growth promoters, biofertilizers, and biocontrol agents of pests and pathogens (Samal et al., 2023). As a result, diazotrophs release hydrolytic enzymes, penetrate the host membrane, produce pathogen-related proteins (PR proteins) and secondary metabolites, use host cellular components as a source of nutrition, and eliminate phytopathogens (Singh et al., 2021). Diazotrophs exhibit antagonistic behavior against a variety of phytopathogens, either by secreting mycolytic enzymes like chitinases, β-1,3-glucanases, cellulase, and protease or by producing antimicrobial metabolites such as peptaibiotics, pyrones, siderophore, ergosterols, terpenoids, and toxins as a weapon to disrupt the cell integrity and break down the polysaccharide cell wall (Li et al., 2017; Guo et al., 2020).

Diazotrophs' potential for biocontrol has drawn a lot of interest lately. *Azotobacter vinelandii*, *Streptomyces albus*, and *Trichoderma harzianum* formed a microbial consortium that reduced *Fusarium oxysporum* in tomatoes and boosted fruit output (Kawicha et al., 2023). Endophytic Streptomyces species improve nitrogen cycling while opposing foliar diseases and parasitic nematodes (Worsley et al., 2020). Recent research has shown that diazotrophs can promote a variety of plant development (Mir et al., 2022). When evaluated in rice (*Oryza sativa* L.), the chosen strains were able to increase phytohormone (IAA) production and nutrient absorption (P and Zinc solubilization capability). Isolated diazotrophs that considerably generate HCN, ACC-diaminase, antifungal secondary metabolites, and cell wall-degrading enzymes against *Rhizoctonia solani* and *Fusarium oxysporum* are listed in this order.

Rice plants responded with increased plant height, biomass (fresh and dry matter of shoots and roots), and chlorophyll content, as well as successful microbial colonization. The cyanobacterium Nostoc sp. shown antifungal efficacy against oomycetes and phytopathogenic fungi in a suspension (Yusupova et al., 2025). The microbial consortia that were created had the ability to mobilize P and break down polysaccharides, which helped to restore fertility in deteriorated soil and hydroponic wastewater. In post-mining environments, genetically engineered consortia enhance plant establishment and nutrient availability (Li et al., 2025).

12.4 Major Constrains and Future Prospectus

1. Microbial diazotrophs are distinct since they are linked to the plant and occasionally particular to its location and tissue. However, developing formulation and technology that targets certain host stages and colonization is challenging.
2. The existence and abundance of uncultured endophytic diazotrophs were revealed by advanced molecular techniques, and it is extremely difficult to cultivate these microorganisms in the lab and employ them for large-scale applications.
3. The development of uncultured microbiome pools is necessary for sustainable agriculture, and media optimization of new nitrogen-fixing bacteria is a difficult challenge.
4. Many diazotrophs generate metabolites linked to a specific host, as well as methods for producing these metabolites in the absence of a host or through laboratory tissue culture for commercial use. It is necessary to determine how genome editing techniques can aid in the development of such technologies.
5. What are the roles of microbial groups and how do they proliferate in plant tissues? Compared to the rhizospheric and epiphytic microbial groups, studying the microbe roles of endophytes diazotrophs is more difficult since they typically collaborate with a variety of individuals. Only a few members of each microbial group can grow readily in a lab. In order to investigate microbial recruitment and function biology in a living system, an integrated approach utilizing molecular and microscopic methods is required.
6. It is necessary to investigate the genes linked to microbial bioactive substances and their relationship with the host. To comprehend the function of genetic material exchange throughout evolution and how these genes integrate with bacterial and fungal chromosomes, genome-wide study is required.

7. To map the position of abiotic factors in plant and microbial diversity and function, plant cytological and genome-wide data on microbial recruitment must be investigated.
8. To determine the link and benefits, the multitrophic interplay of microbial abundance, host plant development, and external elements (light, water, air, temperature, pH, nutrients, minerals, insects, microbial diseases, chemicals, etc.) must be studied concurrently.
9. Understanding the significance of microbes in sustainable farming may be aided by the application of microfluidics techniques in plant microbial interaction and function.

References

Aasfar, A., Bargaz, A., Yaakoubi, K., Hilali, A., Bennis, I., Zeroual, Y., & Meftah Kadmiri, I. (2021). Nitrogen fixing azotobacter species as potential soil biological enhancers for crop nutrition and yield stability. *Frontiers in Microbiology*, *12*, 628379.

Aasfar, A., Meftah Kadmiri, I., Azaroual, S. E., Lemriss, S., Mernissi, N. E., Bargaz, A., Zeroual, Y., & Hilali, A. (2024). Agronomic advantage of bacterial biological nitrogen fixation on wheat plant growth under contrasting nitrogen and phosphorus regimes. *Frontiers in Plant Science*, *15*, 1388775.

Abdelkrim, S., Jebara, S. H., Saadani, O., & Jebara, M. (2018). Potential of efficient and resistant plant growth-promoting rhizobacteria in lead uptake and plant defence stimulation in *Lathyrus sativus* under lead stress. *Plant Biology*, *20*, 857–869.

Abdelwahed, S., Trabelsi, E., Saadouli, I., Kouidhi, S., Masmoudi, A. S., Cherif, A., Mnif, W., & Mosbah, A. (2022). A new pioneer colorimetric micro-plate method for the estimation of ammonia production by plant growth promoting rhizobacteria (PGPR). *Main Group Chemistry*, *21*(1), 55–68.

Al-Karablieh, N., Al-Shomali, I., Al-Elaumi, L., & Hasan, K. (2022). *Pseudomonas fluorescens* NK4 siderophore promotes plant growth and biocontrol in cucumber. *Journal of Applied Microbiology*, *133*, 1414–1421.

Andrade-Bustamante, G., Martínez-Ruiz, F. E., Ortega-García, J., Renganathan, P., Gaysina, L. A., Mahendhiran, M., & Puente, E. O. (2025). Microalgae-based functional foods: A blue-green revolution in sustainable nutrition and health. *Applied Microbiology*, *5*, 39.

Arkani, M., Babaeipour, V., Mohammadi, R., & Deldar, A. (2025). Expression of Vitreoscilla hemoglobin gene in different organisms: a new approach to increase the efficiency of various biological processes. *Preparative Biochemistry & Biotechnology*, 55(8), 955–972.

Arstol, E., & Hohmann-Marriott, M. F. (2019). Cyanobacterial siderophores—physiology, structure, biosynthesis, and applications. *Marine Drugs*, *17*, 281.

Babcock-Jackson, L., Konovalova, T., Krogman, J. P., Bird, R., & Díaz, L. L. (2023). Sustainable fertilizers: Publication landscape on wastes as nutrient sources, wastewater treatment processes for nutrient recovery, biorefineries, and green ammonia synthesis. *Journal of Agricultural and Food Chemistry*, *71*, 8265.

Biological Products Industry Alliance—BPIA. (2023) The listening session report: identifying ambiguities, gaps, inefficiencies, and uncertainties in the coordinated framework for the regulation of biotechnology (p. 27). Biological Products Industry Alliance.

Carvalho, T. L., Balsemão-Pires, E., Saraiva, R. M., Ferreira, P. C., & Hemerly, A. S. (2014). Nitrogen signalling in plant interactions with associative and endophytic diazotrophic bacteria. *Journal of Experimental Botany*, *65*, 5631–5642.

Chen, B., Luo, S., Wu, Y., Ye, J., Wang, Q., Xu, X., Pan, F., Khan, K. Y., Feng, Y., & Yang, X. (2017). The effects of the endophytic bacterium *Pseudomonas fluorescens* Sasm05 and IAA on the plant growth and cadmium uptake of sedum alfredii Hance. *Frontiers in Microbiology, 8*, 269386.

Chen, L., Liu, Y., Wu, G., Veronican Njeri, K., Shen, Q., Zhang, N., & Zhang, R. (2016). Induced maize salt tolerance by rhizosphere inoculation of *Bacillus amyloliquefaciens* SQR9. *Physiologia Plantarum, 158*, 34–44.

Condori, T., Alarcón, S., Huasasquiche, L., García-Blásquez, C., Padilla-Castro, C., Velásquez, J., & Solórzano, R. (2024). Inoculation with *Azospirillum brasilense* as a strategy to reduce nitrogen fertilization in cultivating purple maize (*Zea mays* L.) in the inter-andean valleys of peru. *Microorganisms, 12*, 2107.

Cowan, N., White, S., Olszewska, J., Dobel, A., Sim, G., Eades, L. J., & Skiba, U. (2022). Integration of algae treatment with hydroponic crop waste to reduce impact of nutrient waste streams. *Journal of Sustainable Agriculture and Environment, 1*, 203–215.

Delaporte-Quintana, P., Grillo-Puertas, M., Lovaisa, N. C., Teixeira, K. R., Rapisarda, V. A., & Pedraza, R. O. (2017). Contribution of *Gluconacetobacter diazotrophicus* to phosphorus nutrition in strawberry plants. *Plant and Soil, 419*, 335–347.

El Lateef Gharib, F. A., & Ahmed, E. Z. (2023). *Spirulina platensis* improves growth, oil content, and antioxidant activity of rosemary plant under cadmium and lead stress. *Scientific Reports, 13*, 8008.

El Lateef Gharib, F. A., Abd El Sattar, A. M., & Ahmed, E. Z. (2024). Impact of chlorella vulgaris, nannochloropsis salina, and arthrospira platensis as bio-stimulants on common bean plant growth, yield and antioxidant capacity. *Scientific Reports, 14*, 1398.

Eleazar, F., Jaime, R., Renganathan, P., Gaysina, L. A., Sukhanova, N. V., & Puente, E. O. (2025). Microalgae as functional food ingredients: Nutritional benefits, challenges, and regulatory considerations for safe consumption. *Biomass, 5*, 25.

El-Shazoly, R. M., Aloufi, A. S., & Fawzy, M. A. (2025). The potential use of arthrospira (*Spirulina platensis*) as a biostimulant for drought tolerance in wheat (*Triticum aestivum* L.) for sustainable agriculture. *Journal of Plant Growth Regulation, 44*, 686–703.

Fedorova, E., Pueyo, J. J., & Lucas, M. M. (2018). The symbiosome: Legume and rhizobia co-evolution toward a nitrogen-fixing organelle?. *Frontiers in Plant Science, 8*, 305846.

Filipini, L. D., Pilatti, F. K., Meyer, E., Ventura, B. S., Lourenzi, C. R., & Lovato, P. E. (2021). Application of Azospirillum on seeds and leaves, associated with Rhizobium inoculation, increases growth and yield of common bean. *Archives of Microbiology, 203*, 1033–1038.

Ganusova, E. E., Banerjee, I., Seats, T., & Alexandre, G. (2025). Indole-3-acetic acid (IAA) protects *Azospirillum brasilense* from indoleinduced stress. *Applied and Environmental Microbiology, 91*, e0238424.

Gonçalves, A. L. (2020). The use of microalgae and cyanobacteria in the improvement of agricultural practices: A review on their biofertilising, biostimulating and biopesticide roles. *Applied Sciences, 11*, 871.

Guo, D. J., Singh, R. K., Singh, P., Li, D. P., Sharma, A., Xing, Y. X., Song, X. P., Yang, L. T., & Li, Y. R. (2020). Complete genome sequence of Enterobacter roggenkampii ED5, a nitrogen fixing plant growth promoting endophytic bacterium with biocontrol and stress tolerance properties, isolated from sugarcane root. *Frontiers in Microbiology*, 11, 580081.

Gupta, S.M., Kumar, K., Joshi, R.K., Gupta, S., & Bala, M. (2020) Frankia: A promising N-fixing plant growth promoting rhizobacteria (PGPR) improved drought tolerance in crops at higher altitude. In *Microbiological advancements for higher altitude agro-ecosystems & sustainability* (pp. 411–431). Springer.

Hu, B., & Chu, C. (2020). Nitrogen–phosphorus interplay: Old story with molecular tale. *The New Phytologist, 225*, 1455–1460.

Hu, B., Flemetakis, E., Liu, Z., Hänsch, R., & Rennenberg, H. (2023). Significance of nitrogen-fixing actinorhizal symbioses for restoration of depleted, degraded, and contaminated soil. *Trends in Plant Science, 28*, 752–764.

Imran, A., Hakim, S., Tariq, M., Nawaz, M. S., Laraib, I., Gulzar, U., Hanif, M. K., Siddique, M. J., Hayat, M., Fraz, A., & Ahmad, M. (2021). Diazotrophs for lowering nitrogen pollution crises: Looking deep into the roots. *Frontiers in Microbiology, 12*, 637815.

James, E. K. (2019). Formulation of a highly effective inoculant for common bean based on an autochthonous elite strain of *Rhizobium leguminosarum* bv. phaseoli, and genomic-based insights into its agronomic performance. *Frontiers in Microbiology, 10*, 490875.

Ji, C., Zhang, M., Kong, Z., Chen, X., Wang, X., Ding, W., Lai, H., & Guo, Q. (2021). Genomic analysis reveals potential mechanisms underlying promotion of tomato plant growth and antagonism of soilborne pathogens by *Bacillus amyloliquefaciens* Ba13. *Microbiology Spectrum, 9*, e0161521.

Kaur, T., Devi, R., Kumar, S., Sheikh, I., Kour, D., & Yadav, A. N. (2022). Microbial consortium with nitrogen fixing and mineral solubilizing attributes for growth of barley (*Hordeum vulgare* L). *Heliyon, 8*(4), e09326.

Kawicha, P., Nitayaros, J., Saman, P., Thaporn, S., Thanyasiriwat, T., Somtrakoon, K., Sangdee, K., & Sangdee, A. (2023). Evaluation of soil Streptomyces spp. for the biological control of Fusarium wilt disease and growth promotion in tomato and banana. *The Plant Pathology Journal, 39*, 108–122.

Kerecki, S., Pe´cinar, I., Karliˇci´c, V., Mirkovi´c, N., Kljujev, I., Raiˇcevi´c, V., & Joviˇci´c-Petrovi´c, J. (2022). *Azotobacter chroococcum* F8/2: A multitasking bacterial strain in sugar beet biopriming. *Journal of Plant Interactions, 17*, 719–730.

Kim, M. J., Radhakrishnan, R., Kang, S. M., You, Y. H., Jeong, E. J., Kim, J. G., & Lee, I. J. (2017). Plant growth promoting effect of *Bacillus amyloliquefaciens* H-2-5 on crop plants and influence on physiological changes in soybean under soil salinity. *Physiology and Molecular Biology of Plants, 23*, 571–580.

Kollmen, J., & Strieth, D. (2022). The beneficial effects of cyanobacterial co-culture on plant growth. *Life, 12*, 223.

Koval, E. V., & Ogorodnikova, S. Y. (2021). The prospect of using the cyanobacterium Nostoc muscorum to improve vital activity of barley seedlings by various methods of seed treatment. In In BIO Web of Conferences. (p. 04005). Les Ulis, France: EDP Sciences. Vol 36.

Li, C., Sun, L., Jia, Z., Tang, Y., Liu, X., Zhang, J., & Müller, C. (2025). Microbial inoculants drive changes in soil and plant microbiomes and improve plant functions in abandoned mine restoration. *Plant, Cell and Environment, 48*, 1162–1178.

Li, H. B., Singh, R. K., Singh, P., Song, Q. Q., Xing, Y. X., & Yang, L. T. (2017). Genetic diversity of nitrogen-fixing and plant growth promoting Pseudomonas species isolated from sugarcane rhizosphere. *Frontiers in Microbiology, 8*, 1268.

Li, X., & Xiao, R. (2025). Molecular dialogue in legume-rhizobium symbiosis: Signaling mechanisms and genetic insights. *Rhizosphere, 33*, 101034.

Lim, S. L., Subramaniam, S., Zamzuri, I., & Amir, H. G. (2018). Growth and biochemical profiling of artificially associated micropropagated oil palm plantlets with *Herbaspirillum seropedicae*. *Journal of Plant Interactions, 13*, 173–181.

Mathesius, U. (2022). Are legumes different? Origins and consequences of evolving nitrogen fixing symbioses. *Journal of Plant Physiology, 276*, 153765.

Mevada, P., Bhagat, A., Dave, G., Patel, J., Patel, Y., Gelot, A., & Bhadauria, H. S. (2025). Exploration of plant growth-promoting traits of diazotroph isolates obtained from rhizospheric soil. *Advanced Research, 26*(4), 344–357.

Mir, M. I., Hameeda, B., Quadriya, H., Kumar, B. K., Ilyas, N., Kee Zuan, A. T., El Enshasy, H. A., Dailin, D. J., Kassem, H. S., Gafur, A., & Sayyed, R. Z. (2022). Multifarious indigenous diazotrophic rhizobacteria of rice (*Oryza sativa* L.) Rhizosphere and their effect on plant growth promotion. *Frontiers in Nutrition, 8*, 781764.

Mishra, A., Rajput, S., Gupta, P. S., Goyal, V., Singh, S., Sharma, S., Shukla, S., Singh, A., Shukla, K., & Varma, A. (2021). Role of cyanobacteria in Rhizospheric nitrogen fixation. In In Soil nitrogen ecology. (pp. 497–519). Springer.

Moon, J., Park, Y. J., Choi, Y. B., Truong, T. Q., Huynh, P. K., Kim, Y. B., & Kim, S. M. (2023). Physiological effects and mechanisms of *Chlorella vulgaris* as a biostimulant on the growth and drought tolerance of *Arabidopsis thaliana*. *Plants*, *13*, 3012.

Moreira, V. D., Oliveira, C. E., Jalal, A., Gato, I. M., Oliveira, T. J., Boleta, G. H., Giolo, V. M., Vitória, L. S., Tamburi, K. V., & Filho, M. C. (2022). Inoculation with *Trichoderma harzianum* and *Azospirillum brasilense* increases nutrition and yield of hydroponic lettuce. *Archives of Microbiology*, *204*, 440.

Morillas-España, A., Ruiz-Nieto, Á, Lafarga, T., Acién, G., Arbib, Z., & González-López, C. V. (2022). Biostimulant capacity of Chlorella and Chlamydopodium species produced using wastewater and centrate. *Biology*, *11*, 1086.

Mostafa, M. M., Hammad, D. M., Reda, M. M., & El-Sayed, A. E. (2024). Water extracts of *Spirulina platensis* and *Chlorella vulgaris* enhance tomato *(Solanum lycopersicum* L.) tolerance against saline water irrigation. *Biomass Conversion and Biorefinery*, *14*, 21181–21191.

Nong, Q., Malviya, M. K., Solanki, M. K., Solanki, A. C., Lin, L., Xie, J., Mo, Z., Wang, Z., Song, X., Huang, X., Rai, S., Li, C., & Li, Y.-R. (2022). Sugarcane root transcriptome analysis revealed the role of plant hormones in the colonization of an endophytic diazotroph. *Frontiers in Microbiology*, 13, 924283.

Oliveira, C. E., Jalal, A., Aguilar, J. V., De Camargos, L. S., Zoz, T., Ghaley, B. B., Alarjani, K. M., AbdElgawad, H., & Teixeira Filho, M. C. (2023). Yield, nutrition, and leaf gas exchange of lettuce plants in a hydroponic system in response to *Bacillus subtilis* inoculation. *Frontiers in Plant Science*, *14*, 1248044.

Pal, G., Saxena, S., Kumar, K., Verma, A., Sahu, P. K., Pandey, A., White, J. F., & Verma, S. K. (2022). Endophytic burkholderia: Multifunctional roles in plant growth promotion and stress tolerance. *Microbiological Research*, *265*, 127201.

Pallucchini, M., Franchini, M., Narraidoo, N., Palframan, M. J., Hayes, C. J., Dent, D., Cocking, E. C., Perazzolli, M., Fray, R. G., & Hill, P. J. (2024). *Gluconacetobacter diazotrophicus* AZ0019 requires functional nifD gene for optimal plant growth promotion in tomato plants. *Frontiers in Plant Science*, *15*, 1469676.

Pittol, M., Durso, L., Valiati, V. H., &, Fiuza L. M. (2016). Agronomic and environmental aspects of diazotrophic bacteria in rice fields. *Annals of Microbiology*, 66, 511–527.

Puglisi, I., La Bella, E., Rovetto, E. I., Lo Piero, A. R., & Baglieri, A. (2019). Biostimulant effect and biochemical response in lettuce seedlings treated with a *Scenedesmus quadricauda* extract. *Plants*, *9*, 123.

Rai, S., & Solanki, M. K. (2023). Beneficial endophytic Trichoderma functions in plant health management. In Solanki, M. K., Yadav, M. K., Singh, B. P., Vijai Kumar Gupta, V. J. (Eds.)Eds. *Microbial endophytes and plant growth* . (pp. 233–244). Academic Press.

Rai, S., Singh, P. K., Mankotia, S., Swain, J., & Satbhai, S. B. (2021). Iron homeostasis in plants and its crosstalk with copper, zinc, and manganese. *Plant Stress*, *1*, 100008.

Rai, S., Omar, A. F., Rehan, M., Al-Turki, A., Sagar, A., Ilyas, N., Sayyed, R. Z., & Mirza, H. (2022). Crop microbiome: Their role and advances in molecular and omic techniques for the sustenance of agriculture. *Planta*, *257*, 27.

Rai, S., Ramteke, P.W., Sagar, A., Dhusia, K., & Kesari, S.K. (2019b). Diversity assessment of antagonistic Trichoderma species by comparative analysis of microsatellites. In R. Z. Sayyed, M. S. Reddy, & S. Antonius (Eds.), *Plant growth promoting rhizobacteria (PGPR)* (pp. 233–254). Prospects for Sustainable Agriculture.

Rai, S., Solanki, M. K., Anal, A. K. D. (2020). Modern biotechnological tools: An opportunity to discover complex phytobiomes of horticulture crops. In M. K. Solanki, P. L. Kashyap, R. A. Ansari, & B. Kumari (Eds.), *Microbiomes and plant health: Panoply and their applications* (pp. 85–124). Elsevier.

Rai, S., Solanki, M. K., Solanki, A. C., Samal, S. (2023). Microbial endophytes as probiotics for the plant health: An overview. In M.K. Solanki, M.K. Yadav, B.P. Singh, & V. K. Gupta. Academic Press, pp 269–281.

Rai, S., Solanki, M. K., Solanki, A. C., Surapathrudu, K. (2019a). Biocontrol potential of Trichoderma spp.: Current understandings and future outlooks on molecular techniques. In R. A. Ansari, & I. Mahmood, (Eds.), *Plant health under biotic stress* (pp. 129–160). Springer.

Ramos, A. C., Melo, J., de Souza, S. B., Bertolazi, A. A., Silva, R. A., Rodrigues, W. P., Campostrini, E., Olivares, F. L., Eutrópio, F. J., Cruz, C., & Dias, T. (2020). Inoculation with the endophytic bacterium *Herbaspirillum* seropedicae promotes growth, nutrient uptake and photosynthetic efficiency in rice. *Planta*, 252, 87.

Renganathan, P., Borboa-Flores, J., Rosas-Burgos, E. C., Cárdenas-López, J. L., Murillo-Amador, B., Ortega-García, J., & Rueda-Puente, E. O. (2018). Inoculation of nitrogen-fixing halobacteria in the contribution to tolerance to salt stress in bean tepary. *Revista Mexicana de Ciencias Agricolas*, *9*, 4289–4300.

Renganathan, P., Gaysina, L. A., Jaime, R., Carlos, J., & Omar, E. (2024). Phycoremediated microalgae and cyanobacteria biomass as biofertilizer for sustainable agriculture: A holistic biorefinery approach to promote circular bioeconomy. *Biomass*, 4, 1047–1077.

Romero-Perdomo, F., Abril, J., Camelo, M., Moreno-Galván, A., Pastrana, I., Rojas-Tapias, D., Bonilla, R. (2017). Azotobacter chroococcum as a potentially useful bacterial biofertilizer for cotton (Gossypium hirsutum): Effect in reducing N fertilization. Rev Argent Microbiol, 49(4):377-383.

Safaei, F., Alirezalu, A., Noruzi, P., & Alirezalu, K. (2024). Phytochemical and morpho-physiological response of melissa officinalis L. to different NH4 + to NO3 − ratios under hydroponic cultivation. *BMC Plant Biology*, *24*, 968.

Sagar, A., Rai, S., Sharma, S., Perveen, K., Bukhari, N. A., Sayyed, R. Z., & Mastinu, A. (2024). Molecular characterization reveals biodiversity and biopotential of rhizobacterial isolates of Bacillus Spp. *Microbial Ecology*, 87, 83.

Samal, S., Rai, S., & Upadhaya, R. S. (2023). Endophytic fusarium and their association with plant growth. In Solanki, M. K., Yadav, M. K., Singh, B. P., Vijai Kumar Gupta, V. J. (Eds.)Eds. *Microbial endophytes and plant growth* . (pp. pp. 259–268). Academic Press.

Santini, G., Rodolfi, L., Biondi, N., Sampietro, G., & Tredici, M. R. (2022). Effects of cyanobacterial-based biostimulants on plant growth and development: A case study on basil (*Ocimum basilicum* L). *Journal of Applied Phycology*, *34*, 2063–2073.

Santos, F., Melkani, S., Oliveira-Paiva, C., Bini, D., Pavuluri, K., Gatiboni, L., Mahmud, A., Torres, M., McLamore, E., & Bhadha, J. H. (2024). Biofertilizer use in the United States: Definition, regulation, and prospects. *Applied Microbiology and Biotechnology*, *108*, 511.

Saranraj, P., Al-Tawaha, A. R., Sivasakthivelan, P., Al-Tawaha, A. R., Amala, K., Thangadurai, D. , & Sangeetha, J. (2022) *Azospirillum bioinoculant* technology: Past to current knowledge and future prospects. In *Organic farming for sustainable development* (pp. 51–76). Apple Academic Press.

Silva, L. R., Bento, C., Gonçalves, A. C., Flores-Félix, J. D., Ramírez-Bahena, M. H., Peix, A., & Velázquez, E. (2017). Legume bioactive compounds: Influence of rhizobial inoculation. *AIMS Microbiology*, *3*, 267.

Silva, R., Filgueiras, L., Santos, B., Coelho, M., Silva, M., Vidal, M., Baldani, J. I., & Meneses, C. (2019). Gluconacetobacter diazotrophicus changes the molecular mechanisms of root development in *Oryza sativa* L. growing under water stress. *International Journal of Molecular Sciences*, *21*, 333.

Singh, P., Singh, R. K., Li, H. B., Guo, D. J., Sharma, A., Verma, K. K., & Li, Y. R. (2023). Nitrogen fixation and phytohormone stimulation of sugarcane plant through plant growth promoting diazotrophic Pseudomonas. *Biotechnology and Genetic Engineering Reviews*, *40*(1), 15–35.

Singh, P., Chauhan, P. K., Upadhyay, S. K., Singh, R. K., Dwivedi, P., Wang, J., Jain, D., & Jiang, M. (2022). Mechanistic insights and potential use of siderophores producing microbes in rhizosphere for mitigation of stress in plants grown in degraded land. *Frontiers in Microbiology*, *13*, 898979.

Singh, R. K., Singh, P., Guo, D.-J., Sharma, A., Li, D.-P., Li, X., Verma, K. K., Malviya , M. K., Song, X.-P., Lakshmanan, P., Yang, L.-T. , & Li, Y.-R. (2021) Root-derived endophytic diazotrophic bacteria pantoea cypripedii AF1 and kosakonia arachidis EF1 promote nitrogen assimilation and growth in sugarcane. *Frontiers in Microbiology*, 12, 774707.

Soares, F. S., Soares, A. L., De Souza, S. A., Pinto, V. B., Matiello, L., Da Silva, F. R., Menossi, M., & Apolinário, G. (2023). Fine-tuning of *Arabidopsis thaliana* response to endophytic colonization by *Gluconacetobacter diazotrophicus* PAL5 revealed by transcriptomic analysis. *Plants*, *13*, 1719.

Srikanth, P., Sivakumar, D., Sharma, A., & Kaushik, N. (2025). Recent developments in omics techniques for improving plant abiotic stress using microbes.. *International Journal of Environmental Science and Technology*, *22*, 3787–3810.

Tang, A., Haruna, A. O., Majid, N. M. A., & Jalloh, M. B. (2020). Potential PGPR properties of cellulolytic, nitrogen-fixing, phosphate-solubilizing bacteria in rehabilitated tropical forest soil. *Microorganisms*, *8*(3), 442.

Threatt, S. D., & Rees, D. C. (2023). Biological nitrogen fixation in theory, practice, and reality: A perspective on the molybdenum nitrogenase system. *FEBS Letters*, *597*, 45–58.

Uniyal, S., Singh, P., Singh, R. K., & Tiwari, S. P. (2024). Effects of Nostoc sp. inoculation on the yield and quality of a medicinal plant, *Allium sativum. The Journal of Applied Phycology*, *36*, 3287–3300.

Upadhyay, D., Shukla, K., Mishra, A., Jindal, T., Sharma, S., & Shukla, S. (2021) Molecular aspects and oxygen relations of nitrogen fixation in cyanobacteria. In *Soil nitrogen ecology* (pp. 521–568). Springer.

Verma, P., Bhattacharya, A., Bharti, C., & Arora, N. K. (2024). Diazotrophs for enhancement of productivity of major cereal crops under abiotic stresses. In: N. K. Arora, & B. Bouizgarne (Eds.), *Microbial biotechnology for sustainable agriculture volume 2* (vol. 34). *Microorganisms for Sustainability*. Springer.

Vinci, G., Cozzolino, V., Mazzei, P., Monda, H., Savy, D., Drosos, M., & Piccolo, A. (2018). Effects of *Bacillus amyloliquefaciens* and different phosphorus sources on Maize plants as revealed by NMR and GC-MS based metabolomics. *Plant and Soil*, *429*, 437–450.

Worsley, S. F., Newitt, J., Rassbach, J., Batey, S. F., Holmes, N. A., Murrell, J. C., Wilkinson, B., & Hutchings, M. I. (2020). Streptomyces endophytes promote host health and enhance growth across plant species. *Applied and Environmental Microbiology*, *86*, e01053–20.

Yusupova, A., Kartabayeva, B., Sushchenko, R., Gaysina, K., Renganathan, P., & Gaysina, L. A. (2025). Antifungal potential of cyanobacterium nostoc sp. BCAC 1226 suspension as a biocontrol agent against phytopathogenic fungi and oomycetes. *Applied Microbiology*, *6*, 46.

Zayed, O., Hewedy, O. A., Abdelmoteleb, A., Ali, M., Youssef, M. S., Roumia, A. F., Seymour, D., & Yuan, Z.-C. (2023). Nitrogen journey in plants: From uptake to metabolism, stress response, and microbe interaction. *Biomolecules*, *13*(10), 1443.

Zuluaga, M. Y. A., Lima Milani, K. M., Azeredo Gonc¸alves, L. S., & Martinez de Oliveira, A. L. (2020). Diversity and plant growth-promoting functions of diazotrophic/N-scavenging bacteria isolated from the soils and rhizospheres of two species of Solanum. *PLoS One*, *15*(1), e0227422.

Diazotrophic Plant Growth Promoting Bacteria: A Reliable Tool for Sustainable Agriculture

13

Poonam Kumari and Prafull Kumar

Abstract

Growing global pressures on food production have intensified efforts to identify sustainable strategies for improving nitrogen availability in agricultural systems. While synthetic nitrogen fertilizers remain widely used due to their effectiveness in boosting crop yields, their prolonged application is associated with soil nutrient imbalances, environmental contamination, and declining long-term sustainability. Biological nitrogen fixation (BNF), performed by diazotrophic microorganisms, provides a viable and environmentally responsible alternative by converting atmospheric nitrogen into forms that plants can readily assimilate. A substantial proportion of diazotrophs also express diverse plant growth-promoting traits, classifying them as plant growth-promoting diazotrophs (PGPDs). These microorganisms contribute to plant development and productivity through multiple mechanisms, including phosphate solubilization, siderophore production, phytohormone biosynthesis, and ACC deaminase activity. They also suppress phytopathogens by generating antimicrobial compounds and eliciting induced systemic resistance. Beyond these functions, PGPDs enhance plant tolerance to abiotic stresses such as drought, salinity, and nutrient limitation, thereby supporting crop performance under suboptimal environmental conditions. This chapter provides an in-depth examination of the taxonomic diversity, biochemical capabilities, and ecological relevance of diazotrophic plant growth-promoting bacteria. It discusses their modes of interaction with host

P. Kumari (✉)
SVB College of Agriculture and Research Station, Indira Gandhi Agricultural University, Marra-Patan, Durg, Raipur, Chhattisgarh, India
e-mail: poonam15sep@gmail.com

P. Kumar
Department of Genetics and Plant Breeding, College of Agriculture, Indira Gandhi Agricultural University, Raipur, Chhattisgarh, India

R. K. Singh and P. Singh (eds.), *Diazotrophic Plant Growth-Promoting Microorganisms and Green Energy – Volume I*, https://doi.org/10.1007/978-981-95-9682-9_13

plants, their potential to decrease dependence on chemically synthesized fertilizers, and their contributions to sustainable intensification of agricultural systems. Recent advances in elucidating the molecular, physiological, and ecological dimensions of PGPD activity are highlighted, along with the principal challenges and emerging opportunities associated with their field application. By integrating current research findings, this chapter underscores the central role of diazotrophic microorganisms in driving the transition toward environmentally sustainable and resilient agriculture.

Keywords

Diazotrophs · Sustainability · Plant growth promotion · Phytohormones · Antagonism

13.1 Introduction

Nitrogen is an essential macronutrient required for normal plant growth and development, as it is involved in numerous biochemical and physiological pathways. It constitutes key cellular molecules such as ATP, NADH, NADPH, chlorophyll, amino acids, proteins, a wide range of enzymatic systems, and the nucleic acids DNA and RNA (Zayed et al., 2023). Recognition of nitrogen's central role in plant nutrition dates back to 1836, when the French chemist Jean-Baptiste Boussingault established that the effectiveness of fertilizers was closely associated with their nitrogen content. A major technological shift occurred in 1913 with Fritz Haber's development of a process for the industrial reduction of atmospheric dinitrogen (N_2) into bioavailable forms. The Haber–Bosch technology revolutionized global food production and became a cornerstone of the mid-twentieth-century Green Revolution, ultimately enabling the sustenance of a substantial portion of the world's population. By the 1970s, ammonia synthesis via the Haber–Bosch process exceeded natural terrestrial biological nitrogen fixation, marking a transition in which human activities became the dominant contributor of reactive nitrogen to the biosphere (Galloway, 2013).

Modern agricultural systems have, therefore, become increasingly dependent on synthetic nitrogen fertilizers to maintain high yields. However, fertilizer applications have grown disproportionately relative to crop productivity, revealing a persistent decline in nitrogen use efficiency (Zayed et al., 2023). Only a limited share of applied nitrogen is assimilated by plants, while considerable quantities are lost through leaching, volatilization, and gaseous emissions. Nitrate leaching contaminates freshwater and coastal ecosystems (Melillo, 2021), whereas emissions of nitrous oxide (N_2O) contribute simultaneously to stratospheric ozone depletion (Ravishankara, 2009) and atmospheric greenhouse gas accumulation (Gao & Serrenho, 2023). Additionally, the production of synthetic nitrogen fertilizers is highly energy-intensive and reliant on non-renewable fossil resources (Mandon et al., 2021). Excessive nitrogen fertilization, along with other agrochemical inputs, further undermines soil quality and threatens long-term agricultural sustainability. Given these environmental and economic constraints, expanding the use of biologically fixed nitrogen represents a more sustainable and resource-efficient alternative to continued dependence on mineral nitrogen fertilizers.

Biological nitrogen fixation (BNF) is one of the most important natural processes through which new reactive nitrogen is introduced into terrestrial ecosystems and remains the dominant nitrogen source for leguminous plant species (Wirth et al., 2024). In contrast to industrial nitrogen fixation, BNF proceeds without reliance on fossil fuels or high-energy inputs, making it an inherently clean and environmentally sustainable route for supplying nitrogen to crops. The application of biofertilizers composed of nitrogen-fixing microorganisms—collectively referred to as diazotrophs—offers a means to enhance crop performance while mitigating the ecological and human health impacts associated with excessive use of synthetic nitrogen fertilizers. Efficient management and enhancement of BNF are, therefore, critical for sustaining agricultural productivity and safeguarding ecosystem stability. This chapter explores the taxonomic diversity, functional attributes, and agricultural relevance of diazotrophic plant growth-promoting bacteria (PGPB), with particular emphasis on their contributions to environmentally sustainable, resource-efficient, and resilient farming systems.

13.2 Diazotrophic Bacteria and Their Role in Biological Nitrogen Fixation

Biological nitrogen fixation (BNF) is the enzymatic reduction of atmospheric dinitrogen (N_2) to ammonia by specialized soil microorganisms, providing plants with biologically assimilable nitrogen in exchange for microbial carbon resources (Soumare et al., 2020). Foundational work by Hellriegel and Wilfarth in (1888) first clarified the essential role of nitrogen-fixing symbioses in legumes, a concept later expanded through the comprehensive analyses of Franche et al. (2009). The ability of cyanobacteria to fix atmospheric nitrogen was subsequently demonstrated by Stewart (1969), further broadening the known phylogenetic range of nitrogen-fixing organisms. BNF constitutes the primary natural input of reactive nitrogen in most terrestrial ecosystems (Angus & Grace, 2017) and is a major determinant of the global nitrogen budget. It is estimated to contribute approximately 128 Tg of fixed nitrogen annually, accounting for nearly 60% of the total nitrogen introduced into the biosphere each year (Vitousek et al., 2013). In agricultural systems, diazotrophic microorganisms are responsible for an estimated 30%–50% of total nitrogen inputs (Ormeño-Orrillo et al., 2013), ultimately supplying up to 15% of the nitrogen taken up by crops. The nitrogen derived from BNF is indispensable for the biosphere, as it supports the synthesis of fundamental biomolecules including nucleic acids, amino acids, proteins, and chlorophyll that are required for cellular growth, metabolic regulation, and physiological function in all living organisms (Threatt & Rees, 2023).

The nitrogenase enzyme complex is responsible for catalyzing the ATP-dependent reduction of atmospheric dinitrogen (N_2) to ammonia (NH_3), using protons and electrons as substrates. The overall stoichiometric reaction is:

$$N_2 + 8H^+ + 8e^- + 16ATP \rightarrow 2NH_3 + H_2 + 16ADP + 16Pi$$

This reaction illustrates the high energetic demand of nitrogenase activity. A minimum of 16 ATP molecules is required to reduce a single N_2 molecule, and an additional ~12 ATP are needed for ammonium assimilation and transport processes. Thus, the total energetic expenditure for the fixation and assimilation of one N_2 molecule is ~28 ATP. To sustain this energy-intensive process, nodulated plants must provide substantial carbon resources to their symbiotic partners estimated at roughly 12 g of glucose per gram of nitrogen fixed (Soumare et al., 2020). Despite the significant biological energy demand, biological nitrogen fixation (BNF) remains vastly more energy-efficient than its industrial counterpart, the Haber–Bosch process. Haber–Bosch ammonia synthesis, developed in 1913 requires reaction temperatures of 400°C–500°C and pressures near 200–250 bar. Recent technoeconomic analyses indicate that industrial ammonia production costs range from USD 160–500 t^{-1}, with smaller-scale facilities or high energy and transport requirements pushing costs beyond USD 500 t^{-1} (Renganathan et al., 2025). Energy consumption averages 28–32 GJ t^{-1} NH_3 (≈40 MJ kg^{-1} N), and the process releases approximately 1 t CO_2 for each ton of NH_3 produced (Collado et al., 2024). In comparison, BNF requires only about 16 MJ kg^{-1} N^{-1}, equivalent to roughly 4.1 g C/gram of nitrogen fixed (Holland et al., 2023), underscoring its markedly greater energy efficiency and reduced environmental footprint.

Nitrogenase operates under a critical biochemical limitation: its metal-containing active sites are extremely sensitive to oxygen, which can irreversibly oxidize and deactivate the enzyme. To overcome this constraint, diazotrophic microorganisms have evolved a range of physiological and structural adaptations that minimize oxygen exposure and maintain nitrogenase functionality. Unicellular cyanobacteria such as *Cyanothece* achieve this by temporally separating incompatible processes fixing atmospheric nitrogen only during the night, when oxygenic photosynthesis ceases (Mishra et al., 2021). In contrast, obligate aerobic diazotrophs such as *Azotobacter* spp. rely on respiratory protection, maintaining exceptionally high respiration rates that rapidly consume intracellular oxygen and sustain microoxic conditions compatible with nitrogenase activity (Oelze, 2000). Filamentous cyanobacteria capable of heterocyst differentiation, including *Anabaena* and *Nostoc*, adopt a spatial separation strategy. They generate thick-walled heterocysts that restrict oxygen diffusion and create a protected microenvironment for nitrogenase, isolating it from oxygen produced by adjacent photosynthetic cells (Flores & Herrero, 2010). Additional protective mechanisms include the activity of oxygen-scavenging proteins such as FeSII, the formation of extracellular polysaccharide layers that function as diffusion barriers, and colonization of naturally microaerobic habitats such as biofilms and root nodules (Lery et al., 2010). Together, these diverse strategies underscore the evolutionary importance of safeguarding nitrogenase from oxygen, enabling efficient biological nitrogen fixation across a wide range of ecological niches.

13.3 Diazotrophic Microbes and Plant Interactions: Insights into Biological Nitrogen Fixation

Diazotrophic microorganisms encompass a wide array of phylogenetically and functionally diverse taxa that occupy numerous ecological habitats. This group includes nitrogen-fixing members of the α-, β-, and γ-Proteobacteria (Shin et al., 2016), along with representatives from Cyanobacteria, Firmicutes, and Actinobacteria (Imran et al., 2021). Functionally, diazotrophs are typically organized into three major categories according to the type of interaction they establish with plants: rhizospheric, endophytic, and symbiotic groups (Carvalho et al., 2014). These categories differ in their modes of colonization, degrees of host specificity, and overall contributions to nitrogen fixation and plant growth enhancement. Table 13.1 summarizes representative taxa from each group and outlines the principal mechanisms, by which they facilitate plant growth promotion.

13.3.1 Endophytic Diazotrophs

Endophytic diazotrophs inhabit the internal tissues of plants—including cortical regions, intercellular spaces, and the xylem–phloem continuum—without inducing pathogenic effects (Pal et al., 2022). Occupation of these internal niches provides several ecological and physiological advantages: reduced competition with rhizospheric microbes, access to a steady supply of host-derived nutrients, and improved efficiency of nitrogen transfer due to their close association with the plant vascular system (Carvalho et al., 2014). A diverse range of endophytic nitrogen-fixing genera has been documented in cereal crops such as rice, maize, sorghum, and wheat. Prominent examples include *Herbaspirillum, Gluconacetobacter, Burkholderia, Azoarcus*, and *Azospirillum* (Pal et al., 2022). Species such as *Herbaspirillum seropedicae* and *Burkholderia vietnamiensis* colonize the intercellular spaces of sugarcane tissues and have been shown to enhance nitrogen fixation, particularly when fertilizer inputs are minimal (Carvalho et al., 2014). *Azospirillum brasilense*, a facultative endophyte, exhibits chemotactic responses to root exudates and employs a respiratory protection mechanism—elevating its oxygen consumption rate to maintain nitrogenase activity under aerobic conditions (Upadhyay et al., 2021).

13.3.2 Rhizospheric Diazotrophs

Rhizospheric diazotrophs are free-living or associative nitrogen-fixing bacteria that colonize the root–soil interface, a dynamic zone strongly influenced by plant-derived compounds. Root exudates—including sugars, amino acids, and organic acids—serve as chemoattractants and readily accessible carbon sources that support microbial establishment and activity in this niche (Imran et al., 2021). Many rhizospheric diazotrophs exhibit classic plant growth-promoting rhizobacteria (PGPR) traits, contributing

Table 13.1 Host plant–microbe interactions underlying biological nitrogen fixation

Type of diazotroph	Representative species	Host plants	Plant growth promotion traits	Reference
Endophytic	*Gluconacetobacter diazotrophicus*	Sugarcane Rice Tomato Maize	N-fixation; IAA, GA_3, cytokinin production; phosphate solubilization, biocontrol agent	Arencibia et al. (2006); Rajani et al. (2023); Serbot et al. (2021); Tufail et al. (2021)
	Herbaspirillum seropedicae	Maize Rice	N-fixation, Siderophore production, phytohormone synthesis, ACC deaminase activity, ISR	Waller et al. (2021); Ramos et al. (2020)
	Azoarcus sp.	Rice	Nitrogen fixation; stress tolerance; IAA synthesis	Fernández- Llamosas et al. (2021)
	Burkholderia seminalis	*Arabidopsis*, Pak Choi, Chinese Amaranth, Lettuces, and Other Vegetables	N-fixation; P-solubilization; Fe-siderophore production	Hwang et al.(2021)
	Burkholderia sp.	Maize coffee	N-fixation	Estrada-De Los Santos et al. (2001)
	Enterobacter cloacae and *Klebsiella pneumoniae*	Banana	Biocontrol N-fixation, indole acetic acid production, phosphate solubilization,	Macedo-Raygoza et al. (2019)
Symbiotic	*Rhizobium leguminosarum*	Pea (*Pisum sativum*), lentil	N-fixation	Kaziūnienė et al. (2025)
	Sinorhizobium meliloti	*Medicago*, *Melilotus*, and *Trigonella*	nitrogen fixation	Kearsley et al. (2024)
	Mesorhizobium ciceri	Chickpea (Cicer arietinum)	N-fixation; phosphate solubilization	Wanjofu et al. (2022)
	Bradyrhizobium sp.	Mungbean	Enhanced nodulation, Nitrogen fixation	Favero et al. (2022)

Type of diazotroph	Representative species	Host plants	Plant growth promotion traits	Reference
Rhizosphere/ Associative	*Azospirillum brasilense*	Wheat	Associative nitrogen fixation	Galindo et al. (2022)
	Azotobacter chroococcum	Sugarbeet	Nitrogen, iron, and potassium availability, auxin and 1-aminocyclopropane-1-carboxilic acid deaminase	Kerečki et al. (2022)
	Beijerinckia fluminensis	Litchi	N-fixation	Singh et al. (2024)
	Klebsiella pneumoniae	wheat	N-fixation	Iniguez et al. (2004)
	Pseudomonas stutzeri	Rice, Wheat	N-fixation; nitrate respiration, competitive root colonization	Rediers et al. (2009)
	Paenibacillus durus	Cucumber	N-fixation; stress tolerance; P-solubilization	Li et al. (2022)
	Serratia marcescens	Tea	N-fixation; P-solubilization, biocontrol	Chakraborty et al. (2010)

to plant development through mechanisms such as atmospheric nitrogen fixation, phosphate solubilization, siderophore production, and phytohormone biosynthesis (Renganathan et al., 2025).

Interest in rhizospheric diazotrophs has increased due to their potential to mitigate both the environmental and economic impacts of intensive synthetic nitrogen fertilizer use (Chakraborty et al., 2024). A key biological constraint for free-living nitrogen fixation is its considerable energy requirement (Inomura et al., 2018). Unlike symbiotic diazotrophs—which receive a continuous supply of photosynthetically derived carbon from their hosts—heterotrophic diazotrophs in bulk soil frequently experience carbon limitation. The rhizosphere, however, is an important exception: its abundance of low-molecular-weight carbon compounds enables these microorganisms to meet the energetic demands of nitrogenase activity (Upadhyay et al., 2022). Through their close association with plant roots, rhizospheric diazotrophs enhance nutrient acquisition and contribute to improved plant vigor, productivity, and soil health. Consequently, they are regarded as promising alternatives or complements to mineral nitrogen fertilizers in the transition toward more sustainable agricultural systems. A wide variety of nitrogen-fixing genera—such as *Acetobacter, Arthrobacter, Azoarcus, Azotobacter, Azospirillum, Beijerinckia, Derxia, Enterobacter, Klebsiella, Pseudomonas*, and *Zoogloea*—have been isolated from the rhizosphere of diverse crop species (Thiebaut et al., 2022; Ladha et al., 2022; Imran et al., 2021; Naqqash et al., 2022).

Beyond nitrogen fixation, these free-living and associative diazotrophs contribute to plant growth through multiple PGPR-mediated processes, including enhanced nutrient solubilization, modulation of phytohormone levels, activation of plant defense pathways, synthesis of antioxidants and growth regulators, facilitation of osmotic balance, and improved tolerance to biotic and abiotic stresses. Given that many globally important cereal crops—such as maize, rice, wheat, sorghum, and sugarcane—lack nodulating capabilities, the development of bioinoculants based on nonsymbiotic diazotrophic bacteria has become increasingly relevant (Thiebaut et al., 2022). A deeper understanding of the ecological dynamics and functional capabilities of rhizospheric diazotrophs is, therefore, vital not only for clarifying nitrogen-cycle processes but also for advancing sustainable and low-input agricultural practices (Zhao et al., 2024).

13.3.3 Symbiotic Diazotrophs

Symbiotic nitrogen fixation occurs through mutualistic interactions between plants and diazotrophic bacteria that give rise to specialized root or stem nodules, within which nitrogenase operates under carefully regulated microaerobic conditions (Li & Xiao, 2025). Inside these nodules, the host plant provides photosynthetically derived carbon sources and maintains a controlled internal environment that supports bacterial metabolism and ensures the sustained functioning of nitrogenase (Cowan et al., 2022). In legumes, these symbioses are predominantly formed with members of the order *Rhizobiales*, including *Rhizobium, Bradyrhizobium, Ensifer, Mesorhizobium*, and *Azorhizobium*. The interaction begins when root-secreted flavonoids serve as signaling molecules that attract compatible rhizobia and activate the expression of *nod* genes. The resulting Nod factors mediate highly specific recognition events, initiate infection thread formation, and promote cortical cell division, ultimately driving nodule organogenesis (Li & Xiao, 2025). Although legumes represent the primary hosts for rhizobial symbioses, a small number of nonleguminous plants—most notably *Parasponia andersonii*—also form nodule-forming associations with *Rhizobium* spp., providing an evolutionary bridge between rhizobial and actinorhizal symbiotic systems (Mathesius et al., 2022). In a separate lineage of symbioses, many nonleguminous woody angiosperms form nitrogen-fixing nodules through partnerships with filamentous actinobacteria of the genus *Frankia*. These actinorhizal symbioses occur in species such as *Casuarina, Alnus, Myrica, Elaeagnus*, and *Hippophae* (Gupta et al., 2020). Nodules induced by *Frankia* contain specialized vesicles that protect nitrogenase from oxygen and sporangia that facilitate microbial propagation (Scaria et al., 2022). Collectively, rhizobial and actinorhizal symbiotic systems play essential ecological and agronomic roles. By enhancing biological nitrogen inputs, they contribute to soil fertility, support afforestation initiatives, and enable the reclamation and restoration of nitrogen-deficient landscapes.

13.4 Diazotrophs as Multifunctional Bioinoculants: Mechanisms of Plant Growth Promotion Beyond Nitrogen Fixation

Beyond their ability to fix atmospheric nitrogen, diazotrophic bacteria contribute to plant growth through numerous direct and indirect mechanisms, placing them among the most functionally versatile groups of plant growth-promoting bacteria. These mechanisms include the synthesis of phytohormones, siderophores, hydrogen cyanide (HCN), and various antimicrobial compounds; the solubilization of key mineral nutrients such as phosphorus, zinc, and potassium; and the activity of 1-aminocyclopropane-1-carboxylic acid (ACC) deaminase, which reduces ethylene accumulation and alleviates stress responses in plants (Imran et al., 2021). Several of these pathways are depicted in Fig. 13.1. Diazotrophs possess a competitive advantage in environments characterized by high carbon availability but limited nitrogen, conditions commonly found in the rhizosphere. This ecological niche favors the selective enrichment of nitrogen-fixing microorganisms, enabling them to exert substantial influence on root development and overall plant performance (Dalton & Kraner, 2007). Given their multifunctional benefits, microbial inoculants—particularly those based on diazotrophic strains—represent a promising avenue for improving crop productivity and soil quality. Their application has been shown to enhance yields, promote plant vigor, and support more sustainable agricultural systems capable of meeting the nutritional demands of a growing global population.

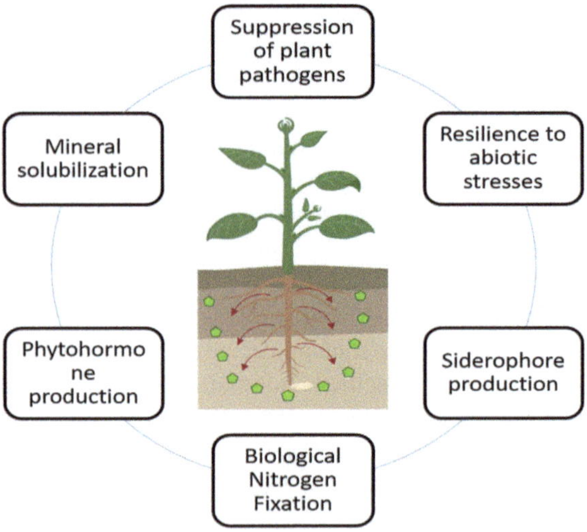

Fig. 13.1 Schematic representation of multifaceted mechanisms of diazotrophic PGPR in plant growth promotion

13.4.1 Phytohormone-Mediated Plant Growth Promotion by Diazotrophs

Phytohormones—also known as plant growth regulators (PGRs)—are central to coordinating plant growth, development, and adaptive responses. They influence numerous physiological processes, including seedling establishment, root proliferation, shoot and root elongation, and overall plant architectural patterning. One of the principal ways in which plant growth-promoting rhizobacteria (PGPR) enhance plant performance is through auxin-mediated modification of root architecture. Auxin-induced expansion of the root system increases the effective absorptive surface area, thereby improving the acquisition of water and essential nutrients and ultimately supporting greater biomass production (Hmaeid et al., 2019).

In addition to their nitrogen-fixing capability, many diazotrophs synthesize a broad suite of phytohormones—including indole-3-acetic acid (IAA), gibberellins, cytokinins, and abscisic acid—that regulate key developmental processes. These hormones influence seed germination, root system expansion and nutrient uptake efficiency, vascular tissue differentiation, shoot and floral development, and whole-plant biomass accumulation (Renganathan et al., 2025). By colonizing roots and producing these regulatory compounds, diazotrophic bacteria stimulate root elongation, lateral root formation, and root hair proliferation, thereby enabling plants to access larger nutrient pools within the soil (Imran et al., 2021). Secondary roots and root hairs, which provide favorable microhabitats, often serve as the primary sites of bacterial attachment and establishment (Imran et al., 2021).

A wide spectrum of diazotrophic and PGPR genera—including *Aeromonas, Azotobacter, Azospirillum, Bacillus, Bradyrhizobium, Burkholderia, Enterobacter, Mesorhizobium, Pseudomonas, Rhizobium*, and *Sinorhizobium*—are known to produce

IAA through multiple biosynthetic routes (Imran et al., 2021; Fukami et al., 2018b; Dobbelaere et al., 2003). Field-based evidence substantiates the role of microbial phytohormones in crop enhancement. Fukami et al. (2018b) reported that *Azospirillum* inoculation increased maize root volume and shoot height and enhanced tillering in wheat, with these improvements attributed to bacterial phytohormone production. Similarly, *Bradyrhizobium* strains isolated as rice endophytes promoted plant growth through a combination of biological nitrogen fixation, ACC deaminase activity, and IAA synthesis (Greetatorn et al., 2019). In maize, inoculation of 27 genotypes with *A. brasilense* Ab-V5 elevated endogenous IAA concentrations, resulting in increased plant biomass, larger and more branched root systems, and improved nutrient uptake efficiency (Zeffa et al., 2019).

Microbial phytohormones frequently act in a manner analogous to exogenously applied plant hormones, influencing endogenous hormonal balance and modulating multiple developmental pathways. IAA plays particularly central roles in root and shoot growth, cell expansion, vascular differentiation, bacterial colonization, and the regulation of plant defense responses. It also facilitates cell-wall loosening, promotes cell division, and shapes overall root architecture (Etesami et al., 2025). Recent work by Naqqash et al. (2022) showed that two *Azospirillum* strains produced substantial quantities of IAA, directly enhancing nutrient and water acquisition. By expanding root surface area, IAA production facilitates bacterial colonization and nutrient cycling, ultimately improving soil nutrient availability to the plant (Hakim et al., 2021).

13.4.2 Phosphate Solubilization by Diazotrophic Microorganisms: Mechanisms and Implications for Plant Nutrition

Phosphorus (P) is an essential macronutrient required for optimal plant growth and development, typically constituting approximately 0.2% of plant dry biomass (Alori et al., 2017). Adequate P availability is fundamental for a range of core biochemical and physiological processes, including the synthesis of nucleic acids and membrane phospholipids, ATP generation, and the functioning of photosynthesis and biological nitrogen fixation in legumes (Bargaz et al., 2021). Although mineral P fertilizers are frequently applied to correct soil P deficiencies, their efficiency is often constrained by the rapid conversion of applied P into insoluble, immobilized, or strongly adsorbed forms through diverse abiotic and biotic soil reactions. In addition, excessive P application increases the risk of nutrient losses via runoff, erosion, and leaching, thereby contributing to groundwater contamination and eutrophication of aquatic environments (Khan et al., 2023). Thus, developing sustainable strategies that enhance P-use efficiency while mitigating environmental impacts is essential for modern crop production.

Diazotrophic microorganisms contribute significantly to nutrient cycling and improved plant nutrient acquisition, particularly for macronutrients and micronutrients such as phosphorus (P), zinc (Zn), and potassium (K). Their contribution is largely mediated through the solubilization or mobilization of insoluble mineral forms. Phosphate-solubilizing bacteria (PSB) facilitate this process primarily by excreting organic acids that lower the surrounding pH and convert insoluble phosphate (PO_4

$^{3-}$) into plant-accessible forms such as HPO_4^{2-} and $H_2PO_4^-$. This acidification can be substantial, with localized pH reductions from neutral values (~7.0) to as low as 2.0 (Suleman et al., 2018). A wide spectrum of organic acids—including gluconic, citric, malic, oxalic, fumaric, malonic, tartaric, glyoxylic, propionic, butyric, glutaric, and adipic acids—participate in P solubilization, with gluconic acid being the most commonly reported (Joshi et al., 2023). Gluconic acid is produced through the direct oxidation of glucose by glucose dehydrogenase, an enzyme encoded by the *gcd* gene and dependent on the cofactor pyrroloquinoline quinone (PQQ) (Deppenmeier et al., 2002).

The majority of characterized *pqq* gene clusters occur in α-, β-, and γ -Proteobacteria, particularly among Gram-negative taxa (Joshi et al., 2023). Bacterial genera frequently associated with PQQ-dependent phosphate solubilization include *Acinetobacter, Azotobacter, Beijerinckia, Bradyrhizobium, Burkholderia, Erwinia, Gluconacetobacter, Klebsiella, Gluconobacter, Methylobacillus, Methylobacterium, Mycobacterium, Pseudomonas, Rhizobium, Streptomyces*, and *Xanthomonas* (Joshi et al., 2023). Environmental and nutritional factors—including elevated glucose availability and higher concentrations of insoluble phosphate—strongly regulate PQQ and glucose dehydrogenase synthesis (An & Moe, 2016).

Several diazotrophic and plant growth–promoting strains, such as *Azospirillum brasilense, Azotobacter chroococcum, Bacillus spp., Pseudomonas fluorescens, Anabaena variabilis*, and *Tolypothrix*, are known producers of gluconic, citric, and oxalic acids. These acids mobilize mineral-bound phosphate through chelation, thereby improving plant P uptake (Renganathan et al., 2024; Condori et al., 2024). In potato, inoculation with phosphate-solubilizing *Azospirillum* strains TN03 and TN09 significantly enhanced shoot and root length, as well as total biomass compared with uninoculated controls (Naqqash et al., 2022). Likewise, Estrada et al. (2013) demonstrated that *Herbaspirillum* and *Burkholderia* strains exhibiting strong phosphate-solubilizing activity—evidenced by halo formation on agar media—improved rice grain yield and increased plant N-use efficiency from applied fertilizers.

13.4.3 Role of Diazotrophic Microbes in Iron Acquisition

Iron (Fe) is an essential micronutrient required by nearly all organisms due to its involvement as a cofactor in a wide array of metalloproteins that drive key metabolic pathways, including mitochondrial respiration, photosynthetic electron transport, amino acid biosynthesis, and sulfur and nitrogen assimilation (Molnár et al., 2023). Despite being the fourth most abundant element in the Earth's crust, Fe deficiency is widespread in plants. This apparent paradox arises because most soil Fe occurs as highly insoluble oxides and hydroxides, especially under neutral to alkaline pH conditions, which severely limits its biological accessibility (Molnár et al., 2023).

Plant growth–promoting rhizobacteria (PGPR) alleviate Fe limitation primarily through the production of siderophores. Under well-aerated soil conditions, iron is predominantly present as Fe^{3+} within oxyhydroxide complexes that exhibit extremely low solubility, on the order of 10^{-18} M at pH 7, rendering it unavailable to plants and microorganisms (Neilands, 1995). To overcome this constraint, many rhizosphere-associated bacteria synthesize siderophores—low-molecular-weight, high-affinity Fe-chelating compounds produced in response to iron scarcity.

Siderophores encompass several structural classes, including hydroxamates, catecholates, and carboxylates. Their biosynthesis is typically induced in alkaline and calcareous soils where Fe solubility is most restricted (Khoso et al., 2024). These compounds solubilize and sequester Fe from mineral phases, forming stable Fe–siderophore complexes that can be taken up by microbial cells. As microbes acquire iron via these complexes, they simultaneously increase Fe bioavailability within the rhizosphere, indirectly enhancing plant Fe nutrition and supporting a wide range of physiological functions (Timoveeva et al., 2022). Siderophores also chelate other micronutrients—such as Mo, Mn, Co, and Ni thereby improving their mobility and uptake in plant–soil systems.

A diverse array of bacterial species are recognized producers of siderophores, including *Bradyrhizobium, Azotobacter, Azospirillum, Bacillus, Enterobacter, Pseudomonas, Rhizobium, Serratia*, and *Streptomyces* (Verma et al., 2013), as well as *Klebsiella, Kosakonia, Methylobacterium, Nocardia, Pantoea, Paenibacillus*, and *Rhodococcus* (Timoveeva et al., 2022). In addition to enhancing Fe acquisition, siderophore-producing PGPR contribute significantly to plant disease suppression. By tightly binding Fe, they limit its availability to phytopathogens, inhibiting pathogen growth and improving plant health. This competitive advantage also helps beneficial microbes maintain ecological dominance within the rhizosphere.

Given the strong Fe-binding affinity and high stability of siderophores, their accumulation in soil can markedly enhance Fe mobility, especially in arid, alkaline, or calcareous soils where Fe availability is highly constrained. For example, siderophore-producing *Burkholderia* strains exhibit strong in vitro antagonism toward phytopathogenic fungi, with pathogen suppression linked to reduced Fe availability demonstrating siderophore production as a key biocontrol trait (Loaces et al., 2011).

13.4.4 ACC Deaminase Producing Diazotrophs: Mitigating Stress Ethylene and Promoting Growth

Ethylene is a gaseous phytohormone produced by all higher plants, capable of eliciting physiological responses at concentrations below 1.0 ppm ($1.0\,\mu L\,L^{-1}$). Even at these trace levels, ethylene regulates a broad range of developmental and metabolic processes, including seed germination, fruit ripening and maturation, senescence, initiation of root and leaf primordia, formation of root hairs and nodules, modulation of root elongation, and induction of epinastic growth (Sun et al., 2019; Shahid et al., 2023). Ethylene biosynthesis begins with the conversion of methionine and ATP into S-adenosyl methionine (SAM) catalyzed by SAM synthetase. SAM is subsequently transformed into 1-aminocyclopropane-1-carboxylic acid (ACC) by ACC synthase, after which ACC oxidase converts ACC into ethylene. This terminal step also generates hydrogen cyanide and carbon dioxide. Multiple gene isoforms encoding ACC synthase and ACC oxidase have been documented in plants examined to date. Transcription of ACC synthase is strongly induced by numerous abiotic and biotic stress factors—including drought, salinity, flooding, heavy metals, agrochemicals, and pathogen infection—resulting in elevated ethylene synthesis (Orozco-Mosqueda et al., 2020). Consequently, increased ethylene accumulation is a hallmark of generalized plant stress responses.

Plant growth–promoting bacteria (PGPB) producing ACC deaminase and indole-3-acetic acid (IAA) play a pivotal role in moderating ethylene biosynthesis and sustaining plant growth. The coordinated interaction between bacterial ACC deaminase activity and microbial IAA production is central to this regulation (Nascimento et al., 2019). Many PGPB secrete IAA, which supplements plant-derived IAA and can induce ACC synthase gene expression, resulting in elevated ACC levels and increased ethylene production. High ethylene concentrations suppress IAA signal transduction and restrict IAA-dependent growth processes. However, when ACC deaminase–producing bacteria are present, they cleave ACC into α-ketobutyrate and ammonia, thereby reducing the pool of ACC available for ethylene formation. This suppression of ethylene biosynthesis alleviates feedback inhibition and restores IAA-driven growth responses (Glick, 2014). In this way, ACC deaminase–producing PGPB function as an efficient biological sink for ACC, mitigating stress-induced ethylene accumulation and promoting root and shoot development under adverse environmental conditions.

ACC deaminase activity is widely distributed among plant-associated bacteria, underscoring its ecological significance in plant–microbe interactions and ethylene regulation. The rhizosphere is a rich reservoir for isolating microorganisms that express ACC deaminase and other plant growth–promoting traits (Wu et al., 2019). Phylogenetic analyses of the **acdS** gene demonstrate its presence across numerous bacterial lineages, including Actinobacteria, α-, β-, and γ-Proteobacteria, Deinococcus, and Firmicutes (Singh et al., 2022). In rhizobia, ACC deaminase contributes to successful legume–rhizobium symbiosis and indirectly supports biological nitrogen fixation (Singh et al., 2022). Various rhizobial genera—including *Rhizobium, Sinorhizobium, Agrobacterium, Mesorhizobium*, and *Phyllobacterium*—harbor ACC deaminase activity (Nascimento et al., 2019). Many other PGPR taxa, such as *Arthrobacter, Achromobacter, Bacillus, Burkholderia, Citrobacter, Enterobacter, Micrococcus, Pseudomonas, Ralstonia*, and *Serratia*, also exhibit this trait (Acuna et al., 2019; Duan et al., 2013; Kang et al., 2019).

Bacteria capable of degrading ACC are, therefore, critical contributors to plant stress tolerance. By preventing excessive ethylene accumulation, they alleviate negative effects associated with salinity, drought, heavy metals, and organic pollutants, ultimately enhancing plant growth, physiological stability, and resilience.

13.4.5 Diazotrophs as Biocontrol Agents: Indirect Growth Promotion by Suppression of Phytopathogens

Diazotrophic bacteria also enhance plant performance indirectly by suppressing phytopathogens and plant-parasitic pests, thereby reducing the damage associated with disease pressure. These microbes produce a broad spectrum of antagonistic

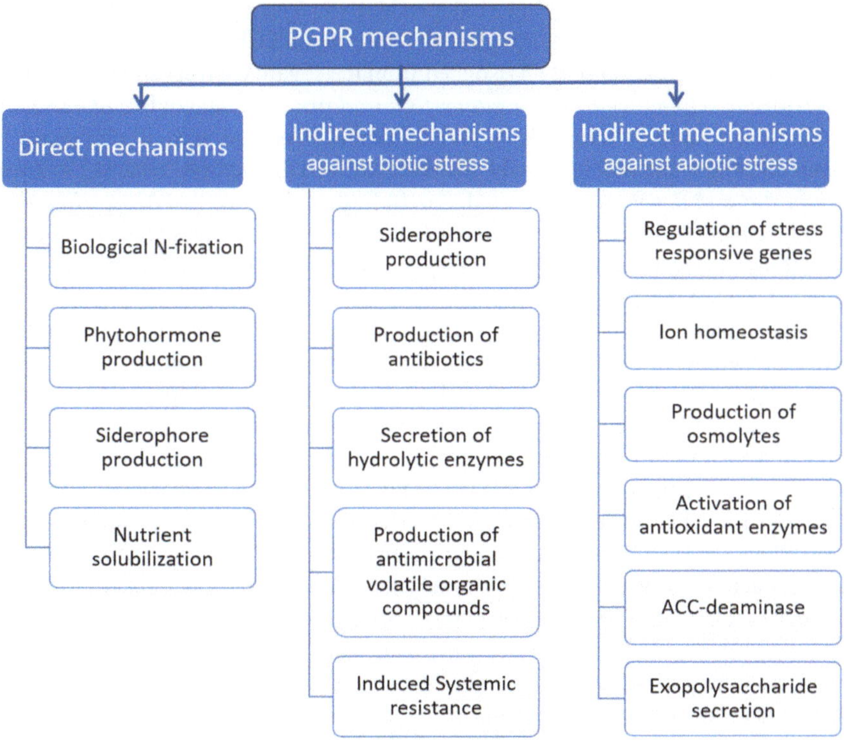

Fig. 13.2 Strategies employed by diazotrophs to enhance plant growth under biotic and abiotic stresses

metabolites—including antibiotics, lytic enzymes, toxins, siderophores, and antimicrobial volatile organic compounds—and can additionally activate induced systemic resistance (ISR) in the host plant. Through these complementary mechanisms, diazotrophs limit pathogen establishment and proliferation, contributing significantly to plant health and protection (Fig. 13.2) (Dobbelaere et al., 2003).

Siderophores produced by biocontrol bacteria generally exhibit much higher iron-binding affinities than those synthesized by fungal pathogens, giving beneficial bacteria a competitive advantage in iron acquisition. By sequestering the majority of bioavailable Fe, these microbes restrict iron access for pathogenic fungi and thereby suppress their growth. Ji et al. (2014) demonstrated that endophytic diazotrophs, including *Klebsiella pneumoniae*, *Bacillus subtilis*, and *Microbacterium* sp., inhibited *Rhizoctonia solani* through siderophore-mediated competition. In addition to siderophores, several bacterial strains release secondary metabolites such as hydrogen cyanide (HCN) and ammonia, both of which have been widely implicated in the suppression of fungal diseases (Martínez-Viveros et al., 2010; Olanrewaju et al., 2017). Ammonia is produced predominantly via nitrite ammonification in diazotrophic bacteria (Geetha et al., 2014), contributing to plant growth while also functioning as an antimicrobial compound either independently or synergistically with HCN and siderophores (Allu et al., 2014). Singh et al. (2021) reported that *Pantoea dispersa* AA7 and *Enterobacter asburiae* BY4, isolated from the sugarcane rhizosphere, inhibited *Sporisorium scitamineum* and *Ceratocystis paradoxa*, with antagonism linked to the production of siderophores, HCN, and ammonia. Likewise, Ji et al. (2014) documented strong antagonistic activity of rice-associated diazotrophs against *Fusarium oxysporum* and *R. solani*. Araújo et al. (2010) further observed that *Azospirillum* sp., *Burkholderia brasilensis*, and *Herbaspirillum seropedicae* suppressed *Fusarium* infections in rice seed inoculation assays, resulting in improved germination and reduced disease incidence.

Beyond these classical antagonistic mechanisms—such as siderophore secretion, HCN and ammonia production, emission of volatile and non-volatile antimicrobial compounds, and synthesis of cell wall–degrading enzymes—endophytic bacteria can also stimulate plant defense pathways via induced systemic resistance (ISR). ISR, triggered by nonpathogenic microorganisms, originates at the initial site of colonization but subsequently primes the entire plant, enhancing resistance responses in distal tissues (van Loon & Bakker, 2005). Once activated, ISR may confer long-term protection (van Loon, 2007). According to van Loon (2007), plant growth-promoting bacteria activate ISR through four primary mechanisms: (i) developmental escape, through accelerated or improved plant growth; (ii) physiological tolerance, mitigating the severity of disease symptoms; (iii) environmental interactions, wherein antagonistic microbes limit pathogen growth in the rhizosphere; and (iv) biochemical resistance, involving reinforcement of cell walls, accumulation of phytoalexins and PR proteins, and heightened readiness for rapid defense activation. Plants perceive a diverse range of microbial signals—including siderophores, lipopolysaccharides, exopolysaccharides, volatile organic compounds, flagellin, salicylic acid, and pyocyanins—which serve as ISR elicitors in both rhizospheric and endophytic bacteria (Gupta et al., 2012). These elicitors enable plants to inhibit pathogen establishment at early stages (Akram et al., 2015), highlighting antagonistic endophyte colonization as a key driver of systemic resistance.

PGPB from the genera *Azospirillum*, *Pseudomonas*, and *Bacillus* are among the most common ISR inducers (Fukami et al., 2018a). Transcriptomic analyses revealed that *Azospirillum* sp. strain B510, associated with rice cultivar "Nipponbare," induced one PR gene while repressing several others, whereas *A. lipoferum* strain 4B activated a broader suite of defense-related genes in "Nipponbare" compared with its native host cultivar "Cigalon" (Drogue et al., 2014). In *Arabidopsis thaliana*, inoculation with *A. brasilense* Sp245 stimulated PR gene expression (Spaepen et al., 2014), and similar PR gene activation was observed in maize inoculated with *A. brasilense* Ab-V5 and Ab-V6 (Fukami et al., 2018b). Yasuda et al. (2009) further reported that rice plants inoculated with *Azospirillum* sp. B510 exhibited enhanced resistance to *Magnaporthe oryzae* and *Xanthomonas oryzae*. This ISR response proceeded through a salicylic acid (SA)-independent pathway, as evidenced by the absence of both SA accumulation and PR protein induction.

13.4.6 Diazotrophs in Enhancing Crop Resilience to Abiotic Stress

The intensification of abiotic stresses under accelerating climate change poses a significant threat to global agricultural productivity. Rising temperatures, recurrent drought episodes, progressive soil salinization, and an increased frequency of extreme weather events collectively undermine crop performance and exacerbate food insecurity (Calanca, 2017). Global projections indicate that by 2050, climate change could reduce crop yields by 20%–30%, with the most severe impacts anticipated in regions already constrained by water scarcity and extreme thermal environments (Kumar et al., 2022). These stressors impair key physiological and metabolic processes including photosynthesis, nutrient uptake, osmotic regulation, and water-use efficiency ultimately compromising both yield quantity and nutritional quality. As climate variability becomes increasingly erratic, enhancing plant tolerance to environmental perturbations has become a major priority for sustainable agricultural development (Kumari et al., 2022).

Plant-associated microbiomes provide an ecologically sustainable and low-input strategy for mitigating the adverse effects of abiotic stresses while contributing to long-term global food security (Shah et al., 2019; Ullah et al., 2025). Diazotrophic plant growth-promoting rhizobacteria (PGPR) employ multiple physiological and biochemical mechanisms to enhance plant resilience under stress conditions, as outlined in Fig. 13.2. Innovative microbial-based technologies—including plant biostimulants, biofertilizers, and synthetic microbial communities (SynComs)—are receiving growing attention for their ability to optimize plant health and stress adaptation (Compant, 2025). Biofertilizers formulated with beneficial microorganisms such as nitrogen-fixing and phosphate-solubilizing bacteria, as well as microalgae, enhance soil nutrient availability, promote nutrient-use efficiency, and improve soil fertility, collectively contributing to improved plant growth, productivity, and environmental sustainability (Alzate Zuluaga, 2024).

13.4.6.1 Diazotroph-Mediated Alleviation of Salinity Stress in Crops

Soil salinization, characterized by the excessive accumulation of soluble salts, particularly sodium chloride significantly disrupts plant physiological processes and poses a major threat to agricultural productivity and long-term soil sustainability. Elevated concentrations of salts such as Na^+ lower the soil water potential, thereby limiting water availability in the rhizosphere and restricting plant water uptake (Van Zelm et al., 2020). This leads to osmotic or water-deficit stress, which decreases cell turgor, disturbs plant water relations, and reduces water-use efficiency (Huang et al., 2017). Salt stress also induces stomatal closure and impairs chlorophyll and photosynthetic apparatus integrity. Furthermore, ionic and osmotic imbalances trigger the excessive formation of reactive oxygen species (ROS), including H_2O_2, $^1O_2$, O_2^-, and $\cdot OH$, due to perturbations in cellular metabolism (Gill & Tuteja, 2010). These ROS cause oxidative damage to lipids, nucleic acids, proteins, and carbohydrates, thereby disrupting redox homeostasis and compromising cellular stability (Gill & Tuteja, 2010).

Soil microbial communities play an important role in mitigating the detrimental impacts of salinity. Beneficial microorganisms synthesize compatible solutes such as glycerol and betaine that assist plant cells in maintaining osmotic balance and structural integrity. They also contribute to ion homeostasis by promoting the selective exclusion or intracellular sequestration of toxic ions such as Na^+, thus reducing ion-induced injury and preserving metabolic function (Kumar et al., 2020). Several halotolerant plant growth-promoting rhizobacteria (PGPR)—including Acinetobacter, Azotobacter, Bacillus spp., Serratia spp., Pseudomonas spp., and Rhizobium spp.—have been documented to enhance plant growth under both optimal and saline conditions via mechanisms such as nitrogen fixation, phosphate solubilization, siderophore production, and phytohormone synthesis (Ullah et al., 2025).

Among these PGPR, species of *Azospirillum*, particularly *A. brasilense*, are among the most extensively studied for ameliorating salt-induced stress (Fasciglione et al., 2015). Hamdia et al. (2004) demonstrated that inoculation of maize with *Azospirillum* spp. improved ionic homeostasis by reducing Na^+ uptake and enhancing K^+ and Ca^{2+} accumulation. This shift in ion balance lowered proline levels—a common physiological marker of stress—and resulted in improved plant growth. Similarly, Fukami et al. (2018b) reported that maize inoculated with *A. brasilense* Ab-V6 exhibited enhanced biomass accumulation and reduced harmful solute buildup, accompanied by the upregulation of genes associated with antioxidant defense. More recently, Degon et al. (2023) showed that *A. brasilense* significantly improved rice performance under saline conditions by modulating gene networks involved in abscisic acid signaling, nutrient transport, salt stress responses, antioxidant regulation, calcium signaling, and Na^+/K^+ homeostasis. Collectively, these studies highlight the multifaceted roles of *Azospirillum* in enhancing plant resilience to salinity through coordinated physiological and molecular mechanisms.

13.4.6.2 Diazotroph-Mediated Mitigation of Drought Stress in Crops

Drought is one of the most severe constraints on global crop productivity, and its impacts are projected to intensify under accelerating climate change. Water deficit disrupts stomatal conductance, thereby lowering the internal CO_2/O_2 ratio in leaves, suppressing photosynthetic efficiency, and ultimately reducing biomass accumulation (Fukami et al., 2018a). Under prolonged or intense drought stress, plants experience oxidative damage due to excessive production of reactive oxygen species (ROS) (Hasanuzzaman et al., 2020). To mitigate this, plants activate a suite of antioxidant enzymes including catalase (CAT), peroxidases (POX), superoxide dismutase (SOD), glutathione peroxidase (GPX), and ascorbate peroxidase (APX), which collectively detoxify ROS and maintain redox homeostasis (Simova-Stoilova et al., 2008). Additional drought-mitigating mechanisms include osmotic adjustment, sustained root functionality, stabilization of cellular membranes, and the accumulation of compatible solutes such as proline, glycine betaine, and trehalose (Ngumbi & Kloepper, 2016).

Rhizosphere-associated microbial communities play a pivotal role in enhancing plant drought resilience. These microorganisms strengthen plant stress tolerance through multiple mechanisms, including improved nutrient acquisition, stimulation of root growth, phytohormone synthesis, production of ACC deaminase, exopolysaccharide (EPS) formation, induction of antioxidant defense pathways, and modulation of stress-responsive gene expression (Ullah et al., 2025). Microbial inoculation also enhances water-use efficiency by promoting root proliferation and increasing the accumulation of osmoprotectants such as proline and trehalose (Vidal et al., 2022).

Endophytic diazotrophs such as *Gluconacetobacter diazotrophicus* have been shown to alleviate drought and nitrogen deficiency in maize (*Zea mays* L.), improving biomass accumulation, chlorophyll content, nitrogen uptake, and water-use efficiency (Tufail et al., 2021). The involvement of *Azospirillum* spp. in drought tolerance is similarly well established, with effects partly attributed to their synthesis of abscisic acid (ABA), which promotes stomatal closure and reduces transpirational water loss (Cohen et al., 2015). In addition to ABA-mediated effects, *Azospirillum* stimulates the accumulation of osmolytes such as free amino acids and soluble sugars, helping maintain cellular hydration under water scarcity. These bacteria also enhance root system architecture, promoting increased root branching, higher root biomass, and greater root hair density, thereby improving soil water acquisition (Hungria et al., 2015). These morphological and physiological changes are strongly linked to phytohormones, particularly indole-3-acetic acid (IAA), produced by plant growth-promoting bacteria (Saharan & Nehra, 2011). Saeed et al. (2016) demonstrated that inoculation of canola seeds with *A. lipoferum* increased germination rate, root surface area, chlorophyll concentration, and water potential under drought stress. Similarly, Curá et al. (2017) reported that maize inoculated with *A. brasilense* or *Herbaspirillum seropedicae* exhibited enhanced desiccation tolerance, associated with modulation of ABA and ethylene signaling pathways.

13.5 Bioinoculants in Sustainable Agriculture: Harnessing Multifaceted Diazotrophic Plant Growth Promoting Bacteria for Crop Productivity

A central component of sustainable agriculture, particularly, efforts to maintain crop productivity while decreasing reliance on chemical fertilizers is the strategic use of beneficial microorganisms. In this context, reshaping plant-associated microbiomes and applying bioinoculants enriched with diazotrophic plant growth–promoting bacteria (PGPB) have emerged as promising strategies (Thiebaut et al., 2022). Numerous studies demonstrate that inoculation with diazotrophic bacteria enhances plant performance through multiple growth-promoting mechanisms (summarized in Table 13.2). Their use can substantially reduce, and in some cases replace, synthetic nitrogen fertilizer inputs, thereby improving crop yields while supporting environmentally sustainable production systems.

The expansion of these practices, however, requires careful consideration of environmental and human health safeguards. Concerns persist regarding the functional reliability and biosafety of microbial inoculants, particularly because some bacterial isolates from plant rhizospheres can pose risks to mammals, including humans (Martínez-Hidalgo et al., 2019). To address these issues, biosafety assessment frameworks such as the Environmental and Human Safety Index (EHSI) have been proposed to evaluate candidate strains prior to field application (Vílchez et al., 2016). Importantly, several widely used PGPB, including *Azospirillum* and *Azotobacter*, are

Table 13.2 Multitrait diazotrophs as plant growth promoting bioinoculants

Diazotrophs	PGPR traits	Effect on plant	Plant	Reference
Azospirillum baldaniorum Sp245 and *A. brasilense* ARG2	Ammonia (NH_3) scavenging	Increased plant's N budget	Wheat	Dias et al. (2024)
Stenotrophomonas maltophilia, Enterobacter sp., *Bacillus* sp., *Ochrobactrum haematophilum, Pseudomonas aeruginosa*	Production of IAA, ACC-deaminase, P-solubilization antagonistic activity, lytic enzyme production and biofilm formation	Higher germination percentage, seed vigor index and total dry biomass	Rice	Mir et al. (2022)
Burkholderia cepacia	Nitrogen fixation, production of ACC-deaminase, iron chelators (pyoverdin and pyochelin), growth-promoting substances, including acetoin, 2,3-butanediol and isoprene, Biocontrol activity (as production of chitinase, phenazine and pyrrolnitrin)	Enhanced plant biomass and grain yield	Rice	Li et al. (2022)
Herbaspirillum, Azospirillum	Production of indolic compounds, ACC deaminase, Biofilm, exopolysaccharides, volatile organic compounds	Higher shoot dry biomass, relative water content of leaves under water deficit condition	Perennial ryegrass	Cortés-Patiño et al. (2021)
Pseudomonas koreensis CY4 and P. entomophila CN11	Inducing Nitrogen uptake and defense-related gene expression in crop	Enhanced nitrogen fixation, defense activity, and PGP attributes in Plants	Sugarcane	Singh et al. (2021)
Brevundimonas sp.	P-solubilization	Enhanced potato growth and stimulated nitrogen uptake	Potato	Naqqash et al. (2022)
Azotobacter sp. *strain Avi2*	Biofilm production	Higher yield and photosynthetic rates	Rice	Banik et al. (2019)
Azospirillum brasilense, Rhizobium tropici	Induced systemic tolerance in plants	Protection from the negative effects of saline stress	Maize	Fukami et al. (2018b)

Diazotrophs	PGPR traits	Effect on plant	Plant	Reference
Azospirillum brasilense	Nitrogen fixation, synthesis of phytohormones, induction of plant stress tolerance and defense genes	Plant growth promotion	Maize	Fukami et al. (2017)
Herbaspirillum and Burkholderia	P-solubilization, N-fixation	Higher grain yield and enhanced Nitrogen use efficiency	Rice	Estrada et al. (2013)
Herbaspirillum sp., *Paenibacillus sp. Bacillus* sp., *Serratia* sp., *Xanthomonas* sp.,	ACC-deaminase activity, production of indole 3-acetic acid (IAA), siderophores and ammonia (NH_3), P and zinc (Zn) solubilization and sulfur (S) oxidation,	Enhanced plant growth	Rice and Canola	Islam et al. (2009)

classified as non-pathogenic and fall under Risk Group 1/Biosafety Level 1 (BSL-1), indicating they are generally safe for agricultural use (Estrada et al., 2013). This strengthens the rationale for integrating diazotrophic bioinoculants as sustainable alternatives or supplements to conventional fertilizers.

Despite their proven contributions to improved stress tolerance and enhanced plant adaptability under fluctuating environmental conditions, significant knowledge gaps persist. A deeper mechanistic understanding of plant–microbe interactions, especially the regulatory networks governing these associations, is vital for improving the consistency and effectiveness of microbial inoculants. Advances in monitoring gene expression "in real time" in both plants and associated diazotrophs will be critical for identifying key molecular pathways that can be optimized to enhance bioinoculant performance.

13.6 Harnessing Diazotrophic Rhizobacteria in Agriculture: Prospects and Practical Challenges

Integrating beneficial microorganisms into agricultural systems has become a central strategy for advancing sustainable crop production while reducing reliance on chemical fertilizers, particularly nitrogen-based inputs. In this context, plant-associated microbiomes and bioinoculants formulated with diazotrophic bacteria have gained considerable attention for their capacity to enhance nutrient-use efficiency, promote plant growth, and support the long-term economic and environmental sustainability of farming systems.

Despite substantial progress in the development, formulation, and commercialization of microbial inoculants, achieving consistent and scalable performance under field conditions remains a persistent challenge (Mitter et al., 2021). Microbial strains that exhibit strong growth-promoting effects under laboratory or greenhouse conditions frequently fail to reproduce these benefits across diverse soil types, cropping systems, and climatic environments (Naamala & Smith, 2020). This discrepancy is largely attributable to the stark contrast between controlled experimental settings characterized by aseptic, homogeneous conditions and the complexity and heterogeneity inherent to agricultural soils (Mitter et al., 2021). Once introduced into natural environments, inoculated strains must compete with well-established native microbial communities, which can rapidly outcompete or suppress the introduced organisms, leading to poor persistence or suboptimal colonization (Batista & Singh, 2024). Environmental stressors such as drought, heat, salinity, low soil fertility, or contaminants may further diminish the performance of strains that previously showed robust efficacy in controlled environments. Practical constraints including limited shelf life, incompatibility with agrochemicals or farm machinery, inconsistent application recommendations, and logistical difficulties in transport and storage also hinder successful field deployment (Batista & Singh, 2024).

Overcoming these challenges will require targeted advancements in multiple stages of microbial product development. Improvements in upstream production processes such as optimized fermentation systems, novel bioreactor configurations, and cost-effective growth media—are essential to enhance yield, stability, and scalability. Equally important is the implementation of rigorous quality control and quality assurance frameworks to ensure batch-to-batch consistency and product reliability (Díaz-Rodríguez et al., 2025). This includes continuous monitoring of parameters such as cell viability, population density, strain identity, and formulation stability throughout storage and distribution (Neuhoff et al., 2024).

From a field-performance standpoint, a more comprehensive understanding of interactions among plant growth-promoting microorganisms (PGPMs), host plants, phytopathogens, and resident soil microbiota is needed. These interactions shape inoculant persistence, functional expression, and long-term impacts on soil health and ecosystem stability. The application of next-generation sequencing, metagenomics, transcriptomics, metabolomics, and other multiomics platforms will be critical for elucidating the composition, dynamics, and functional attributes of plant-associated microbiomes. Such integrated insights will support the design of improved microbial formulations, optimization of application timing and delivery methods, and more effective incorporation of bioinoculants into holistic crop management systems.

13.7 Conclusions

Diazotrophic plant growth-promoting rhizobacteria (PGPR) represent a pivotal resource for advancing sustainable agriculture. Their capacity to fix atmospheric nitrogen, coupled with multiple plant growth–promoting functions such as phytohormone synthesis, nutrient solubilization, modulation of ethylene and other stress-related pathways, and activation of induced systemic resistance enables them to substantially enhance plant performance. These benefits extend across both optimal growth conditions and a wide range of abiotic and biotic stresses, including drought, salinity, and pathogen attack. However, despite the strong potential demonstrated under controlled laboratory and greenhouse conditions, field performance of diazotrophic PGPR remains inconsistent. This variability is influenced by complex interactions among the introduced strains, resident soil microbiota, edaphic factors, climatic conditions, and host plant genotypes. Overcoming these challenges requires a deeper mechanistic understanding of plant–microbe–soil interactions. Emerging multi-omics tools including metagenomics, transcriptomics, proteomics, and metabolomics combined with functional genomics and systems-level analyses, will be essential for identifying the microbial genes, regulatory networks, and metabolic pathways that underpin successful colonization, nitrogen fixation, and stress alleviation. Future research should prioritize the development of more resilient and reliable microbial formulations, the rational design of synthetic microbial consortia that mimic natural microbiome functions, and the integration of microbial inoculants with precision agriculture technologies. Such strategies will enhance strain survival, activity, and consistency under field conditions. Ultimately, the effective deployment of diazotrophic PGPR can reduce dependence on synthetic nitrogen fertilizers, improve crop productivity, and contribute to the establishment of resilient and environmentally sustainable agroecosystems.

References

Acuña, J. J., Campos, M., de la Luz Mora, M., Jaisi, D. P., & Jorquera, M. A. (2019). ACCD producing rhizobacteria from an Andean Altiplano native plant (*Parastrephia quadrangularis*) and their potential to alleviate salt stress in wheat seedlings. *Applied Soil Ecology, 136*, 184–190.

Akram, W., Anjum, T., & Ali, B. (2015). Searching ISR determinant/s from *Bacillus subtilis* IAGS174 against *Fusarium* wilt of tomato. *Biological Control, 60*(2), 271–280. https://doi.org/10.1007/s10526-014-9636-1.

Allu, S., Kumar, N. P., & Audipudi, A. V. (2014). Isolation, biochemical and PGP characterization of endophytic *Pseudomonas aeruginosa* isolated from chilli red fruit antagonistic against chilli anthracnose disease. *International Journal of Current Microbiology and Applied Sciences, 3*(2), 318–329.

Alori, E. T., Glick, B. R., & Babalola, O. O. (2017). Microbial phosphorus solubilization and its potential for use in sustainable agriculture. *Frontiers in Microbiology, 8*, 971. https://doi.org/10.3389/fmicb.2017.00971.

Alzate Zuluaga, M. Y., Fattorini, R., Cesco, S., & Pii, Y. (2024). Plant-microbe interactions in the rhizosphere for smarter and more sustainable crop fertilization: The case of PGPR-based biofertilizers. *Frontiers in Microbiology, 15*, 1440978.

An, R., & Moe, L. A. (2016). Regulation of pyrroloquinoline quinone-dependent glucose dehydrogenase activity in the model rhizosphere-dwelling bacterium *Pseudomonas putida* KT2440. *Applied and Environmental Microbiology, 82*(16), 4955–4964. https://doi.org/10.1128/AEM.00813-16.

Angus, J., & Grace, P. (2017). Nitrogen balance in Australia and nitrogen use efficiency on Australian farms. *Soil Research, 55*, 435–450. https://doi.org/10.1071/SR16325.

Araújo, A. E. D. S., Rossetto, C. A. V., Baldani, V. L. D., & Baldani, J. I. (2010). Rice seed germination and vigour as affected by the inoculation with diazotrophic bacteria. *Ciência e Agrotecnologia, 34*(5), 932–939.

Arencibia, A. D., Vinagre, F., Estevez, Y., Bernal, A., Perez, J., Cavalcanti, J., Santana, I., & Hemerly, A. S. (2006). *Gluconacetobacter diazotrophicus* elicits a sugarcane defense response against a pathogenic bacteria *Xanthomonas albilineans*. *Plant Signaling & Behavior, 1*(5), 265–273. https://doi.org/10.4161/psb.1.5.3390.

Banik, A., Dash, G. K., Swain, P., Kumar, U., Mukhopadhyay, S. K., & Dangar, T. K. (2019). Application of rice (*Oryza sativa* L.) root endophytic diazotrophic *Azotobacter* sp. strain Avi2 (MCC 3432) can increase rice yield under greenhouse and field condition. *Microbiological Research, 219*, 56–65. https://doi.org/10.1016/j.micres.2018.11.005.

Bargaz, A., Elhaissoufi, W., Khourchi, S., Benmrid, B., Borden, K. A., & Rchiad, Z. (2021). Benefits of phosphate solubilizing bacteria on belowground crop performance for improved crop acquisition of phosphorus. *Microbiological Research, 252*, 126842. https://doi.org/10.1016/j.micres.2021.126842.

Batista, B. D., & Singh, B. K. (2024). Realities and hopes in the application of microbial tools in agriculture. *Microbial Biotechnology, 14*(4), 1751–7915. https://doi.org/10.1111/1751-7915.13866.

Calanca, P. P. (2017). Effects of abiotic stress in crop production. In *Quantification of climate variability, adaptation and mitigation for agricultural sustainability*. (pp. 165–180). Springer International Publishing.

Carvalho, T. L., Balsemão-Pires, E., Saraiva, R. M., Ferreira, P. C., & Hemerly, A. S. (2014). Nitrogen signalling in plant interactions with associative and endophytic diazotrophic bacteria. *Journal of Experimental Botany, 65*(20), 5631–5642. https://doi.org/10.1093/jxb/eru292.

Chakraborty, S., Venkataraman, S., Infante, F., Pfleger, B. F., & Ané, J. M. (2024). Scripting a new dialogue between diazotrophs and crops. *Trends in Microbiology, 32*(6), 577–589.

Chakraborty, U., Chakraborty, B. N., & Chakraborty, A. P. (2010). Influence of *Serratia marcescens* TRS-1 on growth promotion and induction of resistance in *Camellia sinensis* against *Fomes lamaoensis*. *Journal of Plant Interactions, 5*(4), 261–272. https://doi.org/10.1080/17429140903551738.

Cohen, A. C., Bottini, R., Pontin, M., Berli, F. J., Moreno, D., Boccalandro, H., Travaglia, C. N., & Piccoli, P. N. (2015). Azospirillum brasilense ameliorates the response of Arabidopsis thaliana to drought mainly via enhancement of ABA levels. Physiologia Plantarum, 153, 79–90. https://doi.org/10.1111/ppl.12221.

Collado, L., Pizarro, A. H., Barawi, M., García-Tecedor, M., Liras, M., & de la Peña O'Shea, V. A. (2024). Light-driven nitrogen fixation routes for green ammonia production. Chemical Society Reviews, 53(23), 11334–11389. https://doi.org/10.1039/D3CS01075A.

Compant, S., Cassan, F., Kostić, T., Johnson, L., Brader, G., Trognitz, F., & Sessitsch, A. (2025). Harnessing the plant microbiome for sustainable crop production. *Nature Reviews Microbiology*, *23*, 9–23.

Condori, T., Alarcon, S., Huasasquiche, L., Garcia-Blasquez, C., Padilla-Castro, C., Velasquez, J., & Solorzano, R. (2024). Inoculation with *Azospirillum brasilense* as a strategy to reduce nitrogen fertilization in cultivating purple maize (*Zea mays* L.) in the Inter-Andean Valleys of Peru. *Microorganisms*, *12*, 2107. https://doi.org/10.3390/microorganisms12102107

Cortés-Patiño, S., Vargas, C., Álvarez-Flórez, F., Bonilla, R., & Estrada-Bonilla, G. (2021). Potential of *Herbaspirillum* and *Azospirillum* consortium to promote growth of perennial ryegrass under water deficit. *Microorganisms*, *9*(1), 91. https://doi.org/10.3390/microorganisms9010091.

Cowan, N., White, S., Olszewska, J., Dobel, A., Sim, G., Eades, L. J., & Skiba, U. (2022). Integration of algae treatment with hydroponic crop waste to reduce impact of nutrient waste streams. *Journal of Sustainable Agriculture and Environment*, *1*, 203–215.

Curá, J. A., Franz, D. R., Filosofía, J. E., Balestrasse, K. B., & Burgueño, L. E. (2017). Inoculation with Azospirillum sp. and Herbaspirillum sp. bacteria increases the tolerance of maize to drought stress. Microorganisms, 5, 41. https://doi.org/10.3390/microorganisms5030041.

Dalton, D. A., & Kramer, S. (2007). Nitrogen-fixing bacteria in non-legumes. In S. S. Gnanamanickam (Ed.), *Plant-associated bacteria* (pp. 39–58). Springer. https://doi.org/10.1007/978-1-4020-4538-7_3

Degon, Z., Dixon, S., Rahmatallah, Y., Galloway, M., Gulutzo, S., Price, H., Cook, J., Glazko, G., & Mukherjee, A. (2023). *Azospirillum brasilense* improves rice growth under salt stress by regulating the expression of key genes involved in salt stress response, abscisic acid signaling, and nutrient transport, among others. *Frontiers in Agronomy*, *5*, 1216503. https://doi.org/10.3389/fagro.2023.121650.

Deppenmeier, U., Hoffmeister, M., & Prust, C. (2002). Biochemistry and biotechnological applications of *Gluconobacter* strains. *Applied Microbiology and Biotechnology*, *60*(3), 233–242. https://doi.org/10.1007/s00253-002-1114-5.

Dias, T., Azmaliyev, K., Melo, J., Santos, A. M., Correia, P., & Cruz, C. (2024). The diazotrophic bacteria *Azospirillum baldaniorum* and *A. brasilense* improve wheat seedlings' nitrogen budget through ammonia scavenging. *Applied Soil Ecology*, *204*, 105737. https://doi.org/10.1016/j.apsoil.2024.105737.

Díaz-Rodríguez, A. M., Parra Cota, F. I., Cira Chávez, L. A., García Ortega, L. F., Estrada Alvarado, M. I., Santoyo, G., & de los Santos-Villalobos, S. (2025). Microbial inoculants in sustainable agriculture: Advancements, challenges, and future directions. *Plants*, *14*(2), 191. https://doi.org/10.3390/plants14020191.

Dobbelaere, S., Vanderleyden, J., & Okon, Y. (2003). Plant growth-promoting effects of diazotrophs in the rhizosphere. *Critical Reviews in Plant Sciences*, *22*(2), 107–149. https://doi.org/10.1080/713610853.

Drogue, B., Sanguin, H., Chamam, A., Mozar, M., Llauro, C., Panaud, O., Prigent-Combaret, C., Picault, N., & Wisniewski-Dyé, F. (2014). Plant root transcriptome profiling reveals a strain-dependent response during *Azospirillum*-rice cooperation. *Frontiers in Plant Science*, *5*, 607. https://doi.org/10.3389/fpls.2014.00607.

Duan, J., Jiang, W., Cheng, Z., Heikkila, J. J., & Glick, B. R. (2013). The complete genome sequence of the plant growth-promoting bacterium *Pseudomonas* sp. UW4. *PLoS One, 8*(3), e58640. https://doi.org/10.1371/journal.pone.0058640.

Estrada, G. A., Baldani, V. L. D., de Oliveira, D. M., Urquiaga, S., & Baldani, J. I. (2013). Selection of phosphate-solubilizing diazotrophic *Herbaspirillum* and *Burkholderia* strains and their effect on rice crop yield and nutrient uptake. *Plant and Soil*, *369*(1), 115–129. https://doi.org/10.1007/s11104-012-1550-7.

Estrada-De los Santos, P., Bustillos-Cristales, R., & Caballero-Mellado, J. (2001). *Burkholderia*, a genus rich in plant-associated nitrogen fixers with wide environmental and geographic distribution. *Applied and Environmental Microbiology*, *67*(6), 2790–2798. https://doi.org/10.1128/AEM.67.6.2790-2798.2001.

Etesami, H. (2025). The dual nature of plant growth-promoting bacteria: Benefits, risks, and pathways to sustainable deployment. *Current Research in Microbial Sciences*, *9*, 100421. https://doi.org/10.1016/j.crmicr.2025.100421.

Fasciglione, G., Casanovas, E. M., Quillehauquy, V., Yommi, A. K., Goni, M. G., Roura, S. I., & Barassi, C. A. (2015). *Azospirillum* inoculation effects on growth, product quality and storage life of lettuce plants grown under salt stress. *Scientia Horticulturae*, *195*, 154–162. https://doi.org/10.1016/j.scienta.2015.09.015.

Favero, V. F., de Carvalho, R. H., Leite, A. B. C., dos Santos, D. M. T., de Freitas, K. M., Boddey, R. M., Xavier, G. R., Rumjanek, N. G., & Urquiaga, S. (2022). *Bradyrhizobium* strains from Brazilian tropical soils promote increases in nodulation, growth and nitrogen fixation in mung bean. *Applied Soil Ecology*, *175*, 104461. https://doi.org/10.1016/j.apsoil.2022.104461.

Fernández-Llamosas, H., Díaz, E., & Carmona, M. (2021). Motility, adhesion and c-di-GMP influence the endophytic colonization of rice by *Azoarcus* sp. CIB. *Microorganisms*, *9*(3), 554. https://doi.org/10.3390/microorganisms9030554.

Flores, E., & Herrero, A. (2010). Compartmentalized function through cell differentiation in filamentous cyanobacteria. *Nature Reviews Microbiology*, *8*(1), 39–50. https://doi.org/10.1038/nrmicro2242.

Franche, C., Lindström, K., & Elmerich, C. (2009). Nitrogen-fixing bacteria associated with leguminous and non-leguminous plants. *Plant and Soil*, *321*(1–2), 35–59. https://doi.org/10.1007/s11104-008-9833-8.

Fukami, J., Cerezini, P., & Hungria, M. (2018a). *Azospirillum*: Benefits that go far beyond biological nitrogen fixation. *AMB Express*, *8*(1), 73. https://doi.org/10.1186/s13568-018-0608-1.

Fukami, J., Ollero, F. J., Megías, M., & Hungria, M. (2017). Phytohormones and induction of plant stress tolerance and defense genes by seed and foliar inoculation with *Azospirillum brasilense* cells and metabolites promote maize growth. *AMB Express*, *7*, 153. https://doi.org/10.1186/s13568-017-0453-7.

Fukami, J., de la Osa, C., Ollero, F. J., Megías, M., & Hungria, M. (2018b). Co-inoculation of maize with *Azospirillum brasilense* and *Rhizobium tropici* as a strategy to mitigate salinity stress. *Functional Plant Biology*, *45*(3), 328–339. https://doi.org/10.1071/FP17167.

Galindo, F. S., Pagliari, P. H., Fernandes, G. C., Rodrigues, W. L., Boleta, E. H. M., Jalal, A., Céu, E. G. O., Lima, B. H. d, Lavres, J., & Teixeira Filho, M. C. M. (2022). Improving sustainable field-grown wheat production with *Azospirillum brasilense* under tropical conditions: A potential tool for improving nitrogen management. *Frontiers in Environmental Science*, *10*, 821628. https://doi.org/10.3389/fenvs.2022.821628.

Galloway, J. N., Leach, A. M., Bleeker, A., & Erisman, J. W. (2013). A chronology of human understanding of the nitrogen cycle. *Philosophical Transactions of the Royal Society B: Biological Sciences*, *368*(1621), 20130120. https://doi.org/10.1098/rstb.2013.0120.

Gao, Y., & Cabrera Serrenho, A. (2023). Greenhouse gas emissions from nitrogen fertilizers could be reduced by up to one-fifth of current levels by 2050 with combined interventions. *Nature Food*, *4*(2), 170–178. https://doi.org/10.1038/s43016-023-00698-w.

Geetha, J., Venkatesham, E., Hindumathi, A., & Bhadraiah, B. (2014). Isolation, screening and characterization of plant growth promoting bacteria and their effect on *Vigna radiata* (L.) R. Wilczek. *International Journal of Current Microbiology and Applied Sciences*, *3*(6), 799–803.

Gill, S. S., & Tuteja, N. (2010). Reactive oxygen species and antioxidant machinery in abiotic stress tolerance in crop plants. *Plant Physiology and Biochemistry, 48*(12), 909–930. https://doi.org/10.1016/j.plaphy.2010.08.016.

Glick, B. R. (2014). Bacteria with ACC deaminase can promote plant growth and help to feed the world. *Microbiological Research, 169*(1), 30–39. https://doi.org/10.1016/j.micres.2013.09.009.

Greetatorn, T., Hashimoto, S., Sarapat, S., Tittabutr, P., Boonkerd, N., Uchiumi, T., & Teaumroong, N. (2019). Empowering rice seedling growth by endophytic *Bradyrhizobium* sp. SUTN9-2. *Letters in Applied Microbiology, 68*(3), 258–266. https://doi.org/10.1111/lam.13114.

Gupta, G., Panwar, J., Akhtar, M. S., & Jha, N. P. (2012). Endophytic nitrogen fixing bacteria as biofertilizer. In E. Lichtfouse (Ed.), *Sustainable agriculture reviews* (Vol. 1, pp. 183–221). Springer. https://doi.org/10.1007/978-94-007-5449-2_8

Gupta, S. M., Kumar, K., Joshi, R. K., Gupta, S., & Bala, M. (2020). Frankia: A promising N-fixing plant growth promoting rhizobacteria (PGPR) improved drought tolerance in crops at higher altitude. In *Microbiological Advancements for Higher Altitude Agro-Ecosystems & Sustainability* (pp. 411–431). Springer. https://doi.org/10.1007/978-981-15-1987-9_20

Hakim, S., Naqqash, T., Nawaz, M. S., Laraib, I., Siddique, M. J., Zia, R., Mirza, M. S., & Imran, A. (2021). Rhizosphere engineering with plant growth-promoting microorganisms for agriculture and ecological sustainability. *Frontiers in Sustainable Food Systems, 5*, 617157. https://doi.org/10.3389/fsufs.2021.617157.

Hamdia, M. A. E. S., Shaddad, M. A. K., & Doaa, M. M. (2004). Mechanisms of salt tolerance and interactive effects of *Azospirillum brasilense* inoculation on maize cultivars grown under salt stress conditions. *Plant Growth Regulation, 44*(2), 165–174. https://doi.org/10.1007/s10725-004-3131-0.

Hasanuzzaman, M., Bhuyan, M. H. M. B., Zulfiqar, F., Raza, A., Mohsin, S. M., Mahmud, J. A., Fujita, M., & Fotopoulos, V. (2020). Reactive oxygen species and antioxidant defense in plants under abiotic stress: Revisiting the crucial role of a universal defense regulator. Antioxidants, 9(8), 681. https://doi.org/10.3390/antiox9080681.

Hellriegel, H., & Wilfarth, H. (1888). Untersuchungen über die Stickstoffnahrung der Gramineen und Leguminosen. Beilageheft zu der Zeitschrift des Vereins für die Rübenzucker-Industrie des Deutschen Reichs, Berlin, Germany 234 pp.

Hmaeid, N., Wali, M., Mahmoud, O. M.-B., Pueyo, J. J., Ghnaya, T., & Abdelly, C. (2019). Efficient rhizobacteria promote growth and alleviate NaCl-induced stress in the plant species *Sulla carnosa. Applied Soil Ecology, 133*, 104–113. https://doi.org/10.1016/j.apsoil.2018.09.017.

Holland, B. L., Matthews, M. L., Bota, P., Sweetlove, L. J., Long, S. P., & diCenzo, G. C. (2023). A genome-scale metabolic reconstruction of soybean and Bradyrhizobium diazoefficiens reveals the cost–benefit of nitrogen fixation. New Phytologist, 240, 744–756. https://doi.org/10.1111/nph.19203.

Huang, L., He, B., Han, L., Liu, J., Wang, H., & Chen, Z. (2017). A global examination of the response of ecosystem water-use efficiency to drought based on MODIS data. *Science of the Total Environment, 601–602*, 1097–1107. https://doi.org/10.1016/j.scitotenv.2017.05.241.

Hungria, M., Nogueira, M. A., & Araujo, R. S. (2015). Soybean seed co-inoculation with Bradyrhizobium spp. and Azospirillum brasilense: a new biotechnological tool to improve yield and sustainability. American Journal of Plant Sciences, 6, 811–817. https://doi.org/10.4236/ajps.2015.66087.

Hwang, H.-H., Chien, P.-R., Huang, F.-C., Hung, S.-H., Kuo, C.-H., Deng, W.-L., Chiang, E.-P. I., & Huang, C.-C. (2021). A plant endophytic bacterium, *Burkholderia seminalis* strain 869T2, promotes plant growth in *Arabidopsis, Pak Choi, Chinese Amaranth, Lettuces*, and other vegetables. *Microorganisms, 9*(8), 1703. https://doi.org/10.3390/microorganisms9081703.

Imran, A., Hakim, S., Tariq, M., Nawaz, M. S., Laraib, I., Gulzar, U., Hanif, M. K., Siddique, M. J., Hayat, M., Fraz, A., & Ahmad, M. (2021). Diazotrophs for lowering nitrogen pollution crises: Looking deep into the roots. *Frontiers in Microbiology, 12*, 637815. https://doi.org/10.3389/fmicb.2021.637815.

Iniguez, L. A., Dong, Y., & Triplett, E. W. (2004). Nitrogen fixation in wheat provided by *Klebsiella pneumoniae* 342. *Molecular Plant-Microbe Interactions*, *17*(10), 1078–1085. https://doi.org/10.1094/MPMI.2004.17.10.1078.

Inomura, K., Bragg, J., Riemann, L., & Follows, M. J. (2018). A quantitative model of nitrogen fixation in the presence of ammonium. *PLoS One*, *13*(11), e0208282. https://doi.org/10.1371/journal.pone.0208282.

Islam, M. T., Deora, A., Hashidoko, Y., Rahman, A., Ito, T., & Tahara, S. (2009). Characterization of plant growth-promoting traits of free-living diazotrophic bacteria and their inoculation effects on growth and nitrogen uptake of crop plants. *Soil Science and Plant Nutrition*, *55*(1), 91–101. https://doi.org/10.1111/j.1747-0765.2008.00325.x.

Ji, S. H., Gururani, M. A., & Chun, S.-C. (2014). Isolation and characterization of plant growth promoting endophytic diazotrophic bacteria from Korean rice cultivar. *Microbiological Research*, *169*(1), 83–98. https://doi.org/10.1016/j.micres.2013.06.003.

Joshi, S., Gangola, S., Jaggi, V., Sharma, P., Pathak, R. K., & Sharma, A. (2023). Functional characterization and molecular fingerprinting of potential phosphate solubilizing bacterial candidates from *Shisham rhizosphere*. *Scientific Reports*, *13*, 7003. https://doi.org/10.1038/s41598-023-33217-9.

Kang, S. M., Shahzad, R., Bilal, S., Khan, A. L., Park, Y. G., Lee, K. E., & Lee, I. J. (2019). Indole-3-acetic-acid and ACC deaminase producing *Leclercia adecarboxylata* MO1 improves *Solanum lycopersicum* L. growth and salinity stress tolerance by endogenous secondary metabolites regulation. *BMC Microbiology*, *19*(1), 80. https://doi.org/10.1186/s12866-019-1450-9.

Kaziūnienė, J., Gegeckas, A., Lapinskienė, L., Razbadauskienė, K., Mažylytė, R., & Supronienė, S. (2025). Competitiveness and nitrogen fixation efficiency analysis of *Rhizobium leguminosarum* strains in different field pea (*Pisum sativum* L.) genotypes. *Agriculture*, *15*(16), 1784. https://doi.org/10.3390/agriculture15161784.

Kearsley, J. V. S., Sather, L. M., & Finan, T. M. (2024). *Sinorhizobium (Ensifer) meliloti* . *Trends in Microbiology*, *32*(5), 516–518. https://doi.org/10.1016/j.tim.2024.03.004.

Kerečki, S., Pećinar, I., Karličić, V., Mirković, N., Kljujev, I., Raičević, V., & Jovičić-Petrović, J. (2022). *Azotobacter chroococcum* F8/2: A multitasking bacterial strain in sugar beet biopriming. *Journal of Plant Interactions*, *17*(1), 719–730. https://doi.org/10.1080/17429145.2022.2091802.

Khan, F., Siddique, A. B., Shabala, S., Zhou, M., & Zhao, C. (2023). Phosphorus plays key roles in regulating plants' physiological responses to abiotic stresses. *Plants*, *12*(15), 2861. https://doi.org/10.3390/plants12152861.

Khoso, M. A., Wagan, S., Alam, I., Hussain, A., Ali, Q., Saha, S., Poudel, T. R., Manghwar, H., & Liu, F. (2024). Impact of plant growth-promoting rhizobacteria (PGPR) on plant nutrition and root characteristics: current perspective. Plant Stress, 11, 100341. https://doi.org/10.1016/j.stress.2023.100341.

Kumar, A., Singh, S., Gaurav, A. K., Srivastava, S., & Verma, J. P. (2020). Plant growth-promoting bacteria: Biological tools for the mitigation of salinity stress in plants. *Frontiers in Microbiology*, *11*, 1216. https://doi.org/10.3389/fmicb.2020.01216.

Kumar, L., Chhogyel, N., Gopalakrishnan, T., Hasan, M. K., Jayasinghe, S. L., Kariyawasam, C. S., Kogo, B. K., & Ratnayake, S. (2022). Climate change and future of agri-food production. In *Future foods* (pp. 49–79). Elsevier. https://doi.org/10.1016/B978-0-323-91034-8.00011-0

Kumari, V. V., Banerjee, P., Verma, V. C., Sukumaran, S., Chandran, M. A. S., Gopinath, K. A., Venkatesh, G., Yadav, S. K., Singh, V. K., & Awasthi, N. K. (2022). Plant nutrition: An effective way to alleviate abiotic stress in agricultural crops. *International Journal of Molecular Sciences*, *23*(15), 8519. https://doi.org/10.3390/ijms23158519.

Ladha, J. K., Peoples, M. B., Reddy, M. B., Biswas, J. C., Bennett, A., Jat, M. L., & Krupnik, T. J. et al. (2022). Biological nitrogen fixation and prospects for ecological intensification in cereal-based cropping systems. *Field Crops Research*, 283, 108541. 10.1016/j.fcr.2022.108541.

Lery, L. M., Bitar, M., Costa, M. G., Rössle, S. C., & Bisch, P. M. (2010). Unraveling the molecular mechanisms of nitrogenase conformational protection against oxygen in diazotrophic bacteria. *BMC Genomics, 11*(Suppl 5), S7. https://doi.org/10.1186/1471-2164-11-S5-S7.

Li, X., & Xiao, R. (2025). Molecular dialogue in legume-rhizobium symbiosis: Signaling mechanisms and genetic insights. *Rhizosphere, 33*, 101034. https://doi.org/10.1016/j.rhisph.2025.101034.

Li, Z., Henawy, A. R., Halema, A. A., Fan, Q., Duanmu, D., & Huang, R. (2022). A wild rice rhizobacterium *Burkholderia cepacia* BRDJ enhances nitrogen use efficiency in rice. *International Journal of Molecular Sciences, 23*(18), 10769. https://doi.org/10.3390/ijms231810769.

Li, Q., Liu, S., Li, Y., Hao, T., & Chen, S (2022). Nitrogen fixation by *Paenibacillus polymyxa* WLY78 is responsible for cucumber growth promotion. *Plant and Soil*, 473, 507–516. https://doi.org/10.1007/s11104-022-05307-6.

Loaces, I., Ferrando, L., & Scavino, A. F. (2011). Dynamics, diversity and function of endophytic siderophore-producing bacteria in rice. *Microbial Ecology, 61*(4), 606–618. https://doi.org/10.1007/s00248-010-9780-9.

van Loon, L. C. (2007). Plant responses to plant growth-promoting rhizobacteria. *European Journal of Plant Pathology, 119*(3), 243–254. https://doi.org/10.1007/s10658-007-9165-1.

Macedo-Raygoza, G. M., Valdez-Salas, B., Prado, F. M., Prieto, K. R., Yamaguchi, L. F., Kato, M. J., Canto-Canché, B. B., Carrillo-Beltrán, M., Di Mascio, P., White, J. F., & Beltrán-García, M. J. (2019). *Enterobacter cloacae*, an endophyte that establishes a nutrient-transfer symbiosis with banana plants and protects against the black Sigatoka pathogen. *Frontiers in Microbiology, 10*, 804. https://doi.org/10.3389/fmicb.2019.00804.

Mandon, K., Nazaret, F., Farajzadeh, D., Alloing, G., & Frendo, P. (2021). Redox regulation in diazotrophic bacteria in interaction with plants. *Antioxidants, 10*(6), 880. https://doi.org/10.3390/antiox10060880.

Martínez-Hidalgo, P., Maymon, M., Pule-Meulenberg, F., & Hirsch, A. M. (2019). Engineering root microbiomes for healthier crops and soils using beneficial, environmentally safe bacteria. *Canadian Journal of Microbiology, 65*(2), 91–104. https://doi.org/10.1139/cjm-2018-0315.

Martínez-Viveros, O., Jorquera, M. A., Crowley, D. E., Gajardo, G., & Mora, M. L. (2010). Mechanisms and practical considerations involved in plant growth promotion by rhizobacteria. *Journal of Soil Science and Plant Nutrition, 10*(3), 293–319. https://doi.org/10.4067/S0718-95162010000100006.

Mathesius, U. (2022). Are legumes different? Origins and consequences of evolving nitrogen fixing symbioses. *Journal of Plant Physiology, 276*, 153765. https://doi.org/10.1016/j.jplph.2022.153765.

Melillo, J. M. (2021). Disruption of the global nitrogen cycle: A grand challenge for the twenty-first century. *Ambio, 50*(4), 759–763. https://doi.org/10.1007/s13280-020-01448-y.

Mir, M. I., Hameeda, B., Quadriya, H., Kumar, B. K., Ilyas, N., Kee Zuan, A. T., El Enshasy, H. A., Dailin, D. J., Kassem, H. S., Gafur, A., & Sayyed, R. Z. (2022). Multifarious indigenous diazotrophic rhizobacteria of rice (*Oryza sativa* L.) rhizosphere and their effect on plant growth promotion. *Frontiers in Nutrition, 8*, 781764. https://doi.org/10.3389/fnut.2021.781764.

Mishra, A., Rajput, S., Gupta, P. S., Goyal, V., Singh, S., Sharma, S., Shukla, S., Singh, A., Shukla, K., & Varma, A. (2021). Role of cyanobacteria in rhizospheric nitrogen fixation. In A. Varma, R. K. Tripathi, & S. Prasad (Eds.), *Soil nitrogen ecology* (pp. 497–519). Springer. https://doi.org/10.1007/978-3-030-79792-2_20

Mitter, E. K., Tosi, M., Obregon, D., Dunfield, K. E., & Germida, J. J. (2021). Rethinking crop nutrition in times of modern microbiology: Innovative biofertilizer technologies. *Frontiers in Sustainable Food Systems, 5*, 29. https://doi.org/10.3389/fsufs.2021.00029.

Molnár, Z., Solomon, W., Mutum, L., & Janda, T. (2023). Understanding the mechanisms of Fe deficiency in the rhizosphere to promote plant resilience. *Plants, 12*(10), 1945. https://doi.org/10.3390/plants12101945.

Naamala, J., & Smith, D. L. (2020). Relevance of plant growth-promoting microorganisms and their derived compounds in the face of climate change. *Agronomy*, *10*(8), 1179. https://doi.org/10.3390/agronomy10081179.

Naqqash, T., Malik, K. A., Imran, A., Hameed, S., Shahid, M., Hanif, M. K., Majeed, A., Iqbal, M. J., Qaisrani, M. M., & van Elsas, J. D. (2022). Inoculation with *Azospirillum* spp. acts as the liming source for improving growth and nitrogen use efficiency of potato. *Frontiers in Plant Science*, *13*, 929114. https://doi.org/10.3389/fpls.2022.929114.

Nascimento, F. X., Tavares, M. J., Franck, J., Ali, S., Glick, B. R., & Rossi, M. J. (2019). ACC deaminase plays a major role in *Pseudomonas fluorescens* YsS6 ability to promote the nodulation of Alpha- and Betaproteobacteria rhizobial strains. *Archives of Microbiology*, *201*(6), 817–822. https://doi.org/10.1007/s00203-019-01658-w.

Neilands, J. B. (1995). Siderophores: Structure and function of microbial iron transport compounds. *Journal of Biological Chemistry*, *270*(45), 26723–26726. https://doi.org/10.1074/jbc.270.45.26723.

Neuhoff, D., Neumann, G., & Weinmann, M. (2024). Testing plant growth promoting microorganisms in the field—A proposal for standards. *Frontiers in Plant Science*, *14*, 1324665. https://doi.org/10.3389/fpls.2023.1324665.

Ngumbi, E., & Kloepper, J. (2016). Bacterial-mediated drought tolerance: Current and future prospects. Applied Soil Ecology, 105, 109–125. https://doi.org/10.1016/j.apsoil.2016.04.009.

Oelze, J. (2000). Respiratory protection of nitrogenase in Azotobacter species: Is a widely held hypothesis unequivocally supported by experimental evidence?. *FEMS Microbiology Reviews*, *24*(3), 321–333. https://doi.org/10.1111/j.1574-6976.2000.tb00544.x.

Olanrewaju, O. S., Glick, B. R., & Babalola, O. O. (2017). Mechanisms of action of plant growth promoting bacteria. *World Journal of Microbiology and Biotechnology*, *33*(11), 197. https://doi.org/10.1007/s11274-017-2364-9.

Ormeño-Orrillo, E., Hungria, M., & Martinez-Romero, E. (2013). Dinitrogen-fixing prokaryotes. In E. Rosenberg, E. F. DeLong, S. Lory, E. Stackebrandt, & F. Thompson (Eds.), *The prokaryotes* (pp. 427–451). Springer. https://doi.org/10.1007/978-3-642-30194-0_61

Orozco-Mosqueda, M. C., Glick, B. R., & Santoyo, G. (2020). ACC deaminase in plant growth-promoting bacteria (PGPB): An efficient mechanism to counter salt stress in crops. *Microbiological Research*, *235*, 126439. https://doi.org/10.1016/j.micres.2020.126439.

Pal, G., Saxena, S., Kumar, K., Verma, A., Sahu, P. K., Pandey, A., White, J. F., & Verma, S. K. (2022). Endophytic *Burkholderia*: Multifunctional roles in plant growth promotion and stress tolerance. *Microbiological Research*, *265*, 127201. https://doi.org/10.1016/j.micres.2022.127201.

Rajani, G., Latha, P. C., Sundaram, R. M., Bandeppa, A., Phule, A. S., Prasad Babu, K. V., Barbadikar, K. M., Prasad Babu, M. B. B., Mandal, P. K., & Surekha Rani, H. (2023). Effect of plant growth promoting endophytic bacterium *Gluconacetobacter diazotrophicus* on germination attributes and seedling growth of rice varieties under *in vitro* conditions. *International Journal of Plant & Soil Science*, *35*(20), 62–71. https://doi.org/10.9734/ijpss/2023/v35i203786.

Ramos, A. C., Melo, J., de Souza, S. B., Bertolazi, A. A., Silva, R. A., Rodrigues, W. P., Campostrini, E., Olivares, F. L., Eutrópio, F. J., Cruz, C., & Dias, T. (2020). Inoculation with the endophytic bacterium *Herbaspirillum seropedicae* promotes growth, nutrient uptake and photosynthetic efficiency in rice. *Planta*, 252, 87. https://doi.org/10.1007/s00425-020-03496-x.

Ravishankara, A. R., Daniel, J. S., & Portmann, R. W. (2009). Nitrous oxide (N_2O): The dominant ozone-depleting substance emitted in the 21st century. *Science*, *326*(5949), 123–125. https://doi.org/10.1126/science.1176985.

Renganathan, P., Puente, E. O., Sukhanova, N. V., & Gaysina, L. A. (2024). Hydroponics with microalgae and cyanobacteria: Emerging trends and opportunities in modern agriculture. *Biotechnology*, *13*(2), 27. https://doi.org/10.3390/biotech13020027.

Renganathan, P., Astorga-Eló, M., Gaysina, L. A., Puente, E. O. R., & Sainz-Hernández, J. C. (2025). Nitrogen fixation by diazotrophs: A sustainable alternative to synthetic fertilizers in hydroponic cultivation. *Sustainability*, *17*(13), 5922. https://doi.org/10.3390/su17135922.

Saeed, Q., Xiukang, W., Haider, F. U., Kučerik, J., Mumtaz, M., Holatko, J., Naseem, M., Naveed, M., & Brtnicky, M. (2016). Azospirillum brasilense and Azospirillum lipoferum improve growth, physiology and antioxidant status of canola under drought stress. Plant Physiology and Biochemistry, 109, 593–602.

Saharan BS, Nehra V (2011). Plant growth promoting rhizobacteria: a critical review. Life Sci Med Res 2011:LSMR-21.

Scaria, S. S., & Ravi, L. (2022). Symbiotic associations of Frankia in actinorhizal plants. In *Microbial Symbionts* (pp. 397–416). Academic Press. https://doi.org/10.1016/B978-0-323-91074-0.00029-4

Shah, K., Chaturvedi, V., & Gupta, S. (2019). Climate change and abiotic stress-induced oxidative burst in rice. In M. Hasanuzzaman, M. Fujita, K. Nahar, & J. K. Biswas (Eds.), *Advances in rice research for abiotic stress tolerance* (pp. 505–535). Elsevier. https://doi.org/10.1016/B978-0-12-814332-2.00022-3

Shahid, M., Singh, U. B., Khan, M. S., Singh, P., Kumar, R., Singh, R. N., Kumar, A., & Singh, H. V. (2023). Bacterial ACC deaminase: Insights into enzymology, biochemistry, genetics, and potential role in amelioration of environmental stress in crop plants. *Frontiers in Microbiology*, *14*, 1132770. https://doi.org/10.3389/fmicb.2023.1132770.

Shin, W., Islam, R., Benson, A., Joe, M. M., Kim, K., Gopal, S., Sa, T., & others. (2016). Role of diazotrophic bacteria in biological nitrogen fixation and plant growth improvement. *Korean Journal of Soil Science and Fertilizer*, *49*(1), 17–29. https://doi.org/10.7745/KJSSF.2016.49.1.017.

Simova-Stoilova, L., Demirevska, K., Petrova, T., Tsenov, N., & Feller, U. (2008). Antioxidative protection in wheat varieties under severe recoverable drought at seedling stage. Plant, Soil and Environment, 54, 529–536.

Singh, P., Singh, R. K., Li, H.-B., Guo, D.-J., Sharma, A., Lakshmanan, P., Malviya, M. K., Song, X.-P., Solanki, M. K., Verma, K. K., Yang, L.-T., & Li, Y.-R. (2021). Diazotrophic bacteria *Pantoea dispersa* and *Enterobacter asburiae* promote sugarcane growth by inducing nitrogen uptake and defense-related gene expression. *Frontiers in Microbiology*, *11*, 600417. https://doi.org/10.3389/fmicb.2020.600417.

Singh, R. P., Ma, Y., & Shadan, A. (2022). Perspective of ACC-deaminase producing bacteria in stress agriculture. *Journal of Biotechnology*, *352*, 36–46. https://doi.org/10.1016/j.jbiotec.2022.06.003.

Singh, M., Kumar, S., Singh, D. K., Ranjan, T., & Pandey, A. K. (2024). Isolation and characterization of a novel nitrogen fixer *Beijerinckia fluminensis* strain BAUMS11 from Litchi chinensis rhizosphere. *Journal of Applied and Natural Science*, 16(3), 1196–1201. 10.31018/jans.v16i3.5331.

Soumare, A., Diedhiou, A. G., Thuita, M., Hafidi, M., Ouhdouch, Y., Gopalakrishnan, S., & Kouisni, L. (2020). Exploiting biological nitrogen fixation: A route towards a sustainable agriculture. *Plants*, *9*(8), 1011. https://doi.org/10.3390/plants9081011.

Spaepen, S., Bossuyt, S., Engelen, K., Marchal, K., & Vanderleyden, J. (2014). Phenotypical and molecular responses of *Arabidopsis thaliana* roots as a result of inoculation with the auxin-producing bacterium *Azospirillum brasilense*. *New Phytologist*, *201*(3), 850–861. https://doi.org/10.1111/nph.12590.

Srebot, M. S., Tano, J., Carrau, A., Ferretti, M. D., Martínez, M. L., Orellano, E. G., & Rodriguez, M. V. (2021). Bacterial wilt biocontrol by the endophytic bacterium *Gluconacetobacter diazotrophicus* in *Río Grande* tomato cultivar. *Biological Control*, *162*, 104728. https://doi.org/10.1016/j.biocontrol.2021.104728.

Stewart, W. D. P. (1969). Biological and ecological aspects of nitrogen fixation by free-living microorganisms. *Proceedings of the Royal Society of London. Series B. Biological Sciences*, *172*(1029), 367–388. https://doi.org/10.1098/rspb.1969.0058.

Suleman, M., Yasmin, S., Rasul, M., Yahya, M., Atta, B. M., & Mirza, M. S. (2018). Phosphate solubilizing bacteria with glucose dehydrogenase gene for phosphorus uptake and beneficial effects on wheat. *PLoS One*, *13*(9), e0204408. https://doi.org/10.1371/journal.pone.0204408.

Sun, L. R., Zhao, Z. J., & Hao, F. S. (2019). NADPH oxidases, essential players of hormone signaling in plant development and response to stresses. *Plant Signaling & Behavior, 14*(8), 1657343. https://doi.org/10.1080/15592324.2019.1657343.

Thiebaut, F., Urquiaga, M. C.dO., Rosman, A. C., da Silva, M. L., & Hemerly, A. S. (2022). The impact of non-nodulating diazotrophic bacteria in agriculture: Understanding the molecular mechanisms that benefit crops. *International Journal of Molecular Sciences, 23*(19), 11301. https://doi.org/10.3390/ijms231911301.

Threatt, S. D., & Rees, D. C. (2023). Biological nitrogen fixation in theory, practice, and reality: A perspective on the molybdenum nitrogenase system. *FEBS Letters, 597*(1), 45–58.

Timofeeva, A. M., Galyamova, M. R., & Sedykh, S. E. (2022). Bacterial siderophores: Classification, biosynthesis, perspectives of use in agriculture. *Plants, 11*(22), 3065. https://doi.org/10.3390/plants11223065.

Tufail, M. A., Touceda-González, M., Pertot, I., & Ehlers, R.-U. (2021). *Gluconacetobacter diazotrophicus* Pal5 enhances plant robustness status under the combination of moderate drought and low nitrogen stress in *Zea mays* L. *Microorganisms, 9*(4), 870. https://doi.org/10.3390/microorganisms9040870.

Tufail, M. A., Touceda-González, M., Pertot, I., & Ehlers, R. U. (2021). *Gluconacetobacter diazotrophicus* Pal5 enhances plant robustness under combined moderate drought and low nitrogen stress in *Zea mays* L. *Microorganisms, 9*(4), 870. https://doi.org/10.3390/microorganisms9040870.

Ullah, F., Ali, S., Siraj, M., Akhtar, M. S., & Zaman, W. (2025). Plant microbiomes alleviate abiotic stress-associated damage in crops and enhance climate-resilient agriculture. *Plants, 14*(12), 1890. https://doi.org/10.3390/plants14121890.

Upadhyay, S. K., Srivastava, A. K., Rajput, V. D., Chauhan, P. K., Bhojiya, A. A., Jain, D., Chaubey, G., Dwivedi, P., Sharma, B., & Minkina, T. (2022). Root exudates: Mechanistic insight of plant growth promoting rhizobacteria for sustainable crop production. *Frontiers in Microbiology, 13*, 916488. https://doi.org/10.3389/fmicb.2022.916488.

Upadhyay, D., Shukla, K., Mishra, A., Jindal, T., Sharma, S., & Shukla, S. (2021). Molecular aspects and oxygen relations of nitrogen fixation in cyanobacteria. In M. Khan, A. Zaidi, & R. Wani (Eds.), *Soil nitrogen ecology* (pp. 521–568). Springer. https://doi.org/10.1007/978-3-030-57636-8_20

van Loon, L. C., & Bakker, P. (2005). Induced systemic resistance as a mechanism of disease suppression by rhizobacteria. In Z. A. Siddiqui (Ed.), PGPR: Biocontrol and biofertilization (pp. 39–66). Springer. https://doi.org/10.1007/1-4020-4152-7_2

Verma, J. P., Yadav, J., Tiwari, K. N., & Kumar, A. (2013). Effect of indigenous Mesorhizobium spp. and plant growth promoting rhizobacteria on yields and nutrients uptake of chickpea (*Cicer arietinum* L.) under sustainable agriculture. *Ecological Engineering, 51*, 282–286.

Vidal, C., Gonzalez, F., Santander, C., Perez, R., Gallardo, V., Santos, C., Aponte, H., Ruiz, A., & Cornejo, P. (2022). Management of rhizosphere microbiota and plant production under drought stress: A comprehensive review. *Plants, 11*(19), 2437. https://doi.org/10.3390/plants11192437.

Vílchez, J. I., Navas, A., González-López, J., Arcos, S. C., & Manzanera, M. (2016). Biosafety test for plant growth-promoting bacteria: Proposed environmental and human safety index (EHSI) protocol. *Frontiers in Microbiology, 6*, 1514. https://doi.org/10.3389/fmicb.2015.01514.

Vitousek, P. M., Menge, D. N., Reed, S. C., & Cleveland, C. C. (2013). Biological nitrogen fixation: Rates, patterns and ecological controls in terrestrial ecosystems. *Philosophical Transactions of the Royal Society B: Biological Sciences, 368*(1621), 20130119. https://doi.org/10.1098/rstb.2013.0119.

Waller, S., Wilder, S. L., Schueller, M. J., Housh, A. B., Scott, S., Benoit, M., Powell, A., Powell, G., & Ferrieri, R. A. (2021). Examining the effects of the nitrogen environment on growth and N_2-fixation of endophytic *Herbaspirillum seropedicae* in maize seedlings by applying ^{11}C radiotracing. *Microorganisms, 9*(8), 1582. https://doi.org/10.3390/microorganisms9081582.

Wanjofu, E. I., Venter, S. N., Beukes, C. W., Steenkamp, E. T., Gwata, E. T., & Muema, E. K. (2022). Nodulation and growth promotion of chickpea by *Mesorhizobium* isolates from diverse sources. *Microorganisms*, *10*(12), 2467. https://doi.org/10.3390/microorganisms10122467.

Wirth, S. B., Braun, J., Heinke, J., Ostberg, S., Rolinski, S., Schaphoff, S., Stenzel, F., von Bloh, W., Taube, F., & Müller, C. (2024). Biological nitrogen fixation of natural and agricultural vegetation simulated with LPJmL 5.7.9. *Biogeosciences*, *22*(6), 1234–1256. https://doi.org/10.5194/bg-22-1234-2025.

Wu, Y., Zhou, J., Li, C., & Ma, Y. (2019). Antifungal and plant growth promotion activity of volatile organic compounds produced by *Bacillus amyloliquefaciens*. *Microbiology Open*, *8*(7), e813. https://doi.org/10.1002/mbo3.813.

Yasuda, M., Isawa, T., Shinozaki, S., Minamisawa, K., & Nakashita, H. (2009). Effects of colonization of a bacterial endophyte, *Azospirillum* sp. B510, on disease resistance in rice. *Bioscience, Biotechnology, and Biochemistry*, *73*(12), 2595–2599. https://doi.org/10.1271/bbb.90402.

Zayed, O., Hewedy, O. A., Abdelmoteleb, A., Ali, M., Youssef, M. S., Roumia, A. F., Seymour, D., & Yuan, Z. C. (2023). Nitrogen journey in plants: From uptake to metabolism, stress response, and microbe interaction. *Biomolecules*, *13*(10), 1443. https://doi.org/10.3390/biom13101443.

Zeffa, D. M., Perini, L. J., Silva, M. B., de Sousa, N. V., Scapim, C. A., Oliveira, A. L. M., Amaral Júnior, A. T., & Gonçalves, L. S. A. (2019). Azospirillum brasilense promotes increases in growth and nitrogen use efficiency of maize genotypes. PLoS One, 14, e0215332.

Van Zelm, E., Zhang, Y., & Testerink, C. (2020). Salt tolerance mechanisms of plants. *Annual Review of Plant Biology*, *71*, 403–433. https://doi.org/10.1146/annurev-arplant-050718-100005.

Zhao, Q., Wang, J., He, Q., & Yin, H. (2024). Carbon type and quantity regulate soil free-living nitrogen fixation through restructuring diazotrophic community. *Applied Soil Ecology*, *202*, 105586. https://doi.org/10.1016/j.apsoil.2024.105586.

Diazotrophic Plant Growth Promoting Microorganisms Are a Key to Soil Health and Productivity

14

Deepshikha Kushwaha, Prashant Katiyar and Shalini Rai

Abstract

Soil health is the foundation of sustainable agriculture, and plant productivity is closely tied to the nutrient dynamics and microbial ecology of the soil. Among the diverse group of microbial populations that interact with plants, diazotrophic microorganisms, which are free-living organisms found in soil and responsible for converting atmospheric nitrogen into available forms that are beneficial to plant growth, development, and health, thereby enhancing crop yield and quality. This chapter highlights the importance of multiple plant growth-promoting soil microorganisms (PGPSMs), including bacteria, fungi, cyanobacteria, and protozoa, which are abundant in fertile soils and have garnered significant attention for their ability to fix atmospheric nitrogen and enhance plant growth. These beneficial microbes not only reduce the dependence on synthetic fertilizers but also improve soil structure, nutrient cycling through mineralization can positively impact the growth, development, and health of crop plants. However, some crucial key factors, such as nitrogen fixation, nutrient solubilization, and increasing crop resilience. This chapter concluded that PGPSMs offer a promising avenue for enhancing both the sustainability and productivity of crop cultivation, laying the groundwork for future research and practical applications in agriculture and environmental sustainability.

D. Kushwaha
Centre for Microbiology, Ewing Christian College, University of Allahabad, Prayagraj, UP, India

P. Katiyar
Department of Biotechnology Engineering, Goel Institute of Technology and Management, Lucknow, UP, India

S. Rai (✉)
Department of Microbiology, DSMNRU, Lucknow, UP, India
e-mail: shalinimicro09@gmail.com

R. K. Singh and P. Singh (eds.), *Diazotrophic Plant Growth-Promoting Microorganisms and Green Energy – Volume I*, https://doi.org/10.1007/978-981-95-9682-9_14

Keywords

Diazotrophic microorganisms · Nutrient cycling · Nitrogen fixation · Nutrient solublization

14.1 Introduction

Modern agriculture is confronted with an increasingly complex and urgent challenge: the need to simultaneously increase food production to feed a rapidly growing global population while minimizing the environmental degradation that often accompanies agricultural practices. As the world's population continues to swell, projected to reach nearly 10 billion by 2050 (United Nations, 2019), there is mounting pressure on the agricultural sector to meet the demands for food, while also addressing the detrimental effects of farming on the environment. A significant factor contributing to these environmental challenges is the widespread use of chemical fertilizers, particularly nitrogen-based fertilizers. These synthetic fertilizers have been key to enhancing crop yields (Tilman et al., 2002), but their excessive application has led to a cascade of negative environmental consequences, including soil acidification, water contamination, and the emission of harmful greenhouse gases in the atmosphere (Galloway et al., 2008).

One of the main problems with nitrogen fertilizers is their tendency to leach into the soil and water systems, leading to contamination of groundwater and surface water bodies (Canfield et al., 2010). This process not only harms aquatic ecosystems but also contributes to the phenomenon of eutrophication, where an overabundance of nutrients causes algal blooms that deplete oxygen levels in water, killing aquatic life. Furthermore, the continuous use of nitrogen-based fertilizers has been linked to soil acidification, a condition where soil pH drops, making it less fertile and disrupting the delicate balance of soil nutrients. Additionally, the production and use of synthetic nitrogen fertilizers are responsible for a significant portion of agricultural greenhouse gas emissions, contributing to climate change. It is evident that while nitrogen fertilizers have increased agricultural productivity, they have also created a host of environmental challenges that threaten the sustainability of farming systems (Foley et al., 2011).

In response to these mounting concerns, researchers and practitioners in the agricultural field are exploring more sustainable alternatives to synthetic fertilizers. One of the most promising avenues is the use of diazotrophic plant growth-promoting microorganisms (PGPMs), a group of beneficial microbes that can naturally enhance nitrogen availability in soils (Bhattacharyya & Jha, 2012). Diazotrophs, as these microbes are called, represent a novel and ecologically responsible solution to the nitrogen challenge. These microorganisms are capable of fixing atmospheric nitrogen (N_2), a gas that makes up about 78% of Earth's atmosphere, into a form that plants can use ammonia (NH_3). This biological process, known as biological nitrogen fixation, allows diazotrophs to supply plants with a crucial nutrient that is essential for growth and development, without the need for synthetic fertilizers.

Diazotrophs, apart from other nitrogen-fixing organisms, such as those found in the roots of legumes, are their ability to exist as free-living or associative microorganisms, independent of specific plant hosts (Rothballer et al., 2008). While legumes typically harbor nitrogen-fixing bacteria (such as *Rhizobium*) in specialized root nodules, many non-leguminous crops, including cereals, vegetables, and other staple crops, rely on free-living or associative diazotrophs for their nitrogen needs. These microbes are often found in the plant rhizosphere, the soil zone surrounding plant roots, or within plant tissues as endophytes, microorganisms that live inside plant cells without causing harm. In both cases, diazotrophs can provide a steady and natural supply of nitrogen to plants, promoting growth and productivity while reducing the need for chemical fertilizers.

The interaction between plants and diazotrophs is a complex and highly beneficial relationship. In the rhizosphere, the microorganisms engage in a mutualistic association with the plant, where the plant supplies the microbes with carbon compounds, primarily sugars, which are essential for the microbes' survival. In return, the diazotrophs convert atmospheric nitrogen into ammonia, which the plants can absorb and use for various physiological processes, such as the synthesis of proteins and chlorophyll (Bhattacharyya & Jha, 2012). This symbiotic exchange enhances plant growth, reduces fertilizer dependency, and improves soil fertility over time. Moreover, some species of diazotrophs are able to live within the plant tissues themselves as endophytes, where they continue to fix nitrogen and supply it directly to the plant, without the need for the plant to interact directly with the soil (Compant et al., 2010). The potential of diazotrophs to revolutionize agricultural practices is immense, particularly in non-leguminous crops that are less well-studied in terms of their microbial interactions. Studies have shown that the use of diazotrophic microorganisms can significantly reduce the need for synthetic nitrogen fertilizers while improving soil health and plant productivity (Bashan & de-Bashan, 2010). For instance, field trials have demonstrated that inoculating non-leguminous crops with diazotrophs can increase crop yields in nitrogen-poor soils, while simultaneously improving soil quality by enhancing microbial diversity and activity. These microbes contribute to the stabilization of soil organic matter, the cycling of other essential nutrients, and the overall health of the soil ecosystem. As a result, diazotrophic microorganisms offer an integrated approach to sustainable farming that benefits both crop productivity and environmental health. Beyond nitrogen fixation, diazotrophs also play a crucial role in maintaining soil fertility through other mechanisms. For example, many diazotrophic species produce plant growth hormones such as auxins and cytokinins, which promote root growth, enhance nutrient uptake, and improve plant resilience to environmental stressors like drought or high temperatures. Additionally, some diazotrophs can degrade organic matter in the soil, releasing essential nutrients like phosphorus, potassium, and sulfur, thereby improving overall soil nutrient availability. These additional benefits further underscore the importance of diazotrophs in creating a holistic, sustainable agricultural system. The use of diazotrophic microorganisms also aligns with the growing emphasis on agroecology, an approach to farming that seeks to integrate ecological principles into agricultural practices. Agroecology emphasizes biodiversity, soil health, and the reduction of chemical inputs in favor of natural processes. By leveraging the power of beneficial microbes, farmers can reduce their reliance on synthetic fertilizers, reduce environmental pollution, and create farming systems that are both productive and ecologically sustainable. Moreover, as the global agricultural sector faces increasing scrutiny over its environmental impact, the adoption of microbial inoculants, such as diazotrophs, offers a promising way forward in building more resilient and sustainable food systems. However, despite the promising potential of diazotrophs, there are still several challenges that need to be addressed before they can be widely adopted in agricultural systems. One of the primary hurdles is the variability in the effectiveness of different diazotrophic species under diverse environmental conditions. While some species may thrive in certain soils or climates, others may be less effective, making it necessary to tailor microbial inoculants to specific crop types, soil types, and environmental conditions (Arora et al., 2020). Additionally, large-scale commercialization of diazotrophic products requires further research and development to ensure their consistency, efficacy, and cost-effectiveness.

Another challenge is the need for proper management and integration of diazotrophic inoculants into existing farming practices. Farmers will need to be educated on the benefits and limitations of these microorganisms, as well as how to effectively integrate them into their cropping systems. Additionally, more research is needed to better understand the complex interactions between diazotrophs, plants, and other soil microorganisms, as well as the long-term effects of their use on soil ecosystems (Bashan et al., 2014).

14.2 Diazotrophic (PGPSMs) and Their Role

Diazotrophic (PGPSMs) refer to a group of beneficial microorganisms that promote plant growth through multiple mechanisms, with nitrogen fixation being a central function. They include bacteria such as *Azospirillum, Azotobacter, Herbaspirillum, Gluconacetobacter, Burkholderia,* and *Klebsiella.* These organisms inhabit various ecological niches: free-living in soil, associated with the rhizosphere, or existing as endophytes within plant tissues (Smith et al., 2023; Wang & Liu, 2022c; Zhang et al., 2021; Patel & Kaur, 2023). The main significant role of diazotrophic microorganisms is highlighted below (Fig. 14.1):

- The biological and ecological roles of diazotrophic PGPSMs.
- Their sig*nifi*cance in sustainable and organic farming.
- The wide-ranging diversity within this group, which affects their adaptability, efficiency, and application in different agroecosystems.

14.2.1 General Biological Nitrogen Fixation (BNF) Mechanism

The key attribute of diazotrophs is their ability to carry out biological nitrogen fixation (BNF) (Smith et al., 2023). This process is catalyzed by the enzyme nitrogenase, which reduces atmospheric nitrogen to ammonia under suitable anaerobic or microaerophilic

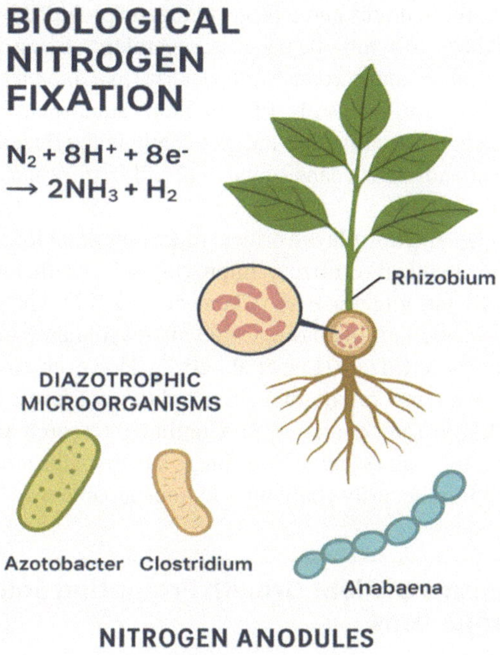

Fig. 14.1 Diazotrophic microorganisms can fix atmospheric nitrogen into ammonia, which is utilized by plants to control their nitrogen deficiencies

conditions, following the given reaction mechanism: $N_2 + 8H^+ + 8e^- \rightarrow 2NH_3 + H_2$ (Wang & Liu, 2022b). However, nitrogen fixation is energy-intensive, requiring ATP and is tightly regulated by the availability of oxygen and nitrogenous compounds in the environment (Zhang et al., 2021). Other than this, biological nitrogen fixation (BNF) is a vital natural process by which certain microorganisms can convert atmospheric nitrogen (N_2) into ammonia (NH_3), a form that plants can readily absorb and utilize for plant growth (Patel & Kaur, 2023). Nitrogen is an essential macronutrient for all living organisms, playing a crucial role in the formation of amino acids, nucleic acids, and chlorophyll. Although nitrogen gas makes up about 78% of Earth's atmosphere, it is an inert and unavailable to most organisms in its gaseous form (Cheng et al., 2021). BNF provides a sustainable and eco-friendly alternative to synthetic nitrogen fertilizers, which are often costly and environmentally harmful (Yang et al., 2023).

The key agents of BNF are diazotrophic microorganisms, including both free-living and symbiotic bacteria and archaea (Zhou et al., 2022). Some of the most well-known symbiotic nitrogen-fixing bacteria are *Rhizobium, Bradyrhizobium*, and *Frankia*, which form nodules on the roots of leguminous plants such as peas, beans, and clover (Kim et al., 2022). Inside these nodules, the bacteria convert atmospheric nitrogen into ammonia through the action of the enzyme nitrogenase, a complex and oxygen-sensitive enzyme system (Smith et al., 2023).

The fixed nitrogen is then assimilated by the plant, while the microorganism benefits from a supply of carbohydrates and a protected niche provided by the host (Patel & Kaur, 2023). In addition to symbiotic bacteria, several free-living diazotrophs such as *Azotobacter, Clostridium*, and certain cyanobacteria (like *Anabaena* and *Nostoc*) also contribute to nitrogen fixation in soils (Fig. 14.1) and aquatic environments (Wang & Liu, 2022b). These microbes function independently in the rhizosphere or in natural ecosystems like wetlands and oceans, enhancing soil fertility and ecosystem productivity (Zhang et al., 2021).

Biological nitrogen fixation plays a critical role in sustainable agriculture by reducing dependence on chemical fertilizers, improving soil health, and promoting environmentally friendly farming practices (Cheng et al., 2021). The global push toward regenerative and organic farming has renewed interest in harnessing nitrogen-fixing microorganisms as biofertilizers (Lee et al., 2022). However, environmental conditions such as soil pH, temperature, moisture, and oxygen levels significantly affect the efficiency of BNF (Yang et al., 2023). Continued research and development in microbial biotechnology are essential to optimize the use of diazotrophs in agriculture and address global food security challenges in a sustainable way (Zhou et al., 2022).

14.3 Mechanisms of Plant Growth Promoting Soil Microorganisms

Plant Growth-Promoting Soil Microorganisms (PGPSMs) play a crucial role in boosting up the plant health and productivity through a variety of direct and indirect mechanisms (Kumar et al., 2023b) (Fig. 14.2). These beneficial microbes, including bacteria such as *Rhizobium*, *Azospirillum*, *Pseudomonas*, *Bacillus* and fungi like *Trichoderma* and *Mycorrhizae*, colonize the rhizosphere the region of soil surrounding plant roots and establish symbiotic relationships that support plant development and soil fertility (Singh et al., 2021). One of the primary direct mechanisms employed by PGPSMs is biological nitrogen fixation, where diazotrophic microbes convert atmospheric nitrogen into ammonia, a usable form for plants (Smith et al., 2023). Other direct mechanisms include phosphate solubilization, in which insoluble phosphorus compounds in soil are converted into soluble forms through the secretion of organic acids and enzymes, enhancing phosphorus availability (Wang & Liu, 2022b).

PGPSMs also produce phytohormones such as indole-3-acetic acid (IAA), gibberellins, and cytokinins that stimulate root elongation, cell division, and overall plant growth (Yang et al., 2021b). In addition, PGPSMs contribute to siderophore production, where they secrete iron-chelating compounds that bind and make iron more available to plants, particularly in iron-deficient soils (Zhang et al., 2022). Many PGPSMs also facilitate the uptake of other essential nutrients like potassium and zinc (Li, Wu et al., 2023). On the other hand, indirect mechanisms involve improving plant resilience against pathogens and environmental stress (Patel & Kaur, 2023). PGPSMs suppress soil-borne diseases by producing antibiotics, antifungal compounds, and hydrolytic enzymes that inhibit or destroy pathogenic microbes (Chen et al., 2022). They also induce systemic resistance in plants by triggering plant defense pathways, preparing

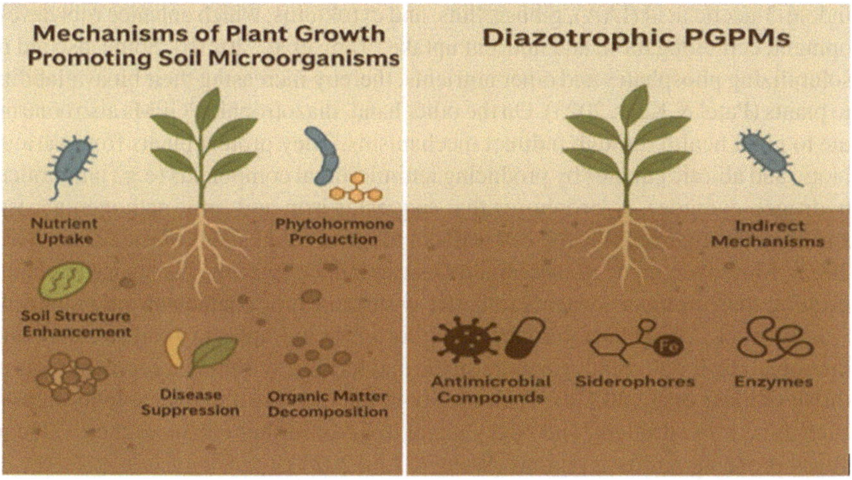

Fig. 14.2 (A) Direct approach: Diazotrophs help in decomposing the organic matter to enhance the soil structure, produce phytohormones involved in the enhancement of plant health by uptake of nutrients into the soil, and suppress the diseases. (B) Indirect approach: Diazotrophs produce some antimicrobial compounds, enzymes, and small iron chelating molecules that increase the plant's health during stress conditions

them for potential pathogen attacks (Li et al., 2022). Furthermore, some PGPSMs can degrade harmful organic pollutants or heavy metals, thereby detoxifying the soil environment (Wang & Liu, 2021). Through these combined actions, PGPSMs significantly improve plant tolerance to abiotic stresses such as drought, salinity, and nutrient deficiency (Yang et al., 2022).

The diverse and synergistic mechanisms of PGPSMs not only promote healthy plant growth but also contribute to sustainable agriculture, reducing the need for chemical fertilizers and pesticides (Kumar et al., 2023b). Harnessing these mechanisms effectively through biofertilizers and biocontrol agents is essential for advancing environmentally friendly and resilient farming systems in the face of global climate and food security challenges (Singh et al., 2021).

These plant growth-promoting microorganisms (PGPMs) are a specialized group of soil microorganisms, capable of fixing atmospheric nitrogen and enhancing plant growth through a combination of direct and indirect mechanisms (Zhou et al., 2023), as shown in Fig. 14.2. These beneficial microbes, including genera such as *Rhizobium*, *Azospirillum*, *Azotobacter*, *Herbaspirillum*, and some cyanobacteria, play a vital role in sustainable agriculture by naturally supplying nitrogen and promoting overall plant health (Patel & Kaur, 2023). The direct mechanisms primarily involve biological nitrogen fixation (BNF), where diazotrophs convert inert atmospheric nitrogen (N_2) into ammonia (NH_3), a form that plants can readily absorb and utilize for protein synthesis and growth (Wang & Liu, 2022c). This process reduces the need for synthetic nitrogen fertilizers, making agriculture more eco-friendly (Smith et al., 2023). In addition, many diazotrophic PGPMs produce phytohormones such as indole-3-acetic acid (IAA), gibberellins, and cytokinins, which enhance root development, cell elongation, and nutrient uptake (Yang et al., 2021b). Some also aid in solubilizing phosphates and other nutrients, thereby increasing their bioavailability to plants (Patel & Kaur, 2023). On the other hand, diazotrophic PGPMs also contribute to plant health through indirect mechanisms. They protect plants from various biotic and abiotic stresses by producing antimicrobial compounds (e.g., antibiotics, hydrogen cyanide), siderophores that sequester iron and limit pathogen access, and enzymes that degrade the cell walls of harmful fungi and bacteria (Zhou et al., 2022). Moreover, many of these microbes can induce systemic resistance in plants, priming their immune systems to better withstand future attacks from pathogens (Li, Wu et al., 2023). Diazotrophs also play a role in improving soil structure and microbial diversity, creating a more stable and fertile environment for plant roots to thrive (Kumar et al., 2023b). Their ability to tolerate and alleviate abiotic stresses such as salinity, drought, and heavy metal toxicity further enhances their value in stress-prone agricultural regions (Wang & Liu, 2021).

14.4 Impact on Soil Health

Healthy soils support a diverse microbial community, efficient nutrient cycling, and robust plant growth (Kumar et al., 2023b). Diazotrophic microorganisms play a crucial role in maintaining and enhancing soil health, primarily through their ability to fix atmospheric nitrogen and contribute to nutrient cycling (Zhang et al., 2022). These microorganisms, including genera such as *Rhizobium*, *Azospirillum*, *Azotobacter*, and cyanobacteria, convert inert atmospheric nitrogen (N_2) into ammonia (NH_3), a bioavailable form that enriches the soil with essential nutrients (Patel & Kaur, 2023). This natural nitrogen input reduces the dependency on chemical fertilizers, which, when overused, can lead to soil acidification, nutrient imbalance, and microbial disruption (Smith et al., 2022). By improving nitrogen content in the soil, diazotrophs enhance soil fertility, supporting the growth of healthy crops and increasing microbial biodiversity (Wang & Liu, 2023b).

Beyond nitrogen fixation, diazotrophs also contribute to soil structure and stability (Yang et al., 2021b). Their activities stimulate root growth, leading to increased organic matter input in the form of root exudates and plant residues (Li et al., 2022). This organic matter enhances soil aggregation, water retention, and aeration (Zhou et al., 2023). Additionally, many diazotrophic microorganisms secrete extracellular polysaccharides and other biofilms that help bind soil particles, improving texture and reducing erosion (Kumar et al., 2023b). These microbes also foster a more diverse and balanced soil microbial community (Patel & Kaur, 2023). By producing antimicrobial compounds and siderophores, diazotrophs suppress harmful pathogens and promote beneficial microbial interactions, creating a biologically active and disease-resistant soil environment (Cheng et al., 2022). Moreover, their ability to function under stress conditions like drought or salinity makes them vital in rehabilitating degraded or marginal soils (Li, Wu et al., 2023). In essence, diazotrophic microorganisms are key contributors to long-term soil

Table 14.1 List of the plant growth-promoting substances secreted by diazotrophs and their role

Diazotrophs	PGPR growth substances	Effects on the plant	References
Azospirillum sp.	Siderophore production, production of Indole acetic acid, improve K aquitions and N_2 fixation, Antibiotics and anti-fungal compounds	Increase the length, number, weight of lateral roots and root hairs, improve Zn in the plant and grain, suppress the Rizoctonia *solani* and *Fusarium oxysporum*	Sivasakthivelan et al. (2023), dos Santos Ferreira et al. (2022)
Acinetobacter sp.	Siderophore, IAA, phosphate solubilization	Salinity tolerance, increase the *nifH* expression	Tafaroji et al. (2022)
Azotobacter sp.	IAA, Gibberellin, ACC deaminase, Kinetin, N_2 fixation, improve K aquitions, exo polysaccharide, antioxidant activity	Increase the length, number, weight of lateral roots and root hairs, improve Zn in the plant and grain, stress tolerance, enhance the plant defense system, enhance nutrient availability	Abdel Latef et al. (2021), Bhadrecha et al. (2023), Danish et al. (2022), Jain et al. (2021)
Bradyrhizobium sp.	IAA, improve K aquitions and N_2 fixation, Siderophore, NH_3, HCN, exopolysaccharides	Influence the lateral root development, growth promotion, cell differentiation, improve plant health in iron-deficient soil, activate defense signals during biotic stress, abiotic stress resistance, enhance nutrient availability	Kiruthika and Kumar (2021), Fagorzi et al. (2018)
Brevundimonas sp.	N_2 fixation, IAA production, ACC deaminase, antifungal compound	Influence the lateral root development, growth promotion, cell differentiation, suppress the growth of *Fusariumredolens* and *Fusarium yellows*	Naqqash et al. (2020), Bekkar et al. (2024), Ratnaningsih et al. (2023)
Brevibacillus sp.	IAA production, Zn resistant	Influence the lateral root development, growth promotion, cell differentiation	Li et al. (2022), Xiang et al. (2023)
Burkholderia sp.	IAA production, ACC deaminase, Siderophore production, phosphate solubilization, stress tolerance	Influence the lateral root development, growth promotion, cell differentiation, improve plant health in iron-deficient soil	Han et al. (2021)

(Continued)

Table 14.1 (Continued)

Diazotrophs	PGPR growth substances	Effects on the plant	References
Bacillus sp.	Siderophore production, IAA, HCN, phosphate solubilization, antifungal compounds production	Influence the lateral root development, growth promotion, cell differentiation, improve plant health in iron-deficient soil, plant development, biotic stress	Wani et al. (2007), Xiang et al. (2023), Mahapatra et al. (2022)
Mesorhizobium sp.	Siderophore production, IAA, HCN, exopolysaccharides	Influence the lateral root development, growth promotion, cell differentiation, improve plant health in iron-deficient soil, seed dormancy, plant development, defense against pathogen and pests, enhance nutrient availability.	Yadav et al. (2022)
Rhizobium sp.	Siderophore production, ACC deaminase, IAA, HCN, exo-polysaccharides, cd stress	Influence the lateral root development, growth promotion, cell differentiation, improve plant health in iron-deficient soil, seed dormancy, plant development, defense against pathogen and pests, enhance nutrient availability	Nofal and Attia (2023), Banjare et al. (2023), Alinia et al. (2022)
Flavobacterium sp.	IAA, Siderophore production	Influence the lateral root development, growth promotion, cell differentiation, improve plant health in iron-deficient soil	Feng et al. (2023)
Pseudomonas sp.	Exo-polysaccharides, Siderophore production, IAA, HCN, phosphate solubilization, 2,3-dihydroxybenzoic acid, Salicylic acid, Antifungal compounds	Enhance nutrient availability, improve soil health, improve plant health in iron-deficient soil, influence the lateral root development, growth promotion, cell differentiation, seed dormancy, defense against pathogen and pests, protect against abiotic stress, suppress the growth of *Penicillium italicum*	Pereira (2019), Parvin et al. (2023), Wang et al. (2021)

(Continued)

Table 14.1 (Continued)

Diazotrophs	PGPR growth substances	Effects on the plant	References
Klebsiella oxytoca	Nitrogenase activity, phosphate solubilization, IAA production	N_2 fixation, influence the lateral root development, growth promotion, cell differentiation	Khalifa and Aldayel (2022)
Enterobacter sp.	HCN, exo-polysaccharides, IAA, siderophore production, ACC deaminase, phosphate solubilization,	Seed dormancy, plant development, defense against pathogen and pests, enhance nutrient availability, influence the lateral root development, growth promotion, cell differentiation, improve plant health in iron-deficient soil protect against abiotic stress	Singh et al. (2022), Saikia et al. (2023)
Xanthomonas sp.	IAA	Increase the length, number, weight of lateral roots and root hairs, cell differentiation	He et al. (2022)
Spingomonas sp.	IAA	Increase the length, number, weight of lateral roots and root hairs, cell differentiation	Asaf et al. (2020)
Stenotrophomonas maltophilia	IAA, nitrogenase activity, ACC deaminase, phosphate solubilization	Increase the length, number, weight of lateral roots and root hairs, cell differentiation, N_2 fixation, help in cell growth	Li et al. (2023)
Serratia marcescens	Siderophore production, IAA, HCN	Improve plant health in iron-deficient soil, influence the lateral root development, growth promotion, cell differentiation, protection against biotic stress, growth promotion, nutrient enhancement	Sriwati et al. (2023)

sustainability, promoting nutrient cycling, microbial diversity, and ecosystem resilience in agricultural and natural landscapes listed in Table 14.1 (Smith et al., 2022; Yang et al., 2021b).

14.4.1 Enriching Soil Nitrogen Pools

Through consistent nitrogen inputs, diazotrophs improve the soil's nitrogen balance, reducing the need for external fertilizer inputs (Kumar et al., 2023b). However, diazotrophic microorganisms play a vital role in enriching soil nitrogen pools by converting atmospheric nitrogen (N_2) into ammonia (NH_3), a form that plants can absorb and utilize (Zhang et al., 2022). This process, known as biological nitrogen fixation (BNF), is carried out by a variety of free-living and symbiotic microbes, including species from the genera *Rhizobium*, *Azospirillum*, *Azotobacter*, and certain cyanobacteria (Wang & Liu, 2023b; Singh et al., 2022). Unlike synthetic fertilizers, which can lead to nutrient leaching and environmental pollution, nitrogen fixed by diazotrophs is released gradually into the soil, ensuring a more sustainable and balanced nutrient supply (Li et al., 2022). Symbiotic diazotrophs, particularly those associated with leguminous plants, form specialized root nodules where nitrogen fixation occurs efficiently, directly enriching the rhizosphere with bioavailable nitrogen (Patel & Kaur, 2023). Free-living diazotrophs, though less efficient individually, contribute significantly over time, especially in diverse cropping systems or organic farming environments (Zhou et al., 2023). The nitrogen added by these microbes not only supports plant growth but also benefits soil microbial communities by improving overall nutrient cycling (Cheng et al., 2022). Incorporating or encouraging diazotrophic microorganisms in agricultural practices enhances soil fertility, reduces dependence on chemical inputs, and contributes to long-term soil health (Wang & Liu, 2023b). Their presence ensures a continuous, eco-friendly supply of nitrogen, making them essential players in sustainable and regenerative agriculture (Kumar et al., 2023b).

14.4.2 Diazotrophic Microorganisms Enhance the Microbial Biodiversity

The presence of microbial biodiversity encourages microbial interactions that lead to a dynamic and resilient soil ecosystem (Patel & Kaur, 2023). The diazotrophic microorganisms significantly contribute microbial biodiversity especially targeted in the soil ecosystems through their multifunctional roles in nutrient cycling and plant-microbe interactions (Zhang et al., 2022). These nitrogen-fixing microbes, including species from genera such as *Rhizobium*, *Azospirillum*, *Azotobacter*, and *Frankia*, improve soil fertility by converting atmospheric nitrogen into ammonia, thus enriching the soil with bioavailable nitrogen (Wang & Liu, 2023a). This added nitrogen not only supports plant growth but also stimulates the activity and growth of other soil microorganisms, creating a more dynamic and diverse microbial environment (Li, Wu et al., 2023). Diazotrophs also produce various bioactive compounds such as phytohormones, siderophores, and organic acids, which modify the rhizosphere environment and promote the colonization of other beneficial microbes (Cheng et al., 2022).

By improving root architecture and increasing root exudation, diazotrophic microbes indirectly support microbial diversity by providing energy sources and signaling molecules that attract a broader range of soil organisms (Zhou et al., 2023). Furthermore, their presence can suppress soil pathogens through competitive exclusion and antimicrobial compound production, allowing beneficial microbial populations to thrive (Patel & Kaur, 2023). In agroecosystems, the introduction or encouragement of diazotrophs fosters a healthier, more balanced soil microbial community, leading to improved nutrient availability, plant resilience, and ecosystem stability (Kumar et al., 2023b). Thus, diazotrophic microorganisms are key players in maintaining and enhancing microbial biodiversity, essential for sustainable and regenerative agriculture (Wang & Liu, 2023a).

14.4.3 Diazotrophic Microorganisms, A maintainers of Soil Structure

Diazotrophic microorganisms play a crucial role in improving and maintaining soil structure, which is fundamental for healthy plant growth and sustainable agriculture. These nitrogen-fixing microbes, such as *Azospirillum*, *Azotobacter*, *Rhizobium*, and certain cyanobacteria, contribute to soil structure enhancement both directly and indirectly (Zhang et al., 2022). The primary benefits are the production of extracellular polysaccharides (EPS) and biofilms, which help in the binding of soil particles together into stable aggregates. These stable aggregates improve soil porosity, aeration, and water retention, creating a more favorable environment for plant roots and other soil organisms. In addition, diazotrophs stimulate the root growth and branching through the secretion of plant hormones like indole-3-acetic acid (IAA). This enhanced root system leads to increased organic matter input in the form of root exudates and residues, which further contribute to soil aggregation and microbial activity. The organic matter also serves as a food source for other beneficial microbes, promoting a thriving soil ecosystem. Furthermore, by improving the availability of nitrogen through biological nitrogen fixation, diazotrophs enhance overall soil fertility, which supports better plant cover and reduces erosion (Singh et al., 2022), as represented in Fig. 14.3.

14.4.4 Applications of Diazotrophic Microorganisms in Sustainable Agriculture

In the face of increasing environmental concerns and the need for eco-friendly farming practices, diazotrophic microorganisms emerge as vital allies. These nitrogen-fixing microbes contribute to sustainable agriculture by enhancing soil fertility, reducing dependency on chemical fertilizers, and promoting plant growth through natural means. Leveraging their biological potential offers a pathway to resilient, productive, and environmentally responsible farming systems.

Diazotrophic microorganisms found in association with crops like legumes, sugarcane, and cereals enhance soil fertility, improve plant growth, and contribute to long-term soil health (Patel & Kaur 2023). Their use supports environmental conservation by minimizing nutrient runoff and greenhouse gas emissions. Integrating diazotrophs into agricultural practices offers a sustainable alternative to chemical inputs, aligning productivity with ecological balance and paving the way for climate-resilient and sustainable food systems.

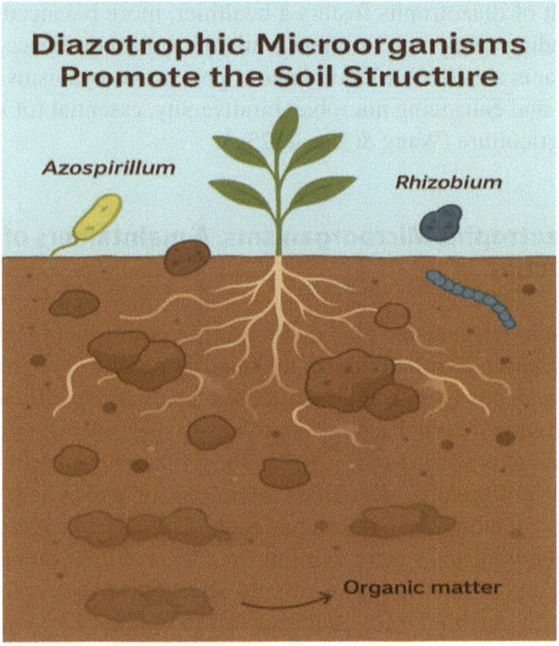

Fig. 14.3 Diazotrophic microbes improve the organic matter of soil, which enhances the porosity, water retention, and fertility of soil, which is beneficial for soil

14.4.5 Diazotroph-Based Bio-Fertilizers

Diazotroph-based bio-fertilizers provide an eco-friendly and sustainable alternative to synthetic nitrogen fertilizers by harnessing the natural ability of certain microorganisms to fix atmospheric nitrogen. Unlike chemical fertilizers, which are energy-intensive to produce and can lead to environmental issues such as water pollution and greenhouse gas emissions, diazotrophic biofertilizers offer a biological solution that promotes soil health and long-term agricultural sustainability(Kumar et al., 2023a). These biofertilizers contain beneficial nitrogen-fixing bacteria such as *Azospirillum* and *Azotobacter*, which colonize the rhizosphere or the interior of plant roots, converting atmospheric nitrogen (N_2) into ammonia (NH_3), a form that plants can readily absorb and utilize for growth. *Azospirillum* is particularly effective in association with cereals like maize, wheat, and rice, enhancing root development and nutrient uptake through the production of phytohormones and nitrogen fixation. Similarly, *Azotobacter* is commonly applied to vegetable crops and non-leguminous plants, where it not only fixes nitrogen but also produces vitamins, growth promoters, and antifungal compounds (Cheng et al., 2022). The use of such inoculants reduces dependency on chemical fertilizers, lowers production costs for farmers, and minimizes negative environmental impacts. As interest in regenerative and organic farming grows, the adoption of diazotroph-based biofertilizers represents a key strategy in building resilient and environmentally sound agricultural systems (Zhang et al., 2022) (Fig. 14.4).

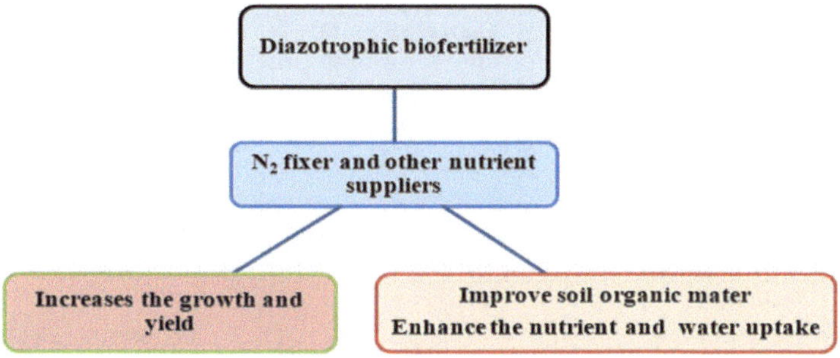

Fig. 14.4 Effect of diazotrophic biofertilizer on plant and soil health

14.5 Integrated Nutrient Management

Combining diazotrophs with organic matter and reduced levels of chemical fertilizers results in a synergistic approach that significantly enhances nutrient availability, uptake, and overall crop productivity. Diazotrophic microorganisms, which naturally fix atmospheric nitrogen into plant-usable forms, work more efficiently when supported by organic matter such as compost, farmyard manure, or crop residues. Organic matter improves soil structure, increases microbial activity, and provides essential nutrients and carbon sources that stimulate the growth and activity of beneficial microbes, including diazotrophs. When small amounts of chemical fertilizers are added to this mix, they serve as an immediate source of nutrients while the microbial populations establish themselves and begin biological nitrogen fixation. This integrated nutrient management approach reduces the total dependency on synthetic fertilizers, thereby minimizing environmental issues like nitrate leaching and soil degradation. Moreover, it enhances the efficiency of nutrient cycling in the soil and promotes better root development, which further facilitates the absorption of both macro and micronutrients.

Studies have shown that such combinations can lead to improved plant vigor, higher yields, and better crop quality, particularly in cereals and vegetables. This method not only supports sustainable farming but also helps farmers to lower input costs while maintaining or even increasing agricultural productivity in an eco-friendly manner, as represented in Fig. 14.5.

14.5.1 Crop-Specific Applications

Crop-specific applications of diazotrophic microorganisms in rice, maize, wheat, and sugarcane have shown significant promise in enhancing productivity, soil health, and sustainability. These crops, though not naturally capable of forming symbiotic nitrogen-fixing nodules like legumes, benefit greatly from the use of free-living or associative diazotrophs such as *Azospirillum*, *Azotobacter*, *Gluconacetobacter*, and *Herbaspirillum*. For instance, in rice cultivation, diazotrophs like *Azospirillum* and

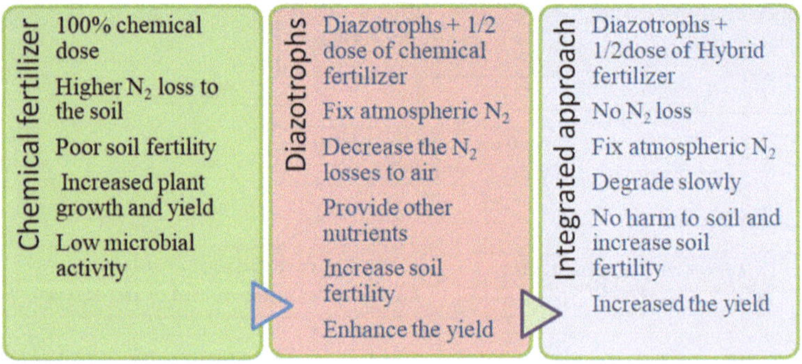

Fig. 14.5 Benefits of the application of integrated nutrient management in soils for sustainable agriculture in comparison with diazotrophs and chemical fertilizers

Azoarcus colonize the rhizosphere and root interior, contributing to biological nitrogen fixation and producing plant growth-promoting hormones that improve tillering and grain filling. In maize, inoculation with *Azospirillum brasilense* has been widely studied and shown to enhance root growth, nutrient uptake, and yield, often reducing the need for nitrogen fertilizers by up to 25%–30%. Wheat responds well to *Azotobacter* and *Azospirillum* inoculation, which boosts nitrogen availability, improves seedling vigor, and enhances grain protein content (Cheng et al., 2021). For sugarcane, *Gluconacetobacter diazotrophicus* is a key endophytic diazotroph that efficiently fixes nitrogen within plant tissues, significantly supporting growth and sugar content without heavy reliance on synthetic inputs. Tailoring diazotroph application based on crop needs allows for more efficient nitrogen management, reduced input costs, and improved environmental outcomes in intensive farming systems (Cheng et al., 2022).

14.6 Recent Molecular Insights

Recent studies have sharpened our understanding of how *nif* genes are tuned to cellular and environmental cues: structural and functional work shows that canonical *Nif*A–*Nif*L pairs integrate redox, energy, and fixed-nitrogen signals so that sigma 54 (σ54-) dependent activation of *nif* operons occurs only when nitrogen is limiting and oxygen/redox conditions permit nitrogenase assembly. Central to many proteobacterial systems is the two-component transcriptional activator-anti-activator pair *Nif* A (activator) and *Nif* L (inhibitor). Under nitrogen limitation and favorable redox/oxygen status, *Nif* A (usually in conjunction with the σ^{54} RNA polymerase holoenzyme) binds upstream activator sequences of *nif* structural operons and initiates transcription (Schmitz et al., 2002). In contrast, when ammonium is abundant, or O_2 is present, *Nif* L senses these signals—via FAD cofactor redox changes, P II-family proteins (e.g., GlnK/GlnB) and other metabolic cues—and inhibits *Nif*A by direct protein–protein interaction, thereby preventing *nif* gene transcription (Yan et al., 2010). Upstream of this system, the global nitrogen regulatory system (Ntr B-Ntr C pathway) controls expression of *nif* LA itself in many free-living bacteria by sensing the internal glutamine/α-ketoglutarate ratio via P II proteins and uridylylation cycles (Shi et al., 2016). New evidence reveals an alternative control layer in proteobacteria that lack *Nif* L — here the PII family protein (GlnK) can bind *Nif* A to form an inhibitory complex under ammonium-replete conditions, providing a direct, reversible brake on *nif* transcription (Boyer et al., 2023). Complementing these mechanistic studies, genome-wide mapping of Gln R binding in Gram-positive diazotrophs (e.g., Paenibacillus) has clarified a distinct regulatory architecture in which Gln R directly links nitrogen status to *nif* promoter activity, explaining how some species control *nif* expression without *Nif* L. Together, these recent structural, biochemical, and omics studies reveal both conserved themes (signal integration to protect energy-intensive, O_2-sensitive nitrogenase) and lineage-specific innovations (*Nif* L-independent PII–*Nif* A regulation; Gln R-mediated control), a synthesis that is guiding engineering efforts to rewire or transplant *nif* regulons for agriculture and synthetic biology (Wang et al., 2023). In this order, Nichio et al. (2025) demonstrated the diversity, distribution, and evolutionary relationships among diazotrophic organisms associated with the *Nif* core (*Nif* H, *Nif* D, *Nif* K, *Nif* E, *Nif* N, and *Nif* B).

14.7 Challenges and Limitations

Despite their significant potential to enhance sustainable agriculture, the widespread adoption of diazotrophic plant growth-promoting microorganisms (PGPMs) still faces several critical hurdles. One of the primary challenges is the lack of awareness and technical knowledge among farmers. Many agricultural communities, especially in developing regions, are not familiar with the benefits of diazotrophs or the correct methods of application. As a result, they continue to rely heavily on conventional chemical fertilizers, which offer immediate results despite their long-term environmental and economic drawbacks.

Another major obstacle is the variability in the field performance of diazotrophic PGPMs. While these microorganisms have shown promising results under controlled or experimental conditions, their effectiveness can vary significantly depending on soil type, climate, crop variety, and farming practices. In some cases, the introduced microorganisms may fail to establish or compete with native microbial populations, reducing their efficacy. Quality control and standardization issues in biofertilizer production also hinder large-scale adoption. Commercial formulations may contain low viable cell counts or be contaminated with ineffective or harmful microbes. Without proper regulation and quality assurance mechanisms, farmers may lose trust in the products if they do not observe consistent benefits. Furthermore, limited shelf life and sensitivity to storage conditions pose logistical challenges. Many diazotroph-based products require specific temperature and moisture conditions to remain viable, which can be difficult to maintain in rural or resource-limited settings. This limits their accessibility and reliability. Policy and institutional support for microbial biofertilizers remains insufficient in many countries. Subsidies, extension services, and incentives are still predominantly directed toward chemical inputs, leaving little room for the promotion and integration of biological alternatives. Lastly, research gaps in strain development, formulation technologies, and delivery mechanisms continue to restrict progress. More efforts are needed to identify robust, high-performing strains adaptable to diverse agro-climatic conditions. Overcoming these barriers will require a coordinated effort involving scientific research, farmer education, government policy support, and private sector innovation to fully harness the benefits of diazotrophic PGPMs in mainstream agricultural systems.

14.8 Future Prospects: Diazotrophic Microorganisms

The future prospects of diazotrophic plant growth-promoting microorganisms (PGPMs) are highly promising, especially when explored through the lenses of genomic and metagenomic tools, biotechnological advancements, and policy and educational initiatives. As global agriculture shifts toward sustainability, these microbial allies offer a powerful alternative to chemical fertilizers, but their full potential is yet to be realized. At the genomic and metagenomic level, cutting-edge tools are revolutionizing our understanding of diazotrophs. Whole-genome sequencing allows researchers to identify genes responsible for nitrogen fixation, phytohormone production, and stress resistance. Metagenomics provides a deeper insight into microbial communities in the rhizosphere, enabling the discovery of previously unknown diazotrophic species and gene functions without the need for culturing. These tools also help in identifying microbial interactions, functional pathways, and environmental adaptability, paving the way for the development of more efficient and crop-specific microbial inoculants. In terms of biotechnological enhancements, synthetic biology and genetic engineering offer new possibilities to improve the efficacy of diazotrophic PGPMs. For instance, genes involved in nitrogen fixation can be transferred or enhanced to create more robust strains capable of surviving harsh field conditions. Engineered microbes can be designed to have multiple plant growth-promoting traits, including phosphate solubilization, drought tolerance, and pathogen resistance, in addition to nitrogen fixation. Furthermore, formulation technologies such as nano-encapsulation, slow-release carriers, and multi-strain consortia are being developed to ensure the stability, shelf-life, and effectiveness of microbial biofertilizers under various agro-climatic conditions. However, scientific progress alone is not sufficient. At the policy level, governments need to recognize the long-term benefits of microbial inputs and create supportive frameworks. This includes offering subsidies for biofertilizers, setting quality control standards, encouraging public–private partnerships, and funding research and development. Policies that integrate microbial solutions into national agricultural programs can accelerate adoption and reduce the environmental footprint of farming. Equally important is the role of education and awareness. Farmers, extension workers, and agribusiness stakeholders must be trained in the correct use, handling, and benefits of diazotrophic PGPMs. Integrating microbiology and sustainable agriculture modules into academic curricula and farmer training programs can create a new generation of agriculturists who are well-versed in biological approaches. In conclusion, the integration of genomics, biotechnology, supportive policies, and education holds the key to unlocking the full potential of diazotrophic PGPMs. With a coordinated effort, these microorganisms can become a cornerstone of sustainable, resilient, and productive agricultural systems worldwide.

14.9 Conclusion

Diazotrophic plant growth-promoting microorganisms hold immense promise for sustainable agriculture by enhancing soil health, reducing dependence on synthetic inputs, and increasing crop productivity. By integrating these beneficial microbes into agricultural practices, turn into more resilient and environmentally friendly food production systems. The path ahead lies in interdisciplinary research, effective policy support, and widespread farmer engagement to unlock the full potential of these microscopic allies. Diazotrophic plant growth-promoting microorganisms enhance the nitrogen availability in soils. These microbes help reduce the need for chemical fertilizers, improve soil health, and promote long-term agricultural productivity. However, the integration of diazotrophs into farming systems could play a key role in achieving the delicate balance between increasing food production and minimizing environmental degradation. Additionally, further research and investment are needed to unlock their full potential and ensure that these soil microorganisms can be effectively harnessed to meet the demands of a growing global population. With continued innovation and adoption, diazotrophic microorganisms could become a cornerstone of sustainable agriculture, paving the way for a more resilient and eco-friendly food system .

References

Abdel Latef, A. A. H., Omer, A. M., Badawy, A. A., Osman, M. S., & Ragaey, M. M. (2021). Strategy of salt tolerance and interactive impact of *Azotobacter chroococcum* and/or *Alcaligenes faecalis* inoculation on canola (*Brassica napus* L.). *Plants Grown in Saline Soil Plants*, *10*(1), 110. https://doi.org/10.3390/plants10010110.

Alinia, M., Kazemeini, S. A., Dadkhodaie, A., Sepehri, M., Mahjenabadi, V. A. J., Amjad, S. F., Poczai, P., El-Ghareeb, D., Bassouny, M. A., & Abdelhafez, A. A. (2022). Co-application of ACC deaminase-producing rhizobial bacteria and melatonin improves salt tolerance in common bean (*Phaseolus vulgaris* L.) through ion homeostasis. *Scientific Report*, *12*(1), 22–105. https://doi.org/10.1038/s41598-022-26084-3.

Arora, N. K., Fatima, T., Mishra, I., Verma, S., & Verma, R. (2020). Microbe-based technologies for sustainable agriculture. *Applied Soil Ecology*, *151*, 103537. https://doi.org/10.1016/j.apsoil.2019.103537.

Asaf, S., Numan, M., Khan, A. L., & Al-Harrasi, A. (2020). *Sphingomonas:* From diversity and genomics to functional role in environmental remediation and plant growth. *Critical Reviews in Biotechnology*, *40*(2), 138–152. https://doi.org/10.1080/07388551.2019.1709793.

Banjare, U., Patel, A., Pandey, A., Kumar, S., Singh, R. K., Masukar, P., Singh, R. K., & Gupta, S. K. (2023). Biochemical and molecular evaluation of *Rhizobium* spp. and its growth promotion studies with Lentil (Lens culinaris Medik. L.). *Journal of Pure Applied Microbiology*, *17*(1), 155–166. https://doi.org/10.22207/jpam.17.1.04.

Bashan, Y., & de-Bashan, L. E. (2010). How the plant growth-promoting bacterium *Azospirillum* promotes plant growth: A critical assessment. *Advances in Agronomy*, *108*, 77–136. https://doi.org/10.1016/S0065-2113(10)08002-8.

Bashan, Y., de-Bashan, L. E., Prabhu, S. R., & Hernandez, J. P. (2014). Advances in plant growth-promoting bacterial inoculant technology: Formulations and practical perspectives (1998–2013). *Plant and Soil*, *378*, 1–33. https://doi.org/10.1007/s11104-013-1956-x.

Bekkar, A. A., & Zaim, S. (2024). Newly isolated *Brevundimonas naejangsanensis* as a biocontrol agent against *Fusarium redolens* the causal of *Fusarium* yellows of chickpea. *Folia Microbiology*, *69*, 835–846. https://doi.org/10.1007/s12223-023-01126-z.

Bhadrecha, P., Singh, S., & Dwibedi, V. (2023). A plant's major strength in rhizosphere: The plant growth promoting rhizobacteria. *Archives of Microbiology*, *205*(165), 1–25. https://doi.org/10.1007/s00203-023-03502-2.

Bhattacharyya, P. N., & Jha, D. K. (2012). Plant growth-promoting rhizobacteria (PGPR): Emergence in agriculture. *World Journal of Microbiology and Biotechnology*, *28*, 1327–1350. https://doi.org/10.1007/s11274-011-0979-9.

Boyer, N. R., Tokmina-Lukaszewska, M., Batista, M. B., Mus, F., Dixon, R., Bothner, B., & PetersJ, W. (2023). Structural insights into redox signal transduction mechanisms in the control of nitrogen fixation by the *Nif*LA system. *Proceedings of the National Academy of Sciences of the United States of America*, *120*(30), 1–9. https://doi.org/10.1073/pnas.2302732120.

Canfield, D. E., Glazer, A. N., & Falkowski, P. G. (2010). The evolution and future of Earth's nitrogen cycle. *Science*, *330*(6001), 192–196. https://doi.org/10.1126/science.1186120.

Chen, L., Zhao, X., & Zhang, Y. (2022). Antimicrobial activity of plant growth-promoting rhizobacteria against soil-borne pathogens. *Microbial Pathogenesis*, *96*, 129–139.

Cheng, L., Zhang, X., & Zhou, S. (2021). The role of nitrogen-fixing bacteria in environmental nitrogen cycling. *Environmental Microbiology*, *48*(6), 472–481.

Cheng, L., Zhang, X., & Zhou, S. (2022). Role of nitrogen-fixing microbes in nutrient cycling and soil health. *Microbial Ecology in Agriculture*, *45*(5), 249–262.

Compant, S., Clément, C., & Sessitsch, A. (2010). Plant growth-promoting bacteria in the rhizo- and endosphere of plants: Their role, colonization, mechanisms involved and prospects for utilization. *Soil Biology and Biochemistry*, *42*(5), 669–678. https://doi.org/10.1016/j.soilbio.2009.11.024.

Danish, M., Shahid, M., Zeyad, M. T., Bukhari, N. A., Al-Khattaf, F. S., Hatamleh, A. A., & Ali, S. (2022). *Bacillus mojavensis*, a metal-tolerant plant growth-promoting bacterium, improves growth, photosynthetic attributes, gas exchange parameters, and Alkalo-Polyphenol contents in silver nanoparticle (Ag-NP)-treated *Withania somnifera* L.(Ashwagandha). *ACS Omega*, *7*(16), 13878–13893. https://doi.org/10.1021/acsomega.2c00262.

Fagorzi, C., Checcucci, A., DiCenzo, G.C., Debiec-Andrzejewska, K., Dziewit, L., Pini, F. & Mengoni, A. (2018). Harnessing rhizobia to improve heavy-metal phytoremediation by legumes. *Genes (Basel)*, 9 (11), 542. https://doi.org/10.3390/genes9110542.

Feng, Q., Cao, S., Liao, S., Wassie, M., Sun, X., Chen, L., & Xie, Y. (2023). *Fusarium* equiseti-inoculation altered rhizosphere soil microbial community, potentially driving perennial ryegrass growth and salt tolerance. *Science of Total Environment*, *871*, 162153. https://doi.org/10.1016/j.scitotenv.2023.162153.

Foley, J. A., Ramankutty, N., Brauman, K., Cassidy, E. S., Gerber, J., Johnston, M., Mueller, N. D., Connell, C. O. ', Ray, D. K., West, P. C., Balzer, C., Bennett, E. M., Carpenter, S. R., Hill, J., Monfreda, C., Polasky, S., Rockström, J., Sheehan, J., Siebert, S., & Zaks, D. P. M. (2011). Solutions for cultivated planet. *Nature*, *478*(7369), 337–342. https://doi.org/10.1038/nature10452.

Galloway, J. N., Townsend, A. R., Erisman, J. W., Bekunda, M., Cai, Z., Freney, J. R., & Sutton, M. A. (2008). Transformation of the nitrogen cycle: Recent trends, questions, and potential solutions. *Science*, *320*(5878), 889–892. https://doi.org/10.1126/science.1136674.

Han, L., Zhang, H., Xu, Y., Li, Y., & Zhou, J. (2021). Biological characteristics and salt-tolerant plant growth-promoting effects of an ACC deaminase-producing *Burkholderia pyrrocinia* strain isolated from the tea rhizosphere. *Archives of Microbiology*, *203*, 2279–2290. https://doi.org/10.1007/s00203-021-02204-x.

He, Y., Wang, H., Liu, Y., Zuo, M., Wang, X., Shen, Y., Li, G., Hu, Y., Zhao, M., & Yang, Y. (2022). Distribution and variation of soil bacterial community of two lead-zinc tailings in Qinling

Mountains. *Geomicrobiology Journal*, *40*, 1–11. https://doi.org/10.1080/01490451.2022.2124330.

Jain, D., Sharma, J., Kaur, G., Bhojiya, A. A., Chauhan, S., Sharma, V., Suman, A., Mohanty, S. R., & Maharjan, E. (2021). Phenetic and molecular diversity of nitrogen fixating plant growth promoting *Azotobacter* isolated from semiarid regions of India. *Biomedical Research International*, *2021*, 1–9. https://doi.org/10.1155/2021/6686283.

Khalifa, A., & Aldayel, M. (2022). Isolation and characterization of Klebsiella oxytoca from the rhizosphere of Lotus corniculatus and its biostimulating features. *Brazilian Journal of Biology*, *82*, e266395. https://doi.org/10.1590/1519-6984.266395.

Kim, Y., Lee, J., & Park, H. (2022). Symbiotic nitrogen fixation in leguminous plants: Mechanisms and potential for agricultural sustainability. *Frontiers in Plant Science*, *13*(4), 876–890.

Kiruthika, S., & Arunkumar, M. (2021). A comprehensive study on IAA production by *Bradyrhizobium japonicum* and *Bacillus subtilis* and its effect on *Vigna radiata* plant growth. Indian. *Journal of Agriculture Research*, *55*(5), 570–576. https://doi.org/10.18805/IJARe.A-5521.

Kumar, S., Patel, R., & Singh, N. (2023a). Microbial nitrogen fixation in soil: Importance for sustainable farming. *Environmental Microbiology Reports*, *15*(1), 102–115.

Kumar, S., Patel, R., & Singh, N. (2023b). Role of plant growth-promoting microorganisms in sustainable agriculture. *Microorganisms*, *11*(3), 212–225.

Lee, S., Cho, M., & Kim, J. (2022). Nitrogen fixation: A critical process for plant growth and ecosystem stability. *Soil Biology and Biochemistry*, *62*, 290–299.

Li, H., Zhang, Y., & Li, X. (2022). Nitrogen cycling by diazotrophs: Implications for sustainable agriculture. *Soil Biology & Biochemistry*, *58*(7), 448–459.

Li, H., Zhang, Y., & Li, X. (2023). Mechanisms of nutrient acquisition by plant growth-promoting microorganisms. *Frontiers in Microbiology*, *14*, 1223–1236.

Li, J., Wu, H., Pu, Q., Zhang, C., Chen, Y., Lin, Z., Hu, X., & Li, O. (2023). Complete genome of *Sphingomonas paucimobilis* ZJSH1, an endophytic bacterium from *Dendrobium officinale* with stress resistance and growth promotion potential. *Archives of Microbiology*, *205*, 132. https://doi.org/10.1007/s00203-023-03459-2.

Mahapatra, S., Yadav, R., & Ramkrishna, W. (2022). *Bacillus subtilis* impact on plant growth, soil health and environment: Dr. Jekyll and Mr. Hyde. *Journal of Apllied. Microbiology*, *132*(5), 3543–3562. https://doi.org/10.1111/jam.15480.

Naqqash, T., Imran, A., Hameed, S., Shahid, M., Majeed, A., Iqbal, J., Hanif, M. K., Ejaz, S., & Malik, K. A. (2020). First report of diazotrophic Brevundimonas spp. as growth enhancer and root colonizer of potato. *Scientific Report*, *10*, 12893. https://doi.org/10.1038/s41598-020-69782-6.

Nichio, B. T.dL., Chaves, R. B. R., Pedrosa, F.dO., & Raittz, R. T. (2025). Exploring diazotrophic diversity: Unveiling Nif core distribution and evolutionary patterns in nitrogen-fixing organisms. *BMC Genomics*, *26*, 81.

Nofal, M. M., & Attia, M. S. (2023). Management of root-knot nematode (*Meloidogyne incognita*) and improve defensive ability of cucumber plant using *Serratia marcescens*, salicylic acid and fosthiazate (melotherin) through adaptive processes and antioxidant capacity. *Menoufia Journal of Plant Protection*, *8*, 31–45. https://doi.org/10.21608/mjapam.2023.197242.1011.

Parvin, N., Mukherjee, B., Roy, S., & Dutta, S. (2023). Characterization of plant growth promoting rhizobacterial strain Bacillus cereus with special reference to exopolysaccharide production. *Journal of Plant Nutrition*, *46*(11), 1–13.

Patel, R., & Kaur, H. (2023). Biological nitrogen fixation and its role in agriculture: The potential of diazotrophic bacteria as biofertilizers. *Journal of Applied Microbiology*, *15*(3), 122–135.

Pereira, F. (2019). Rhizobacteria as Bioprotectants against stress conditions. Plant Growth promoting Rhizobacteria for sustainable stress management. *Rhizobacteria Abiotic Stress Management*, *1*, 157–177. https://doi.org/10.1007/978-981-13-6536-2.

Ratnaningsih, H. R., Noviana, Z., Dewi, T. K., Leokito, S., Wiyono, S., Gafur, A., & Antonius, S. (2023). IAA and ACC deaminase producing-bacteria isolated from the rhizosphere of pineapple plants grown under different abiotic and biotic stresses. *Heliyon 26*, *9*(6), e16306. https://doi.org/10.1016/j.heliyon.2023.e16306.

Saikia, J., Kotoky, R., Debnath, R., Kumar, N., Gogoi, P., Yadav, A., & Saikia, R. (2023). De novo genomic analysis of Enterobacter asburiae EBRJ12, a plant growth-promoting rhizobacteria isolated from the rhizosphere of *Phaseolus vulgaris L. Journal of Applied Microbiology*, *134*(2), lxac090. https://doi.org/10.1093/jambio/lxac090.

dos Santos Ferreira, N., Coniglio, A., Puente, M., Sant'Anna, F. H., Maroniche, G., Garcia, J., Molina, R., Nievas, S., & Volpiano, C. G. A. (2022). Genome-based reclassification of Azospirillum brasilense Az39 as the type strain of Azospirillum argentinense sp. nov. *International Journal of Systmatic Evolutionary Microbiology*, *72*(8), 005475. https://doi.org/10.1099/ijsem.0.005475.

Schmitz, R. A., Klopprogge, K., & Grabbe, R. (2002). Regulation of nitrogen fixation in Klebsiella pneumoniae and Azotobacter vinelandii: *Nif*L, transducing two environmental signals to the *nif* transcriptional activator *Nif*A. *Journal of Molecular Microbiology and Biotechnology*, *4*(3), 235–42.

Shi, H. w, Wang, L. y, Li, X. x, Xiao-meng, L., Hao, T. -y, He, X. -j, & Chen, S. -f (2016). Genome-wide transcriptome profiling of nitrogen fixation in *Paenibacillus* sp. WLY78. *BioMed Central Microbiology*, *16*, 25. https://doi.org/10.1186/s12866-016-0642-6.

Singh, J., Kumar, P., & Sharma, R. (2021). Synergistic interactions of PGPSMs for sustainable crop production. *Plant Soil & Environment*, *67*(10), 563–575.

Singh, J., Kumar, P., & Sharma, R. (2022). Diazotrophic bacteria and their role in agricultural systems. *Plant Soil & Environment*, *68*(8), 520–531.

Singh, R. P., Pandey, D. M., Jha, P. N., & Ma, Y. (2022). ACC deaminase producing rhizobacterium *Enterobacter cloacae* ZNP-4 enhance abiotic stress tolerance in wheat plant. *PLOS ONE*, *17*(5), e0267127. https://doi.org/10.1371/journal.pone.0267127.

Sivasakthivelan, P., Saranraj, P., Sayyed, R., Arivukkarasu, K., Kokila, M., Manigandan, M., & Seifi, S. (2023). Inoculant production and formulation of Azospirillum Species. In *Plant growth promoting microorganisms of arid region* (pp. 423–455). Springer. https://doi.org/10.1007/978-981-19-4124-5_19.

Smith, A., Turner, C., & Martinez, G. (2023). Diazotrophic bacteria and nitrogen fixation: Insights into sustainable agricultural practices. *Microbial Ecology in Agriculture*, *55*(2), 114–130.

Smith, J., Turner, C., & Martinez, G. (2022). Effects of synthetic nitrogen fertilizers on soil acidification and microbial diversity: A review. *Agricultural Systems*, *112*(9), 775–787.

Sriwati, R., Maulidia, V., Intan, N., Oktarina, H., Khairan, K., Skala, L., & Mahmud, T. (2023). Endophytic bacteria as biological agents to control fusarium wilt disease and promote tomato plant growth. *Physiology and Molecular Plant Pathology*, *125*, 101994. https://doi.org/10.1016/j.pmpp.2023.101994.

Tafaroji, S. H., Abtahi, A., Jafarinia, M., & Ebadi, M. (2022). The effect of plant growth-promoting rhizobacteria producing ACC-deaminase, IAA, siderophore and phosphate solubilization on growth indices, chlorophyll, proline and protein in alfalfa at different levels of salinity. *Plant Production*, *45*(3), 375–384. https://doi.org/10.22055/ppd.2022.40265.2018.

Tilman, D., Cassman, K. G., Matson, P. A., Naylor, R., & Polasky, S. (2002). Agricultural sustainability and intensive production practices. *Nature*, *418*(6898), 671–677. https://doi.org/10.1038/nature01014.

United Nations (2019). World population prospects 2019: Highlights. Department of Economic and Social Affairs, Population Division. https://population.un.org/wpp/Publications.

Wang, Q., & Liu, Z. (2021). Biological nitrogen fixation in soils: Mechanisms, benefits, and environmental impact. *Agricultural Systems*, *28*(6), 701–713.

Wang, Q., & Liu, Z. (2022a). Mechanisms of nitrogen fixation in free-living and symbiotic diazotrophs. *Journal of Soil Microbiology*, *44*(8), 674–682.
Wang, Q., & Liu, Z. (2022b). Nitrogen fixation by PGPSMs and its contribution to soil fertility. *Soil Microbiology and Plant Nutrition*, *48*(7), 555–566.
Wang, Q., & Liu, Z. (2022c). Nitrogen fixation by plant growth-promoting rhizobacteria and their role in nutrient cycling. *Soil Biology & Biochemistry*, *54*(7), 677–688.
Wang, Q., & Liu, Z. (2023a). Diazotrophic microorganisms and their contribution to soil fertility and microbial biodiversity. *Soil and Plant Science Journal*, *38*(6), 320–332.
Wang, Q., & Liu, Z. (2023b). The contribution of diazotrophs to soil fertility and plant growth promotion. *Soil Microbiology and Plant Nutrition*, *52*(2), 94–105.
Wang, T., Zhao, X., Wu, X., & Chen, S. (2023). Genome-wide mapping of GlnR-binding sites reveals the global regulatory role of GlnR in controlling the metabolism of nitrogen and carbon in *Paenibacillus polymyxa* WLY78. *BioMed Central Genomics*, *24*, 85. https://doi.org/10.1186/s12864-023-09147-1.
Wang, Z., Zhong, T., Chen, K., Du, M., Chen, G., Chen, X., Wang, K., Zalan, Z., Takacs, K., & Kan, J. (2021). Antifungal activity of volatile organic compounds produced by *Pseudomonas fluorescens* ZX and potential biocontrol of blue mold decay on postharvest citrus. *Food Control*, *120*(2021), 107499. https://doi.org/10.1016/j.foodcont.2020.107499.
Wani, P. A., Khan, M. S., & Zaidi, A. (2007). Chromium reduction, plant growth-promoting potentials, and metal solubilizatrion by *Bacillus* sp. isolated from alluvial soil. *Current Microbiology*, *54*(3), 237–243. https://doi.org/10.1007/s00284-006-0451-5.
Xiang, H., He, Y., Wang, X., Wang, J., Li, T., Zhu, S., Zhang, Z., Xu, X., & Ww, Z. (2023). Identification and characterization of siderophilic biocontrol strain SL-44 combined with whole genome. *Environmental Science and Pollution Research*, *30*, 62104–62120. https://doi.org/10.1007/s11356-023-26272-2.
Yadav, P., Ramesh, C., & Pareek, N. (2022). Plant Growth Promoting Mesorhizobia as a Potential Inoculant for Chickpea (*Cicer arietinum* L.): A Review. *Bhartiya Krishi Anusandhan Patrika*, *37*(4), 328–333. https://doi.org/10.18805/BKAP553.
Yan, Y., Ping, S., Peng, J., Han, Y., Li, L., Yang, J., Dou, Y., Li, Y., Fan, Y., Li, D., Zhan, Y., Chen, M., Lu, W., Zhang, W., Cheng, Q., & Lin, M. (2010). Global transcriptional analysis of nitrogen fixation and ammonium repression in root-associated Pseudomonas stutzeri A1501. *BioMed Central Genomics*, *11*, 11. https://doi.org/10.1186/1471-2164-11-11.
Yang, Y., Zhao, S., & Liu, T. (2022). Environmental factors affecting biological nitrogen fixation efficiency. *Frontiers in Soil Science*, *22*(8), 223–234.
Yang, Y., Zhao, S., & Liu, T. (2023). The impact of environmental factors on biological nitrogen fixation in agricultural systems. *Soil and Plant Nutrition*, *49*(7), 845–860.
Yang, Z., Wang, F., & Liu, H. (2021a). Soil microbial communities and their role in improving soil structure and stability. *Soil and Plant Science Journal*, *32*(4), 789–801.
Yang, Z., Wang, F., & Liu, H. (2021b). The role of phytohormones in promoting plant growth via PGPSMs. *Journal of Plant Growth Regulation*, *40*(4), 789–802.
Zhang, Y., Chen, L., & Wei, F. (2021). Diazotrophic microorganisms in agriculture and their ecological importance. *Journal of Agricultural Science*, *79*(2), 234–245.
Zhang, Y., Chen, L., & Wei, F. (2022). Siderophore production by plant growth-promoting bacteria and its role in iron acquisition. *Microbial Ecology*, *65*(4), 478–490.
Zhou, T., Liu, X., & Hu, Z. (2022). The interaction between nitrogen-fixing bacteria and plants: Mechanisms and applications in sustainable farming. *Applied Soil Ecology*, *36*(5), 298–310.
Zhou, T., Liu, X., & Hu, Z. (2023). Extracellular polysaccharides of diazotrophic bacteria in soil stabilization. *Applied Soil Ecology*, *82*(7), 201–213.

15 Sustainable Sustenance: Leveraging Biological Nitrogen Fixation in Agriculture

Rupali Dutt, Charu Gupta and Mahendra K. Gupta

Abstract

Green energy technologies are crucial for lowering dependency on fossil fuels and promoting sustainable development. Nitrogen fixation is an important green energy process in which both biologically and chemically convert atmospheric nitrogen (N_2) into bioavailable ammonia (NH_3). Nitrogen, a key component for plant growth, necessitates sustainable nitrogen fixation solutions to diminish reliance on energy-intensive chemical fertilizers, thereby promoting green agriculture and environmental sustainability. However, the intricate mechanisms involved in this process are mainly mediated by nitrogenase enzymes in biological systems or by sophisticated electrochemical and photochemical techniques in industrial settings. Furthermore, artificial techniques like the Haber-Bosch process depend on high temperatures and pressures and use a lot of energy. These techniques cause a detrimental effect on agriculture and the environment. To overcome this problem, Biological Nitrogen Fixation (BNF) uses specialized microorganisms and operates in ambient conditions. Therefore, Nitrogen fixation has applications in greenhouse gas emissions reduction, biofertilizer development, sustainable agriculture, and reducing reliance on synthetic fertilizers. Green energy solutions can provide a substantial contribution to food security, environmental preservation, and the shift to sustainable development by combining environmentally friendly and BNF techniques.

R. Dutt · C. Gupta · M. K. Gupta (✉)
Microbiology Research Lab., School of Studies in Botany, Jiwaji University, Gwalior, Madhya Pradesh, India
e-mail: mkgsac@yahoo.com

R. K. Singh and P. Singh (eds.), *Diazotrophic Plant Growth-Promoting Microorganisms and Green Energy – Volume I*, https://doi.org/10.1007/978-981-95-9682-9_15

Keywords

Biological nitrogen fixation · green energy · agriculture · environment

15.1 Introduction

Globally, the use of inorganic nitrogen fertilizers has disrupted ecosystems (Soumare et al., 2020; Zheng et al., 2019; Usha et al., 2018; Yang et al., 2015). This supports the growing call to encourage sustainable farming and agroforestry methods and lessen the systematic use of inorganic N fertilizers (Shah and Wu, 2019; Araujo et al., 2012). Utilizing biological nitrogen fixation (BNF) seems to be one of the other strategies for lowering the amount of N fertilizers used in agriculture and, consequently, their detrimental effects on the environment. Actually, BNF is a natural process that converts atmospheric nitrogen (N_2) into NH4+, a simple, soluble, nontoxic form that plant cells use to synthesize a variety of biomolecules. One of the main ways that plants obtain nitrogen is through nitrogen fixation, which is also a crucial process for the ecosystem's distribution of this nutrient (Sur et al., 2010; Saikia & Jain, 2007).

Bacteria and archaea are the only prokaryotes that can fix nitrogen biologically. There are several types of bacteria involved, including cyanobacteria linked to cycads, symbiotic bacteria like Rhizobium in legumes, Frankia in actinorhizal plants, and free-living bacteria from genera like *Azotobacter, Azospirillum, Bacillus*, or *Clostridium* (Ininbergs et al., 2011; Ravikumar et al., 2007). Nitrogen fixation in archaea is still limited to methanogens, which are species that create methane (Welte, 2018).

Among the main nutrients found in soil, nitrogen (N) is special since it comes from the atmosphere and is primarily mediated by biological activities and the water cycle in an ecosystem. Fowler et al. (2015) predicted that a relatively modest portion of the atmosphere's huge, well-mixed, but biologically inaccessible pool of di-nitrogen (N_2) gas (4×109 Tg N) is transformed each year into reactive or physiologically available pools of N. These changes were thought to be caused by (a) a natural non-biological process (lightning 5 Tg N), (b) biological N_2 fixation (BNF: 128 Tg N for natural terrestrial ecosystems, 120 Tg N for marine and aquatic ecosystems, and 60 Tg N in agricultural ecosystems), (c) industrial fixation of ammonia (NH_3) through the Haber-Bosch process (120 Tg N), and (d) burning fossil fuels (40 Tg N). According to Fowler et al. (2015), 220 Tg N of the 473 Tg N fixed from atmospheric N_2 was thought to have come from human activity.

15.1.1 Biological Fundamentals of Nitrogen Fixation

Nitrogen Cycle: The most important step in the N cycle is nitrogen fixation. Chemically triple-bonded N-N is how dinitrogen enters the biosphere. Regardless of whether the triple bond is broken by BNF or the Haber-Bosch process, around 103 kJ mol^{-1} is needed to transform it into NH_3. When it comes to BNF, nitrogenase activity quickly transforms the NH_3 into ammonium (NH_4^+), which is then used for the biochemical synthesis of proteins and amino acids that are momentarily stored in microbial or plant biomass. These organic N sources contribute to the soil organic N (SON) pool during senescence and death, with some of that N being subsequently mineralized by soil microorganisms and released as nitrate (NO_3^-), an accessible form of nitrogen that can be absorbed by plant roots and other soil microbial community members.

N is not entirely recycled by organisms like other elements. Instead, it is constantly lost through a variety of loss mechanisms, such as: (a) The anaerobic denitrification of NO_3-mediated by microbes resulting in the release of strong greenhouse gases like nitrous oxide (N_2O) from the soil, (b) the volatilization of NH_3 from compost, alkaline soils, and animal urine patches, (c) NO_3^-leaching to groundwater beyond the crop rooting zone, and (d) the loss of organic and inorganic forms of nitrogen in runoff water and erosion. As a result, BNF may play a significant role in restocking organic N reserves in a plant-soil system. The soil is in a state of dynamic flux and functions as both a source and a sink of nitrogen. A soil system's capacity to hold nitrogen depends on how well its net gains and losses are balanced. Thus, the fundamental mechanisms behind the N-cycle can be regarded as BNF and the numerous N loss pathways.

15.2 Nitrogen Fixation

15.2.1 Discovery

In 1886, Hellriegel and Wilfarth discovered biological nitrogen fixation when they found that certain legumes could use nitrogen gas (N_2) from nodules on their roots (Franche et al., 2009). Beijerinck made the initial isolation of the N-fixing bacterial strain *Rhizobium leguminosarum* 2 years later in 1888 (Beijerinck, 1901). Winogradsky reported isolating *Clostridium pasteurianum* in 1893 (Winogradsky, 1895), and Guthrie presented the idea of inoculating legumes with N-fixing rhizobia to farmers in New South Wales, Australia, in 1896 (Gurthie, 1896). Beijerinck discovered and characterized the aerobic heterotroph *Azotobacter* five years later in 1901.

In 1905, Hawkesbury Agricultural College in Australia conducted the first field trials utilizing a rhizobial and field pea culture. Later, in 1914, farmers received rhizobial inoculants (Bullard et al., 2005). In 1953, Johanna Döbereiner's research led to the discovery of a new class of nitrogen-fixing endophytes: *Azotobacter paspali*, which is linked to Bahia grass (*Paspalum notatum*), and *Beijerinckia fluminensis*, which is linked to sugarcane (Baldani & Baldani, 2005; Dobereiner, 1966). *Fabio Pedrosa* created a nitrogen-free medium known as NFb (Fb for *Fabio Pedrosa*) at the end of 1975 in order to isolate *Spirillum* species. Thus, the semi-solid NFb medium has been used to isolate two species of *Azospirillum* (*A. lipoferum* and *A. brasilense*) (Baldani & Baldani, 2005; Dobereiner, 1976). These two types of bacteria are non-symbiotic and heterotrophic, with an average annual fixation rate of 20–30 kg N/ha (Kizilkaya, 2009). Stewart demonstrated nitrogen fixation in cyanobacteria (previously known as blue-green algae) in 1979 using the Nitrogen-15 (15 N) tracing technique (Stewart,

1969). Only lately have separate findings of diazotrophic growth in two distinct methanogenic archaea—*Methanococcus thermolithotrophicus* (Leigh, 2000; Belay et al., 1984) and *Methanosarcina barkeri* (Leigh, 2000; Murray & Zinder, 1984) —highlighted the ability of archaea to fix nitrogen. Nitrogen fixation in *M. barkeri* and *M. thermolithotrophicus* has been verified by the 15 N tracing technique and the acetylene reduction assay (ARA) technique, respectively. At least thirteen genera in the prokaryotics group are currently recognized to fix nitrogen.

15.2.2 Nitrogen-Fixing Organisms

There are three types of nitrogen-fixing organisms: symbiotic, associative, and free-living N fixers. The rhizospheres of legume and non-legume plants contain the final two categories (Soumare et al., 2020; Mus et al., 2016; Santi et al., 2013) (Fig. 15.1).

15.2.3 Free Living Nitrogen Fixers

Numerous heterotrophic bacteria inhabit the soil and fix large amounts of nitrogen without coming into close contact with other living things. This category of nitrogen-fixing bacteria includes species of *Azotobacter, Klebsiella, Clostridium*, and *Bacillus*. These creatures have to find their own energy supply, usually through oxidizing organic molecules that are generated by other species or from decomposition. Certain free-living organisms can use inorganic substances as a source of energy because they are chemolithotrophs. Free-living microorganisms fix nitrogen by acting as anaerobes or microaerophiles since oxygen can block nitrogenase as shown in Table 15.1. Their contribution to the rates of nitrogen fixation worldwide is typically regarded as negligible due to the lack of adequate carbon and energy supplies for these species. But according to a recent study conducted in Australia on an intense

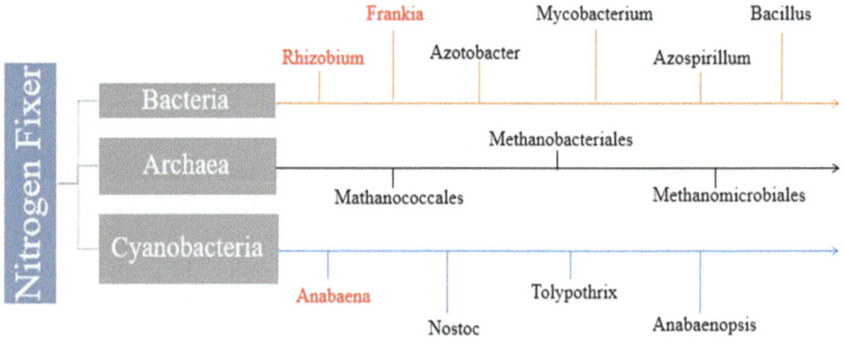

Fig. 15.1 The three classes of organisms that fix nitrogen include a few major genera. Red indicates genera that contain symbiotic nitrogen-fixing species, while black indicates orders or genera that contain free-living nitrogen-fixing species (Soumare et al., 2020)

wheat rotation farming system, free-living microorganisms helped the cropping system meet its long-term nitrogen needs by contributing 20 kilos per hectare annually (30%–50% of the total needs) (Vadakattu & Patterson 2006). This system's limited tillage and maintenance of wheat stubble created the ideal high-carbon, low-nitrogen environment for the free-living organisms' maximum activity.

15.2.4 Associative Nitrogen Fixers

Several members of the Poaceae (grasses), including agronomically significant cereal crops like rice, wheat, corn, oats, and barley, can create close interactions with *Azospirillum* species. Significant amounts of nitrogen are fixed by these bacteria in the host plants' rhizosphere. There have been reports of 52 mg N_2 g^{-1} malate efficacies (Stephan et al. 1981). A number of variables affect the amount of nitrogen fixation, such as soil temperature (*Azospirillum* species prefer more temperate and/or tropical climates), the host plant's capacity to create an oxygen-low rhizosphere, the availability of host photosynthates for the bacteria, the bacteria's competitiveness, and nitrogenase efficiency (Vlassak & Reynders, 1981).

Table 15.1 Key genera of free-living and symbiotic diazotrophs of agricultural significance and examples of associated host plant species

Heterotrophic		Phototrophic	
Free living	Symbiotic (host)	Free living	Symbiotic (host)
Aerobic	*Azorhizobium*	*Cyanobacteria*	Anabaena (Azolla)
Azotobacter	*Sesbania*	*Gloeocapsa*	
Azospirillum	*Bradyrhizobium*	*Trichodesmium*	
Azoarcus	(peanut, soybean, vigna)	*Nostoc*	
Beijerinckia	*Mesorhizobium*	*Anabaena*	
Rhizobium	(Chickpea, milk vetch)	Calothrix	
Facultative anaerobic	*Rhizobium*	Photosynthetic bacteria	
Bacillus	(Common beans, peas, lentils, vetch, Medicago, sesbania)	*Rhodospirillum*	
Klebsiella	*Sinorhizobium*	*Rhodopseudomonas*	
Enterobacter	(soybean, sesbania)	*chromatium*	
Erwinia		*Chlorobium*	
Anaerobic			
Clostridium			
Desulfobirio			

Source: Ladha et al. (2022)

15.2.5 Symbiotic Nitrogen Fixers

Numerous microorganisms work in tandem with a host plant to fix nitrogen symbiotically. The nitrogen-fixing microbe uses the sugars produced by photosynthesis from the plant to give the energy it needs to fix nitrogen. The bacterium gives the host plant fixed nitrogen for growth in return for these carbon sources. The cyanobacterium *Anabena azollae* and the water fern *Azolla's* symbiotic relationship is one instance of this kind of nitrogen fixation. Cavities created at the base of *Azolla* fronds are colonized by *Anabaena*. Significant amounts of nitrogen are fixed by the cyanobacteria in specialized cells known as heterocysts. In Southeast Asian wetland paddies, this symbiosis has been utilized as a biofertilizer for at least a millennium. During the growing season, *Azolla* "blooms," which blankets rice paddies, can fix up to 600 kg N ha^{-1} yr^{-1}.

Actinorhizal plants and shrubs, like alder (*Alnus sp.*), and the actinomycete *Frankia* have a symbiotic relationship. These North American native plants usually do well in conditions with low nitrogen levels. Being the pioneer species in successional plant communities, they are frequently the most prevalent non-legume nitrogen fixers in many regions. Among the various ecosystems where actinorhizal plants can be found are alpine, xeric, chaparral, forest, glacial till, riparian, coastal dune, and arctic tundra habitats.

The interactions between legumes and the *Rhizobium* and *Bradyrhizobium* bacteria are by far the most significant nitrogen-fixing symbiotic partnerships, despite the fact that the symbiotic partners mentioned above are significant to the global ecology of nitrogen fixation. In agricultural systems, alfalfa, beans, clover, cowpeas, lupines, peanuts, soybeans, and vetches are major legumes. Soybeans account for 68% of the world's total legume production and are produced on 50% of the land dedicated to legumes worldwide (Vance, 2001) (Fig. 15.2).

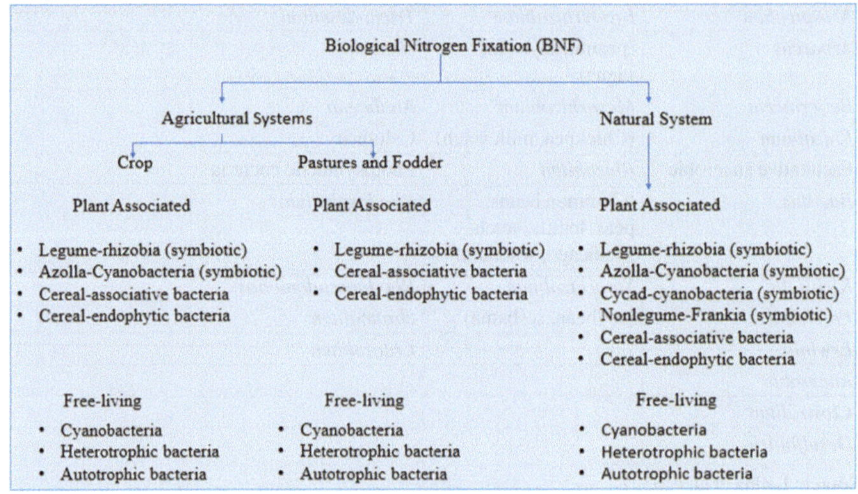

Fig. 15.2 Nitrogen-fixing organisms are found in agricultural and natural systems (Wagner, 2011)

The process of biological nitrogen fixation transforms di-nitrogen (N_2) into NH_4^+, which is largely used by plants. N_2 is combined with the hydrogen ions from water to complete the reaction. The oxidation of N_2 to nitrate (NO_3) by fire or lightning makes N_2-fixation more than just a physiologically mediated process. Of the net nitrogen fixed annually, lightning produces around 1% ammonia (Igarashi & Seefeldt, 2003). BNF is an essential source of nitrogen for all living things, including prokaryotes and eukaryotes. For the creation of proteins, nucleic acids, and other organic nitrogenous substances, this N is the primary ingredient. Since 16 ATP molecules are required to break down one N_2 molecule, biological nitrogen fixation is an energy costly process. To transport and assimilate NH_4^+, an extra 12 ATP molecules are needed, for a total of 28 ATP molecules. For the nodulating plants to benefit 1 g N in part, they must give their bacterial partners 12 g of glucose (Buscot & Varma , 2005). Nevertheless, this method is still less costly in terms of energy than the 1913 Haber-Bosch process. For the Haber-Bosch process to yield the equivalent amount of nitrogen, temperatures between 400 and 500 degrees Celsius and pressures between 200 and 250 bars are needed (Gilchrist & Benjamin, 2017).

The majority of nitrogen-fixing bacteria have nitrogenase, which catalyzes N_2 fixation. Nitrogenase is an enzyme complex that has two metal components: dinitrogenase reductase (Fe protein) and dinitrogenase MoFe (molybdenum-iron protein), which acts as the catalytic component. The nif genes, which code for MoFe dinitrogenase (nif D and nif K) and Fe dinitrogenase reductase (nif H) (Nonaka et al., 2019; Seefeldt et al., 2018; Buren & Rubio, 2018; Dai et al., 2014), encode these two metal components. Nif genes encode a number of regulatory proteins involved in nitrogen fixation in addition to nitrogenase. Three distinct kinds of nitrogenase exist, depending on the needs of iron (Fe), vanadium (V), or molybdenum (Mo). For the reduction of the substrate, each nitrogenase has an active site that is made up of a complex metal group known as the FeV-cofactor, FeFe-cofactor, and FeMo-cofactor, which stand for V-nitrogenase, Fe-nitrogenase, and Mo-nitrogenase, respectively (Harris et al., 2018). Because nitrogenase is so sensitive to oxygen, it is inactivated in aerobic environments regardless of its form. In fact, oxygen inhibits the processes involved in nitrogen fixation and assimilation and deactivates and kills nitrogenase (Berman-Frank et al., 2005). While Mo-nitrogenase is marginally less sensitive to oxygen, Fe-nitrogenase and V-nitrogenase are extremely sensitive (Fay, 1992).

In nitrogen fixation, the Fe protein links and hydrolyzes two ATP molecules, transferring one electron to the 8Fe–7 S cluster from the 4Fe–4 S cluster within the Fe protein. Subsequently, the electrons are conveyed to the Mo–Fe cofactor to engage in the reduction of the N2 molecule. N2 possesses a stable intramolecular triple bond, rendering the reduction of N2 to NH3 by nitrogenase a highly challenging reaction that necessitates surmounting the activation energy barrier of catalysis. High-energy electrons contained within the Mo–Fe cofactor serve as reducing agents to surmount the activation energy barrier of catalysis in the presence of substantial quantities of ATP and protons [16]. V-Fe and Fe-Fe nitrogenases can also use this catalytic process, but they require additional high-energy electrons, ATP, and protons to break through the energy barrier. While V-Fe and Fe-Fe nitrogenases require 24 to 32 ATP molecules to immobilize a molecule of N2, Mo-Fe nitrogenase requires at least 16 ATP molecules (Zhang et al., 2023).

15.3 Nodule Formation in Legume

The bacteria *Bradyrhizobium or Rhizobium* colonize the root system of the host plant, causing the roots to develop nodules that will hold the bacteria. Next, the bacteria start fixing the nitrogen that the plant needs. The plant can develop nitrogen-fortified leaves that can be recycled throughout the plant when it has access to the fixed nitrogen. As a result, the plant can produce more nitrogen-rich seeds by increasing its photosynthetic capacity. When legumes are grown in soil that is low in nitrogen, the results of not nodding them can be quite drastic. The resulting plants usually produce relatively little seed, have little nitrogen content, and are chlorotic. An organized connection between the bacterium and the host plant is demonstrated by the nodulation process (Napoli & Hubbell 1975, Kamst et al. 1998; van Rhijn & Vanderleyden 1995, Cheng & Walker 1998). The process starts when flavonoids produced by the roots of the host legume attract the rhizobia. The bacteria then start to cling to root hairs, which are extensions of root epidermal cells, in legumes like alfalfa, clover, and soybeans (others, including lupines and peanuts, create nodules in other ways). Actually, the attachment mechanism consists of two steps. The bacteria first bind utilizing rhicadhesin, a protein that binds $Ca^{2+.}$ A more robust attachment, including lectins and/or cellulose fibrils and fimbriae produced by the host plant and bacteria, respectively, occurs after the bacteria have accumulated and anchored themselves to the surface of the root hair as shown in Table 15.2.

The rhizobia then produce chemicals known as Nod factors, which the host legume detects and causes the colonized root hairs to curl, creating a shepherd's crook. Once inside the root hairs, rhizobia usually develop into an infection thread, which is a tubular structure. The bacteria cause cortical cell divisions that result in the development of a nodule after they have reached the root itself. In the process of producing the nodule, the bacteria are liberated into the plant cells and encased in a membrane formed from the plant. In order to create bacteroid—large, irregularly shaped branching cells—the

Table 15.2 Major cross-inoculation groups of legumes

Crop	Nodulating bacteria
Alfalfa	*Sinorhizobium melilotii*
Beans	*Rhizobium leguminosarum biovar phaseoli and Rhizobium tropici*
Clover	*Rhizobium leguminosarum biovar trifolii*
Lotus	*Mesorhizobium loti*
Peas	*Rhizobium leguminosarum biovar viceae*
Soybean	*Bradyrhizobium japonicum, Bradyrhizobium elkanii, Rhizobium fredii*
Sesbania	*Azorhizobium caulinodans*

Source: Wagner (2011)

bacteria then lose their cell walls and undergo a significant shift in cell morphology. For their energy requirements, they are then totally reliant on the host plant. In exchange, the bacteria fix nitrogen for the plant.

The synthesis of leghemoglobin is another illustration of the complex interaction between the rhizobia and the host legume (Appleby 1984). The structure and function of this molecule are comparable to those of hemoglobin, which is present in human blood. It binds and controls the oxygen levels in the nodule and is only generated in fully functional root nodules. The amount of nitrogen fixation in the nodule would be inhibited by free oxygen in the cytoplasm of the nodule cell since the enzyme nitrogenase is oxygen-sensitive. It appears that leghemoglobin carries just enough oxygen for the rhizobia to perform cellular respiration, without preventing nitrogenase from doing its job.

15.4 Non-Symbiotic BNF in Maize, Rice, and Wheat

During cereal cropping, non-symbiotic N_2 fixation occurs in the soil, on the plant-soil surface, and within the plant. These are known as autochthonous (indigenous) BNF systems, and they are made up of cyanobacteria and heterotrophic and phototrophic bacteria that are found in soil, plants, and floodwater (Roger and Ladha, 1992; Giller, 2001; Reddy et al., 2002; Bageshwar et al., 2017; Van Deynze et al., 2018).

The majority of rice is grown in submerged soils, which also support free-living heterotrophic diazotrophs and phototrophic (cyanobacteria) BNF. Under ideal circumstances, 20–40 kg N ha^{-1} $crop^{-1}$ can be fixed by cyanobacteria (Roger and Watanabe, 1986). Since it was discovered that cyanobacteria play a significant role in providing nitrogen gains during floods, numerous inoculation experiments have been carried out utilizing either native or cultured cyanobacteria as potential methods to increase soil fertility and rice grain output.

15.5 Symbiotic BNF (Aquatic Green Manures) in Rice Farming

The aquatic habitat of rice lowlands makes them ideal for aquatic legumes like *Sesbania* and *Aeschynomene spp.* that form symbioses with heterotrophic and phototrophic rhizobia, as well as the water fern *Azolla*, which is home to the symbiotic N2-fixing cyanobacteria *Anabaena azollae* (Ladha et al., 1992). Because they are not common in agriculture, these allochthonous (exogenous) BNF systems made up of *Azolla* and legumes must be added to rice fields in order to give the crop more nitrogen (Reddy et al., 2002).

Typically, *azolla* is cultivated as a cover crop and inoculated with or sans rice to be added to the soil as a top dressing when growing rice. Under ideal circumstances, significant amounts of BNF can be fixed, up to 70% of which becomes available to the rice crop upon incorporation (Roger and Ladha, 1992), and up to 99% of *Azolla* N can be obtained from the atmosphere (Yoneyama et al., 1997). During a growth cycle of roughly 40 days, a single standing crop of *Azolla* can accumulate between 20- and 146-kilogram N ha^{-1} (average 70 kg ha^{-1}), and the N_2-fixing rate can vary between 0.4 and 3.6 kg N ha^{-1} d^{-1} (average 2 kg N ha^{-1} d^{-1}) (Becker et al., 1995).

Green manured aquatic legumes have been found to fix up to 458 kg N ha^{-1}, but BNF commonly varied between 100 and 180 kg N ha^{-1} (Becker et al., 1990). As the plants grew older, the percentage of Ndfa in well-nodulated *S. rostrata* and *S. cannabina* rose from 50%–75% at 25 days after sowing, 70%–95% at 45–55 days, and nearly 100% by 65 days. The amount of N_2 fixed was larger for *S. rostrata* due to higher N accumulation in biomass, even if the percentage Ndfa tended to be similar in both Sesbania species. Because *A. afraspera* is less sensitive to photoperiod than *S. rostrata*, it accumulates more nitrogen and produces more BNF during the shorter days of the year.

15.6 Factors Affecting Biological Nitrogen Fixation

The following variables have a significant impact on growth, survival, and metabolic activity of nitrogen-fixing bacteria and plants, as well as their capacity to form symbiotic relationships: Environmental stress, Temperature, Water status, Soil fertility, pH fluctuations, and Availability of nutrients.

15.6.1 Environmental Stress

Understanding the interactions between macro and micro-symbionts at the cellular and molecular levels is crucial to comprehending the impact of environmental stress. Environmental disturbances have independent effects on the nitrogen-fixation and nodulation processes. Bacterial nodulation gene activation and repression are tightly regulated processes that significantly impact host specificity and environmental sensitivity.

As a global regulatory gene, the nodD gene promotes the transcription of other inducible nodulation genes in conjunction with plant flavonoid-signal molecules. Plant fertility, pH, and nod factors are among the environmental factors that affect the production of the flavonoid nod-gene inducers, which are specific for a certain legume-Rhizobium interaction. Repressor proteins are also involved in the regulation of nod-genes. It is commonly known that the symbiotic nitrogen fixation (SNF)-dependent legumes are particularly sensitive to salt in terms of nodule initiation, development, and function.

15.6.2 Soil Water Content

The growth of soil microorganisms is influenced by soil water through mass movement, diffusion, and nutrient concentration processes. Soil water and pore space are connected; soils with more pore space and larger pores hold less water. Rhizobia and the majority of soil microorganisms can therefore grow more readily in soil aggregates with fewer interior pore spaces. Rhizobia population declines during the dry season are probably the cause of poor legume nodulation in desert soils. Rhizobia have developed osmotic stress adaptation mechanisms, mostly through the intracellular buildup of organic and inorganic solutes. For instance, *R. meliloti* builds up suitable solutes such as $K+$, glutamate, proline, proline betaine, trehalose, etc. to overcome the growth inhibition caused by osmotic stress.

15.6.3 Soil Nutrient

The symbiotic relationship, as well as the independent growth and survival of both partners, are greatly influenced by soil nutrients. Additionally, as legumes age, fixation tends to decline, primarily due to the corresponding rise in soil N. When N was administered to the soil surface or to 0–20 cm of the top soil, a negative exponential connection between N fertilizer rate and N_2 fixation was found.

15.6.4 Soil Temperature

The survival and persistence of rhizobial strains in soils are significantly impacted by temperature. As an illustration, cowpea rhizobia from West Africa's hot, dry Sahel savannah grow best at 37°C, while over 90% of the strains identified from this area thrived at 40°C. It seems that the impact of temperature on rhizobia depends on both soil and strain. It has also been demonstrated that very high root temperatures affect infection, N_2-fixation capacity, and legume growth. They also have a significant impact on the interaction between particular strains and cultivars.

15.6.5 Soil pH

The symbiotic bacteria and the host plant are both impacted by the pH of the soil. The pH of the soil has an impact on the forage plants' ability to absorb nutrients from it. Generally speaking, the ideal conditions for the forage plant's maximum uptake are found in soil with a pH between 6.0 and 7.0. Thus, keeping the pH of the soil within this range tends to encourage the best possible plant growth and BNF in a variety of forage crops. If the pH of the soil is below 6.0 and it is cost-effective to do so, limestone can be applied to raise the pH and, consequently, the BNF.

15.6.6 Soil Fertility

An overabundance of soil nitrates or a shortage of a vital nutrient that restricts plant growth and development are frequently linked to the reduction of BNF brought on by soil fertility. BNF will generally be lowered by high soil nitrate levels. Forage crops that can use BNF are therefore rarely advised to get fertilizer N. It is necessary to perform soil tests if there is any indication of nutrient shortages in the soil.

15.6.7 Climate

Almost every climatic factor that influences crop growth rate will also have an impact on BNF. In particular, drought, lack of sunshine, and low air/soil temperatures will probably, at least temporarily, lower BNF. Although forage producers currently have no control over climate change, they can adopt certain management measures to lessen its effects.

15.7 Role of Diazotrophs in Plant Resilience Under Abiotic and Biotic Stress Conditions

There is an increasing need for sustainable and environmentally friendly technologies that assist plants in overcoming these challenges in order to maintain agricultural productivity. By improving plant resilience through a variety of direct and indirect processes, plant growth-promoting rhizobacteria (PGPRs) have become important. These bacteria are widely recognized for improving soil quality and fertility, leading to higher agricultural output through biotic activities that promote nutrient turnover and sustainability. The defense mechanisms employed by PGPR are essential for increasing plant productivity and moving their use into commercialization. In order to increase plant resilience, PGPR uses both direct and indirect methods. Mineral solubilization and the synthesis of hormones that regulate growth, such as gibberellic acid (GA3) and indole-3-acetic acid (IAA), are examples of direct systems. Indirect mechanisms include the release of antibiotics, secondary metabolites, and siderophores to suppress infections (Sindiswa et al., 2025).

15.8 Importance of BNF

The improvement of soil fertility is one of the main ecological effects of biological nitrogen fixation. Certain plants have a competitive edge in many habitats, particularly those with low nitrogen availability, due to their capacity to establish symbiotic interactions with nitrogen-fixing bacteria. The capacity of leguminous plants, including soybeans, peas, and clover, to develop nodules on their roots that are home to bacteria that fix nitrogen is especially well-known. These plants receive their ammonia from the bacteria, which act as a direct source of nitrogen for the plants, fostering their development and growth. The entire ecosystem is impacted by the increased nitrogen availability in soils brought about by biological nitrogen fixing.

Since the nitrogen fixed by bacteria is made available to them through processes like mineralization and breakdown, non-leguminous plants also benefited indirectly. By encouraging the growth of nitrogen-demanding species and changing the competitive dynamics between various plant species, this nitrogen availability affects the composition and structure of plant communities. Additionally, by encouraging the growth of nitrogen-fixing plants, biological nitrogen fixation helps to preserve biodiversity.

In ecosystems, these plants, which are frequently pioneer species, foster the growth of other plant species. Their organic matter enriches the soil as they develop and die, increasing soil fertility even further. Plant communities flourish as a result of this process, which boosts ecological stability and variety. Biological nitrogen fixation affects plant tissues' nutritional value in addition to plant development. The availability of nitrogen, a crucial component of proteins, has a direct impact on the amount of protein in plants. For example, legume crops are prized for their high protein content as well as their capacity to fix nitrogen.

Agricultural systems utilize crop rotation, intercropping, and cover crops to maximize nitrogen availability and advance sustainable agriculture, thereby leveraging the ecological relevance of biological nitrogen fixation. Biological nitrogen fixation has consequences for aquatic settings in addition to its function in terrestrial ecosystems. Water bodies can have nitrogen fixed by some cyanobacteria, also referred to as blue-green algae (Masuda, 2024).

15.9 Conclusion

Biological Nitrogen Fixation (BNF) is fundamental to sustainable agriculture, connecting ecological integrity with food security. By utilizing the inherent ability of microorganisms to transform atmospheric nitrogen into forms accessible to plants, biological nitrogen fixation diminishes reliance on energy-intensive synthetic fertilizers, alleviates greenhouse gas emissions, and rejuvenates soil fertility. The incorporation of legumes, efficient inoculants, and sustainable management approaches improves nutrient cycling and production in agricultural systems. To fully grasp its potential, the implementation of BNF necessitates concerted efforts, including scientific advancements to enhance microbial efficiency, and governmental frameworks that promote sustainable nitrogen management. A synergistic strategy that integrates traditional ecological knowledge, contemporary biotechnology, and adaptive management can elevate biological nitrogen fixation from a specialized ecological process to a worldwide sustainability solution.

Ultimately, utilizing BNF transcends mere fertilizer substitution; it entails reengineering agricultural systems to operate in concert with nature. This technique makes sustainable subsistence possible, ensuring resilient food production, environmental preservation, and economic viability in future generations.

References

Appleby, C. A. (1984). Leghemoglobin and Rhizobium respiration. *Annual Review of Plant Physiology*, *35*, 443–478.

Araujo, A. S. F., Leite, L. F. C., Iwata, B. D. F., Lira, Jr, M. D. A., Xavier, G. R., & Figueiredo, M. D. V. B. (2012). Microbiological process in agroforestry systems. A review. *Agronomy for Sustainable Development*, *32*(1), 215–226.

Bageshwar, U. K., Srivastava, M., Pardha-Saradhi, P., Paul, S., Gothandapani, S., Jaat, R. S., Shankar, P., Yadav, R., Biswas, D. R., Kumar, P. A., & Padaria, J. C. (2017). An environmentally friendly engineered Azotobacter strain that replaces a substantial amount of urea fertilizer while sustaining the same wheat yield. *Applied and Environmental Microbiology*, *83*(15), e00590-17.

Baldani, J. I., & Baldani, V. L. (2005). History on the biological nitrogen fixation research in graminaceous plants: special emphasis on the Brazilian experience. *Anais da Academia Brasileira de Ciências*, *77*, 549–579.

Becker, M., Ladha, J. K., & Ali, M. (1995). Green manure technology: Potential, usage, and limitations. A case study for lowland rice. *Plant and Soil*, *174*(1), 181–194.

Becker, M., Ladha, J. K., & Ottow, J. C. G. (1990). Growth and N2-fixation of two stem-nodulating legumes and their effect as green manure on lowland rice. *Soil Biology and Biochemistry*, *22*(8), 1109–1119.

Beijerinck, M. W. (1901). Uber oligonitrophile mikroben. *Zentralbl. Bakterol. Parasitenkd. Infektionskr. Hyg. Abt. II.*, *7*, 561–582.

Belay, N., Sparling, R., & Daniels, L. (1984). Dinitrogen fixation by a thermophilic methanogenic bacterium. *Nature*, *312*(5991), 286–288.

Berman-Frank, I., Chen, Y. B., Gerchman, Y., Dismukes, G. C., & Falkowski, P. G. (2005). Inhibition of nitrogenase by oxygen in marine cyanobacteria controls the global nitrogen and oxygen cycles. *Biogeosciences Discussions*, *2*(2), 261–273.

Bullard, G. K., Roughley, R. J., & Pulsford, D. J. (2005). The legume inoculant industry and inoculant quality control in Australia: 1953–2003. *Australian Journal of Experimental Agriculture*, *45*(3), 127–140.

Buren, S., & Rubio, L. M. (2018). State of the art in eukaryotic nitrogenase engineering. *FEMS Microbiology Letters*, *365*(2), fnx274.

Buscot, F., & Varma, A. (2005). Microorganisms in Soils: Roles in Genesis and Functions (3, p. 422). Springer Science Business Media.

Cheng, H. P., & Walker, G. C. (1998). Succinoglycan is required for initiation and elongation of infection threads during nodulation of alfalfa by Rhizobium meliloti. *Journal of Bacteriology*, *180*(19), 5183–5191.

Dai, Z., Guo, X., Yin, H., Liang, Y., Cong, J., & Liu, X. (2014). Identification of nitrogen-fixing genes and gene clusters from metagenomic library of acid mine drainage. *PLOS ONE*, *9*(2), e87976.

Döbereiner, J. (1966). Azotobacter paspali sp. n., uma bactéria fixadora de nitrogênio na rizosfera de Paspalum Pesquisa Agropecuária Brasileira, (pp. 357–365).

Döbereiner, J. (1976). Fixação de nitrogênio atmosférico em gramíneas tropicais. In *Proceedings of the Congresso Brasileiro de Ciência do Solo, Colombia, Cali*, (Vol. 15, pp. 593–602).

Fay, P. (1992). Oxygen relations of nitrogen fixation in cyanobacteria. *Microbiological Reviews*, *56*(2), 340–373.

Fowler, D., Steadman, C. E., Stevenson, D., Coyle, M., Rees, R. M., Skiba, U. M., Sutton, M. A., Cape, J. N., Dore, A. J., Vieno, M., Simpson, D., Zaehle, S., Stocker, B. D., Rinaldi, M., Facchini, M. C., Flechard, C. R., Nemitz, E., Twigg, M., Erisman, J. W., Butterbach-Bahl, K., & Galloway, J. N. (2015). Effects of global change during the 21st century on the nitrogen cycle. *Atmospheric Chemistry and Physics*, *15*, 13849–13893.

Franche, C., Lindström, K., & Elmerich, C. (2009). Nitrogen-fixing bacteria associated with leguminous and non-leguminous plants.

Gilchrist, M., & Benjamin, N. (2017). From atmospheric nitrogen to bioactive nitrogen oxides. In *Nitrite and nitrate in human health and disease* . (pp. 11–19). Springer International Publishing.

Giller, K. E. (2001). *Nitrogen fixation in tropical cropping systems* . CAB Publishing.

Gurthie, F. B. (1896). Inoculation of the soil for leguminous crops. *The Agricultural Gazette of New South Wales*, *7*, 690–694.

Harris, D. F., Lukoyanov, D. A., Shaw, S., Compton, P., Tokmina-Lukaszewska, M., Bothner, B., Kelleher, N., Dean, D. R., Hoffman, B. M., & Seefeldt, L. C. (2018). Mechanism of N2 reduction catalyzed by Fe-nitrogenase involves reductive elimination of H2. *Biochemistry, 57*(5), 701–710.

Igarashi, R. Y., & Seefeldt, L. C. (2003). Nitrogen fixation: The mechanism of the Mo-dependent nitrogenase. *Critical Reviews in Biochemistry and Molecular Biology, 38*(4), 351–384.

Ininbergs, K., Bay, G., Rasmussen, U., Wardle, D. A., & Nilsson, M. C. (2011). Composition and diversity of nifH genes of nitrogen-fixing cyanobacteria associated with boreal forest feather mosses. *New Phytologist, 192*(2), 507–517.

Kamst, E., Spaink, H.P., & Kafetzopoulos, D. (1998). Biosynthesis and secretion of rhizobial lipo-chitin-oligosaccharide signal molecules. *Plant-Microbe Interactions, 29*, 29–71.

Kizilkaya, R. (2009). Nitrogen fixation capacity of Azotobacter spp. strains isolated from soils in different ecosystems and relationship between them and the microbiological properties of soils. *Journal of Environmental Biology, 30*(1), 73–82.

Ladha, J. K., Peoples, M. B., Reddy, P. M., Biswas, J. C., Bennett, A., Jat, M. L., & Krupnik, T. J. (2022). Biological nitrogen fixation and prospects for ecological intensification in cereal-based cropping systems. *Field Crops Research, 283*, 108541.

Ladha, J. K., Pareek, R. P., & Becker, M. (1992). Stem-nodulating legume-Rhizobium symbiosis and its agronomic use in lowland rice. In *Advances in soil science: Volume 20* . (pp. 147–192). Springer New York.

Leigh, J. A. (2000). Nitrogen fixation in methanogens: The archaeal perspective. *Current Issues in Molecular Biology, 2*(4), 125–131.

Masuda, Y. (2024). Environmental importance of biological nitrogen fixation. *Journal of Microbial & Biochemical Technology, 16*, 599.

Murray, P. A., & Zinder, S. H. (1984). Nitrogen fixation by a methanogenic archaebacterium. *Nature, 312*(5991), 284–286.

Mus, F., Crook, M. B., Garcia, K., Garcia Costas, A., Geddes, B. A., Kouri, E. D., Paramasivan, P., Ryu, M. H., Oldroyd, G. E., Poole, P. S., & Udvardi, M. K. (2016). Symbiotic nitrogen fixation and the challenges to its extension to nonlegumes. *Applied and Environmental Microbiology, 82*(13), 3698–3710.

Napoli, C. A., & Hubbell, D. H. (1975). Ultrastructure of Rhizobium-induced infection threads in clover root hairs. *Applied Microbiology, 30*(6), 1003–1009.

Nonaka, A., Yamamoto, H., Kamiya, N., Kotani, H., Yamakawa, H., Tsujimoto, R., & Fujita, Y. (2019). Accessory proteins of the nitrogenase assembly, NifW, NifX/NafY, and NifZ, are essential for diazotrophic growth in the nonheterocystous cyanobacterium Leptolyngbya boryana. *Frontiers in Microbiology, 10*, 495.

Ravikumar, S., Kathiresan, K., Alikhan, S. L., Williams, G. P., & Gracelin, N. A. A. (2007). Growth of Avicennia marina and Ceriops decandra seedlings inoculated with halophilic azotobacters. *Journal of Environmental Biology, 28*(3), 601.

Reddy, P. M., James, E. K., & Ladha, J. K. (2002). *Nitrogen fixation in rice* (pp. 421–445). Elsevier Science. ISBN 9780444509659.

Roger, P. A., & Watanabe, I. (1986). Technologies for utilizing biological nitrogen fixation in wetland rice: potentialities, current usage, and limiting factors. In *Nitrogen economy of flooded rice soils* . (pp. 39–77). Springer.

Roger, P. A., & Ladha, J. K. (1992). Biological N2 fixation in wetland rice fields: Estimation and contribution to nitrogen balance. *Plant and Soil, 141*, 41–55.

Saikia, S. P., & Jain, V. (2007). Biological nitrogen fixation with non-legumes: An achievable target or a dogma? *Current Science, 92*, 317–322.

Santi, C., Bogusz, D., & Franche, C. (2013). Biological nitrogen fixation in non-legume plants. *Annals of Botany, 111*(5), 743–767.

Seefeldt, L. C., Hoffman, B. M., Peters, J. W., Raugei, S., Beratan, D. N., Antony, E., & Dean, D. R. (2018). Energy transduction in nitrogenase. *Accounts of Chemical Research, 51*(9), 2179–2186.

Shah, F., & Wu, W. (2019). Soil and crop management strategies to ensure higher crop productivity within sustainable environments. *Sustainability, 11*(5), 1485.

Sindiswa, K., Augustine, I. D., Nkosingiphile, N., Khethiwe, N., Sanele, S., Siyabonga, N., Gcinokuhle, N., Arun, G., Marshall, K., Ashwil, K., Lee-Ann, N., & Mbukeni, N. (2025). Optimizing plant resilience with growth-promoting Rhizobacteria under abiotic and biotic stress conditions. *Plant Stress*, *17*, 100949. ISSN 2667-064X.

Soumare, A., Diedhiou, A. G., Thuita, M., Hafidi, M., Ouhdouch, Y., Gopalakrishnan, S., & Kouisni, L. (2020). Exploiting biological nitrogen fixation: A route towards a sustainable agriculture. *Plants*, *9*(8), 1011.

Stephan, M. P., Pedrosa, F. O., & Dobereiner, J. (1981). Physiological studies with Azospirillum spp.

Stewart, W. D. P. (1969). Biological and ecological aspects of nitrogen fixation by free-living micro-organisms. *Proceedings of the Royal Society of London. Series B. Biological Sciences*, *172*(1029), 367–388.

Sur, S., Bothra, A.K., & Sen, A. (2010). Symbiotic nitrogen fixation – A bioinformatics perspective.

Usha, B. (2018). Agriculture and the dark side of chemical fertilizers. *Environmental Analysis & Ecology Studies*, *1*, 3.

Vadakattu, G., & Paterson, J. (2006). Free-living bacteria lift soil nitrogen supply. *Farming Ahead*, *169*, 40.

Van Deynze, A., Zamora, P., Delaux, P. M., Heitmann, C., Jayaraman, D., Rajasekar, S., Graham, D., Maeda, J., Gibson, D., Schwartz, K. D., & Berry, A. M. (2018). Nitrogen fixation in a landrace of maize is supported by a mucilage-associated diazotrophic microbiota. *PLOS Biology*, *16*(8), e2006352.

van Rhijn, P., & Vanderleyden, J. (1995). The Rhizobium-plant symbiosis. *Microbiological Reviews*, *59*(1), 124–142.

Vance, C. P. (2001). Symbiotic nitrogen fixation and phosphorus acquisition. Plant nutrition in a world of declining renewable resources. *Plant Physiology*, *127*(2), 390–397.

Vlassak, K., & Reynders, L. (1981). Agronomic aspects by biological dinitrogen fixation by Azospirillum spp.

Wagner, S. C. (2011). Biological nitrogen fixation. *Nature Education Knowledge*, *3*(10), 15.

Welte, C. U. (2018). Revival of archaeal methane microbiology. *mSystems*, *3*(2), 10–1128.

Winogradsky, S. (1895). Recherches sur l'assimiltion de l'azote libre de l'atmosphëre par les microbes.

Yang, X., & Fang, S. (2015). Practices, perceptions, and implications of fertilizer use in East-Central China. *Ambio*, *44*(7), 647–652.

Yoneyama, T., Muraoka, T., Kim, T. H., Dacanay, E. V., & Nakanishi, Y. (1997). The natural 15N abundance of sugarcane and neighbouring plants in Brazil, the Philippines and Miyako (Japan). *Plant and Soil*, *189*(2), 239–244.

Zhang, W., Chen, Y., Huang, K., Wang, F., & Mei, Z. (2023). Molecular mechanism and agricultural application of the NifA-NifL system for nitrogen fixation. *International Journal of Molecular Sciences*, *24*(2), 907.

Zheng, M., Zhou, Z., Luo, Y., Zhao, P., & Mo, J. (2019). Global pattern and controls of biological nitrogen fixation under nutrient enrichment: A meta-analysis. *Global Change Biology*, *25*(9), 3018–3030.

Challenges in Enhancing Biological Nitrogen Fixation in Non-Leguminous Crops

16

Bhupendra Prasad, Reena Antony and Alina Hussain

Abstract

Associations of diazotrophs with non-leguminous crops are beneficial for agriculture as they increase the nitrogen (N_2) status of soil. New technologies increase possibilities to make such types of associations effectively. Spontaneous associations provide a potential benefit through nitrogen (N_2) fixation, and it is a rarely manageable part of agriculture. For raising the nitrogen content in dry land areas, these associations give a wonderful result. If we make an association of nodules with the roots of cereal crops like wheat and rice with free-living diazotrophs like *azospirilla, rhizobia* symbiotically, we raise the N_2 status. Free-living diazotrophs such as *Azospirillum, Azotobacter*, and certain strains of *Rhizobium* can form beneficial relationships with the roots of cereal crops like wheat, maize, and rice. These microorganisms not only fix atmospheric nitrogen but also help in root growth, improve water and nutrient uptake, and enhance overall plant strength and health. Encouraging such biofertilizer-based practices reduces the dependence on chemical fertilizers, lowers environmental pollution, and supports sustainable agricultural productivity.

Keywords

Biological nitrogen fixation · Nitrogen fixations · Diazotrophic association · Benefits of diazotrophs

B. Prasad (✉) · R. Antony · A. Hussain
Department of Microbiology, Career College, Bhopal, India
e-mail: bhupendra.prasad@careercollegeindia.com

R. K. Singh and P. Singh (eds.), *Diazotrophic Plant Growth-Promoting Microorganisms and Green Energy – Volume I*, https://doi.org/10.1007/978-981-95-9682-9_16

16.1 Introduction

For the plant's growth, nitrogen is an important nutrient, and it plays a fundamental role in the formation of amino acids, nucleic acids, and chlorophyll. Despite the presence of 78% nitrogen in the atmosphere, most plants cannot consume nitrogen in its gaseous form (N_2). To overcome this, chemical fertilizers like urea and ammonium nitrate are used by farmers to supply nitrogen, but these fertilizers can damage the environment, soil degradation, groundwater pollution, and greenhouse gas emissions and are even quite expensive. For the benefit of BNF farmers and scientists are studying forward to look for ways through natural and biological ways to supply nitrogen to plants mostly through the process of Biological Nitrogen Fixation (BNF).

Pulses and the bean family of legume Crops like soybean, pea, and lentil grow from nodules on their roots in nature with the help of *rhizobium* bacteria. These nodules fix a vast amount of nitrogen from the atmosphere. But some crops like rice, wheat, and maize (which are non-legume) do not form these nodules naturally. If we associate with these non-legume Crops and engage them with nitrogen-fixing bacteria, then it could revolutionize agriculture, more in dry and low-fertility soils.

On the other hand, Biological Nitrogen Fixation (BNF) gives us a sustainable and eco-friendly solution by allowing microorganisms to convert atmospheric nitrogen into ammonia, which is plant available in nature. Through this process, legumes are extensively exploited, but the main challenge remains the same as in non-leguminous crops such as rice, wheat, and maize. To address the global nitrogen crisis and ensure the security of the future, we develop sustainable methods to enhance BNF in these crops, which are extremely essential for the future.

16.2 What Is Biological Nitrogen Fixation?

A process where atmospheric nitrogen (N_2) converts into ammonia (NH_3), a useful form for plants through a symbiotic relationship between plants and microorganisms like diazotrophs, is called BNF (Biological Nitrogen Fixation)

16.2.1 Types of Biological Nitrogen Fixation

1. Symbiotic

Nitrogen Fixation for the bacteria living in close association, mostly in root nodules with the host plant, like *Rhizobium* with legumes (peas, beans, clover) and *Frankia* with non-legume plants (e.g., *Alnus*, *Casuarina*).
To convert atmospheric nitrogen into ammonia, the bacteria from the root nodules use nitrogenase enzyme, which the plant uses for growth, and in exchange, the plant gives the bacteria shelter and carbohydrates.

2. Free-Living

Bacteria living in the soil independently carry out nitrogen fixation without the association of plants.

Example:
Aerobic bacteria: *Azotobacter.*

- Anaerobic bacteria: *Clostridium.*
- Cyanobacteria (blue-green algae): *Anabaena, Nostoc.*
- Aerobic bacteria: *Azotobacter.*

These microorganisms give nitrogen in the soil, which is replenished by the plants.

16.2.2 Diazotrophic Associations in Non-Legumes

Non-leguminous crops cannot form nodules, but they can still absorb and use certain nitrogen-fixing bacteria in many different ways:

(a) Free-living bacteria, e.g., *Azotobacter*.
These are independent, free-living bacteria found in soil for nitrogen fixation.
(b) Associative bacteria, e.g., *Azospirillum* with cereal roots.
They live in close association with the plant roots, in some cases attached.
(c) Endophytic diazotrophs, e.g., *Gluconacetobacter* in sugarcane.
They absorb deep inside the plant's root tissue to form nodules to help nitrogen fixation.

16.2.3 Benefits of Diazotrophs for Non-Leguminous Crops

There are many diazotrophs that do not form nodules, and these bacteria can be useful in many different ways:

- Nitrogen supply- It fixes a small yet significant quantity of nitrogen
- Plant growth- some of the diazotrophs generate auxins and gibberellins, which restore root growth.
- Better root development- more roots and root hairs in plants help absorb more nutrients and water.
- Stress tolerance- plants with bacteria in dry land can tolerate stress better.

Farmers of dry land struggle more due to low soil fertility. Using BNF biofertilizers can improve crop yield even with low cost and low rainfall.

16.2.4 Importance of BNF in Non-Leguminous Crops

Non-leguminous cereals like rice and wheat contributes a major portion of the global food supply. They are cultivated across billions of hectares and feed more than half of the world's population.

Inducing Biological Nitrogen Fixation in these crops can extensively reduce the dependency on synthetic fertilizers, abundantly in the poor resource regions. In accordance with the (FAO) Food and Agriculture Organization, global nitrogen fertilizer consumption increases up to 110 million tons annually on average, while there is only a fraction that is utilized effectively by plants.

The rest is seeped down into the groundwater bodies, and some even escape into the atmosphere as nitrogen oxides, which are potent greenhouse gas. Integrating nitrogen-fixing bacteria like *Azospirillum, Azotobacter, and Gluconacetobacter* into non-legume systems helps in lowering these environmental footprints as it also improves the resilience of crops to withstand stress-related conditions such as drought and nutrient deficiency.

16.3 Molecular Mechanisms of Biological Nitrogen Fixation

The molecular mechanism of Biological Nitrogen Fixation (BNF) includes a complex set of biochemical and genetic interactions between microorganisms and host plants. At the center of this process is the nitrogenase enzyme complex, which catalyzes the reduction of atmospheric nitrogen (N_2) to ammonia (NH_2). Nitrogenase is composed of two metalloproteins: dinitrogenase reductase and dinitrogenase, which require reducing power and high amounts of ATP (adenosine triphosphate), typically given by ferredoxin or flavodoxin.

Nitrogenase is highly oxygen sensitive, so the reaction mechanisms occur under severe anaerobic conditions. Diazotrophs have evolved protective mechanisms such as high respiration rates, heterocyst formation (in cyanobacteria), and oxygen-scavenging enzymes to survive in any microaerobic environments. Genetically, nitrogen fixation is governed by nif genes, organized in clusters encoding structural, regulatory, and accessory proteins. The working of these genes is highly regulated by nitrogen availability, energy status, and environmental signals.

Recent studies using transcriptomics (studies the complete set of RNA transcripts in a cell or organism, which indicates which genes are actively being expressed) and Proteomics (studies the complete set of proteins, which are the functional products of those expressed genes, including their modifications and interactions) have revealed cross-communication between plant root exudates and bacterial gene activation. Flavonoids secreted by plant roots can induce the expression of bacterial nodulation (nod) genes, initiating signal transduction cascades that lead to attachment, colonization, and symbiotic establishment. In plants with no legumes, synthetic engineering of signaling pathways aims to mimic this communication and giving chance to endophytic bacteria like Gluconacetobacter and Azospirillum to form pseudo-symbiotic relationships.

The complex biochemical processes of BNF depend on a rather sophisticated orchestration of molecular events, primarily by the nitrogenase enzyme complex. This multicomponent system is not a single enzyme but elaborates molecular machinery necessary for high-energy input and precisely regulated redox reactions to take place. The two indispensable metalloproteins are dinitrogenase (MoFe protein) and dinitrogenase reductase (Fe protein) that, while operating cyclically, pass electrons in successive steps under strictly anaerobic or microaerobic conditions to reduce dinitrogen (N_2) to two molecules of ammonia (NH_2).

Thus, the overall reaction can be expressed as:

$$N_2 + 8H^+ + 8e^- + 16ATP \rightarrow 2NH_3 + H_2 + 16ADP + 16Pi$$

This reaction emphasizes the extremely high energetic cost of BNF, about 16 molecules of ATP are consumed per N_2 reduced. The Fe protein is an ATP-dependent electron donor that transfers electrons to the MoFe protein, which contains the iron-molybdenum cofactor FeMo-co as the active catalytic site where nitrogen binding and reduction occur. The FeMo-co represents an extremely complex metal cluster ($MoFe_7S_9C$-homocitrate) that sets up the proper coordination environment for the activation of inert N≡N bonds-a process that is thermodynamically unfavorable under ambient conditions.

In diazotrophic bacteria like *Azotobacter vinelandii* and *Rhizobium leguminosarum*, other accessory proteins, such as NifH, NifD, and NifK, provide essential roles in the stabilization of nitrogenase structure and redox homeostasis. More specifically, chaperone proteins are involved in FeMo-co assembly, involving the nifEN operon, to accurately position molybdenum and iron-sulfur clusters. Observations of mutant analyses and transcriptomic studies identify that an absence of these assembly factors leads to a non-functional nitrogenase, establishing their essentiality (Fig. 16.1).

Oxygen sensitivity is a critical factor that controls nitrogenase functionality. Even traces of O_2 irreversibly damage the FeMo-co and stop nitrogen fixation. Diazotrophs protect themselves by various means, which include high respiratory activity, microaerophilic cell differentiation, and

At the genetic level, nitrogen fixation is strictly controlled by the expression network of nif, fix, and vnf genes. The gene nifA encodes a transcriptional activator that induces the transcription of structural genes when nitrogen becomes limiting. Conversely, nifL acts as a repressor of the same genes when ammonia is available in sufficient amounts. This regulatory cascade is a means for energy conservation by shutting off the nitrogen-fixing machinery when fixed sources of nitrogen are available. There is now great awareness of the complexity of the cross-talk between environmental signals-oxygen, ammonium, and carbon status-and their corresponding regulatory proteins such as NtrC, GlnK, and FnrN. Such regulatory interplay allows the diazotroph to dynamically adjust nitrogenase expression according to ecological circumstances.

In cereals that are not legumes, research on molecular engineering to mimic this signaling has focused on expressing key bacterial nif gene clusters in plant organelles, such as chloroplasts or mitochondria, which provide a naturally low-oxygen and ATP-rich microenvironment suitable for nitrogenase function. Recent success using

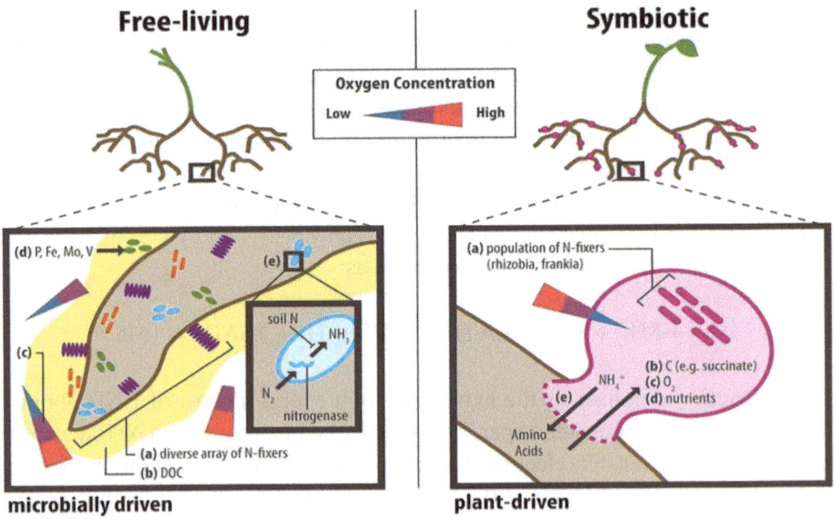

Fig. 16.1 Showing the structure and oxygen protection mechanisms of nitrogenase complex in free-living vs. symbiotic diazotrophs, the production of oxygen-binding molecules like leghemoglobin or hemoglobins in symbiotic systems. In free-living bacteria, such as Azotobacter, rapid respiration and the production of extracellular polysaccharides form oxygen-limited microzones around the cell. In cyanobacteria, specialized thick-walled cells called heterocysts separate nitrogen fixation spatially from photosynthesis, maintaining enzyme integrity

synthetic biology approaches to partly express functional nitrogenase in yeast and tobacco mitochondria encourages the prospect of future transgenic cereals able to fix nitrogen autonomously. While this is still highly experimental, the successful partial expression of nitrogenase is a significant breakthrough in the field of plant–microbe symbiotic engineering.

For example, metabolomic profiling of inoculated maize roots reveals increased levels of amino acids, polyamines, and organic acids that correspond to improved bacterial colonization and nitrogen fixation efficiency. In this line, proteomic data have shown stress-responsive proteins expressed by *Azospirillum brasilense*, which are induced during successful plant-root interaction and highlight the basis of metabolic reprogramming underlying successful diazotrophic association. These "multi-omics" datasets not only clarify molecular mechanisms but also aid in designing superior bacterial strains for biofertilizer applications. Besides, computational modeling and AI-based simulations can predict the dynamics of nitrogen flux under different environmental conditions. This kind of predictive modeling, if combined with field-level data on optimal bacterial inoculation densities, carrier compositions, and soil moisture thresholds, could ultimately lead to maximum efficiency in BNF. Coupling molecular understanding with data-driven modeling thus represents the next frontier in nitrogen fixation research.

16.4 Environmental and Ecological Impact of BNF

Biological nitrogen fixation contributes much to improving soil fertility and, at the same time, contributes considerably to environmental sustainability. The use of BNF-based bio-fertilizers reduces greenhouse gas emissions, nitrate leaching, and the eutrophication of aquatic systems by replacing chemical fertilizers. Recent data from the FAO estimated that excessive fertilizer application is responsible for almost 12% of global anthropogenic nitrous oxide (N_2O) emissions.

BNF enhances soil microbial diversity by supporting beneficial microbial networks that promote nutrient cycling, organic matter decomposition, and disease suppression. Inclusion of nitrogen-fixing systems in agricultural landscapes enhances ecological resilience and carbon sequestration. For example, BNF coupled with no-till increases soil organic carbon by 15%–25% within a decade, as reported in numerous long-term field experiments.

In addition, diazotrophic associations enhance tolerance of plants to drought and salinity by modulating osmolyte accumulation and antioxidant enzyme activity. This dual role for nutrient enrichment and stress mitigation underlines the ecological importance of nitrogen-fixing microbes in agroecosystems.

BNF performs dual ecological functions, enriching soil fertility while mitigating various environmental challenges associated with intensive agricultural practices. At a core level, BNF is a natural biogeochemical process that contributes to the global nitrogen cycle and replenishes soil nitrogen pools without energy-intensive industrial inputs like synthetic fertilizers. This process improves nutrient cycling and productivity while reducing the anthropogenic footprint associated with reactive nitrogen pollution.

Among the major ecological benefits of BNF is a reduction in N_2O emissions, a greenhouse gas nearly 300 times more potent compared with CO_2, due to its global warming potential. Excessive application of nitrogen fertilizers increases microbial denitrification, a process that releases N_2O and thus contributes to atmospheric pollution. On the other hand, diazotrophic microorganisms fix nitrogen in a controlled biological form, with minimal gaseous losses. According to recent models presented by the Intergovernmental Panel on Climate Change (IPCC 2024), widespread substitution of chemical nitrogen fertilizers with BNF-based biofertilizers might reduce agricultural N_2O emissions by 15%–25% globally.

Biological nitrogen fixation also affects the biochemical health of soils through its influence on the rhizosphere, the biologically active soil zone around plant roots. Diazotrophic bacteria excrete exopolysaccharides, enzymes, and organic acids, which help them modify soil structure and improve aggregate stability. Improved soil porosity allows greater soil aeration and water infiltration, processes that are highly important for root respiration and microbial activity. Over time, the levels of organic carbon in soils build up, thereby promoting long-term carbon sequestration. This is particularly important under conservation agriculture systems such as no-till or minimal tillage, where microbial nitrogen inputs synergize with soil carbon accumulation to enhance overall soil organic matter content.

Another key ecological role is the contribution of BNF to biodiversity. To begin with, by lessening the dependence on synthetic fertilizers, which often alter soil pH and microbial diversity, BNF encourages a balanced soil microbiome. Symbiotic and associative diazotrophs, for example, *Azospirillum spp., Herbaspirillum spp., and Burkholderia spp.*, form microhabitats for other beneficial microorganisms, such as PSB, AMF, and PGPR, among others. This complex microbial network enhances nutrient mobilization, suppresses soil-borne pathogens through competitive exclusion, and leads to secondary metabolite production.

BNF acts as an ecological stabilizer in fragile ecosystems such as drylands, saline regions, and degraded soils. Nitrogen-fixing cyanobacteria and free-living diazotrophs colonize barren soils, initiating primary succession by improving the fertility of those soils with time. Crust-forming cyanobacteria, such as *Nostoc commune* and *Anabaena sp.*, enhance the cohesion of the soil and minimize erosion through the formation of a biocrust that traps moisture and organic particles. In sub-Saharan Africa, studies in the arid zones of Rajasthan have shown that microbial crusts similar to such increase nitrogen availability by 20–40 kg N/ha annually and restore productivity on nutrient-poor soils. This ecological restoration potential underlines BNF's role as a tool not only for agriculture but for environmental rehabilitation.

From an aquatic ecosystem perspective, the role of BNF is no less important. The runoff of chemical fertilizers frequently promotes eutrophication, or excessive nutrient loading, which stimulates algal blooms and oxygen depletion in the water. By replacing or reducing fertilizer use, BNF indirectly reduces eutrophication pressures. Some cyanobacteria, although natural nitrogen fixers occurring in water bodies, may pose harm when the nutrient levels are unnaturally high. Thus, the balance of nitrogen by controlled BNF helps in sustainable aquatic biodiversity, besides minimizing the development of harmful blooms.

At the macro ecological scale, the integration of nitrogen-fixing systems into agricultural landscapes strengthens ecological resilience. Fields with active diazotrophic populations show increased drought tolerance and more efficient nutrient use. This is a result of physiological adaptations due to microbial interaction, including the induction of antioxidant enzyme activities like superoxide dismutase, catalase, and peroxidase, and osmolyte accumulation such as proline and glycine betaine. Plants inoculated with *Azospirillum brasilense* or *Gluconacetobacter diazotrophicus* showed increased levels of stress-responsive metabolites that, even under water-limited conditions, maintain photosynthetic performance. These mechanisms underline how BNF indirectly fortifies crops against climatic stresses.

Besides, the long-term ecological monitoring of multiple cropping systems, including rice–wheat rotations, sugarcane monocultures, and maize–soybean intercrops, revealed improved nutrient cycling efficiency due to biofertilizer-based nitrogen fixation. The recovery rate of nitrogen could be over 60%–70% in systems with BNF, while it may be up to 30–40% in systems dominated by chemical fertilizers.

This efficiency is reflected in the less leaching of nitrates into groundwater and low losses due to volatilization, which translates into an ecological and economic advantage. A broader ecological implication of BNF lies in its contribution to planetary nitrogen homeostasis. The Haber-Bosch process, although revolutionary for agriculture, added nearly 120 Tg (teragrams) of reactive nitrogen annually to global ecosystems-a rate far beyond natural biological inputs. The imbalance between nitrogen fixation and denitrification has resulted in "nitrogen saturation," manifesting as acid rain, soil acidification, and aquatic toxicity. BNF, being a self-regulating bioprocess, maintains equilibrium via feedback control. The microbial nitrogenase activity naturally falls once soil ammonia has achieved sufficient levels, avoiding over-fertilization. Thus, BNF acts as a natural moderator in the nitrogen cycle, balancing agricultural productivity with ecological integrity. Finally, the ecological value of BNF encompasses more than nutrient supply-it embodies the concept of circular bioeconomy. In transforming atmospheric nitrogen into plant-available forms through biological means, BNF is in tune with sustainable cycling of resources and low-carbon agriculture. Its incorporation into regenerative models of agriculture will guarantee the long-term ecosystem services of maintaining soil fertility, enhancing biodiversity, and sequestrating carbon. In short, the impact of Biological Nitrogen Fixation upon the environment and ecology surpasses that of conventional nutrient management. It forms one of the cornerstones of sustainable agricultural intensification, balancing productivity with ecological stability. When this is integrated within global climate and conservation strategies, BNF holds the potential for a fundamental shift in how agriculture interacts with the environment, from exploitation to regeneration.

16.5 Challenges for Enhancing BNF in Non-Leguminous Crop

There are many obstacles that can be clearly seen in using BFN:

- No nodules in cereals—like legumes, cereals do not form any root nodules, so the nitrogen fixation rate stays low.
- Low bacterial survival—instigating bacteria often cannot compete with native microorganisms previously present in the soil.
- Short-term colonization—these microorganisms may not stay long enough in the roots or inside the plant.
- Environmental stress- high temperature, lack of organic matter, dry soil reduces bacterial activity.
- Inconsistent results—oftentimes we observe that when induced bacteria work well in some regions, while in others it fails due to soil differences.

Given the potential of BNF in non-leguminous crops is abundant, and the process through which multiple Biological and technical barriers can be modified. Firstly, these cereals have inherited the lack of specialized root structures, which have nodules that are required for the synthesis of symbiotic nitrogen fixation. Secondly, the nitrogenase enzyme complex, which is found in diazotrophs, is highly sensitive to oxygen, which limits the activity in aerobic root conditions.

However, non-legume plants also exhibit a weak signaling mechanism that attains and maintains beneficial bacteria. Other restrictions include competition between induced strains of soil microbes and native strains of soil microbes, which in turn reduces survival rates under unfavorable climate conditions, and unpredictability in field performance across the regions. For example, *Azospirillum* inoculation might show significant yield benefits in tropical soils but minimal effects in temperate regions due to variations in soil microbiota and temperature.

For that reason, an interdisciplinary approach including molecular genetics, soil ecology, and agronomy is essential to get better control of these obstacles.

Integration of BNF into crops other than legumes is among the most enduring scientific and agronomic challenges of modern plant-microbe biotechnology, notwithstanding five decades of progress in research on nitrogen fixation. Unlike legumes, cereals and other non-legume plants lack the anatomical and genetic machinery for the establishment of classical root nodules, the basis of efficient symbiotic nitrogen fixation. Thus, the process of nitrogen fixation in non-legumes is commonly limited by structural, molecular, and environmental bottlenecks that hinder the formation, survival, and functionality of nitrogen-fixing microbial populations.

The basic barrier is the absence of nodule-forming genes and their signaling pathways in non-leguminous plants. The nodulation process is highly specialized in its development, mediated by flavonoids from plants and bacterial nod factors; these active molecules stimulate root cortical cell division and the organogenesis of nodules. Cereals specifically lack receptor-like kinases, such as NFR1 and NFR5, and downstream transcriptional regulators like NSP1 and NIN, which prevent nodule initiation in rice, wheat, and maize. These plants do host associative and endophytic diazotrophs, but most such interactions are transient and involve no specialized nitrogen-fixing organs. Synthetic biology approaches try to reintroduce such genetic modules or mimic them, but stable integration and coordinated expression are a technical challenge.

A second and equally serious restraint involves the oxygen sensitivity of the nitrogenase enzyme complex; the iron-molybdenum cofactor that catalyzes nitrogen reduction is irreversibly inactivated in the presence of oxygen. Whereas legumes circumvent this problem with specialized oxygen-binding molecules such as leghemoglobin, non-leguminous crops have no apparent internal regulation mechanisms. Free-living and associative diazotrophs thus depend on various alternative methods, including high respiratory rates, formation of extracellular polysaccharide matrices, and microaerophilic niches in root tissue. Under field conditions, however, with fluctuating oxygen and moisture levels, these microbial environments are highly unstable. Indeed, this environmental variability often results in inconsistent nitrogen fixation performance across regions and seasons.

Another major challenge is the energy demand of nitrogen fixation. In every reaction cycle, the conversion of atmospheric nitrogen to ammonia requires approximately 16 molecules of ATP, together with an adequate supply of reducing power (electrons). In non-leguminous associations, the host plant usually provides only limited carbon and energy substrates to the bacteria, with suboptimal activity of nitrogenase as a consequence. For continued BNF, proper energy flux between plant and microbe is essential, but in non-symbiotic systems, this is not well documented. Advanced metabolic modeling studies now indicate that increased carbon flux through the tricarboxylic acid cycle and increased efficiency of bacterial oxidative phosphorylation may enhance nitrogenase functioning, but field application based on such approaches has been limited so far.

The inconsistency of results across various environments is a key bottleneck for field implementation. While controlled greenhouse experiments often report promising yield increases after inoculation with either *Azospirillum* or *Gluconacetobacter*, field trials in different agroecological zones frequently result in variable outcomes. Soil texture, organic matter content, temperature, and moisture are just a few of the factors that make a huge difference in microbial establishment. For example, the same strain of *Azospirillum brasilense* can enhance maize yields by 25% in tropical Brazil and be completely ineffective in the same crop in European temperate soils because of differences in rhizospheric pH and competition from the native microbiome. Such geographical inconsistencies in response complicate standardization and large-scale adoption.

The poor host specificity and signaling compatibility between non-legumes and diazotrophs pose other challenges. The communication between plant roots and bacteria, through such signaling molecules as flavonoids, lipo-chitooligosaccharides, and volatile organic compounds, is generally weak or misinterpreted in the case of non-legume systems. For instance, leguminous plants produce specific flavonoid profiles that precisely activate bacterial nod genes, while cereals emit a more generalized array of root exudates lacking such specificity. Bioengineering of root exudation profiles or the introduction of compatible receptor proteins into non-legumes has been attempted but remains at an early stage.

However, most diazotrophic bacteria colonizing cereal roots are relatively short-lived, thus limiting the potential for sustained nitrogen input during the crop lifecycle. Many associative diazotrophs only form temporary associations, effective in colonizing the root surface or intercellular spaces but not deeply invading vascular tissues. Without systemic colonization, nitrogen fixation remains spatially and temporally restricted. Various advanced imaging techniques, including CLSM (Confocal Laser Scanning Microscopy) and FISH (Fluorescence In Situ Hybridization), have shown that while *Herbaspirillum seropedicae* can invade maize roots, population density declines drastically from the post-flowering stage, suggesting a transient colonization behavior. This underlines the need for more persistent bacterial strains able to sustain their endophytic residency.

From a biochemical point of view, the regulation of nitrogenase gene expression under stress conditions is also limited. Many environmental factors, such as nitrate concentration, ammonium availability, and oxidative stress, generally repress the transcription of nif genes due to the activity of global regulators like NtrC and GlnK. It remains very important to understand how these circuits are regulated to be able to act on them for consistent nitrogen fixation performance under variable agricultural conditions. New CRISPR–Cas-based genetic editing tools are now being used to edit such regulatory elements that will enable diazotrophs to continue with nitrogenase expression in moderately nitrogen-rich soils, an important step toward stabilizing BNF efficiency in non-leguminous crops.

Soil physicochemical properties further influence BNF outcomes. Parameters such as pH, salinity, cation exchange capacity, and micronutrient availability directly affect bacterial metabolism and enzyme stability. For example, deficiencies of molybdenum and iron, key cofactors for FeMo-co synthesis, can suppress nitrogenase activity drastically. Similarly, bacterial colonization is reduced in alkaline soils as a result of altered membrane potential and nutrient uptake. Hence, soil management through organic amendments, biochar incorporation, and micronutrient supplementation becomes imperative in order to support active nitrogen-fixing populations. Another subtle but important limitation involves the measurement and quantification of BNF activity. Traditional approaches, such as ARA and ^{1}N isotope tracing, are reliable under controlled conditions but usually not practical at field-scale monitoring due to their high cost and labor requirements. Newer technologies, including remote-sensing-based spectral indices and nano sensor-assisted soil probes, bring new opportunities for real-time assessment of nitrogen fixation rates. These tools are still in the development phase and need further calibration across diverse agro ecosystems. Finally, farmer awareness, product quality control, and institutional support are formidable sociotechnical challenges. Most farmers have no access to high-quality biofertilizers, and the market regulation of microbial inoculants is poor in several developing regions. The low level of efficacy due to poor storage conditions, expired formulation, or lack of strain recommendations for site-specific applications has often reinforced skepticism about the reliability of biofertilizers. For translation of laboratory successes into the large-scale agricultural practice, strengthening policy frameworks, implementation of strict quality standards for biofertilizer products, and conducting region-specific training programs are required. Overcoming the challenges in the improvement of BNF within non-legumes requires a multidisciplinary approach, combining molecular genetics, microbial ecology, agronomy, and environmental engineering. Improvements will involve not only the development of more robust diazotrophic strains but also holistic agroecosystem designs that favor their survival and function over the long term. The next generation of BNF research should, therefore, focus on combining biotechnology with precision farming and sustainable soil management to unlock the full potential of this natural process.

New Technologies and Scientific Progress

In recent years, many new opportunities have arisen with new scientific tools:

- Genetic engineering—scientist have been trying to transfer the nitrogen fixation gene into crops of rice and wheat.
- Synthetic biology—some scientific researchers have shown ways for a better artificial symbiosis system.
- CRISPR gene editing—this technique helps in modifying crops or bacteria to improve their affinity.
- Better carrier material for biofertilizer—for helping bacteria survive, in this method, they use polymers, peat, or liquid for it to survive longer in seeds.
- Microbiome research—To insert the correct combination of microbiome, studying microbiome is very important (Fig. 16.2).

16.6 Strategies to Improve Association

To improve the association of bacteria more successful many practical strategies of inoculation are used:

- Seed inoculation—before sowing seeds is treated with *azospirillum* or *azotobacter*.
- Selecting important or efficient strains—finding local strains more adapted to the local soil and crop.
- Combination with organic manure—this helps the bacteria survive better.
- Co-inoculation—using diazotrophs with other beneficial microbes like phosphorus solubilizing microorganism.

16.7 Quantitative Evidence and Field Data

Experimental studies across various continents have shown the impact of diazotrophic inoculation on crop yield.

- In Brazil, fields having gene strains with *Azospirillum brasilense* on maize and wheat have shown yield increases ranging from 10% to 30%, along with a 25% reduction in nitrogen fertilizer requirements.

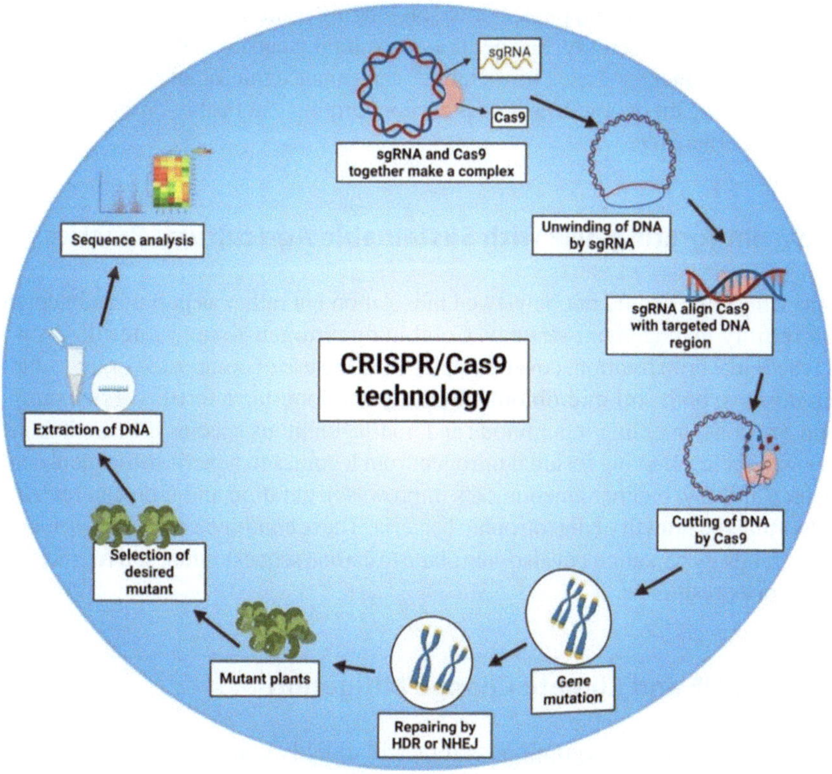

Fig. 16.2 CRISPR gene editing

- In India, the application of cyanobacteria in the field of paddy has resulted in an increase in yield up to 20%, while sugarcane treated with *Gluconacetobacter diazotrophicus* has demonstrated in enhancing sugar content and biomass.
- In dry lands of Africa, integrating Biological Nitrogen Fixation promotes biofertilizers with organic compost, improving sorghum yields by 15%–25%.

These quantitative findings help in understanding the practical feasibility of BNF-based agricultural systems, especially when combined with organic soil amendments.

16.8 Potential Solutions and Strategies for Improvement

To improve the success rate of BNF in non-leguminous crops, several biotechnological and agronomic interventions have been explored. Genetic engineering approaches aim to introduce nodulation and nitrogen fixation genes into cereals. Recent studies have used CRISPR-Cas9 technology to modify key transcription factors in rice and maize, enabling partial expression of nitrogenase-related pathways. Additionally, synthetic biology offers tools to design novel host-microbe interactions that mimic legume symbiosis. From an agronomic perspective, co-inoculation techniques combining nitrogen-fixing and phosphorus-solubilizing bacteria have shown synergistic effects on crop productivity. Furthermore, developing carrier materials such as alginate beads and biochar-based formulations can enhance microbial survival and shelf life. Integrating these practices with precision farming can further optimize nitrogen utilization efficiency.

16.9 Integrating BNF with Sustainable Agricultural Practices

Enhancing BNF should not be viewed in isolation but rather as part of an integrated soil fertility management strategy. Combining nitrogen-fixing biofertilizers with practices like crop rotation, cover cropping, and the use of composted organic matter can improve both soil microbiome diversity and long-term fertility. For example, crop rotations involving leguminous and non-leguminous species create a balanced nitrogen cycle, allowing residual nitrogen from legumes to benefit subsequent cereal crops. Similarly, biochar amendments improve soil aeration and moisture retention, supporting the growth of diazotrophic bacteria. These combined approaches not only enhance nitrogen fixation but also contribute to carbon sequestration, soil regeneration, and climate resilience.

16.10 BNF and Climate Change Mitigation

Increased BNF in non-leguminous crops will directly contribute to climate-smart agriculture. The decreased use of synthetic fertilizers reduces N_2O emissions, which have a global warming potential approximately 300-fold higher than that of CO_2.

Biofertilizer-based nitrogen fixation contributes to climate adaptation by promoting soil carbon sequestration, decreasing energy use associated with fertilizer production, and improving drought and temperature tolerance.

BNF systems also support regenerative agriculture frameworks emphasizing minimal soil disturbance, permanent soil cover, and diversified cropping. These practices promote soil biodiversity and nutrient retention while maintaining ecosystem balance.

16.11 Economic and Policy Perspectives

The extensive use of BNF-based systems can significantly lower the agricultural production costs. Fertilizer costs alone amount to nearly 30–40% of total crop input costs in many developing nations. By integrating biofertilizers, farmers can achieve comparable yields with reduced dependency on external nitrogen inputs. However, large-scale implementation requires supportive policy frameworks, farmer education programs, and quality assurance standards for biofertilizer products. Supportive policy frameworks include incentives to encourage farmers; subsidies on the use of biofertilizers, training modules, and incorporation in sustainable farming under the UN SDGs.

16.12 Future Prospects

The future expectation of BNF is to make cereals from nodules just like legumes- but we're still approaching that criteria. However, with the technology and dedicated study in the field of nitrogen-fixing microbiomes, gene editing, nanotechnology, and improved and much-fertilized biofertilizers, we are moving forward each day.

Further researche is focusing on unwinding the genetic codes of BNF in diazotrophs and host plants. Integrative omics technologies, such as genomics, transcriptomics, and metabolomics, can help identify key genes and regulatory networks responsible for nitrogen fixation. Developing genetically engineered cereal crops capable of maintaining stable associations with diazotrophs represents a long-term goal. In addition to exploring initial microbial diversity through metagenomic sequencing, which led to the discovery of novel nitrogen- fixing strains adapted to specific agro-ecological zones. Eventually, the integration of digital agriculture and AI-based predictive model could enable better management and monitoring of soil nitrogen dynamics and up to the mark biofertilizer applications.

16.13 Future Expects

1. Genetic Engineering for Autonomous Nitrogen Fixation

Synthetic biology and gene editing tools such as CRISPR-Cas9 and TALENs might enable scientists to directly introduce from atmospheric nitrogen without inoculation.

2. Artificial Symbiosis Systems nitrogenous genes into the genomes of cereal crops. This would enable plants to fix nitrogen autonomously. Development Scientists are investigating engineered symbiosis between cereals and nitrogen-fixing bacteria by manipulating plant signaling processes. By replicating legume-*rhizobium* communication (via flavonoid signals or nod factors), it could be feasible to create stable long-term relationships in non-legumes.
3. Nanotechnology-Assisted Nitrogen Fixation

Nano-materials are being evaluated for the purpose of improving the efficiency of microbial biofertilizers through better delivery and protection against stress for bacteria. Nano-carriers might ensure bacterial viability and slow release of nutrients into soil.

4. Microbiome Engineering and Synthetic Consortia

The future can hold designer microbial communities designed to deliver not only nitrogen but several plant advantages—such as phosphorus solubilization, stress resistance, and plant growth promotion. Synthetic consortia can be tailored to suit particular soil and crop conditions.

5. Integrating BNF with Climate-Smart Agriculture

With intensifying climate stresses, BNF-promoting systems may be integrated with climate-resilient practices like no-till agriculture, organic fertilization, and carbon sequestration methods to enhance soil resilience and health to drought.

6. Searching for Novel Diazotrophic Strains

Ongoing metagenomic research is discovering novel and effective nitrogen-fixing bacteria from extreme and understudied environments (e.g., saline soils, arid soils, and wetlands). The novel strains may work better in non-legume systems compared to conventional inoculants.

7. Precision Biofertilizer Technology

Design of precision-prepared biofertilizers able to sense actual soil conditions in real-time through AI-based sensor data could ensure uniform bacterial activity and nutrient supply with minimized wastage and variability.

8. Plant Breeding for Improved BNF Compatibilization

Breeding programs of the future could involve the selection of crop types with root exudates that better attract and promote nitrogen-fixing bacteria. It could dramatically improve the efficiency of microbial colonization.

9. Employment of Omics Technologies

Concurrent use of genomics, proteomics, and metabolomics has the ability to identify the molecular processes underlying successful plant-microbe interactions. This information can inform the design of next-generation inoculants and genetically compatible hosts.

10. Field-Level Validation and Long-Term Trials

There is a need for additional large-scale, multi-location field trials to validate laboratory results and test long-term soil fertility, economic viability, and environmental sustainability of BNF-based systems.

11. Integration with Sustainable Development Goals (SDGs)
Future action towards increasing BNF directly contributes to UN SDGs, particularly Goal 2 (Zero Hunger), Goal 12 (Responsible Consumption and Production), and Goal 13 (Climate Action), highlighting its international importance.

12. Public Awareness and Policy Support
Strong policy environments and public awareness programs are necessary for the promotion of farmers' adoption of BNF-based technologies. Incentives for using biofertilizers and research support will expedite field-level implementation

16.14 Conclusion

Enhancing biological nitrogen fixation (BNF) in non-leguminous crops like rice and wheat holds enormous potential to revolutionize sustainability and is a promising yet challenging task in the field. Diazotrophs can provide some growth Benefits and nitrogen to the plant, but are limited by poor colonization and low nitrogen fixation rates can still be seen. With continuous research and proper field practice, these associations and applications could significantly reduce the use of chemical-based fertilizer, especially in dry land for agriculture. This approach supports sustainable, eco-friendly, and affordable farming. Even with the given challenges, continuous enhancement and advances in molecular biology, microbial ecology, and bio-fertilizer formulation are giving ways for practical knowledge and improvement.

The benefits of diazotrophs, proven by the quantitative field data, have improved crop yields, soil fertility, and reduced fertilizer dependency. Collaboration among plant biologists, soil scientists, and agricultural engineers is vital to achieve scalability. Eventually, Sustainable Development Goals (SDGs) related to food security, environmental protection, and climate action in the (Biological Nitrogen Fixation) BNF-based system play a central role in achieving sustainable development.

Summary Points

- Nitrogen is essential, but chemical fertilizers are costly.
- Non-legumes do not form nodules, but can still benefit from diazotrophs.
- *Azospirillum*, *Azotobacter*, and endophytic bacteria can improve growth and nitrogen availability.
- Challenges include poor survival of bacteria and inconsistent performance.
- New advances like genetic engineering and microbiome studies offer hope.
- Field-based practices like seed inoculation and using local strains improve results.
- BNF in non-leguminous crops like rice and wheat has distinctively reduced dependency on chemical nitrogen fertilizers (Bashan & de-Bashan, 2015; Beatty & Good, 2011; Bashan & de-Bashan, 2010; Behera et al., 2021; Santi et al., 2013; Giller et al., 2024; Li et al., 2024; Galindo et al., 2024; Mahto et al., 2025; Connolly et al., 2025; Misu et al., 2025; Oh et al., 2025).

References

Bashan, Y., & de-Bashan, L. E. (2015). Inoculant preparation and formulations for Azospirillum spp. In *Handbook for Azospirillum: Technical issues and protocols*. Springer International Publishing. (pp. 469–485).

Bashan, Y., & de-Bashan, L. E. (2010). How the plant growth-promoting bacterium Azospirillum promotes plant growth—a critical assessment. *Advances in Agronomy, 108*, 77–136.

Beatty, P. H., & Good, A. G. (2011). Future prospects for cereals that fix nitrogen. *Science, 333*(6041), 416–417.

Behera, B., Das, T. K., Raj, R., & Ghosh, S. (2021). Microbial consortia for sustaining productivity of non-legume crops: Prospects and challenges. *Agricultural Research, 10*(1), 1–14.

Connolly, L. N., Lorenz, N., Maleki, K., Kayafas, N., Dick, R. P., & Mercer, K. L. (2025). Nitrogen fixation rates and aerial root production among maize landraces. *Frontiers in Plant Science, 16*, 1502884.

Galindo, F. S., Pagliari, P., da Silva, E. C., de Lima, B. H., Fernandes, G. C., Thiengo, C. C., Bernardes, J. V. S., Jalal, A., Oliveira, C. E. S., de Sousa Vilela, L., Junior, E. F., Nogueira, T. A. R., do Nascimento, V., Filho, M. C. M. T., & Lavres, J. (2024). Impact of nitrogen fertilizer sustainability on corn crop yield: The role of beneficial microbial inoculation interactions. *BMC Plant Biology, 24*, 268.

Giller, K. E., James, E. K., Ardley, J., & Unkovich, M. (2024). Science losing its ways: Examples from the realm of microbial N2 fixation in cereals and other non-legumes. *Plant and Soil, 511*, 1–24.

Li, P., Huang, Y., Zhu, Y., Tian, M., Zhu, Y., Chen, X., Hu, R., Qin, T., Liu, Y., Peng, S., Yi, Z., Liu, Z., Ao, H., & Li, J. (2024). Mechanism and microbial action of the inoculated nitrogen-fixation bacterium for growth promotion and yield enhancement in rice. *Advanced Biotechnology, 2*, 32.

Mahto, R. K., Singh, R. K., Kumar, A., Yadav, R., Dey, D., Hamwieh, A., & Kumar, R. (2025). Symbiotic nitrogen fixation for sustainable chickpea yield and prospects for genome editing in changing climatic situations. *Frontiers in Plant Science, 16*, 1621191.

Misu, I. J., Kayess, M. O., Siddiqui, M. N., Gupta, D. R., Islam, M. N., & Islam, T. (2025). Microbiome engineering for sustainable rice production: Strategies for biofertilization, stress tolerance, and climate resilience. *Microorganisms, 13*(2), 233.

Oh, J.-H., Kim, E., & Cho, M. (2025). Biofilm formation by rice rhizosphere nitrogen-fixing microorganisms and its effect on rice growth promotion. *Biology, 14*(9), 1249.

Santi, C., Bogusz, D., & Franche, C. (2013). Biological nitrogen fixation in non-legume plants. *Annals of Botany, 111*(5), 743–767.

MIX
Papier aus verantwortungsvollen Quellen
Paper from responsible sources
FSC® C105338

If you have any concerns about our products,
you can contact us on
ProductSafety@springernature.com

In case Publisher is established outside the EU,
the EU authorized representative is:
Springer Nature Customer Service Center GmbH
Europaplatz 3, 69115 Heidelberg, Germany

Printed by Libri Plureos GmbH
in Hamburg, Germany